Supplements to the 2nd Edition of

RODD'S CHEMISTRY OF CARBON COMPOUNDS

ELSEVIER SCIENCE PUBLISHERS B.V.
Sara Burgerhartstraat 25
P.O. Box 211, 1000 AE Amsterdam, The Netherlands

Distributors for the United States and Canada:

ELSEVIER SCIENCE PUBLISHING COMPANY INC.
655, Avenue of the Americas
New York, NY 10010, U.S.A.

Library of Congress Card Number: 64-4605
ISBN 0-444-87399-6

Printed in The Netherlands

Supplements to the 2nd Edition of

RODD'S CHEMISTRY OF CARBON COMPOUNDS

VOLUME I

ALIPHATIC COMPOUNDS
★

VOLUME II

ALICYCLIC COMPOUNDS
★

VOLUME III

AROMATIC COMPOUNDS
★

VOLUME IV

HETEROCYCLIC COMPOUNDS
★

VOLUME V

MISCELLANEOUS
GENERAL INDEX
★

Supplements to the 2nd Edition (Editor S. Coffey) of

RODD'S CHEMISTRY OF CARBON COMPOUNDS

A modern comprehensive treatise

Edited by
MARTIN F. ANSELL
Ph.D., D.Sc. (London) F.R.S.C. C. Chem.
Reader Emeritus, Department of Chemistry,
Queen Mary College, University of London, Great Britain

Supplement to

VOLUME IV HETEROCYCLIC COMPOUNDS

PART K:

Six-membered Heterocyclic Compounds Containing Two or More Hetero-atoms, One or More of which are from Groups II, III, IV, V or VII of the Periodic Table. Heterocyclic Compounds with Seven or More Atoms in the Ring

ELSEVIER
Amsterdam — Oxford — New York — Tokyo   1989

VI

CONTRIBUTORS TO THIS VOLUME

David W. Allen, Ph.D., D.Sc., C.Chem., F.R.S.C.
Department of Chemistry, Sheffield City Polytechnic,
Sheffield S1 1WB

Christopher D. Gabbutt, Ph.D., C.Chem., M.R.S.C.
Department of Chemistry, Lancashire Polytechnic,
Preston PR1 2TQ

John D. Hepworth, B.Sc., Ph.D., C.Chem., F.R.S.C.
Department of Chemistry, Lancashire Polytechnic,
Preston PR1 2TQ

Karen Johnson, B.Sc., M.Sc., Ph.D.,
Department of Chemistry, University College,
University of London, London WC1H 0AJ

Paul D. Lickiss, B.Sc., D.Phil., C.Chem., M.R.S.C.
School of Chemistry and Molecular Science,
University of Sussex, Sussex BN1 9QJ

J. A. Hugh MacBride, M.A., D.Phil.
Department of Chemistry, University of Durham,
Durham DH1 3LE

Timothy J. Mason, B.Sc., Ph.D., C.Chem., F.R.S.C.
Department of Chemistry, Coventry Polytechnic,
Coventry CV1 5FB

John T. Sharp, B.Sc., Ph.D., D.Sc., C.Chem., F.R.S.C.
Department of Chemistry, University of Edinburgh,
Edinburgh EH9 3JJ

Malcolm Sainsbury, Ph.D., D.Sc., C.Chem., F.R.S.C.
Department of Chemistry, The University,
Bath BA2 7AY
(Index)

PREFACE TO SUPPLEMENT IV K

The publication of this volume is another step in the sup-
plementation of the second edition of Rodd's Chemistry of Carbon
Compounds, thus continuing to keep this major work of reference
up-to-date. This supplement covers Chapters 48-56, inclusive,
of the second edition. It covers what may perhaps be best described as a "heterogeneous collection of heterocycles". These
include six-membered heterocyclic compounds with two or more
heteroatoms, one or more of which are from Groups II, III, IV, V
or VII of the Periodic Table, and also heterocycles with seven or
more atoms in the ring. The statement in the preface to Volume
IV K of the second edition that "The volume should be of particular
interest to both organic and inorganic chemists, biochemists, pharma-
cologists and readers interested in any of the biomedical sciences"
applies equally well to this supplement. Compounds reviewed range
from organometallics to physiologically active compounds such as the
benzodioxapines and macrolide antibiotics, from the commercially
important caprolactam to the synthetically useful crown ethers and
theoretically interesting heteroannulenes. Although the chapters in
this book stand on their own, it is intended that each one should be
read in conjunction with the parent chapter in the second edition.

At a time when there are many specialist reviews, monographs
and reports available, there is still in my view a place for a
book such as "Rodd" which gives a broader coverage of organic
chemistry. One aspect of the value of this work is that it
allows the expert in one field to quickly find out what is
happening in other fields of chemistry. On the other hand a
chemist looking for a way into a fresh field of study will find
in "Rodd" an outline of the important aspects of that area of
chemistry together with leading references to other works to
provide more detailed information.

As can sometimes happen with a multi-author project the
arrival of manuscripts was spread over quite a considerable
period of time. I wish to thank those authors who had to wait
patiently for publication. I am grateful to all the contributors
to this supplement as each one provided a carefully compiled
critical survey of their particular subject area. I very much
appreciate the considerable amount of effort which each con-
tributor put in, not only in compiling the material for their
chapter, but also in producing, with the aid of their secretaries

VIII

and artists, very clear camera-ready copy. I also wish to
thank the staff of Elsevier for the help they have given me
and for seeing the production of this volume through to its
final form.

January 1989   Martin F. Ansell

CONTENTS

VOLUME IV K

Heterocyclic Compounds: Six-membered Heterocyclic Compounds Containing Two or More Hetero-atoms, One or More of which are from Groups II, III, IV, V or VII of the Periodic Table. Heterocyclic Compounds with Seven or More Atoms in the Ring

Chapter 48. Compounds with Six-membered Heterocyclic Rings Possessing Two or More Unusual Hetero-atoms
by P.D. LICKISS

X

Chapter 50. Seven-membered Ring Compounds Containing Oxygen or Sulphur in the Ring
by J.T. SHARP

Chapter 51. Seven-membered Ring Compounds Containing Nitrogen in the Ring
by J.T. SHARP

Chapter 52. *Compounds with Larger Heterocyclic Rings: Seven-membered Ring Compounds with Two or More Different Elements in the Ring* by C.D. GABBUTT and J.D. HEPWORTH

Chapter 56. Compounds with Larger Heterocyclic Rings: Compounds with Rings of More than Eight Atoms
by J.A.H. MACBRIDE

XVIII

OFFICIAL PUBLICATIONS

B.P.	British (United Kingdom) Patent
F.P.	French Patent
G.P.	German Patent
Sw.P.	Swiss Patent
U.S.P.	United States Patent
U.S.S.R.P.	Russian Patent
B.I.O.S.	British Intelligence Objectives Sub-Committee Reports
F.I.A.T.	Field Information Agency, Technical Reports of U.S. Group Control Council for Germany
B.S.	British Standards Specification
A.S.T.M.	American Society for Testing and Materials
A.P.I.	American Petroleum Institute Projects
C.I.	Colour Index Number of Dyestuffs and Pigments

SCIENTIFIC JOURNALS AND PERIODICALS

With few obvious and self-explanatory modifications the abbreviations used in references to journals and periodicals comprising the extensive literature on organic chemistry, are those used in the World List of Scientific Periodicals.

LIST OF COMMON ABBREVIATIONS AND SYMBOLS USED

A	acid
Å	Ångström units
Ac	acetyl
a	axial; antarafacial
as, $asymm.$	asymmetrical
at	atmosphere
B	base
Bu	butyl
b.p.	boiling point
C, mC and µC	curie, millicurie and microcurie
c, C	concentration
C.D.	circular dichroism
conc.	concentrated
crit.	critical
D	Debye unit, 1×10^{-18} e.s.u.
D	dissociation energy
D	dextro-rotatory; dextro configuration
DL	optically inactive (externally compensated)
d	density
dec. or decomp.	with decomposition
deriv.	derivative
E	energy; extinction; electromeric effect; Entgegen (opposite) configuration
E1, E2	uni- and bi-molecular elimination mechanisms
E1cB	unimolecular elimination in conjugate base
e.s.r.	electron spin resonance
Et	ethyl
e	nuclear charge; equatorial
f	oscillator strength
f.p.	freezing point
G	free energy
g.l.c.	gas liquid chromatography
g	spectroscopic splitting factor, 2.0023
H	applied magnetic field; heat content
h	Planck's constant
Hz	hertz
I	spin quantum number; intensity; inductive effect
i.r.	infrared
J	coupling constant in n.m.r. spectra; joule
K	dissociation constant
kJ	kilojoule

LIST OF COMMON ABBREVIATIONS

k	Boltzmann constant; velocity constant
kcal	kilocalories
L	laevorotatory; laevo configuration
M	molecular weight; molar; mesomeric effect
Me	methyl
m	mass; mole; molecule; *meta-*
ml	millilitre
m.p.	melting point
Ms	mesyl (methanesulphonyl)
[M]	molecular rotation
N	Avogadro number; normal
nm	nanometre (10^{-9} metre)
n.m.r.	nuclear magnetic resonance
n	normal; refractive index; principal quantum number
o	*ortho-*
o.r.d.	optical rotatory dispersion
P	polarisation, probability; orbital state
Pr	propyl
Ph	phenyl
p	*para-*; orbital
p.m.r.	proton magnetic resonance
R	clockwise configuration
S	counterclockwise config.; entropy; net spin of incompleted electronic shells; orbital state
S_N1, S_N2	uni- and bi-molecular nucleophilic substitution mechanisms
$S_N i$	internal nucleophilic substitution mechanisms
s	symmetrical; orbital; suprafacial
sec	secondary
soln.	solution
symm.	symmetrical
T	absolute temperature
Tosyl	*p*-toluenesulphonyl
Trityl	triphenylmethyl
t	time
temp.	temperature (in degrees centigrade)
tert.	tertiary
U	potential energy
u.v.	ultraviolet
v	velocity
Z	zusammen (together) configuration

LIST OF COMMON ABBREVIATIONS

α	optical rotation (in water unless otherwise stated)
$[\alpha]$	specific optical rotation
αA	atomic susceptibility
αE	electronic susceptibility
ϵ	dielectric constant; extinction coefficient
μ	microns (10^{-4} cm); dipole moment; magnetic moment
μB	Bohr magneton
μg	microgram (10^{-6} g)
λ	wavelength
ν	frequency; wave number
$\chi, \chi d, \chi \mu$	magnetic, diamagnetic and paramagnetic susceptibilities
~	about
(+)	dextrorotatory
(-)	laevorotatory
(±)	racemic
$\ominus$	negative charge
$\oplus$	positive charge

Chapter 48

COMPOUNDS WITH SIX-MEMBERED HETEROCYCLIC RINGS POSSESSING TWO
OR MORE UNUSUAL HETERO ATOMS

P. D. LICKISS

The number of compounds prepared with six-membered rings
containing two or more unusual hetero atoms has grown rapidly.
This has been due mainly to their use as precursors to unusual
organometallic species and to their formation as products from
the trapping of reactive intermediates. No six-membered rings
containing five unusual hetero atoms seem to have been
prepared which is presumably due to a lack of synthetic effort
rather than any particular instability of such compounds.
Rings containing elements from different groups of the
periodic table have been arranged so that the highest group
number has priority, for example, rings containing boron and
silicon are found in the group (IV) section and rings containing germanium and phosphorus are in the group (V) section.

1.   Boron Compounds

(a)   1,4-Diborins

The reaction of potassium in graphite with dibromo-
(methyl)borane in the presence of an internal alkyne affords
the 1,4-dimethyl-1,4-diboracyclohexa-2,5-diene (e.g. 1-5).
Thus, 5-decyne gives (1), b.p. 60-65°/1x10^{-6}mm, and 3- and 4-
decyne give respectively the pairs of isomeric products (2 and
3) and (4 and 5) (S.M. Van Der Kerk *et al.*, Polyhedron, 1984,
3, 271; J. organometallic Chem., 1980, 190, C8).

(1)   R = R¹ = R² = R³ = n-Bu
(2)   R = R² = Et, R¹ = R³ = n-Hexyl
(3)   R = R³ = Et, R¹ = R² = n-Hexyl
(4)   R = R² = n-Pr, R¹ = R³ = n-Pentyl
(5)   R = R³ = n-Pr, R¹ = R² = n-Pentyl

The formation of the 1,4-diborin ring is thought to occur either *via* dimerization of a boracyclopropene (from addition of a carbene analogue, MeB: to the C≡C bond) or *via* haloboration. Several groups, including the original authors, have failed to repeat earlier work in which a mixture of diphenylacetylene, dibromo(phenyl)borane and potassium gave hexaphenyl-1,4-diboracyclohexa-2,5-diene (Van Der Kerk *et al.*, *loc. cit.*).

Much work has been directed towards the use of 1,4-diborins as ligands in transition metal chemistry. A useful route to these compounds involves treatment of 1,1,4,4-tetramethyl-1,4-distannacyclohexa-2,5-diene with dibromo(ferrocenyl)borane to give 1,4-diferrocenyl-1,4-diboracyclohexa-2,5-diene (6), m.p.>300°, as a violet powder in ~ 83% yield (G. E. Herberich and B. Hessner, J. organometallic Chem., 1978, **161**, C36). Treatment of (6) with nickeltetracarbonyl or methanol gives (7), decomp.>180°, and (8) respectively.

FcB[   ]BFc

(6)

FcB[≡]BFc
|
Ni
(CO)$_2$

(7)

MeOB[   ]BOMe

(8)

Fc = $C_5H_5FeC_5H_4$

Compounds (6) and (8) have both been used in the preparation of a variety of transition metal complexes, for example, (6) reacts with Fe, Ru and Os carbonyls to give fair to good yields of complexes in which the 1,4-diborin ring acts as a four electron donor ligand (Herberich and M. M. Kucharska-Jansen, J. organometallic Chem., 1983, **243**, 45). Complexes of Ni, Co, Rh and Pt containing 1,4-diborin rings with a range of substituents have also been prepared starting from (6) and (8) (Herberich and Hessner, Ber., 1982, **115**, 3115). Replacement of both methoxides in complexes containing ligand (8) by Ph, Me, H and Cl at boron is achieved using Ph$_3$Al, Me$_3$Al, H$_3$Al and BCl$_3$ and treatment of the chlorides with Me$_2$NH affords dimethylamino derivatives. The structures of several complexes containing 1,4-diborin ligands have been determined by X-ray crystallography: the air-stable cobalt complex (9), m.p. 122°, (Herberich *et al.*, J. organometallic Chem., 1980, *192*, 421); the triple-decker sandwich compound (10), m.p. >320°, (Herberich *et al.*, Angew. Chem. intern. Edn., 1981, *20*, 472); (11), derived from Ni(CO)$_4$ and (12), m.p. >320°; and (12) (J. A. K. Howard *et al.*, J. chem. Soc. Dalton, 1975, 2466).

The first example, (13) of a 1,4-diborabenzene without substituents on boron has been observed by e.s.r. spectroscopy (H. Bock *et al.*, Ber., 1980, *113*, 3196).

DME=dimethoxyethane   (13)

(b) 1,3,5-Triborinanes

The adamantane-like structure of a product from the pyrolysis of trimethylborane, 2,4,6,8,9,10-hexamethyl-2,4,-6,8,9,10-hexabora-adamantane, (14), has been confirmed by an X-ray crystallographic study (I. Rayment and H. M. M. Shearer, J chem. Soc. Dalton, 1977, 136).

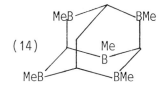

(14)

B-C bond length, mean 1.56 Å
Ring C-B-C angle, mean 117.2°

(c) 1,2,4,5-Tetraborinanes

1,2,4,5-Tetrakis(dimethylamino)-1,2,4,5-tetraborinane (15), m.p. 144°(decomp.), is formed in 29% yield when bis-(cholorodimethylaminoboryl)methane reacts with Na/K alloy. An X-ray crystal structure determination showed that the C_2B_4 ring adopts a chair conformation, with the four boron atoms nearly coplanar (H. Fisch *et al.*, Angew. Chem. intern. Edn., 1984, __23__, 608).

R=Me$_2$N   (15)

2. Compounds of silicon, germanium and tin

(a) Compounds with two hetero atoms

There is interest in reactions of silylenes $R_2Si:$, with alkenes and alkynes to form reactive three membered ring intermediates, dimerizations of silacyclopropenes to 1,4-disilacyclohexadienes receiving particular attention. Many advances have been made in the preparation of compounds with double bonds between the heavier group IV elements. This has led to the preparation of many cyclic compounds containing two adjacent silicon or germanium atoms as precursors of such species. The highly reactive Si=Si and Ge=Ge species may be trapped by dienes, giving rise to products containing a heterocyclic ring.

(i) 1,2-Disilacyclohexanes

The addition of difluorosilylene to ethylene and substituted ethylenes to give 1,1,2,2-tetrafluoro-1,2-disila-cyclohexanes has been the subject of much investigation. Two mechanisms have been proposed, one involving addition of $F_2Si:$ to the alkene to give a silacyclopropane (1) which then dimerizes, and the other involving ·SiF_2SiF_2· addition to one ethylene molecule to give a diradical (2) which then reacts with a second ethylene unit:

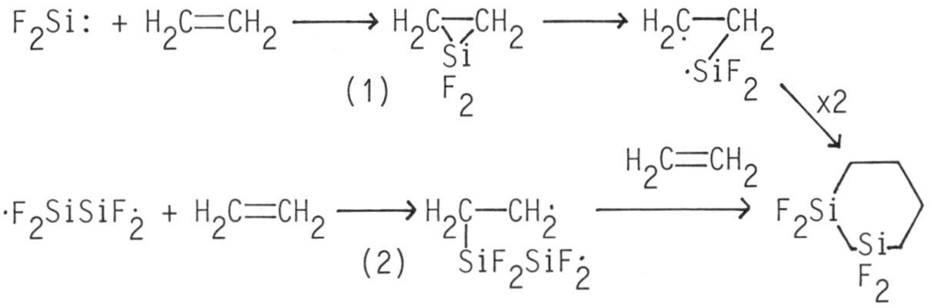

Seyferth and Duncan (J. Amer. chem. Soc., 1978, 100, 7734) found that pure 1,1-difluoro-1-silatetramethylcyclopropane undergoes dimerization at 60° to give 1,1,2,2-tetrafluoro-1,2-disilaoctamethylcyclohexane, m.p. 203-205°, demonstrating that a mechanism involving a silacyclopropane is possible. Examination of the polymeric co-products from co-condensation of F_2Si: with ethylene (W. F. Reynolds et al., Can. J. Chem., 1980, 58, 419) shows that few F_2SiSiF_2 units are present, this also being consistent with a F_2Si: addition pathway. However the reaction of F_2Si: in the gas phase (where monomeric F_2Si: should predominate) with propene gives no 1,2-disilacyclohexane product, but with cycloheptatriene the major volatile product contains an F_2SiSiF_2 unit and it thus seems that both reaction conditions and the alkene greatly influence the reaction mechanism (W. L. Lee et al., J. organometallic Chem., 1986, 302, 23).

Heating of (3-butenyl)-1,1,2,2-tetramethyldisilane in the presence of di-tert-butylperoxide leads to intramolecular cyclization to form 1,1,2,2-tetramethyl-1,2-disilacyclohexane, 23% (T. J. Barton and A. Revis, J. Amer. chem. Soc., 1984, 106, 3802).

(ii)   1,2-Disilacyclohex-3-enes,   1,2-disilacyclohex-4-enes, their benzologues and their germanium and tin analogues

Oxidation by singlet oxygen and subsequent reduction of 1,2-disilacyclohex-4-enes (3, R = H or R = Me) leads to what seem to be the only known 1,2-disilacyclohex-3-enes (4) R = H, b.p. 118°/22 mm (A. Laporterie et al., Nouv. J. De Chim., 1983, 7, 225).

$$\underset{(3)}{\overset{R \quad R}{\underset{Me_2Si-SiMe_2}{\bigcirc}}} \xrightarrow[\text{2)NaBH}_4]{\text{1)}^1O} \underset{(4)}{\overset{R \quad R}{\underset{Me_2Si-SiMe_2}{\bigcirc}}OH}$$

A range of 1,2-disila- and -digerma-cyclohex-4-enes (5) has been made by trapping disilenes or digermenes (usually generated by photolysis or thermolysis) with dienes, as shown in Table 1. Trapping of tetramethyldisilene and tetramethyl-digermene with 1,4-diphenyl-butadiene gives (6) M = Si, 15% m.p. 155°; M = Ge, 17%, m.p. 169° (A. Marchand et al., J. organometallic Chem., 1984, 267, 93).

Table 1.

1,4-Disila(germa)cyclohex-4-enes (5) from diene trapping experiments

Disilene/Digermene	R and R' in diene $H_2C=CRCR'=CH_2$		Yield(%)	Reference
	R	R'		
$Me_2Si=SiMe_2$	H	Me	22	1
$Me_2Si=SiMe_2$	Me	Me	38	2
$(Me_3Si)MeSi=SiMe_2$	Me	Me	20-34	3,4
$t\text{-}Bu_2Si=SitBu_2$	Me	Me	4	5
$(Et_2CH)_2Si=Si(CHEt_2)_2$	Me	Me	19	6
$Et_2Ge=GeEt_2$	Me	Me	5	7
$Ph_2Ge=GePh_2$	Me	Me	65*	7

*Impure

References:
1. T. J. Barton and J. A. Kilgour, J. Amer. chem. Soc., 1976, 98, 7746.
2. H. Sakurai et al., Chem. Letters, 1984, 1379.
3. Sakurai et al., J. Amer. chem. Soc., 1982, 104, 6156.
4. Sakurai et al., Organometallics, 1983, 2, 1484.
5. S. Masamune et al., ibid., 1983, 2, 1464.
6. Masamune et al., J. Amer. chem. Soc., 1983, 105, 6524.
7. P. Riviere et al., J.organometallic Chem., 1984, 264, 193.

(5)
M = Si or Ge

(6)
M = Si or Ge

(7)

Photolysis of peralkylpolystannanes gives rise to dialkyl stannylenes, which react as either R_2Sn: or $\cdot R_2SnSnR_2\cdot$ to give for example, (7, R = Me), 14%, m.p. 78° (W. P. Neumann and A. Schwarz, Angew. Chem. intern. Edn., 1975, 14, 812). In the presence of a mixture of rhenium oxide, aluminium oxide and tetrabutylstannane at 40°, 1,2-diallyltetramethyldisilane loses ethylene to form 1,1,2,2-tetramethyl-1,2-disilacyclohex-4-ene in 55-60% yield (N. V. Ushakov et al., Izvest. Akad. Nauk S.S.S.R., 1981, 2835).

The addition of disilenes or digermenes to cyclic dienes and aromatic compounds leads to a range of bridged compounds which can themselves be used to generate disilenes or digermenes. Disilenes (8) react with anthracene to give the bridged dibenzo products (9), R = R' = Me₃Si, 48%, m.p. 275°; R,R' = Me₃Si,Me, 72%, b.p. 160-180°/0.2 mm; R = R' = Me, 36%; R,R' = Me,Ph, 40% (mixture of *cis* and *trans* isomers); R = R' = Ph, 46% (Y. Nakadaira et al., J. Amer. chem. Soc., 1979, 101, 486). Addition of the pure *cis* or pure *trans* isomer of (8, R,R' = Ph,Me) to anthracene proceeds with retention of stereochemistry to give isomerically pure (9, R,R' = Ph,Me) *cis* m.p. 136-137°, *trans* m.p. 165° (*idem, ibid.*, 1979, 101, 487; Nakadaira et al., Chem. Letters, 1985, 643).

RR'Si≡SiRR' +
(8)

(9)

The reaction of tetramethyldigermene with anthracene gives (10), m.p. 173-175° (decomp.), and that with 1,1-dimethyl-2,5-diphenyl-1-silacyclopentadiene gives (11), 67% (P. Bleckmann *et al.*, Tetrahedron Letters, 1984, 25, 2467; Sakurai *et al.*, Chem. Letters, 1982, 1855). Tetramethyldisilene gives the disilabicyclo[2.2.1]hept-2-ene (12) with cyclopentadiene (Nakadaira *et al.*, Tetrahedron Letters, 1981, 22, 2417), and the 1,2-disilacyclohexadiene (15, see iii below) reacts with maleic anhydride to give (13), m.p. 168-170° (decomp.) (Laporterie *et al.*, Nouv. J. De Chim., 1983, 7, 225).

(10)   (11)   (12)   (13)

(iii)   1,2-Disilacyclohexadienes their benzologues and their germanium analogues

The first 1,2-disilacyclohexadiene (15), b.p. 45°/12 mm, without substituents on carbon has been prepared by dehydration of (14).It reacts with dienophiles such as maleic anhydride (see (ii) above).

(14)   (15)

Treatment of 1,4-dilithio-1,4-diphenyl-1,3-butadiene with 1,2-dichlorotetramethyldigermane gives (16), b.p. 150°/0.07 mm, (Sakurai *et al.*, Chem. Letters, 1982, 1855) and with 1,2-difluoro-1,2-diphenyldimethyldisilane gives a *cis/trans* mixture which can be separated by chromatography to give (17), *cis*, m.p. 114°, and *trans*, m.p. 112-113°, (Nakadaira *et al.*,

J. Amer. chem. Soc., 1979, 101, 487). Heating the 1,1,2,2-tetramethyl analogue of (17) with iron pentacarbonyl gives a product in which the 1,2-disilacyclohexadiene ring is coordinated to an Fe(CO)₃ centre (Nakadaira et al., J. organo-metallic Chem., 1979. 165, 399).

(16)   (17)   (18)   (19)

The reaction of benzyne with (16) and with the isomers of (17) gives respectively (18), 46%, m.p. 145-150°, (Sakurai et al., Chem. Letters, 1982, 1855) and (19) cis, m.p. 192-193°, trans, m.p. 199-200°, (Nakadaira et al., J. Amer. chem. Soc., 1979, 101, 487). Dilithiobiphenyl reacts with 1,2-di-fluorotetramethyldisilane and 1,2-dichlorotetramethyldigermane to give the dibenzo compounds (20, M = Si), m.p. 61° and (20, M = Ge), m.p. 58-60° respectively , (M. Kira et al., J. Amer. chem. Soc., 1983, 105, 7469). Thermolysis of the disila-cyclopropane (21) afforded a quantitative yield of the benz-annelated (22), m.p. 140-143°, (M. Ishikawa et al., J. Amer. chem. Soc., 1985, 107, 7706).

M = Si or Ge   mes = 2,4,6-trimethylphenyl

(20)   (21)   (22)

(iv)   1,3-Disilacyclohexanes and their tin analogues.   1,3-disilacyclohexenes and their benzologues

A 75% total yield of three of the four possible isomers of (23) is obtained when two equivalents of vinyldimethyl-ethoxy-silane react with tert-butyllithium in THF; in hexane the same

reagents give silacyclobutanes (P. R. Jones *et al.*, J. organometallic Chem., 1978, <u>159</u>, 99). A low yield of 1,1,3-tri-
methyl-1,3-disilacyclohexane (24) is obtained when 1-tri-
methylsilyl-3-(dichloro)methylsilylpropane is passed over Na/K
alloy at about 300°; the product is thought to arise from
intramolecular silylene insertion into a C-H bond of one of
the $SiMe_3$ methyl groups (L. E. Gusel'nikov *et al.*, J. organometallic Chem., 1985, <u>292</u>, 189). The 1,3-distannacyclohexanes
(26) are formed when the disodium reagents (25) are treated
with 1,3-dichloropropane in liquid ammonia (26, R = Me), b.p.
120°/12 mm, (K. Jurkschat and M. Gielen, Bull. Soc. chim.
Belg., 1985, <u>94</u>, 299).

$$Me_2Si \qquad Me_2Si \diagup \diagdown SiMeH \qquad H_2C(SnR_2Na)_2 \xrightarrow{Cl(CH_2)_3Cl} R_2Sn \diagup \diagdown Sn^{R_2}$$

R = Me or Ph

(23)   (24)   (25)   (26)

        Pyrolytic methods are useful in the preparation of a
number of 1,3-disilacyclohexenes. Thus, heating 1,2-divinyl-
tetramethyldisilane at 620° gives a low yield of (27, R = Me)
(T. J. Barton and W. D. Wulff, J. organometallic Chem., 1979,
<u>168</u>, 23); heating 1,1-dihalo-1-silacyclopent-3-enes at 680°
leads, *via* silylene insertion into a C-C bond, to (27), R = F,
40%, b.p. 93-94°/750 mm; R = Cl, 45%, b.p. 98-100°/10 mm (E.
A. Chernyshev *et al.*, Doklady Akad. Nauk. S.S.S.R., 1984, <u>276</u>,
1151); head-to-tail dimerization of 1-silatoluene generated
thermolytically affords (28), 14% (C. L. Kreil *et al.*, J.
Amer. chem. Soc., 1980, <u>102</u>, 841; Barton and M. Vuper *ibid.*,
1981, <u>103</u>, 6788); and heating of 1,1,2,2-tetra(2,6-di-
methylphenyl)-1,2-disilacyclopropane at 255° leads via Si-Si
bond scission, to its isomer (29), 54%, (S. Masamune *et al.*,
ibid., 1983, <u>105</u>, 7776).

    (27)   (28)   (29)
                                    R=2,6-dimethylphenyl

(v) 1,4-Disilacyclohexanes

The reactions of tricyclo[4.1.0.0.^{2,7}]hept-1-yllithium with dichlorosilanes and silicon tetrachloride give several unusual polycyclic 1,4-disilacyclohexane derivatives, such as (30), R = Me, m.p. 141° and R = Ph, m.p. 257-260° (decomp.). The structure of (30, R = Me) has been confirmed by an X-ray crystal structure analysis (H.-G. Zoch *et al.*, Ber., 1981, 114, 3896). Transition metal catalysed ring opening of octamethyl-1,2-disilacyclobutane in the presence of allene gives the exocyclic methylene compound (31), 91%, m.p. 76-78° (D. Seyferth *et al.*, J. organometallic Chem., 1984, 271, 337). The highly reactive 6,6-dimethyl-6-silafulvene (32) can be prepared by treating cyclopentadienyldimethylchlorosilane with either methylenetriphenylphosphorane (N. N. Zemlyanskii *et al.*, Zhur. organicheskoi Khim., 1981, 17, 1323) or *tert*-butyllithium (P. R. Jones *et al.*, Organometallics, 1985, 4, 1321) or by thermolysis of allylcyclopentadienyldimethylsilane (Y. Nakadaira *et al.*, Chem. Letters, 1980, 1071; T. J. Barton *et al.*, Tetrahedron Letters, 1981, 22, 7). The fulvene dimerizes and rearranges to give (33), m.p. 106-107.5°, which in solution undergoes ready silyl shifts. The solid state structure has been confirmed by X-ray crystallography, which shows the central ring to have a chair conformation (V. K. Belsky *et al.*, Cryst. Struct. Comm., 1982, 11, 497).

(30)   (31)   (32)   (33)

The bis-silacyclopentadiene (34) undergoes a reversible intramolecular [2+2] cycloaddition to give (35), m.p. 172-173°, (H. Sakurai *et al.*, Organometallics, 1983, 2, 1814).

(34)   (35)

Hydroboration of group (IV) dimethyldi-*iso*-propenyl compounds has given *cis/trans* mixtures of heterocycles, for example (36), containing boron and silicon, or germanium and tin, and these have been used to prepare diols and metallacyclohexanones (J. A. Soderquist and A. Hassner, J. org. Chem., 1983, 48, 1801; Soderquist *et al.*, *ibid.*, 1984, 49, 2565).

M = Si,Ge or Sn   (36)

(vi)   *1,4-Disilacyclohex-2-ene,   1,4-disilacyclohexadienes, their benzologues and their germanium and tin analogues*

Treatment of octamethyl-1,2-disilacyclobutane with a terminal alkyne or an activated internal alkyne in the presence of bis(triphenylphosphine)palladium dichloride provides a good route to the 1,4-disilacyclohexenes (37); for example phenylacetylene gives (37, R = Ph R' = H) and MeO₂CC≡CCO₂Me gives (37, R = R' = MeO₂C), m.p. 81.5-83°, (D. Seyferth *et al.*, J. organometallic Chem., 1984, 271, 337). Thermolysis of 1,2-divinyltetramethyldisilane at 620° gives a low yield of (37, R = R' = H) (Barton and Wulff, J. organometallic Chem., 1979, 168, 23), and thermolysis of the conjugated ene-yne (38) gives a nearly quantitative yield of (39), m.p. 87-88°, (B. C. Berris and K. P. C. Vollhardt, Tetrahedron, 1982, 38, 2911).

(37)   (38)   (39)

Ring opening of 1,2-disilacyclobutenes and silacyclopropenes, either thermally or under transition metal catalysis, provides a useful route for the preparation of a variety of substituted 1,4-disilacyclohexadienes. Heating of 1,1,2,2,3,-4-hexamethyl-1,2-disilacyclobutene with 2-butyne and with 3-hexyne leads respectively to (40, R = Me) and (40, R = Et) (Barton and J. A. Kilgour, J. Amer. chem Soc., 1976, 98, 7746). The dichlorosilacyclopropanes (41) (prepared by :SiCl$_2$ addition to acetylenes) dimerize to give (42) R = H, m.p. 87°; R = Ph, m.p. >350°. In (42, R = H) the four chlorines have been replaced by methyl, hydride and fluoride, b.p.'s 157-158°, 84°, and 62-63°, respectively (E. A. Chernyshev et al., Zhur. obshchei Khim., 1978, 48, 830).

(40)   (41)   (42)

Treatment of silacyclopropenes (43) with bis(triethylphosphine)-palladium dichloride gives the 1,4-disilacyclohexadienes (44) R = Me, m.p. 182-183°; R = Et, m.p. 133-134°; R = tert-Butyl, m.p. 168-169° and R = Ph, m.p. 135°. Where there is a possibility of forming both the cis and the trans isomer, only the trans isomer is found (M. Ishikawa et al., Organometallics 1982, 1, 1473). Analogues of (44), such as (45), m.p. 231-232°, (Ishikawa et al., loc. cit.) and (46), m.p. 80°, have been prepared by similar routes (Ishikawa et al., ibid., 1985, 4, 2040).

(43)   (44)   (45)   (46)

The reactive intermediate hexamethyl-1,4-disilabenzene can be trapped by various unsaturated substrates to give a range of polycyclic compounds. Formation of hexamethyl-1,4-disilabenzene by reduction of (47) with lithium in the presence of anthracene gives (48), m.p. 246-247.5°, which when heated alone at 700° gives a low yield of (49). On photolysis or thermolysis at 600° (48) loses hexamethyl-1,4-disilabenzene (50), which can be trapped with acetylene to give (51, R = H); with hexafluoro-2-butyne to give (51, R = CF₃), and with methanol to give a *cis/trans* mixture of (52) (J. D. Rich and R. West, J. Amer. chem. Soc., 1982, 104, 6884).

The syntheses of a number of novel polyspiro-disilacyclo-hexadienes have been achieved using lithium reagent (53), which with silicon tetrachloride gives (44), m.p. 77-78°, and with 1,1,4,4-tetrachloro-1,4-cyclohexadiene gives (55), m.p. 164-165.5°. The compound, m.p. 195-196°, containing three spiro silicons is prepared in a similar way. A red shift in the uv spectra of these compounds with increasing number of spiro silicons seems to indicate that they relay the conjugation effectively (Sakurai *et al.*, Tetrahedron Letters 1982, 23, 543).

The versatility of dialkynyl metalloids for the preparation of hetero-cyclohexadienes has been exploited further. Addition of dimethylstannane to dimethyldiethynylstannane catalysed by azo-bis-isobutyronitrile affords (56), m.p. 30-31°, (G. E. Herberich and B. Hessner, Z. Naturforsch., 1978, 33B, 180), addition of di-n-butylstannane to dialkynylsilanes catalysed by potassium hydroxide gives (57) R = Me, b.p. 85-88°/0.0004 mm; R = Ph, b.p. 200-210°/8x10⁻⁵ mm (G. Märkl et al., Synthesis 1977, 842). Treatment of 1,1,4,4-tetramethyl-1,4-silastannacyclohexadiene with (dichloro)phenylborane gives (58), b.p. 76°/0.05 mm, which has been used as a ligand in transition metal complexes of iron, cobalt, nickel and platinum; an example is the yellow iron complex (59), m.p. 99°, (Herberich and M. Thönnessen, J. organometallic Chem., 1979, 177, 357; Herberich et al., ibid., 1980, 191, 27; Ber., 1977, 110, 760).

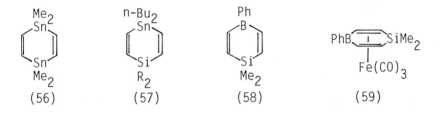

(56)   (57)   (58)   (59)

Tetrachloro-9,10-disilanthrene (60), m.p. 166° is obtained when o-dichlorobenzene is passed over a Si/Cu mixture at 540° (E. A. Chernyshev et al., Zhur. obshchei Khim., 1980, 50, 1790). The dimetallic reagents (61, M = MgCl or Li) react with group (IV) dichlorides to give a range of heterocyclic anthracenes; for example (62) R₂E = Me₂Si, Ph₂Si, Me₂Ge, Me₂Sn and Ph₂Pb, m.p.'s 81-82, 125-126.5, 83-84, 65-66 and 128-129°, respectively. The X-ray crystal structure of (62, R₂E = Me₂Ge or Me₂Sn) shows the central ring to be in a boat conformation (W. Z. McCarthy et al., Organometallics 1984, 3, 255). The crystal structure of (63) (O. A. D'yachenko et al., Zhur. strukt. Khim., 1976, 17, 496) and that of the chromium tricarbonyl complex of (62, R₂E = Me₂Si) (V. A. Sharapov et al., Organometallics, 1984, 3, 1755) also show the central ring to have a boat conformation. A polyfluoro triptycene analogue (64), m.p. 252-253°, has been prepared in a multi-step synthesis starting from 1,2-dibromotetrafluorobenzene (N. A. A. Al-Jabar and A. G. Massey, J. organometallic Chem., 1985, 287, 57).

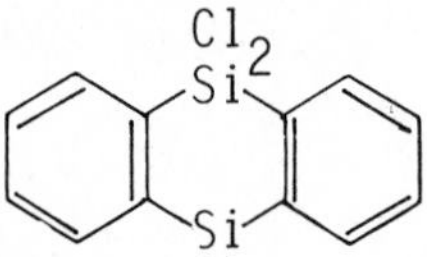

(60) 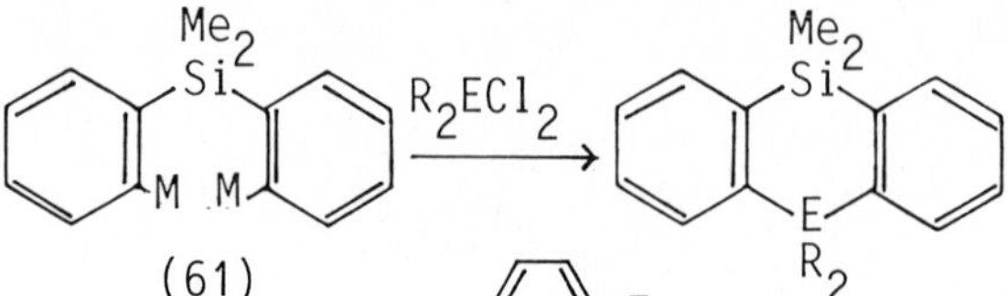　　　　　　　　　(61)

(62)

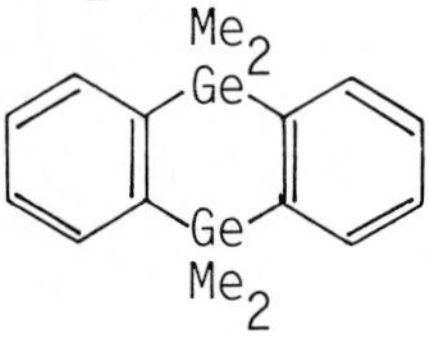

(63) 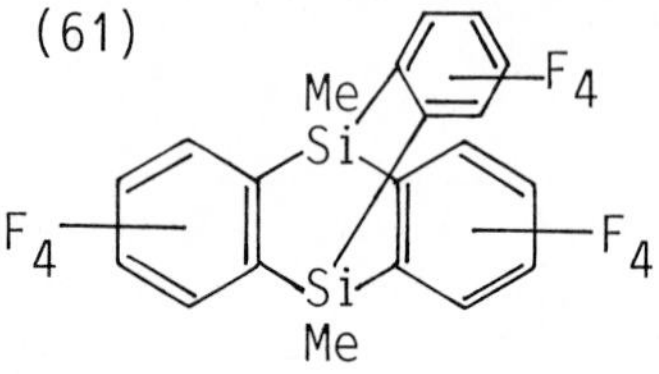　　　　　　　　　(64)

(b) *1,2,3-Trisilacyclohexanes, 1,3,5-trisilacyclohexanes and their tin analogues*

A mixture of 1,1-dichlorotetramethyldisilane and 1,3-bis-(chlorodimethylsilyl)propane reacts in the presence of lithium and triphenylsilyllithium to give (65) (M. Ishikawa *et al.*, J. organometallic Chem., 1985, **292**, 167), and treatment of (66) with activated magnesium yields (67) (Z. Zhang and X. Zhou, Chem. Abstr., 1984, **101**, 151928p).

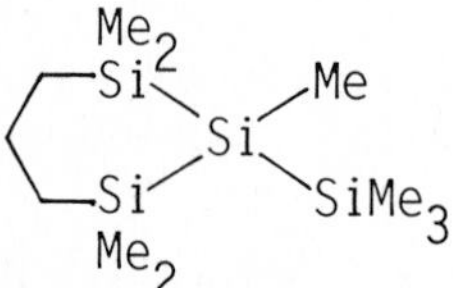

(65) 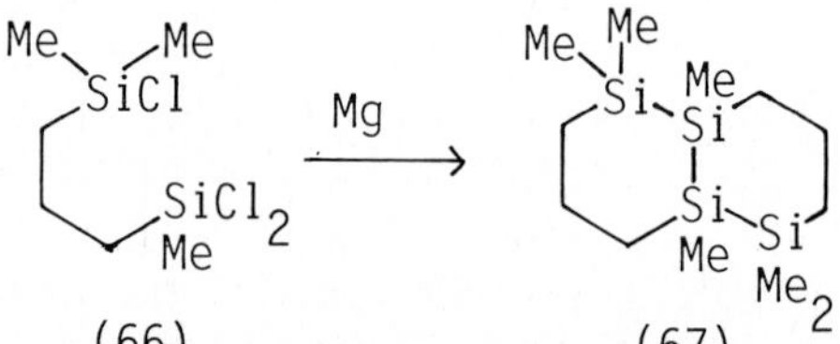　　　　　　　　(66)　　　　　　　　(67)

The high molecular weight products containing from six to nine silicon atoms obtained from the pyrolysis of tetra-methylsilane have been separated by column chromatography or HPLC. Eleven pure compounds and some mixtures of isomers have been isolated, all containing the 1,3,5,7-tetrasilaadamantane unit either, alone, for example, (68), or fused with one or more others, for example (69), m.p. 143-145°, (G. Fritz and K.-P. Wörns, Z. anorg. allg. Chem., 1984, **512**, 103).

(68)   (69)

The passage of dichloromethane or chloroform over a bed of Si/Cu at about 320° gives a variety of Si-chlorinated trisila-cyclohexanes, which on reduction with lithium aluminium hydride gave (70) and (71) (Fritz and A. Wörsching, *ibid.*, 1984, **512**, 131).

(70)   (71)

Reactions of 1,3,5,7-tetrasilaadamantanes have been carried out at either the silicon or carbon atoms of the rings, and a variety of methyl and trimethylsilyl derivatives have been prepared (Fritz and K. Kreilein, *ibid.*, 1977, **433**, 61). The 1,3,5-trisilacyclohexanes can likewise undergo reaction at silicon to give a range of halogen- alkyl- and aryl-substituted products (Fritz *et al.*, *ibid.*, 1981, **478**, 20) or at carbon to give a variety of halogen-, alkyl- or trimethyl-silyl substituted products (Fritz and W. Speck, *ibid*, 1981, **481**, 60; Fritz and Wartanessian, J. organometallic Chem., 1979, **178**, 11). Treatment of mono- di- and tri- bromosilyl-1,3,5-trisilacyclohexanes with a variety of metal carbonyl anions gives products such as (72), m.p. 36°, (73), m.p. 157°, and (74), m.p. 106-109°, (Fritz and K. Hohenberger, Z, anorg. allg. Chem., 1980, **464**, 107).

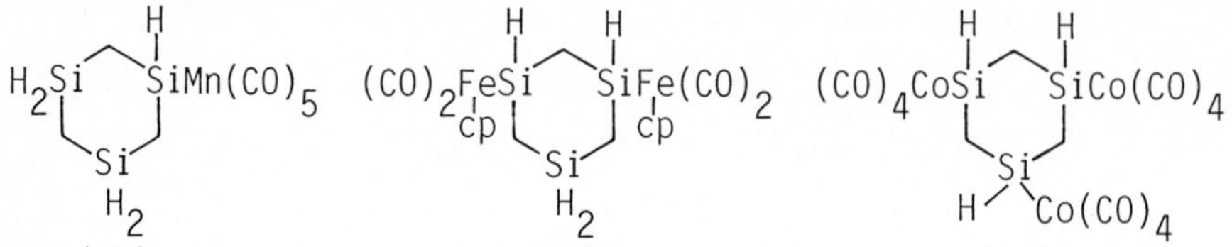

(72) (73) (74)

The ylide trimethyl(silylmethylene)phosphorane is thermally unstable and trimerizes at room temperature to give the cyclic triple ylide (75), m.p. 145°, (H. Schmidbaur and B. Zimmer-Gasser, Z.Naturforsch., 1977, 32B, 603). The mono ylide (76) has been prepared by refluxing a solution of bis[(chlorodimethylsilyl)methyl]dimethylsilane and [bis(trimethylsilyl)methylene]trimethylphosphorane (Schmidbaur and M. Heimann, Ber., 1978, 111, 2696). A 1,3,5-tristannacyclohexane (77), m.p. 184-185° (decomp.), is produced by trimerization of the reactive double bond species 6,6-dimethyl-6-stannafulvene. An X-ray crystallographic study showes that the central ring has a chair conformation (N. N. Zemlyanskii *et al.*, Izvest. Akad. Nauk S.S.S.R., 1981, 2837).

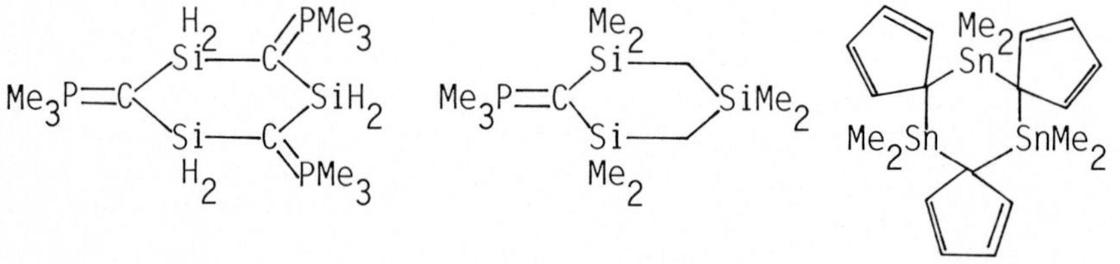

(75) (76) (77)

(c) 1,2,3,4,-Tetrasilacyclohexanes, 1,2,4,5-tetrasilacyclohexanes their tin analogues and related compounds

The 1,2,3,4-tetrasilacyclohexane ring has been formed in several ways. Treatment of a mixture of 1,2-bis(chlorodimethylsilyl)ethane and 1,1-dichlorotetramethyldisilane with lithium yields (78), m.p. 99-100°, (Ishikawa *et al.*, J. organometallic Chem., 1985, 292, 167). Octaethylcyclotetrasilane undergoes ring-opening in the presence of tetrakis(triphenylphosphine)palladium(0) and reacts with alkynes to give 1,2,3,4-tetrasilacyclohex-5-enes (79) (C. W. Carlson and R. West, Organometallics, 1983, **2**, 1801). The benz-annelated compound (80), m.p. 182-184°, can be prepared in low yield by treating o-dichlorobenzene with 1,4-dilithiooctaphenyltetrasilane (E. Hengge and R. Sommer, Monatsh., 1977, **108**, 1413).

$R=R'=CO_2Me$

$R=H, R'=n-Bu$

$R=H, R'=Ph$

(78)　　　　　(79)　　　　　(80)

A range of 1,2,4,5-tetrasila- and tetrastanna-cyclohexanes can be prepared by treating dimetallomethanes (81) with alkali metals; for example, (81, $MR_2X = SiMe_2Br$) reacts with Na/K alloy or lithium powder to give (82, $MR_2 = SiMe_2$) (Fritz and B. Grunert, Z. anorg. allg. Chem., 1981, **473**, 59), and (81, $MR_2X = SnPh_2Br$) reacts with sodium in liquid ammonia to give (82, $MR_2 = SnPh_2$), m.p. 180-182°, (K. Jurkschat and M. Gielen, J. organometallic Chem., 1982, **236**, 69), the X-ray structure of which shows the central ring to be in a chair conformation (J. Meunier-Piret, *ibid.*, 1983, **252**, 289). A decamethyl analogue 1,1,2,2,3,4,4,5,5,6-decamethyl-1,2,4,5-tetrastanna-cyclohexane, m.p. 79-80°, is prepared by a similar route; only the *cis* isomer is formed and the ring has a boat conformation (M. Preut *et al.*, Acta Cryst., 1984, **C40**, 370).

$H_2C\begin{array}{c} MR_2X \\ MR_2X \end{array}$ + Alkali metal $\longrightarrow$

(81)　　　　　(82)

3. Compounds containing phoshorus, arsenic, antimony and bismuth.

(a) Compounds with two hetero atoms

(i)   1,2-Diphosphacyclohexanes, 1,2-diphosphacyclohexenes, and related arsenic, antimony and bismuth compounds

Treatment of ditosylate (1) with 1,2-dilithiodiphenyl-diphosphine gives a mixture of diastereomers of (2), one of which, m.p. 114-116°, can be isolated pure by crystallization. This optically active compound can be used to prepare asymmetric transition metal complexes to be used for asymmetric reductions (S. Y. Zhang *et al.*, Tetrahedron Letters 1981, 22, 3955).   Cyclization of 1,4-bis[(bromo)phenylphosphino]butane with either magnesium or potassium gives a good yield of (3), m.p. 92°, (K. Diemert *et al.*, Ber. 1982, 115, 1947).

Phosphabutadienes undergo spontaneous dimerization to 1,2-diphosphacyclohex-3-enes to give, for example, (4, R^1, R^2, R^3, R^4 = Ph,Ph,OSiMe₃,H) m.p. 118°, (disulphide m.p. 171°); (4, R^1, R^2, R^3, R^4 = cyclohexyl, Ph,OSiMe₃,H), m.p. 105°, (disulphide m.p. 164°) and (4, R^1, R^2, R^3, R^4 = Ph,Ph,H,OSiMe₃), m.p. 124°, (disulphide m.p. 154°) (R. Appel *et al.*, Ber., 1984, 117, 3151). It has been shown by nmr spectroscopy that in solution at low temperature stibabenzene, bismabenzene and *p*-methyl-bismabenzene (5a) are each in equilibrium with the related head-to-head Diels Alder dimer (5b) (A. J. Ashe III *et al.*, J. Amer. chem. Soc., 1982, 104, 5693).

(5a) M = Sb R = H
      M = Bi R = H or Me

The trapping of reactive diphosphene species with dienes leads to the formation of 1,2-diphosphacyclohex-4-enes; for example, di-*tert*-butyldiphosphene gives with 2,3-dimethylbutadiene (6), b.p. 138-139°/2mm and with cyclopentadiene (7, R = t-Bu), b.p. 95°/0.02 mm (J. Escudie *et al.*, J. organometallic Chem, 1982, **228**, C76; C. Couret *et al.*, Organometallics, 1986, 5, 113). Treatment of tetrakis(trimethylsilyl)diaminodiphosphene with cyclopentadiene affords (7, R = $(Me_3Si)_2N$), decomp. 40-45°, (E. Niecke and R. Rüger, Angew. Chem. intern. Edn., 1983, **22**, 155). A mixture of the *cis* and the *trans* disulphides (8) is formed when 1,2-dipotassiodiphenyldiphosphine is treated with 1,4-dichloro-2-butene; the isomers are separable by column chromatography; *cis*-(8) m.p. 183-185°, *trans*-(8) m.p. 198.2-199.5° (T. Kawashima *et al.*, Heterocycles, 1982, **17**, 341). A chromium complex (9) of a 1,2-diarsacyclohex-4-ene is prepared in good yield by treating a reactive intermediate containing an As=As double bond with 2,3-dimethylbutadiene (G. Huttner and I. Jibril, Angew. Chem. intern. Edn., 1984, **23**, 740).

(6)   (7)   (8)   (9)

The *trans*- isomer of disulphide (8) when treated sequentially with *n*- or *sec*-butyllithium and bromotrichloromethane gives the analogous 3-bromide (10), m.p. 165-166°,which in the presence of triethylamine gives the 1,2-diphosphacyclohexadiene (11), m.p. 168-170°, (Kawashima, *et al.*, Heterocycles 1982, **17**, 341).

(10)   (11)

(ii)   1,3-Diphosphacyclohexanes, 1,3-diphosphacyclohexenes and related arsenic compounds

An unusual five-coordinate iron (II) complex (13) is formed in 40% yield when a mixture of carbon disulphide, hydrated iron (II) tetrafluoroborate and the tridentate phosphine (12) is heated in solution (C. Bianchini *et al.*, Angew. Chem. intern. Edn., 1980, 19, 1021). The bis-phosphite (14) when heated with a 1,3-dihalopropane in *p*-xylene solution gives (15), m.p. 84-85°, which on hydrolysis gives the corresponding diacid, m.p. 243-244°, (Z. S. Novikova *et al.*, Zhur. obshchei Khim., 1977, 47, 2636).

(12)   (13)   2(BF$_4$)$^-$

(14)   X = Cl or Br   (15)

The preparation of 1,3-diphosphabenzene *via* 1,3-diphospha-cyclohexanes and cyclohexenes, reported by Märkl (Z. Natur-forsch., 1963, 18B, 1136), could not be repeated by Mastryukova *et al.* (Zhur. obshchei Khim., 1978, 48, 263), who obtained different melting points for the intermediates and found that ring opening occurred in the final deprotonation step rather than formation of a 1,3-diphosphabenzene. The reaction of bis(diphenylphosphino)methane with 1,3-dibromo-propane in refluxing toluene yields (16), m.p. 260° (decomp.), which affords the cyclic carbodiphosphorane (17), m.p. 35° (decomp.) when treated with trimethyl-methylenephosphorane. The carbodiphosphorane (17) gives a methiodide salt m.p. 158° (decomp.) on treatment with methyl iodide, a 1:1 complex with borane and metal bridged dimers with dimethylzinc and dimethylcadmium (H. Schmidbaur and T. Costa, Ber., 1981, 114, 3063).   An intramolecular cyclisation occurs when (18) is treated with sodamide giving a mixture of the *cis*- and the *trans*- cyclic carbodiphosphorane (19)(Schmidbaur and S. Schnatterer, *ibid.*, 1983, 116, 1947).

mes=2,4,6-trimethylphenyl

A tris(phosphonium)methanide salt (21), m.p. >300°, is formed on refluxing a solution of 1,3-dibromopropane and (20) (Schmidbaur *et al.*, *ibid.*, 1983, 116, 3559). Unusual hetero-cyclic noradamantanes (22) R = CO₂Me, m.p. 164°; R = CO₂Et, m.p. 116°, and R = C(O)Ph m.p. 220° (decomp.), are formed when MeC(CH₂AsI₂)₃ is treated with H₂C(COOMe)₂, H₂C(COOEt)₂ and H₂C(COPh)₂ respectively in the prescence of triethylamine (J. Ellermann and M. Lietz, Z. Naturforsch., 1981, 36B, 1532). The unusual nature of the structure (for R = CO₂Et) has been confirmed by an X-ray crystallographic study (G. Thiele *et al.*, *ibid.*, 1984, 39B, 1344).

$Ph_3P\!=\!C(PPh_2)_2$ + $Br(CH_2)_3Br \rightarrow$

(20)   (21)   (22)

(iii) 1,4-Diphosphacyclohexanes

The reaction between primary phosphines and divinyl subs-tituted phosphines catalysed by azo-bis-isobutyronitrile or potassium hydroxide gives a variety of 1,4-diphosphacyclo-hexanes; for example, (23, R = Ph, R' = SiMe₃) is formed in the reaction between phenyldivinylphosphine and trimethyl-silylphosphine (D. M. Schubert and A. D. Norman, Inorg. Chem., 1984, 23, 4130) and (23, R = Et₂N, R' = benzyl), b.p. 122-135°/0.026 mm, is formed in the reaction between diethylaminodivinylphosphine and benzylphosphine (R. B. King and W. F. Masler, J. Amer. chem. Soc., 1977, 99, 4001). The phosphite (24) on treatment with 1,2-dibromoethane then with hydrochloric acid gives (25), m.p. >330° (decomp.), (Y. M. Polikarpov et al., Izvest Akad. Nauk S.S.S.R., 1977, 1188).

(23)　　　　　　　　(24)　　　　　　　　　　　　　　　(25)

(iv)　1,4-Diphosphacyclohexenes, hexadienes and related compounds and their arsenic, antimony and bismuth analogues

Potassium hydroxide/18-crown-6 catalysed cycloadditions of aryl-phosphines, -arsines and -stannanes with pentane-1,4-diynes provide a general route to 1,4-hetero-atom cyclohexadienes and yields are better than those previously obtained using either azo-bis-isobutyronitrile or lithium amide/ammonia as the catalyst. Phenylphosphine and phenylarsine react with (butyl)diethynylphosphine to give (26, M = P), m.p. 53-57°, and (26, M = As), m.p. 47-52°, with (dimethyl)diethynylsilane to give (27, M = P), b.p. 110-115°/4x10^{-4}mm, and (27, M = As), b.p. 25°/4x10^{-4}mm. (*Tert*-Butyl)diethynylphosphine reacts with di-*n*-butylstannane to give (28), b.p. 120-125°/6x10^{-4} mm, (G. Märkl et al., Synthesis 1977, 842). Upon treatment with HCl in acetic acid various aryl substituted phenylethynyldiarylphosphines form cyclic dimers; for example, (29, R = *p*-F), m.p. 231-233°, and R = *o*-Me, m.p. 284-285° (decomp.), (J. C. Williams et al., Phosphorus, 1976, 6, 169). Heating 1-phenyl-2,5-dimethylarsole at 160° generates arsenic atoms which react with diphenylacetylene to give the diarsabicyclo-[2.2.2]-octatriene (30), m.p. 270°, (G. Sennyey and F. Mathey, Tetrahedron Letters, 1981, 22, 4713).

t-Bu　　　　Me$_2$　　　t-Bu　　　　Ph

(26)　　　　(27)　　　　(28)　　　　(29)　　　　　　(30)

The effect of 1,4-diphosphoniacyclohex-2,5-diene salts on the rate of electron transfer at a mercury/dimethylformamide interface has been investigated (T. Kakutani *et al.*, J. Amer. chem. Soc., 1975, 97, 7226); the electrochemical reduction of the salts (R. D. Rieke, *ibid.*, 1977, 99, 6656; E. A. Berdnikov *et al.*, Izvest. Akad. Nauk S.S.S.R., 1977, 803) and their antibacterial activity (F. C. Pearson *et al.*, Chem. Abstr. 1977, 86, 150944a) have also been investigated.

The 1,4-diphosphabenzene (32) prepared by refluxing a hexane solution of (31) is thermally stable but very sensitive to oxygen and shows a large absorption maximum at 282 nm in its uv spectrum suggesting that the ring has some aromatic character. Treatment of (32) with but-2-yne, dimethyl acetylenedicarboxylate and hexafluorobut-2-yne gives the 1,4-diphosphabicyclo[2.2.2]octatrienes (33, R' = Me), m.p. 152.5-153°), (33, R' = CO₂Me), m.p. 108-110°, and (33, R' = CF₃) respectively. With carbon tetrachloride (32) gives (34, X = CCl₂, m.p. 39-39.5°), with sulphur it gives (34, X = S) and with hexamethyl Dewar benzene it gives (35), m.p. 182-184°, (Y. Kobayashi *et al.*, J. Amer. chem. Soc., 1980, 102, 252; 1981, 103, 2465).

A mixture of *endo* and *exo* adducts (36) m.p. 111-112° and (38), m.p. 97-98°, is formed when (32) reacts with cyclobuta-dieneiron tricarbonyl in the presence of cerium (IV) ion. The *exo*-isomer is inert to uv iradiation but the *endo*-isomer undergoes an internal cyclization to yield the hexacyclic product (37), m.p. 201-203°, the structure of which has been confirmed by X-ray crystallography (Kobayashi *et al.*, J. org. Chem., 1979, 44, 4930; D. Schomburg and Kobayashi, Phos. and Sulf., 1981, 10, 17).

(36)   R = CF$_3$ (37)   (38)

Many phosphanthrenes (39) and arsanthrenes (40) having the hetero atoms in either the (III) or the (V) oxidation state have been prepared by use of the chlorides (39, X = Cl) and (40, X = Cl) as precursors; (39) X = Et, m.p. 52-53°; X = I, m.p. 193-196°; X = OH, m.p. >300°; X = Cl$_3$, m.p. 234-236°; X = I$_3$, m.p. 161-164°; X = (O)Cl, m.p. >300°; X = (O)OH, m.p. >300°, (T. V. Kovaleva, Zhur. obshchei Khim., 1977, 47, 1036), (40) X = I, m.p. 202-206°; X = (O)OH, m.p. >360°; (41) X = O, m.p. 196-197°; X = S, m.p. 184-185°; (42) X = O, m.p. 278-280°; X = NPh, m.p. 262-263° (H. Vermeer et al., Tetrahedron, 1979, 35, 155). Polyfluorinated anthracene analogues (43) M = P R = Ph , m.p. 89°, M = As R = Me m.p. 105° and M = As R = Ph, m.p. 118°, are formed in the reaction of 1,2-dilithiotetrafluorobenzene with (dichloro)phenylphosphine, (diiodo)-methylarsine and (di-iodo)phenylarsine respectively (W. R. Cullen and A. W. Wu, J. fluor. Chem., 1976, 8, 183).

(39)   (40)   (41)

(42)   (43)

The range of heterocyclic triptycene analogues has been extended to include both antimony and bismuth derivatives many of which form solvates which may account for some of the discrepancies in melting points recorded for this type of compound. When white phosphorus and *o*-dichlorobenzene are heated together at 325° in the presence of iron (III) chloride and titanium(IV)chloride, 1,6-diphosphatriptycene (44, M = M' = P), m.p. 313-315°, (dioxide m.p. 488-490°, disulphide m.p. 396-399°) is formed (K. G. Weinberg, J. org. Chem., 1975, 40, 3586) The antimony analogue (44, M = M' = Sb), m.p. 254-255°, is formed on heating antimony and *o*-phenylenemercury trimer (N. A. A. Al-Jabar *et al.*, J. organometallic Chem., 1985, 295, 29). Perhalo-triptycene derivatives have been prepared in several ways. When antimony is heated with 1,2-di-iodotetrachloro- or -tetrafluorobenzene the perchloro and perfluoro analogues of (44, M = M' = Sb) are formed (Al-Jabar and A. G. Massey, J. organometallic Chem., 1984, 276, 331). Treatment of 1,2-dilithiotetrafluorobenzene with arsenic (III), antimony (III) or bismuth (III) chloride affords a good yield of the perfluorinated analogue of (44, M = M' = As, Sb or Bi) (Cullen and Wu, J. fluor. Chem., 1976, 8, 183). A lower yielding but more versatile route to perfluoro triptycenes involves the reaction between tris(2-lithiotetra-fluorophenyl) group (V) derivatives and group (V) trichlorides which yields perfluoro derivatives of (44) M = M' = P, m.p. 208-209°; M = M' = As m.p. 206-207°; M = M' = Sb, m.p. 258-259°; M = M' = Bi, m.p. 314°; M = P M' = As, m.p. 214-215°; M = P M' = Sb, m.p. 198-199°; M = As M' = Sb, m.p. 222-223°, (Al-Jabar and Massey, J. organometallic Chem., 1985, 287, 57).

(44)

(b)   Compounds containing three hetero atoms

The thermally unstable unsaturated species (45, M = P or As) trimerize to give 1,3,5-triphospha- and -triarsacyclohex-anes (46) (A. B. Burg, Inorg. Chem., 1983, 22, 2573; J. Grobe and D. Le Van, Angew, Chem. intern. Edn., 1984, 23, 710). Energy barriers for conformational changes in compounds such as (47, for example MR₂ = SiMe₂ or GeEt₂) have been studied by

^{31}P nmr spectroscopy (A. Hauser *et al.*, Phosphorus, 1975, $\underline{5}$, 261; A. Zschunke and I. Nehls, Z. Chem., 1977, $\underline{17}$, 335).

(45)   (46)   (47)

The reaction between bis[bis(trimethylsilyl)methyl] phosphine and (1,5-cyclooctadiene)(2-methylallyl)rhodium(I) affords the novel 1,3,5-phosphadisilacyclohexane rhodium complex (48), m.p. 182°, the structure of which has been confirmed by X-ray crystallography (B. D. Murray *et al.*, Organometallics 1983, $\underline{2}$, 1700; Murray and P. P. Power, *ibid.*,1984, $\underline{3}$, 1199).

$(\eta^4 - \text{COD})\text{Rh}$ (48)   COD=1,5-cyclo-octadiene

The mono and bicyclic ylides (50), b.p. 128°/50 mm, and (49), m.p. 58-59°, are prepared by silylation of a dilithio derivative of an ylide and an ylide respectively (Schmidbaur and M. Heimann, Ber., 1978, $\underline{111}$, 2696).

(49) $\text{Me}_3\text{P}=\text{C(SiMe}_3)_2$ (50)

(c) Compounds containing four hetero atoms

The dimerization of diphosphacyclopropanes, (prepared by various methods), yields 1,2,4,5-tetraphospha-cyclohexanes (51) R = Ph R' = Me; R = t-Bu R' = H, m.p. 141-143°; R = i-Pr R' = H, b.p. 118°/0.01 mm, tetrasulphide m.p. 208° (M. Baudler and B. Carlsohn, Z. Naturforsch., 1977, 32B, 1490; Baudler and F. Saykowski, *ibid.*, 1978, 33B, 1208; Z. S. Novikova *et al.*, Zhur. obshchei Khim., 1979, 49, 471). Treatment of the dilithio reagent (52) with dimethyldichlorosilane gives (53) (C. Couret *et al.*, J. organometallic Chem., 1979, 182, 9), and trapping of the double bonded intermediates (54) yields (55, M = Ge) and (55, M = Sn) although both are thermally unstable and can not be isolated in a pure form (J. Escudie *et al.*, Organometallics, 1982, 1, 1261; Couret *et al.*, J. organometallic Chem., 1981, 208, C3).

R'_2
$RP-C-PR$
$RP-C-PR$
R'_2
(51)

$Me_2SiCH_2CH_2PPh$
$PhPLi \quad Li$
(52)

$\xrightarrow{Me_2SiCl_2}$

Ph Me$_2$
$Me_2Si \begin{array}{c} P-Si \\ C-C \\ H_2 \; H_2 \end{array} PPh$
(53)

$Me_2M=PPh$
(54)

$+$

$Me_2Ge-PPh$
$\boxed{}$

$\longrightarrow$

Ph Me$_2$
$Me_2M \begin{array}{c} P-Ge \\ \end{array} PPh$
(55)

Chapter 49

COMPOUNDS WITH SIX-MEMBERED HETEROCYCLIC RINGS
CONTAINING ONE OR MORE UNUSUAL HETERO ATOMS
TOGETHER WITH NITROGEN OR ELEMENTS OF GROUP VI

D.W. ALLEN

Introduction

The following review articles provide detailed
coverage of some areas of chemistry relevant to
this chapter.
    Spectroscopic information for boron-containing
heterocycles, H. Nöth and B. Wrackmeyer, in "N.m.r.
Spectroscopy of Boron Compounds", Springer Verlag,
New York, 1978.
    The properties of heterocyclic boron compounds as
ligands in transition metal compounds, W. Siebert,
Advances in organometallic Chemistry, 1980, 18, 301.
    Recent developments in the chemistry of 1,3,2-
dioxasila-heterocycles, R.H. Cragg and R.D. Lane,
J. organometallic Chem., 1984, 267, 1.
    The synthesis of element-phosphorus heterocycles,
K. Issleib, Phosphorus Sulfur, 1976, 2, 219.
    The chemistry of dihydrophenophosphazines,
A.I. Bokanov and B.I. Stepanov, Russ. chem. Revs.,
1977, 46, 855.
    Heterocyclic systems containing the P-N-N
linkage, J.P. Majoral, Synthesis, 1978, 557.
    The chemistry of 1,3,2-diheterophosphorinanes,
R.P. Arshinova, Zh. obshch. Khim., 1981, 51, 1007;
R.A. Cherkasov, V.V. Ovchinnikov, M.A. Pudovik, and
A.N. Pudovik, Russ. chem. Revs., 1982, 51, 746;
E.E. Nifant'ev and A.I. Zavalishina, Russ. chem.
Revs., 1982, 51, 921.
    The synthesis and structure of heterocyclic com-
pounds containing Group V and Group VI elements in

the ring, B.A. Arbuzov, R.P. Arshinova, and
N.A. Polezhaeva, Izv. Akad. Nauk SSSR, Ser. Khim.,
1983, 2507.

1. Boron Compounds

(a) Dioxaborins and Oxathiaborins

(i) Benzo-[1,3,2]-dioxaborins
Azeotropic removal of liberated water from the
reaction of o-hydroxybenzyl alcohol with tri-
isopropylborate in benzene leads to 2-isopropoxy-
4H-1,3,2-benzodioxaborin (1), a colourless liquid,
b.p. 72°/0.01 mm. This compound reacts with a wide
variety of hydroxylic compounds with replacement of
the 2-isopropoxy group. Thus the reaction of (1)
with ethylene glycol gives 2,2'-ethylenedioxybis-
(4H-1,3,2-benzodioxaborin) (2), m.p. 70°, (97%).
Similarly, with tributyltin hydroxide, the 2-
tributylstannyloxy ester (3), b.p. 178-180°/0.9 mm.
is formed. (G. Srivastava and P.N. Bhardwaj, Bull.
chem. Soc. Japan., 1978, 51, 524). The benzo-

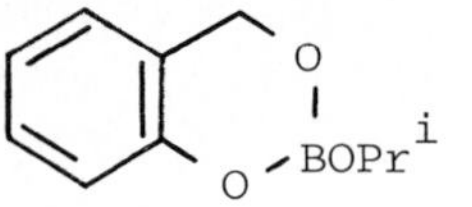

1

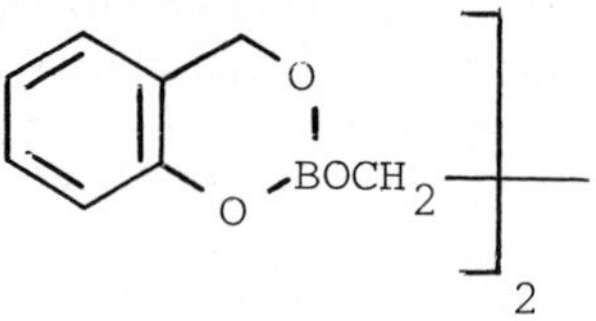

2

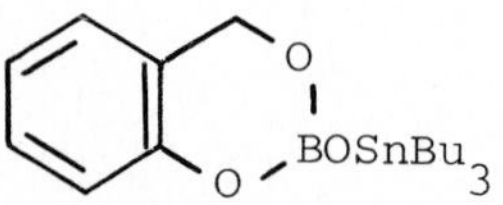

3

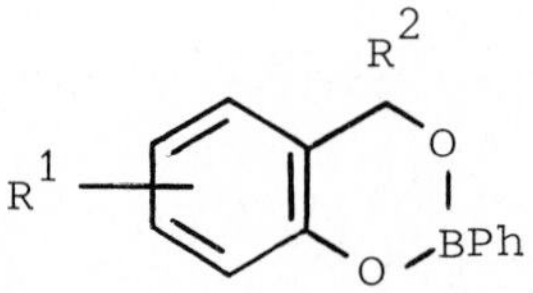

4; R^1 = H, Me, Cl or COR
R^2 = H, Me, n-C_5H_{11}

or Ph

[1,3,2]-dioxaborin system (4) is also formed in the reactions of simple phenols, having a free *ortho* position, with phenylboronic acid and an aldehyde in benzene containing a small amount of propionic acid. These compounds undergo ring cleavage on treatment with either hydrogen peroxide or propan-1,3-diol to form the related *o*-(hydroxy-alkyl)phenol. The overall transformation therefore provides a convenient method for the *ortho*-specific α-hydroxylation of phenols (W. Nagata, K. Okada, and T. Aoki, Synthesis, 1979, 365).

A number of examples have appeared of the formation of various benzo-[1,3,2]-dioxaborins as volatile derivatives of dihydroxycompounds which are of value for the gas-chromatographic detection and identification of such compounds (D.J. Harvey, Biomed. mass Spectrom., 1977, 4, 88; C.F. Poole, S. Singhawangcha, and A. Zlatkis, J. Chromatog., 1979, 186, 307; G. Munro, L.R. Rowe, and M.B. Evans, *ibid.*, 1981, 204, 201).

(ii) Benz-[3,1,2]-oxathiaborins
The reaction of *o*-mercaptobenzyl alcohol with boron trioxide in the presence of an alcohol or phenol in refluxing benzene gives the previously unknown 4*H*-[3,1,2]-benzoxathiaborin system (5), (5, R = Ph, b.p. 68-70°/3 mm.) (A.M. Bernard, *et al.*, J. heterocyclic Chem., 1982, 19, 297)

5

(b) Dioxaborins bearing another unusual heteroatom

(i) 1,3,4,2-Dioxasilaborins
The oxasilaborole (6) readily undergoes a regioselective autoxidative ring-expansion reaction in the presence of tertiary amine oxides with the formation of the 1,3,4,2-dioxasilaborin (7).

(R. Köster and G. Seidl, Angew. Chem., intern. Edn, 1984, <u>23</u>, 155).

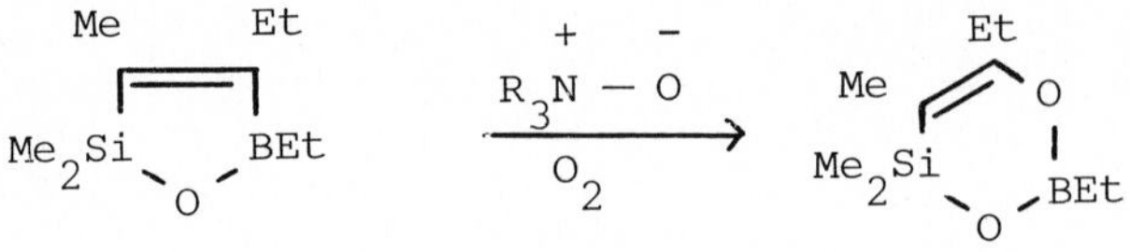

Me   Et

Me$_2$Si — BEt   ⟶ $R_3\overset{+}{N} - \overset{-}{O}$ ⟶   Me   Et

O O_2 Me$_2$Si   O
                                                     BEt
                                                  O

6   7

(ii) 1,3,5,2-Dioxaphosphaborinanes

The chemistry of this system has largely been developed by Russian workers who have prepared a number of compounds, (8, R = H, Me, Pr^i or Ph) by the reactions of bis(hydroxymethyl)phenylphos-phines with boron esters or anhydrides. (B.A. Arbuzov, *et al.*, Izv. Akad. Nauk SSSR, Ser. Khim., 1979, 2349; 1983, 1374). Zwitterionic systems (9), have also been characterised (Arbuzov, O.A. Erastov, and G.N. Nikonov, *ibid.*, 1980, 954).

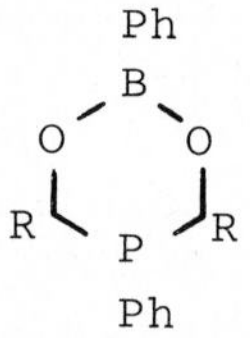

Ph
B
O   O
R   R
P
Ph

8

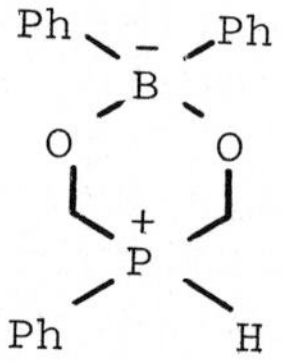

Ph   Ph
B
O   O
$\overset{+}{P}$
Ph   H

9

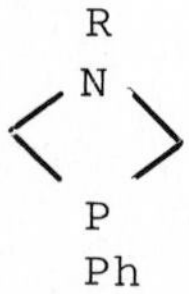

R
N
P
Ph

10

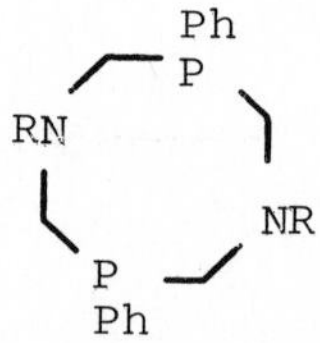

Ph
P
RN
NR
P
Ph

10a

Treatment of these compounds with a primary amine leads to elimination of the boron atom with the formation of heterocyclic aminomethylphosphines containing either a four- (10) or an eight- (10a) membered ring (Arbuzov, *et al.*, *ibid.*, 1980, 735; 1983, 1846).

(c) Oxadiborins

(i) 1,2,6-Oxadiborinanes
The reaction of the bis(chloroborane) (11) with bis(trimethyltin)oxide gives the saturated ring system (12). (W. Haubold, A. Gemmler, and U. Kraatz, Z. anorg. allgem. Chem., 1983, 507, 222)

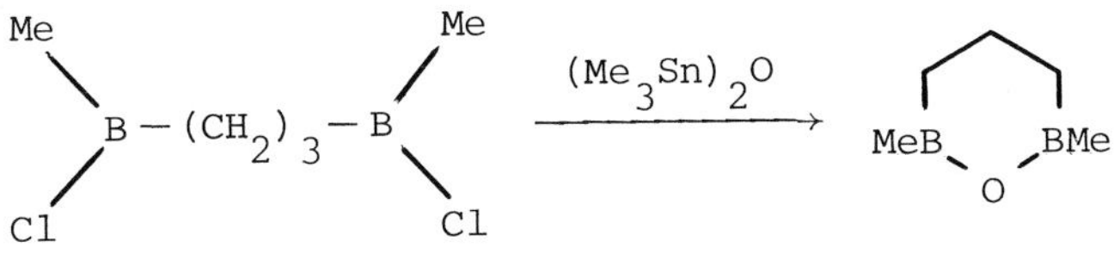

11   12

(ii) 2H-1,2,5- and 2H-1,2,6-Oxadiborins
Both these ring systems are formed by the remarkably facile insertion of carbon monoxide into the diborole (13), which results in a 3:1 mixture of the 2H-1,2,5-oxadiborin (14) and the 2H-1,2,6-oxadiborin (15). (J. Edwin, W. Siebert, and C. Krüger, J. organometallic Chem., 1981, 215 255).

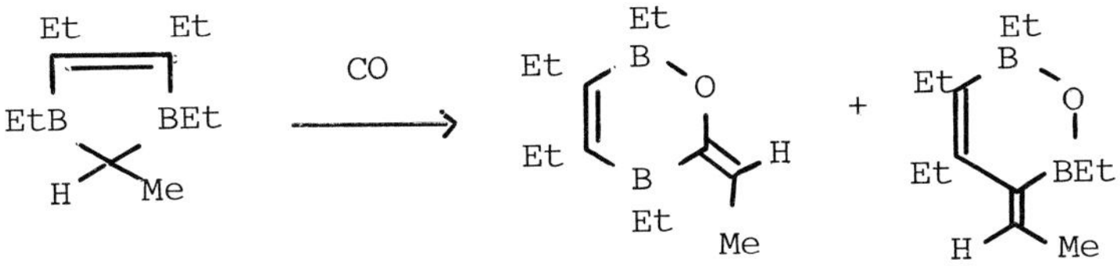

13   14   15

(d) Diazaborines

(i) Perhydro-1,3,2-diazaborines, benzo-[1,3,2]-diazaborines and related oxazaborines
It has been claimed that the reaction of 2-phenyl-1,3,2-diazaboracyclohexane (16) with phenyl isocyanate proceeds with ring-expansion to give a product having the eight-membered ring system (17). (U.W. Gerwath and K.D. Müller, J. organometallic Chem., 1975, 96, C33; Müller and Gerwath, *ibid.*, 1976, 110, 15). However on the basis of the spectroscopic evidence presented in this work, the product with the six-membered ring system (18) is also a distinct possibility (R.H. Cragg and T.J. Miller, *ibid.*, 1985, 294, 1).

Similar comments have been applied to an earlier report of the related reaction of the bicyclic system (19) which is claimed to yield the ring-expansion product (20). (P. Fritz, K. Niedenzu, and J.W. Dawson, Inorg. Chem., 1965, 4, 886).

In an attempt to resolve these issues, Cragg and Miller (*loc. cit.*,) prepared a series of bicyclic systems which contain only tertiary nitrogen atoms and studied their reactions with phenyl isocyanate, the formation of exocyclic products being precluded by the absence of -NH- groups. Thus, the reaction of bis(dimethylamino)phenylborane with 2-(2-hydroxyethyl)piperidine yields the bicyclic system (21), b.p. 100°/0.1 mm, (90%), which with phenyl isocyanate yields the ring-expansion product (22), m.p. 105°. Similarly, the benz-[3,1,2]-oxazaborin-4-one (23), b.p. 160°/0.1 mm. (from the reaction of N-methylanthranilic acid with bis(dimethyl-amino)phenylborane) undergoes ring-expansion on treatment with *p*-toluenesulphonyl isocyanate to yield (24), m.p. 90°. The reaction of hetero-cyclic aminoboranes with phenyl isocyanate thus offers a convenient route to large ring organo-boranes *via* ring-expansion.

21   22

23   24

The reaction of amino-alcohols with boronic acid derivatives or borate esters gives the 1,3,2-oxaza-borine system (25) (Poole, Singhawangcha, and

Zlatkis, J. Chromatog., 1979, _186_, 307; Cragg and
A.F. Weston, J. chem. Soc., Dalton, 1975, 93) and
(26) (H. Piotrowska, _et al._, Roscz. Chem., 1977,
_51_, 1997) and also the 2_H_,4_H_-benz-[3,1,2]-oxaza-
borines (27) (R = Ph, b.p. 210-215°/25 mm.) (A.M.
Bernard, _et al._, J. heterocyclic Chem., 1982, _19_,
297), and (28), m.p. 123° (J.W. Blunt, _et al._,
Austral. J. Chem., 1979, _32_, 1045).

25   26   27

28   29   30

The reactions of _o_-aminobenzoic acid with arylboronic acids give a range of 2_H_-benz-[3,1,2]-
oxazaborin-4-ones (29) (Poole, Singhawangcha, and
Zlatkis, _loc. cit._) A novel route to (30), m.p.
145-147°, is the reaction of aniline with phenyl-
dichloroborane and _p_-nitrobenzaldehyde (T. Toyodo,
K. Sasakura, and T. Sugasawa, Tetrahedron Letters,
1980, _21_, 173).

(ii) Benzo-[2,4,1]-diazaborines
Cyclisation of substituted amidines with trialkyl-
or triaryl- borons provides a satisfactory route
to the 1,2-dihydro- benzo- [2,4,1]-diazaborine
system (31, R = Me or Ph; R^1 = Pr, Bu or Ph)

(V.A. Dorokhov, *et al*. Izv. Akad. Nauk SSSR, Ser. Khim., 1979, 14).

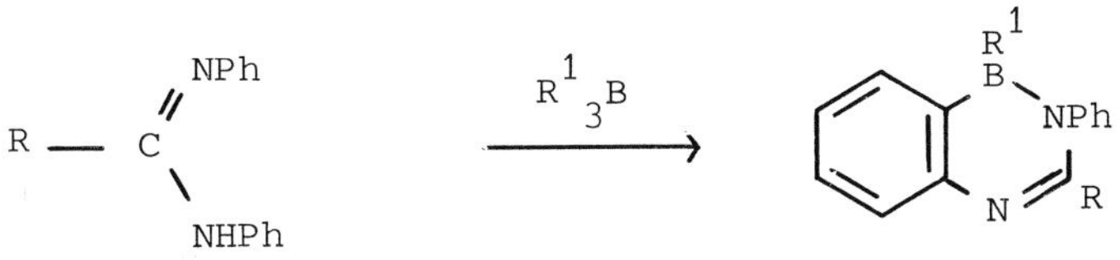

R —— C $\overset{\displaystyle /\!/\text{NPh}}{\underset{\displaystyle \backslash\text{NHPh}}{}}$ $\xrightarrow{R^1{}_3B}$

31

(e) Azadiphosphaborines

A route to the 1,2,6,4-azadiphosphaborine ring system is provided by the deprotonation of the phosphazeno-ylide (32), resulting in the salt (33), which, on subsequent treatment with the chloro-borane-triethylamine adduct, yields the zwitter-ionic system (34) as a colourless sublimable solid, m.p. 105°, (74%) (H. Schmidbaur, *et al*., Ber., 1979, $\underline{112}$, 1448).

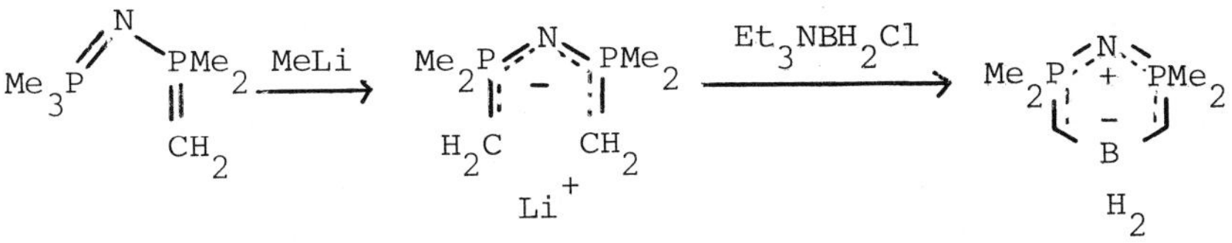

Me$_3$P $\xrightarrow{\text{MeLi}}$ Me$_2$P $\xrightarrow{\text{Et}_3\text{NBH}_2\text{Cl}}$ Me$_2$P

32   33   3 4

(f) Tetrazaborines

Cyclisation of carbohydrazide with the borane-THF (or triethylamine) adduct gives the tetra-hydro-1,2,4,5,3-tetrazaborin-6-one (35), m.p. > 400°, in 95% yield. The related reaction with bis(dimethylamino)phenylborane gives a similarly high yield of (36), m.p. 132-135° (J. Bielawski and

K. Niedenzu, Inorg. Chem., 1980, <u>19</u>, 1090).

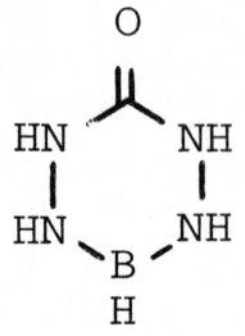

35

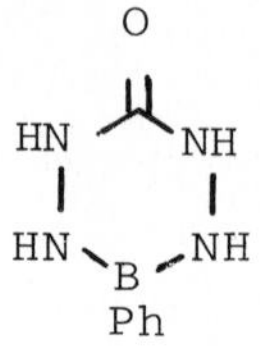

36

(g) Azadiborines

The cyclocondensation of the bis(chloroboranes) (37) with bis(trimethylsilyl)- or bis(trimethyl-stannyl)-methylamine gives the perhydro-1,2,6-azadiborine system (38, R = Me or Cl) (W. Haubold, A. Gemmler, and U. Kraatz, Z. anorg. allgem. Chem., 1983, <u>507</u>, 222).

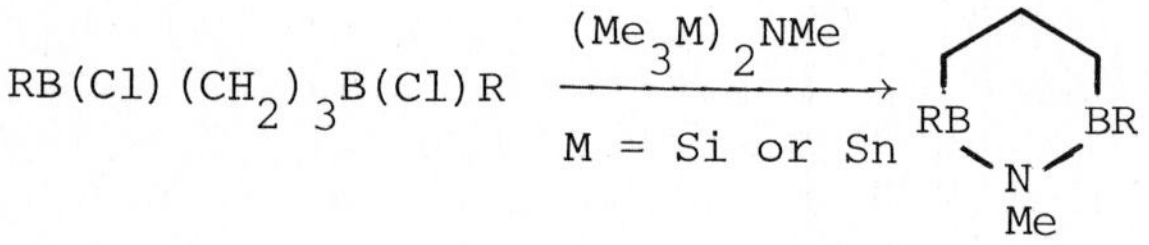

37   38

(h) Diazadiborines

Access to the perhydro-1,2,3,6-diazadiborine ring system is afforded by the cyclocondensation of the bis(chloroborane) (39) with N,N'-bis-[(methyl)(trimethylsilyl)]hydrazine, which gives the perhydro-1,2,3,6-diazadiborine (40) (Haubold,

Gemmler and Kraatz, *loc. cit.*).

RB(Cl)CH$_2$CH$_2$B(Cl)R

$$
\begin{array}{cc}
RB & BR \\
\diagdown & \diagup \\
N & - N \\
Me & Me
\end{array}
$$

39   40

(i) Triazadiborines

(i) 1,3,5,2,4-Triazadiborines and related oxadiaza- and thiadiazadiborines

The reactions of N,N'-disubstituted ureas and thioureas with halogeno-organoboranes leads to these systems. The course of individual reactions is highly dependent on the structure of the urea or thiourea, and also on the reaction conditions (W. Maringgele, J. organometallic Chem., 1981, <u>222</u>, 17; Ber., 1982, <u>115</u>, 3271). In the presence of butyllithium, N,N'-disubstituted ureas and thioureas react with organodihalogenoboranes to give a mixture of the 2,3,4,5-tetrahydro-1,3,5,2,4-triazadiborin-6-ones (or thiones) (41), (41; $R^1 = R^2 = Me$; X = O, b.p. 65°/0.002 mm.) and the 2,3,5,6-tetrahydro-1,3,5,2,6-oxa(or thia)diazadiborin-4-ones (or thiones) (42), (42; $R^1 = R^2 = Ph$; $R^3 = Me$, X = O, m.p. 82-84°), the latter system predominating. The tetrahydrotriazadiborines (41), but <u>not</u> the oxa (or thia) diazadiborines (42) are among the products of the reactions of disubstituted ureas (and thioureas) with organodihalogenoboranes in tetrachloromethane in the absence of a metallating agent. In contrast, the reactions of certain ureas with bromodimethylborane in tetrachloromethane yield the oxadiazadiborin-4-ones (42, X = O), (42; X = O; $R^1 = R^3 = Me$; $R^2 = $ 2,6-dimethylphenyl, m.p. 144-148°) whereas the related reactions of the corresponding thioureas give rise to a mixture of the triazadiborin-6-

thiones (41; X = S) and the 2,3,4,5-tetrahydro-
1,3,5,2,4-thiadiazadiborin-6-imines (43), (43;
R^1 = 2,6-dimethylphenyl; R^2 = Et; R^3 = Me, b.p.
190-210°/0.002 mm.). The latter systems are also
formed in the reactions of thioureas with organodihalogenoboranes in halogenated solvents in the
absence of a metallating agent.

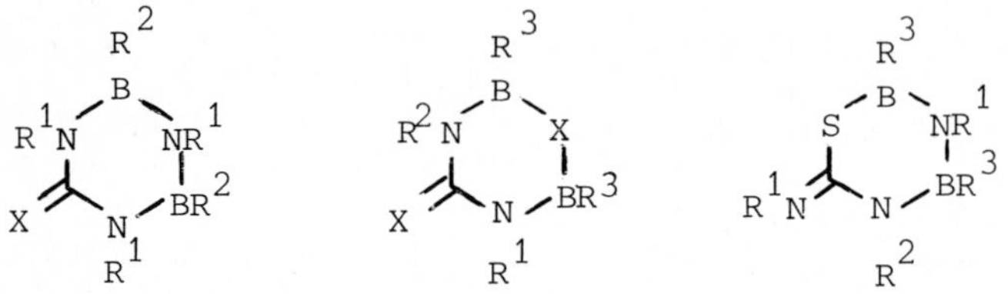

41   42   43

Other routes to the oxadiazadiborine systems
have also been reported. The reaction of *N*-lithio-
N-trimethylsilylacetamide with dibromo(methyl)-
borane gives (44), b.p. 65°/20 mm. (Maringgele
and A. Meller, Ber., 1979, 112, 1595). The
diazadiboretane (45) undergoes ring expansion on
treatment with acrolein or crotonaldehyde to give
(46, R = H), m.p. 109° or (46, R = Me), m.p. 89°
(P. Paetzold, *et al.*, Ber., 1979, 112, 3811).

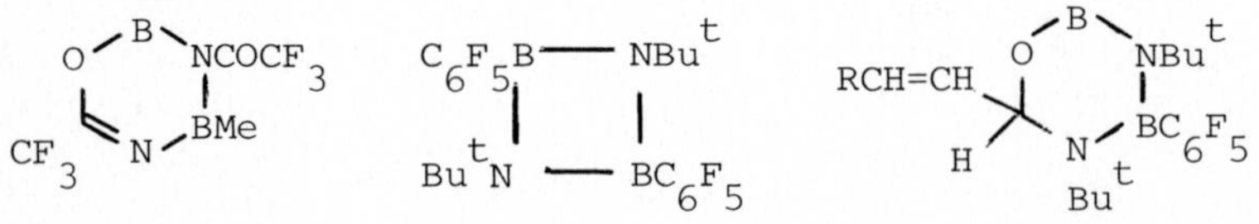

44   45   46

2. Silicon, Germanium, Tin and Lead Compounds

(a) Oxa-silins, -germanins, -stannins, and related
compounds

(i)   1,2-Oxasilins, benz-[1,2]-oxa-silins and
-germanins, benz-[2,1]-oxasilins, and benz-[2,3]-
oxasilins

Interest in the possible existence of extended
π-interactions between vacant 3d orbitals on
silicon and an adjacent π-system has prompted
attempts to prepare members of the 1,2-oxasilin
system. Co-pyrolysis of 2,5-dimethylfuran and
sym-tetramethoxydimethylsilane leads to the 1,2-
oxasilin (1) but in only 3.3% yield. On the basis
of its uv-spectrum in cyclohexane solution,
(λ_{max} 273 nm, ε 8500), it has been concluded that
there is little evidence of the hoped-for
conjugation (M.E. Childs and W.P. Weber, J. org.
Chem., 1976, 41, 1799). A better route to 1,2-
oxasilins is afforded by flash vacuum pyrolysis
of the diene (2) which leads to a 52% yield of
(3), approx. b.p. 80-95°. Attempted Diels-Alder
reaction of the latter with perfluoro-2-butyne
does not give rise to isolable adducts but appears
to result in extrusion of the silanone $Me_2Si=O$,
which can be subsequently trapped by a variety of

| 1 | 2 | 3 |

reagents. A stable Diels-Alder adduct of (3) with
maleic anhydride can be isolated in 80% yield as a
stable crystalline solid, m.p. 124-125°, which also
extrudes the silanone upon either thermolysis or

photolysis. (G. Hussmann, W.D. Wulff, and
T.J. Barton, J. Amer. chem. Soc., 1983, $\underline{105}$, 1263).
   The photochemical reaction of the silabutene (4)
with acetone gives rise to the [4+2] adduct (5)
(P.B. Valkovich, and W.P. Weber, Tetrahedron
Letters, 1975, 2153). In related work, photo-

4   5

chemical insertion of carbonyl compounds into
benzosilacyclobutenes leads to the predominant
formation of the 3,4-dihydro-1H-benz-[2,1]-
oxasilins (6), (6, R^1 = Ph, R^2 = R^3 = Me, m.p.
111-112.5°). In a few cases, the isomeric 1,4-
dihydro-1H-benz-[2,3]-oxasilins (7) are also
formed (K.T. Kang, R. Okazaki, and N. Inamoto,
Bull. Korean chem. Soc., 1984, $\underline{5}$, 32.

6   7

   Cycloaddition of the silaethene $Me_2Si=C(SiMe_3)_2$
with benzophenone gives a quantitative yield of
the unsaturated compound (8), m.p. 81° (N. Wiberg,
G. Preiner, and O. Schieda, Ber., 1981, $\underline{114}$, 3518).
Silicon and germanium analogues of allylphenyl
ethers, $PhOM(Me)_2CH=CH_2$ (M = Si or Ge) undergo

aromatic Cope-type rearrangements on prolonged heating at 170° to form the 3,4-dihydro-2H-benz-[1,2]-oxasilin (9, M=Si), b.p. 93°/15 mm., and -oxagermanin (9, M = Ge) b.p. 110°/12 mm., respectively, (G. Bertrand, P. Mazerolles, and J. Ancelle, Tetrahedron, 1981, 37, 2459).

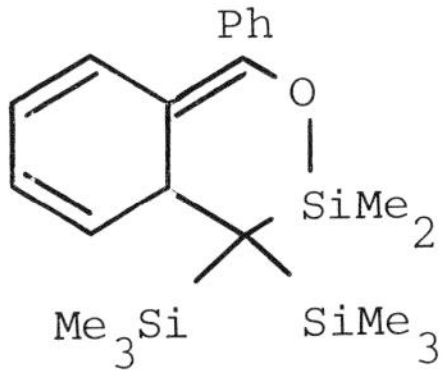

Ph

SiMe$_2$

Me$_3$Si   SiMe$_3$

8

O

MMe$_2$

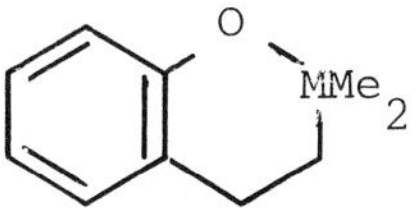

9

(ii)   Perhydro-1,2-oxa- and -thia-silins and -stannins

The chiral 1,2-oxa- and 1,2-thia-silacyclohexanes (10;   X = O or S), (10;   X = O, b.p. 183°/0.3 mm) have been prepared as part of a study of the stereochemical course of nucleophilic substitution at silicon (R.J.P. Corriu, *et al.*, J. organometallic Chem., 1976, 114, 21;   Bull. Soc. chim. Belg., 1980, 89, 783;   Nouv. J. Chim., 1984, 8, 279).   The stannacyclopentane (11) gradually decomposes in moist air to give a mixture of products which include the 1,2-oxastannacyclohexane (12, X = O), m.p. 135-138°.   The yield of

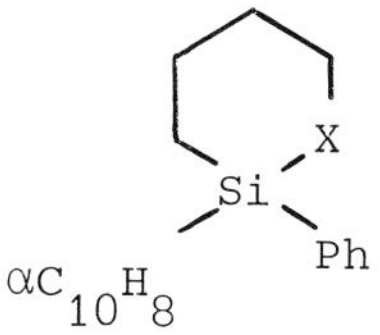

X

Si

Ph

αC$_{10}$H$_8$

10

Sn

Me$_2$

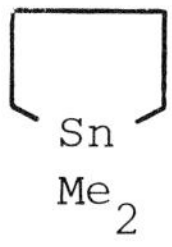

11

X

Sn

Me$_2$

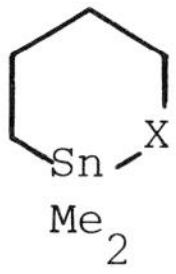

12

this compound is greatly improved if the stannacyclopentane is oxidised with potassium permanganate in acetone solution. The related reaction of (11) with sulphur at 200° leads to the 1,2-thiastannacyclohexane (12; X = S), as a yellow oil, b.p. 87-90°/0.3 mm (E.J. Bulten and H.A. Budding, J. organometallic Chem., 1978, 153, 305; 1979, 166, 339).

(iii) Perhydro-1,3-oxasilins, and the sulphur, selenium and tellurium analogues
An established route to 1,3-oxasila- and 1,3-thiasila-cyclohexane (13; X = O or S), involving the action of aqueous alkali or sodium sulphide on 1,5-dichloro-2-silapentanes (R.J. Fessenden and M.D. Coon, J. org. Chem., 1964, 29, 1607), has been extended to the related reactions with sodium selenide and telluride, giving the 1,3-selena- and -tellura-silacyclohexanes (14; X = Se or Te; R = H or Me)(R. Dedeyne and M.J.O. Anteunis, Bull. Soc. chim. Belg., 1976, 85, 319). N.m.r. studies reveal that all of these ring systems adopt a chair conformation in solution (Anteunis and Dedeyne, Org. mag. Res., 1977, 9, 127).

13   14

(b) Dioxa-germanins, -stannins, and plumbins, and their thia analogues

(i) Benz-[3,1,2]-dioxa- and -oxathia-stannins, and related oxathiaplumbins
The reaction of tin(II) methoxide with salicylic acid and with o-mercaptobenzoic acid in methanol gives, in each case, a high melting associated solid which has been formulated as the 4H-benz-

[3,1,2]-dioxastannin(II)-4-one (15; Z = O) and
the 4H-benz-[3,1,2]-oxathiastannin(II)-4-one
(15; Z = S), respectively (W.D. Honnick and
J.J. Zuckerman, Inorg. Chem., 1978, <u>17</u>, 501). The
related reactions of the above carboxylic acids

15   16   17

with diorganotin(IV) oxides, $(R_2SnO)_n$, yield the
corresponding tin(IV) derivatives (16; Z = O or
S), (16; Z = S, R = Bu, m.p. 185-186°). In the
solid state, these compounds are also highly
associated *via* intermolecular coordination to tin
by carbonyl groups of other molecules, but in
pyridine solution they have been found to be monomeric (*idem. ibid.*, 1979, <u>18</u>, 1437). Similarly,
the reaction of *o*-mercaptobenzoic acid with bis-
(triphenyltin) oxide yields (16; Z = S; R = Ph),
as a white solid, m.p. > 260° (J.F. Vollano, *et al.*,
Inorg. Chem., 1984, <u>23</u>, 3153), and with tin(IV)
acetate, the bis(acetato) derivative (16; Z = S;
R = OAc), m.p. 320°, is formed (R.C. Mehrotra,
G. Srivastava, and E.N. Vasanta, Inorg. Chim. Acta,
1980, <u>47</u>, 125). The ability of the tin atom in
(16; Z = S, R = Me) to act as an electron-
acceptor has been investigated, and a five
coordinate anionic heterocyclic tin derivative
prepared by addition of chloride ion (Vollano,
R.O. Day, and R.R. Holmes, Organometallics, 1984,
<u>3</u>, 750).
    Reactions of *o*-mercaptobenzoic acid with di-
organolead(IV) compounds in chloroform solution
give the related lead heterocycles (17; R = Me or
Ph) (C.D. Hager and F. Huber, Z. Naturforsch.,
1980, <u>35B</u>, 542).

(ii) Perhydro-1,5,2-dioxagermanins and -1,4,2-dithiastannins

Insertion of carbonyl compounds into 1,1-diethyl-1,2-germaoxetanes has given a series of 1,5,2-dioxagermacyclohexanes (18), (18; R = CF$_3$, b.p. 62°/0.7 mm.) (J. Barrau, *et al.*, J. organometallic Chem., 1971, <u>30</u>, C67). On pyrolysis, these compounds give the highly reactive diethyl-germanone, Et$_2$Ge=O, which can be trapped by trimethylene oxide to give the 1,3,2-dioxagermacyclohexane (19) (Barrau, *et al.*, *ibid.*, 1983, <u>246</u>, 227) The 1,4,2-dithiastannacyclohexane (20) has been

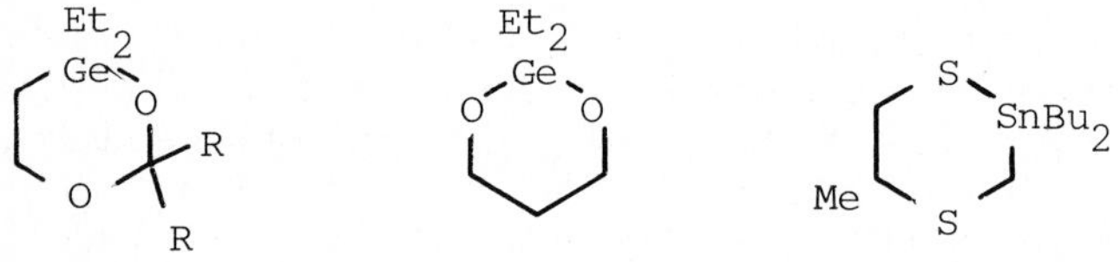

18   19   20

described (G. Domazetis, R.J. Magee, and B.D. James, J. organometallic Chem., 1978, <u>148</u>, 339).

(c) Oxadisilins

(i) Perhydro-1,2,3-oxadisilins (1,2,3-oxadisila-cyclohexanes)

This system is afforded by insertion of the dimethylsilylene, Me$_2$Si:, (generated by photolysis of dodecamethylcyclohexasilane) into the 1,2-oxasilacyclopentanes (21) to give the 1,2,3-oxadisilacyclohexanes (22) (T.Y. Gu and W.P. Weber, J. organometallic Chem., 1980, <u>195</u>, 29).

21   22

*(ii) Perhydro-1,2,5-oxadisilins (1,2,5-oxadi-
silacyclohexanes)*
The reactions of the 1,2-disilacyclobutene (23)
with carbonyl compounds are considered to proceed
via the initial attack of the carbonyl oxygen on
the silicon-silicon bond to give a diradical
intermediate which subsequently ring-closes to give
a mixture of isomeric 1,2,5-oxadisilacyclohexanes
(24) and (25) (C.P. Chin and C.S. Liu, J. organometallic Chem., 1982, 235, 7; Y.L. Lu and Liu,
ibid., 1983, 243, 393).

23   24   25

*(d) Dioxadisilins, dithiadisilins, and dithia-
germanins*

(i) 1,4,2,3-Dioxadisilins
Three syntheses of this system have been
described. The cyclisation of catechol with 1,2-
dichloro-1,1,2,2-tetramethyldisilane affords the
benz-[1,4,2,3]-dioxadisilin (26), m.p. 44°
(E. Hengge, E. Brandstätter, and W. Veil, Monatsh.,

1977, _108_, 1425). Photolysis of dodecamethylcyclohexasilane in the presence of 3,5-di-_t_-
butyl-_o_-benzoquinone gives (27), as an air-

26   27   28

sensitive substance (W. Ando and M. Ikeno, Chem.
Comm., 1979, 655). The cycloaddition of benzil to
the sterically crowded disila-alkene
$But_2Si=SiBut_2$, at -70° in THF, yields the 1,4,2,3-
dioxadisilin (28), m.p. 138-139° (M. Weidenbruch,
A. Schäfer, and K.L. Thom, Z. Naturforsch, 1983,
_38B_, 1695).

_(ii) Perhydro-1,4,2,3-dioxa-(and dithia)-disilins
and germanium analogues_
Cyclocondensation of 1,2-diphenylethane-1,2-diol
with 1,2-dichloro-1,1,2,2-tetramethyldisilane in
the presence of triethylamine gives the 1,4,2,3-
dioxadisilacyclohexane (29), very unstable to air
and water. This compound is considered to be a
key intermediate in the reaction of the sila-
alkene $Me_2Si=SiMe_2$ with benzaldehyde, which gives
rise to _trans_-stilbene and other products derived

29   30   31

from the highly reactive silanone, $Me_2Si=O$. These
products are also formed on pyrolysis of (29) at
500° (T.J. Barton and J.A. Kilgour, J. Amer. chem.
Soc., 1976, 98, 7231). The related dithiadisila-
cyclohexane (30), b.p. 68°/0.1 mm; m.p. 23-24°,
is accessible in 35% yield by cyclocondensation of
the above dichlorodisilane with ethane-1,2-dithiol
in the presence of pyridine. Like the dioxa-
analogue (29), this compound undergoes pyrolysis
to form ethylene and the highly reactive thia-
silanone, $Me_2Si=S$, which can be trapped in various
ways (H.D. Soysa, I.N. Jung, and W.P. Weber, J.
organometallic Chem., 1979, 171, 177).

The dithiadigermacyclohexane (31), b.p. 73°/
0.3 mm, is similarly formed in a cyclocondensation
of ethane-1,2-dithiol with 1,2-bis(diethylamino)-
1,1,2,2-tetramethyldigermane. On photolysis or
pyrolysis, this compound gives rise to both the
highly reactive germylene, $Me_2Ge:$, and the
germathione, $Me_2Ge=S$, characterised as various
adducts (J. Barrau, *et al.*, *ibid*, 1984, 277, 323).

*(iii) Perhydro-1,3,2,4-dioxadisilins and
-dithiadigermanins*

The 1,3,2,4-dioxadisilacyclohexane (32) is among
the products derived from the silanone, $Me_2Si=O$,
formed on pyrolysis of vinyldimethylcarbinoxydi-
methylsilane (T.H. Lane and C.L. Frye, *ibid*, 1979,
172, 213). The related dithiadigermacyclohexane
(33) is formed on pyrolysis of the germathiacyclo-
butane (34), *via* the reactive intermediate $Et_2Ge=S$

32   33   34

(J. Barrau, *et al.*, Syn. React. inorg. met.-org.
Chem., 1980, <u>10</u>, 515).

*(e) Oxa- and dioxa-trisilacyclohexanes and related
systems*

(i) 1,2,3,6-oxatrisilacyclohexanes
Photochemically-generated silylenes, $R_2Si:$,
readily insert into the silicon-oxygen bonds of
the cyclic disiloxane (35), to form the 1,2,3,6-
oxatrisilacyclohexanes (36; R = Me or Ph)
(H. Okinoshima and W.P. Weber, J. organometallic
Chem., 1978, <u>155</u>, 165; D. Seyferth,
D.C. Annarelli, and D.P. Duncan, Organometallics,
1982, <u>1</u>, 1288).

35

36

*(ii) 1,2,4,6-oxatrisilacyclohexanes, 1,3,2,4,6-
dioxatrisilacyclohexanes, and related systems*
Co-pyrolysis of heterocyclic silane precursors
which generate a mixture of the silaethene,
$Me_2Si=CH_2$, and the silathione, $Me_2Si=S$, leads to
the formation of the heterocyclic systems (37) and
(38), together with other products. Treatment of
the pyrolysate with water leads to the evolution
of hydrogen sulphide and the formation of the
related oxasila heterocycles. As yet, none of
these compounds has been fully characterised or,
indeed, isolated in a pure state other than by gas
chromatography (L.E. Gusel'nikov, V.V. Volkova,
V.G. Avakyan, *et al.*, J. organometallic Chem.,
1984, <u>271</u>, 191).

$$\underset{37}{\overset{\displaystyle Me_2Si\overset{S}{\diagdown}SiMe_2}{\underset{\underset{Me_2}{Si}}{\bigsqcup}}}
\qquad
\underset{38}{\overset{\displaystyle Me_2Si\overset{S}{\underset{\displaystyle S}{\diagdown}}SiMe_2}{\underset{\underset{Me_2}{Si}}{\bigsqcup}}}
\qquad
\underset{39}{\overset{\displaystyle Me_2Si\overset{O}{\diagdown}SiMe_2}{\underset{\underset{Cl_2}{Sn}}{\bigsqcup}}}$$

Various 1,2,6,4-oxadisilastannacyclohexanes (e.g., 39), have been described and their reactivity studied with particular reference to the behaviour of the tin atom in the ring (V.F. Mironov, *et al.*, Zh. obshch. Khim., 1972, 42, 631; 1976, 46, 1043; G.I. Magomedov, *et al.*, *ibid.*, 1981, 51, 2380; Koord. Khim., 1983, 9, 351).

(f) Azasilines and azastannines

(i) 1,2-Azasilines, benz-[2,3]-azasilines, and perhydro-1,2-azastannines

Silanimines of the type $Me_2Si=NSiR_3$ (R = Me or But) undero [2+4] cycloaddition reactions with 1,3-dienes to give the partially reduced 1,2-azasilines (40) (N. Wiberg, J. organometallic Chem., 1984, 273, 141). Addition of *N*-trimethyl-silyldiphenylketinimine to the sila-ethene $Me_2Si=C(SiMe_3)_2$ gives the tetrahydro-benz-[2,3]-azasiline (41) (Wiberg, Preiner and Schieda, Ber., 1981, 114, 3518). The 1,2-azastannacyclohexane (42), b.p. 49-51°/0.3 mm, is prepared by the reaction of lithium *t*-butylamide on 4-halogeno-butyl(dimethyl)(halogeno)stannanes, which are easily accessible by the ring-opening of di-methylstannacyclopentanes with bromine or iodine (D. Hänssgen and E. Odenhausen, J. organometallic Chem., 1977, 124, 143). This compound undergoes ring-expansion on treatment with either dimethyl acetylenedicarboxylate or diphenylcarbodiimide to

R–SiMe$_2$
R–NSiR$_3$

40

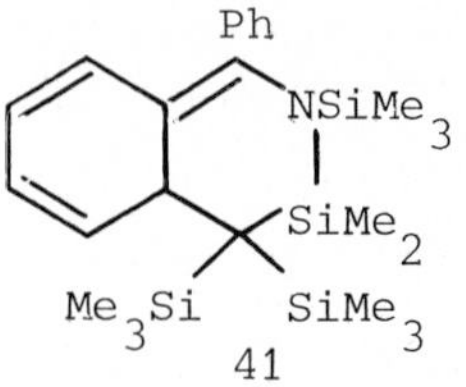

Ph–NSiMe$_3$
SiMe$_2$
Me$_3$Si   SiMe$_3$

41

SnMe$_2$
NBu^t

42

form eight-membered ring systems, the reagent inserting between the tin and nitrogen atoms.

(ii) Perhydro-1,4-azasilines
The reaction of bis(2-bromoethyl)diphenyl-silane with a primary aliphatic amine in the presence of triethylamine, and the reaction of diphenyldivinylsilane with lithium alkylamides, followed by treatment with water, result in the formation of (43, R = alkyl), from which a series of quaternary salts has been prepared on treatment with hydrogen chloride (M. Gerlach, *et al.*, Z. Naturforsch., 1982, **37B**, 657). The mercury-catalysed addition of a primary aromatic amine to diallyldimethylsilane, followed by treatment of the initial adduct with alkaline sodium boro-hydride, yields *N*-aryl systems (e.g. 44; b.p. 75-80°/0.001 mm) as a mixture of geometrical isomers (J. Barluenga, *et al.*, Synthesis, 1982, 414).

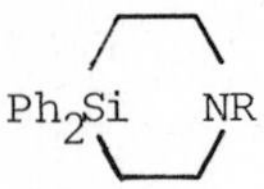

Ph$_2$Si   NR

43

H
Me
Me$_2$Si   NPh
Me
H

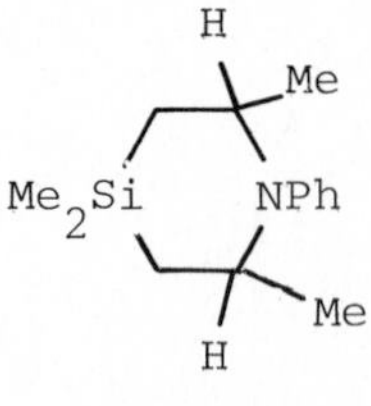

44

(g) Diazasilines and related systems

(i) Benz-[1,2,3]-diazasilines and perhydro-1,3,2-oxazasilines
The tetrahydrobenz-[1,2,3]-diazasiline (45) is formed in the cycloaddition of the silaethene $Me_2Si = C(SiMe_3)_2$ with azobenzene (Wiberg, Preiner and Schieda, *loc. cit.*) Related cycloadditions of silanimines with 2-methyl-2-propenal have given the tetrahydro-1,3,2-oxazasilins (46), (46; R = Bu^t), b.p. 48°/0.01 mm (J. Neeman and U. Klingebiel, Ann., 1980, 1978; Clegg, Klingebiel and Sheldrick, Z. Naturforsch., 1982, **37B**, 423).

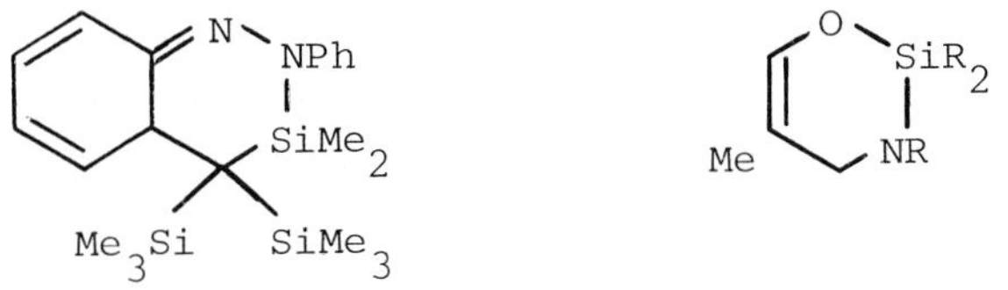

45

46

(ii) Benz-[1,4,2]-oxazasilines and perhydro-1,4,2-oxazasilacyclohexanes
Cyclisation of halogenomethyldimethylchlorosilanes, $ClMe_2SiCHXPh$ (X = Cl or Br), with *o*-aminophenol yields 2,2-dimethyl-3-phenyl-1,2,3,4-tetrahydro-1,4,2-benzoxazasiline (47), m.p. 111-112° (P. Voss, *et al.*, J. prakt. Chem., 1978, **320**,

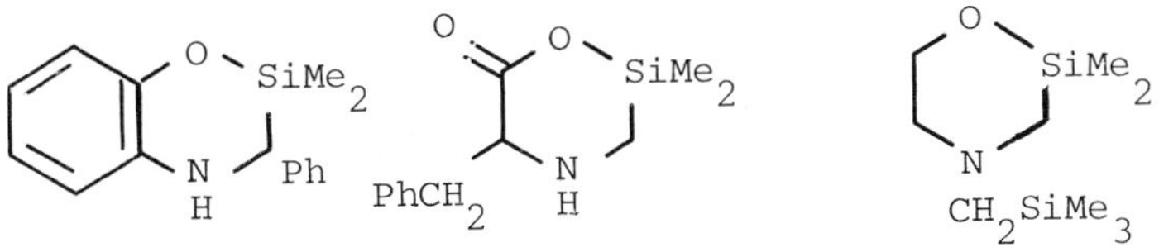

47

48

49

34).

Similar cyclisation reactions of chloromethyl-dimethylchlorosilane with difunctional aliphatic compounds have yielded a variety of new examples of "silamorpholines", such as (48), (C.G. Hammar, Biomed. mass Spec., 1978, 5, 25) and (49) (V.D. Sheludyakov, *et al.*, Zh. obshch. Khim., 1982, 52, 2646; 1984, 54, 1134).

(iii) Perhydro-1,5,2-oxaphospha-silines and -germanines, and perhydro-1,2,6-oxa-(and -thia-) phosphagermanines

Insertion of benzaldehyde into the silicon- (or germanium)- phosphorus bond of the four-membered ring compound (50; Me = Si or Ge) gives rise to the six-membered ring compound (51; M = Si), b.p. 141-145°/0.5 mm (C. Couret, *et al.*, J. organometallic Chem., 1979, 182, 9) and (51; M = Ge), b.p. 151-154°/0.5 mm (J. Escudié, *et al.*, Organometallics, 1982, 1, 1261) as mixtures of diastereoisomers.

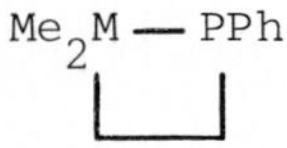

50

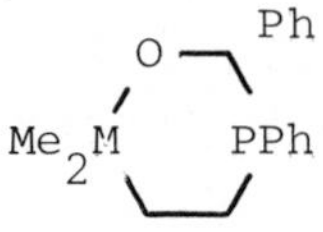

51

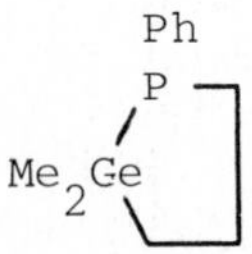

52

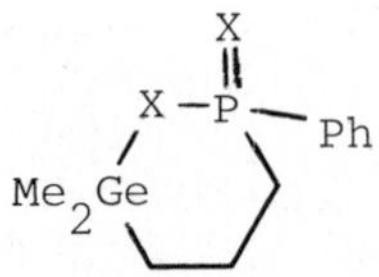

53

The germaphospholane (52) also suffers insertion
into the germanium-phosphorus bond in its reactions
with dioxygen or elemental sulphur, to give,
respectively, (53; X = O), b.p. 147-150°/0.03 mm,
and (53; X = S) m.p. 91° (C. Couret, *et al.*, Rec.
trav. Chim., 1976, 95, 240).

(h) Triazasilines and related compounds

(i) 1,3,5,2-Triazasilines and -triazagermanines
Treatment of *N,N'*-disubstituted ureas with
silicon tetrachloride provides a route to the
1,3,5,2-triazasilines (54), (54; R = Bu, b.p.
111-112°/0.02 mm)(Maringgele, Z. Naturforsch.,
1983, 38B, 71). The reaction of the cyclo-
germazane (55) with phenyl isocyanate and phenyl
isothiocyanate leads to the formation of the tri-
azagerma-dione (56; X = O) and the -dithione (56;
X = S), respectively (G. Lacrampe, *et al.*, Rec.,
J.R. Netherlands chem. Soc., 1983, 102, 21).

54   55   56

(ii) 1,3,5,2-Dioxazasilines and related compounds
The 1,3,5,2-dioxazasiline (57) is formed in the
reaction of *N*-acetylacetamide with bis(penta-
fluorophenyl)dimethylsilane (I.D. Kalikhman, *et al.*,
Izv. Akad. Nauk SSSR, Ser. Khim., 1983, 1515). The
perhydro-1,3,5,2-dioxaphosphasilacyclohexane
system (58, R^1 = alkyl or Ph; R^2 = alkyl) is
easily prepared by the reaction of a bis(hydroxymethyl)phosphine with a dialkyl dichlorosilane
(M.G. Voronkov, *et al.*, Dokl. Akad. Nauk SSSR,
1979, 247, 609). The phosphorus atom in this
system exhibits the normal reactivity pattern of

trivalent phosphorus, forming a phosphine sulphide on reaction with sulphur (Voronkov, *et al.*, Zh. obshch. Khim., 1981, **51**, 2176).

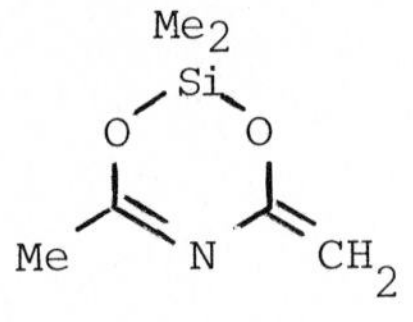

Me₂
Si
O   O
Me   N   CH₂

57

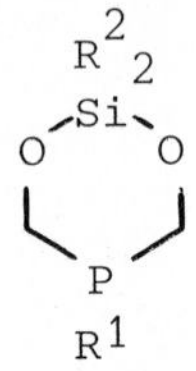

R²₂
Si
O   O
P
R¹

58

Many physicochemical and spectroscopic studies of the conformational properties of this system have also been reported (*inter alia*, I.I. Patsanovskii, *et al.*, Zh. obshch. Khim., 1984, **54**, 1738).

(i) Azadisilines

The perhydro-1,3,5-azadisiline (59), b.p. 70-71°/27 mm., is formed (79%) by the reaction of methylamine with the α,ω-dibromodisilane $BrCH_2Si(Me)_2CH_2Si(Me)_2CH_2Br$ in an autoclave at 137°, (H. Bock, *et al.*, J. organometallic Chem., 1978, **164**, 295). Cycloaddition of the sila-alkene $Me_2Si=C(SiMe_3)_2$ to benzonitrile gives rise

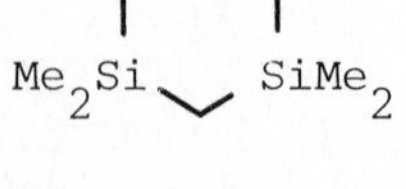

Me
N
Me₂Si   SiMe₂

59

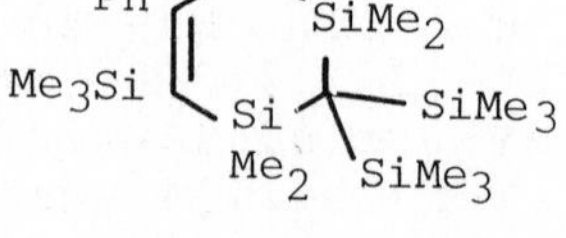

SiMe₃
Ph   N   SiMe₂
Me₃Si   Si   SiMe₃
Me₂   SiMe₃

60

to the tetrahydro-1,2,4-azadisilin (60), m.p.
125° (N. Wieberg, G. Preiner and O. Schieda,
Ber., 1981, 114, 3518).

(j) Diazadisilines, triazadisilines and analogues

   (i) Perhydro-1,4,2,3-diazadisilines and perhydro-1,3,2,4-diazadigermanines
Perhydro-1,4,2,3-diazadisilines have been
prepared by conventional cyclisation reactions
between N,N'-dilithioethylenediamines and 1,2-
dichlorotetramethyldisilane, giving (61), b.p. 46°/
3 mm) (U. Wannagat and G. Eisele, Monatsh., 1978,
109, 1059).

61   62   63

   Thermal decomposition of the germa-azetidine (62)
generates the reactive germa-imine $Ph_2Ge{=}NMe$,
which inserts into the germanium-nitrogen bond of
(62) to yield the perhydro-1,3,2,4-diazadigermanine
(63), b.p. 145°/0.15 mm. (Rivière-Baudet,
P. Rivière, and Satgé, J. organometallic Chem.,
1978, 154, C23; Rivière-Baudet, *et al.*, Rec.
trav. Chim., 1979, 98, 42).

   *(ii) 1,3,5,2,4-Triazadisilines and related
compounds*
Examples of both the 1,3,5,2,4-triazadisiline
system, (64; R = Bu, X = O, b.p. 135°/0.02 mm),
and the 1,3,5,2,6-oxadiazadisiline system, (65;
R = Me; m.p. 44-47°) have been isolated from the
reactions of N,N'-disubstituted ureas with bis-
(diethylamino)dimethylsilane (Maringgele,
Z. Naturforsch., 1983, 38B, 71).

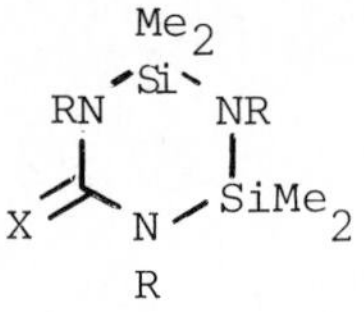

64

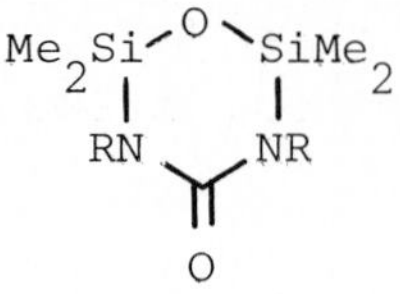

65

1,3,5,2,4-Triazadigermanines, (66), are the initial products of insertion of heterocumulenes or unsaturated dipolar compounds, e.g. phenyl isocyanate, into the cyclogermazanes (55) (Lacrampe *et al.*, *loc. cit.*) Similarly, the cyclodistannazane (67) also undergoes insertion

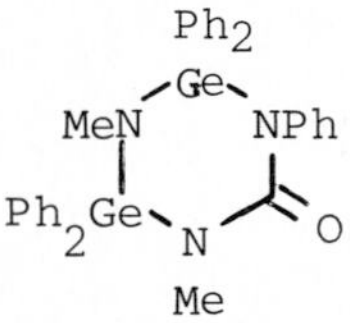

66

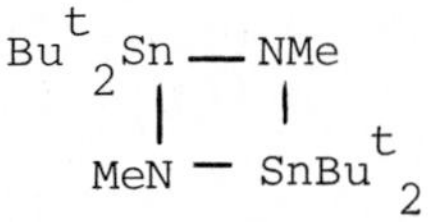

67

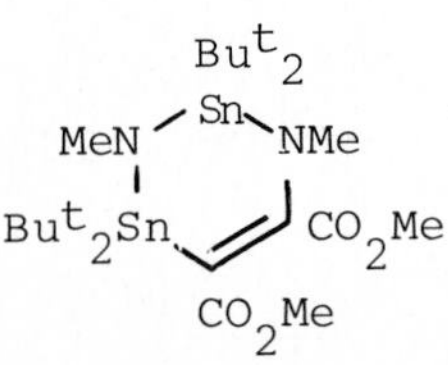

68

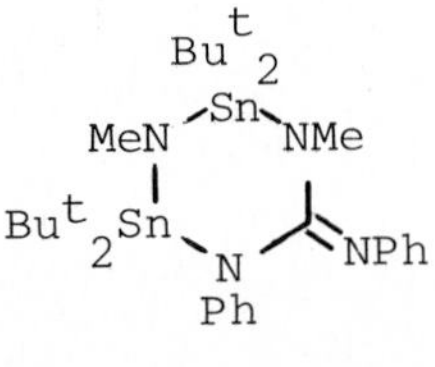

69

reactions on treatment with reactive addends such as dimethyl acetylenedicarboxylate or diphenyl-carbodiimide, to form, respectively, the 1,3,2,4-diazadistannine (68), m.p. 65°, and the 1,3,5,2,4-

triazadistannine (69), m.p. 135-140° (Hänssgen and
I. Pohl, Ber., 1979, 112, 2798).

3. Phosphorus, arsenic, antimony, and bismuth compounds

(a) Oxaphosphorins, Oxarsenins, and their thia analogues

(i) Benz-[1,2]-oxaphosphorins, benz-[2,1]-oxaphosphorins, benz-[2,3]-oxaphosphorins and thia analogues
The 5,6-benz-[e][1,2]-oxaphosphorin (1), an
orange solid, m.p. 196°, is formed in 57% yield in
the reaction of hexaphenylcarbodiphosphorane,
$Ph_3P=C=PPh_3$, with salicylaldehyde in refluxing
toluene (H.J. Bestmann and W. Kloeters, Tetrahedron Letters, 1977, 79). Cyclisation of phenyl
esters of β-alkoxyvinylphosphonic acids with polyphosphoric acid leads to the esters (2; R^1 = H, Me
or Et; R^2 = Et or Ph) in 17-77% overall yield
(K.A. Petrov, *et al.*, Zh. obshch. Khim., 1978, 48,
2667).

| 1 | 2 | 3 |

Perhydrobenz-[1,2]-oxaphosphorins, (3), are
formed on cyclisation of esters of certain 2-(1-
cyclohexenyl)vinylphosphonic acids with chlorine
or bromine (Kh. Angelov, *et al.*, Zh. obshch. Khim.,
1979, 49, 2381; 1982, 52, 181; Tetrahedron
Letters, 1981, 22 359).
Alkyl- and aryl-phosphonic acid esters of the
type (4) undergo a palladium-catalysed cyclisation

to the benz-[2,1]-oxaphosphorin (5; R = Me, Bu or Ph), (5; R = Ph, m.p. 83-84°), on heating in toluene with $(Ph_3P)_4Pd$ in the presence of tri-ethylamine (Y. Xu and J. Zhang, Tetrahedron Letters, 1985, 26, 4771).

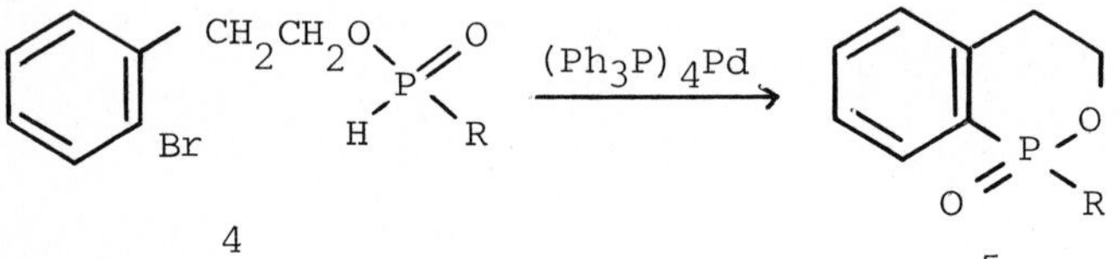

4   5

The seven-membered cyclic ester (6), (obtained from phenyldichlorophosphine and 1,2-bis(hydroxymethyl)benzene in the presence of a base) rearranges on heating at 170° in air to form the benz-[2,3]-oxaphosphorin (7, X = O), b.p. 210°/0.1 mm., in 60% yield. Treatment of this compound with Lawesson's reagent (a phosphinodithioic anhydride derived from the reaction of anisole with P_2S_5) converts it into the corresponding thiophosphoryl compound , (7, X = S), m.p. 77°. However, attempts to reduce the latter to the tri-valent phosphorus system using the nickelocene-allyl iodide procedure developed earlier by

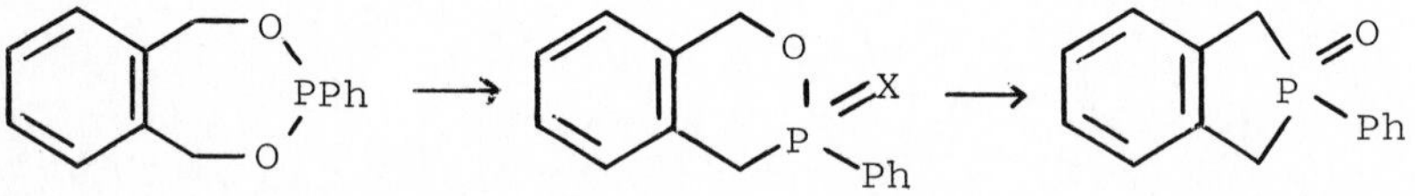

6   7   8

Mathey's group result in a rearrangement to form the isophosphindoline oxide (8) in 38% yield

(F. Mathey and F. Mercier, Chem. Comm., 1980, 191).
Treatment of the dihydropyridine (9) with
Lawesson's reagent leads to the unexpected
formation of the benz-[2,3]-thiaphosphorins (10;
Z = O or S), isolated in poor yield (R. Shabana,
S. Scheibye, *et al.*, Nouv. J. Chim., 1980, $\underline{4}$, 47).

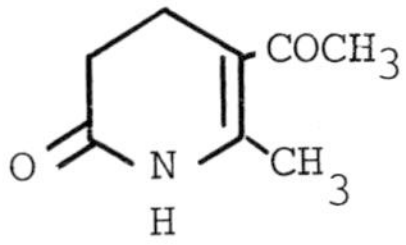

9

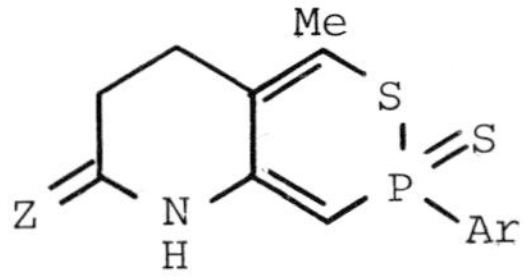

10

(ii) Perhydro-1,3-oxarsenins
The arsinocarboxylic acids (11) undergo cyclisation on treatment with aldehydes in benzene in the
presence of *p*-toluenesulphonic acid to form the
perhydro-1,3-oxarsenins (12, R^1 = Bu or Ph;
R^2 = Me, Et, Pr^i or Ph) (A. Tzschach and
W. Voigtländer, J. organometallic Chem., 1978,
$\underline{155}$, 195).

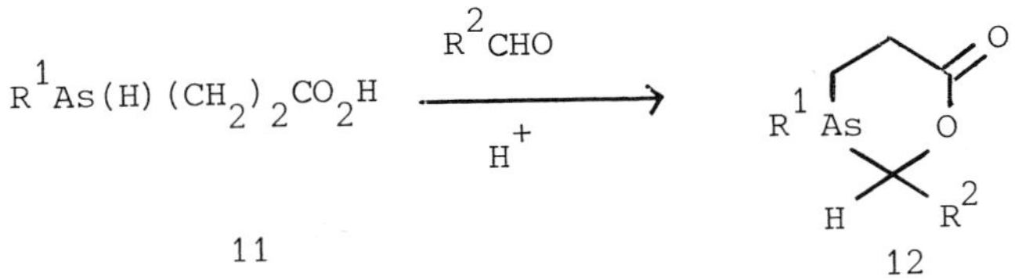

11

12

(iii) Benz-[1,4]-oxaphosphorins
This system is accessible in good yield by the
polyphosphoric acid-induced cyclisation of the
phosphonium salts (13), giving the benz-[1,4]-
oxaphosphorinanium salts (14; R = Me or Ph),
(14; R = Ph, m.p. 277°, 90%)(M.B. Marszak and

M. Simalty, Tetrahedron, 1979, __35__, 775).

13   14

(b) Dioxaphosphorins and related compounds

(i) 4H-1,3,2-Dioxaphosphorins and analogues
   The reaction of acetylacetone with a halogenophosphorus compound in the presence of triethylamine gives a 4H-1,3,2-dioxaphosphorin. Thus,
the reaction of phosphorus trichloride with the
above reagents yields (15; R = Cl; b.p. 28-38°/
0.35 mm;) but in only 20% yield (G.M.L. Cragg,
et al., Chem. Comm., 1977, 569). Similarly, the
related reaction of phenyldichlorophosphine leads
to (15; R = Ph; m.p. 40°) but again in very low
yield (7%) (J. von Seyerl and G. Huttner,
Z. Naturforsch., 1980, 35B, 1373). Thiophosphoryl
derivatives are accessible from the corresponding
reactions of dihalogeno(thiono)phosphorus
compounds (F.S. Mukhametov, *et al.*, Zh. obshch.
Khim., 1983, __53__, 1274). Considerably improved
yields are obtained if the halogenophosphine is

15   16

used in the form of a complex with a transition
metal carbonyl acceptor (Von Seyerl, *et al.*, Ber.,
1979, <u>112</u>, 3637).

Access to the 4*H*-1,3,2-dioxarsenin system (16)
is afforded by the reactions of metal complexes of
dichlorophenylarsine with acetylacetone in the
presence of triethylamine (Von Seyerl, G. Huttner
and K. Kruger, Z. Naturforsch., 1980, <u>35B</u>, 1552).

A number of benzo-4*H*-[1,3,2]-dioxaphosphorins
and their thia analogues have also been described.
The reactions of alkyl *o*-hydroxyphenyl ketones with
phosphorus trichloride and triethylamine afford the
benzo-4*H*-[1,3,2]-dioxaphosphorins (17;   R = Me or
Et), (17, R = Me, b.p. 72-76°/0.3 mm;   45%)
(Cragg, *et al.*, *loc. cit.*)

17

18

The reactions of salicylic acid and of
o-mercaptobenzoic acid (or their esters) with
Lawesson's reagent yields compounds of the benzo-
4*H*-[1,3,2]-oxathiaphosphorin (18;   X = O or S,
Y = O) and -dithiaphosphorin series (18;   X = O
or S;   Y = S) respectively. Thus, salicylic acid
yields (18;   X = Y = O), m.p. 98°, in 72% yield
(A.A. El-Barbary and S.O. Lawesson, Tetrahedron,
1981, <u>37</u>, 2641).

A route to perhydro-1,3,2-oxathiaphosphorins
and -1,3,2-oxaselenaphosphorins (19;   Z = S or Se)
is provided by the reaction of oxetanes with
diaryl esters of thio- and of selena-phosphonic
acids, respectively, (B.A. Arbuzov and
O.N. Nuretdinova, Izv. Akad. Nauk SSSR, Ser.
Khim., 1983, 675). The 1,3,2-dioxabismane (20),
m.p. > 200° decomp., is formed in 43% yield in the

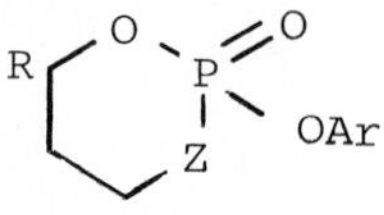

19

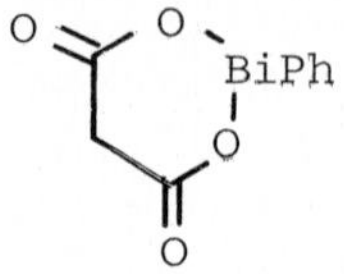

20

reaction of triphenylbismuth with malonic acid
(A. Georgiades and H.P. Latscha, Z. Naturforsch.,
1980, 35B, 1000).

(ii) 2H-1,5,2-Dioxaphosphorins
The reactive phosphene oxide (21), generated by
photolysis of a diphenylphosphoryldiazoalkane,
readily undergoes [4+2] cycloaddition reactions
with carbonyl compounds to give, with other
products, the 2H-1,5,2-dioxaphosphorins (22),
(22; R = Ph, m.p. 182-183°) (M. Regitz, W. Illger,
and G. Mass, Ber., 1978, 111, 705). The perhydro-
1,5,2-dioxaphosphorin (23) is formed in the

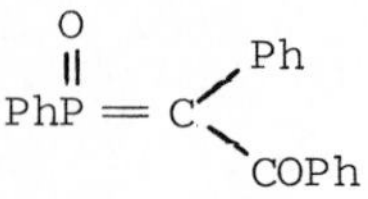

21

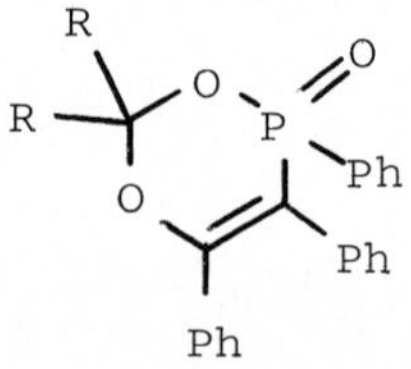

22

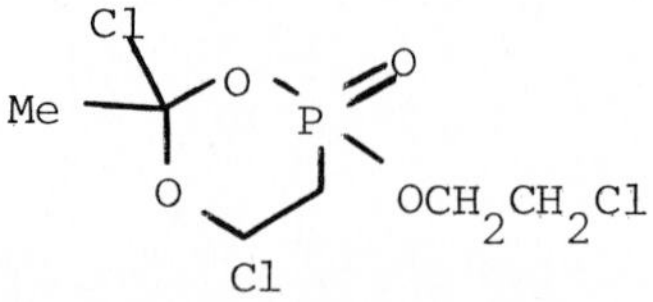

23

reaction of vinyl acetate with phosphorus pentachloride, followed by treatment with ethylene
oxide (A.A. Krolovets, A.F. Kolomiets, and
A.V. Fokin, Izv. Akad. Nauk, SSSR, Ser. Khim.,
1979, 1162).

(c) Oxadiphosphorins and related compounds

(i) 2H-1,2,4-Oxadiphosphorins
Two examples of this ring system, (24) and (25), have been described. The former, m.p. 258-260°, is among the addition products of the phosphene oxide (21) with benzophenone (Regitz *et al.*, *loc. cit.*) The latter is the dimerisation product of the phospha-alkene $Bu^tP=C(CO_2Me)_2$ (O.I. Kolodyazhyni, Zh. obshch. Khim., 1980, 50, 230).

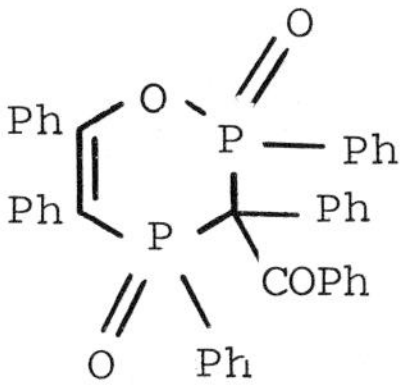

24

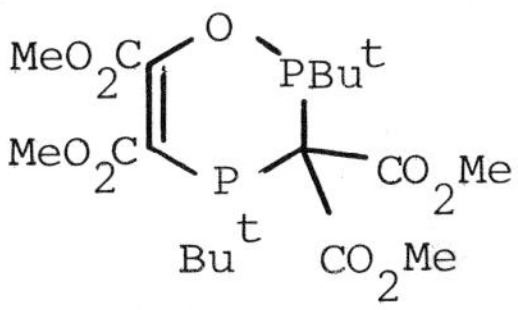

25

(ii) Perhydro-1,2,6-oxadiphosphorins and -thiadiphosphorins
The 1,2,6-oxadiphosphacyclohexane (26), m.p. 198°, has been isolated in very low yield from the products of pyrolysis of the poly(trimethylene-phosphinate) polymers obtained by the treatment of perhydro-1,3,2-dioxaphosphorins with alkyl halides

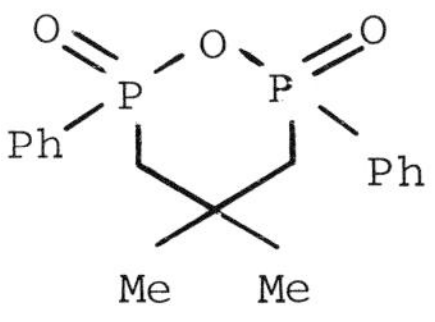

26

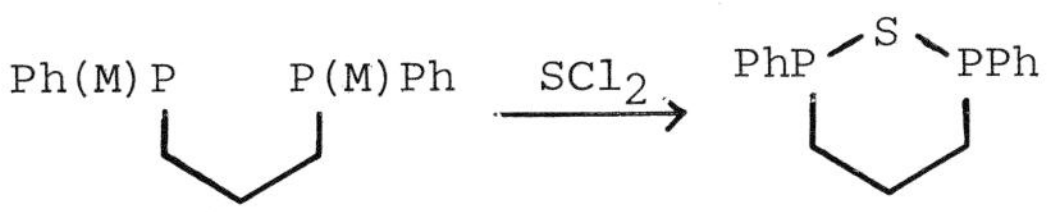

27

28

(G. Singh, J. org Chem., 1979, 44, 1060; Singh, G.S. Reddy, and J.C. Calabrese, *ibid.*, 1984, 49, 5132).

Cyclisation of the bis(metallophosphide) reagent
(27; M = Li or K) with sulphur dichloride provides
a route to the 1,2,6-thiadiphosphacyclohexane (28)
(K. Issleib and W. Böttcher, Syn. React. inorg.
met.-org. Chem., 1976, 6, 179).

(d) Dioxadiphosphorins and thia analogues

Examples of the perhydro-1,3,2,4-dioxadiphospho-
rin and -1,4,2,6-dithiadiphosphorin systems have
been described. The bifunctional diphosphorus
compound $(MeO)_2P(O)CH_2CH_2OPCl_2$, cyclises on
heating to give (29) (M.B. Gazizov, *et al.*, Zh.
obshch. Khim., 1984, 54, 225). The reactions of
trialkylamines with P_4S_{10} yield initially

29   30   31

quaternary ammonium salts of the cyclic dianions
(30), one of which (R = Pr^i) on pyrolysis gives
the bicyclic system (31) (P.J. Retuert, *et al.*,
Z. anorg. allgem. Chem., 1985, 521, 153).

(e) Azaphosphorines and their arsenic analogues

(i) 1,2-Azaphosphorines and related compounds
The iminophosphorane formed from the reaction of
trimethylphosphite with 1-(phenyl)vinyl azide
readily undergoes cycloaddition reactions with
acetylenes substituted with electron-withdrawing
groups to give, initially, the 1,4-dihydro-1,2-
λ^5-azaphosphorines (32), which subsequently
rearrange to form the quasi-aromatic $1,2\lambda^5$-
azaphosphorines (33), (33; R = CO_2Me, m.p. 85-
86°) (T. Kobayashi and M. Nitta, Chem. Letters,

1985, 1459). When the latter is heated with a
further quantity of dimethyl acetylenedicarboxylate in xylene, elimination of benzonitrile occurs
with the formation of the λ^5-phosphorin (34). The
bicyclic system (35) is formed in the reaction of

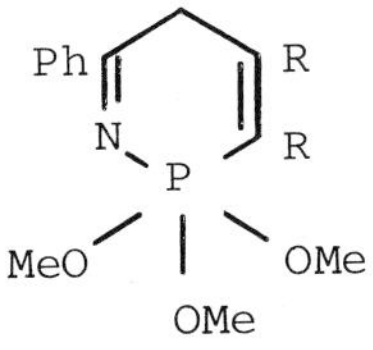

32

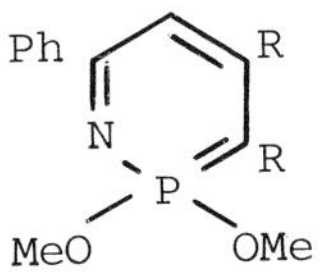

33

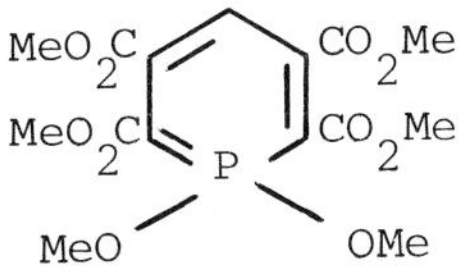

34

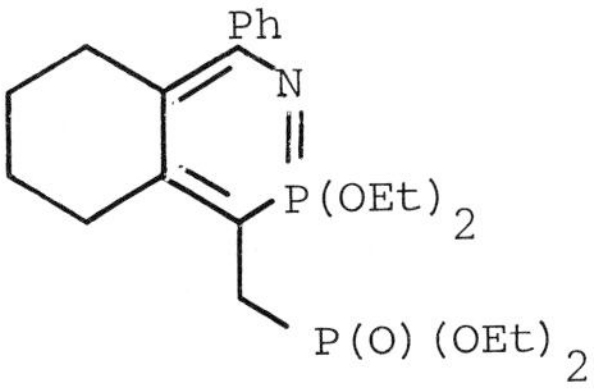

35

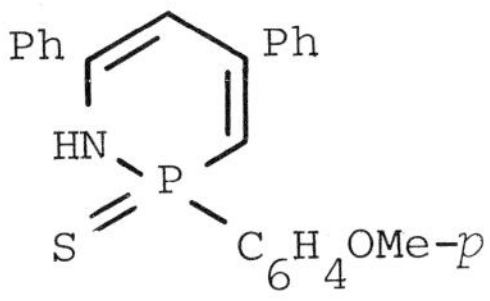

36

a phosphorylated allene with a [phenyl(diethylamino)methylene]aminophosphite (N.G. Khusainova,
et al., Zh. obshch. Khim., 1982, 52, 789). The
1,2-dihydro-1,2-azaphosphorine (36), m.p. 186-
187°, has been isolated in 15% yield from the
reaction of acetophenone oxime with Lawesson's
reagent (R. Shabana, *et al.*, Sulfur Letters,
1984, 2, 223).

(ii) Perhydro-1,2-azaphosphorines
A route to 6-oxo derivatives of this system
involves cyclisation of the phosphonamidates (37)
in the presence of base to give (38; R = H, Me
or allyl), (38; R = H, m.p. 109-112°) (D.G. Hewitt

and M.W. Teese, Austral. J. Chem., 1984, 37, 1631).

$$\text{EtO} - \overset{\overset{\displaystyle O}{\|}}{\underset{\underset{\displaystyle NHR}{|}}{P}} - (CH_2)_3CO_2Et \quad \xrightarrow[\text{Bu}^t\text{OH}]{\text{KOBu}^t}$$

37   38

(iii) 1,3-Azaphosphorines and related systems
Access to the 2,3,4,5-tetrahydro-1,3-azaphosphorine system (39; R^1 = Ph; R^2 = H; R^3 = Ph, b.p. 140-143°/0.08 mm.) is afforded by the reaction of β-ketoalkyl secondary phosphines with a mixture of an aldehyde and ammonia. The related reaction with imines derived from aldehydes and primary amines furnishes the isomeric 1,2,3,4-tetrahydro-1,3-azaphosphorine system (40; R^1 = R^2 = R^4 = R^5 = Ph; R^3 = H; R^6 = Me, m.p. 171-172°)(K. Issleib and P. Von Malotki, Phosphorus, 1973, 3, 141).

39   40

Secondary phosphines of the type $RPH(CH_2)_3NH_2$ (R = Bu^t, Ph or C_6H_{11}) readily combine with carbon disulphide to form thiones, (41; R = Ph, m.p. 128-131°), which on treatment with alkyl halides in the presence of a base undergo alkylation at sulphur to form the tetrahydro-1,3-azaphosphorines

(42), (42; R^1 = Ph; R^2 = Et, b.p. 180-185°/
3.5 mm.). In the absence of a base, alkylation

41   42

occurs at phosphorus (H. Oehme, E. Leissring, and
R. Thamm, J. prakt. Chem., 1978, 320, 600).
   Cyclisation of the phosphinate ester (43) with
ammonia provides a route to the perhydro-1,3-
azaphosphorine (44; b.p. 130°/0.15 mm.) in 36%
yield (Hewitt and Teese, Austral. J. Chem., 1984,
37, 205).

43   44

   A radical-induced cyclisation of diallylamino-
methylphosphine gives a mixture of the two isomeric
bicyclic perhydro-1,3-azaphosphorine systems (45;
b.p. 41-42°/0.8 mm.) and (46). The latter is
easily separated from the mixture *via* its highly-
insoluble adduct with carbon disulphide (Issleib,
U. Kühne, and F. Krech, Phosphorus Sulfur, 1983,
17, 73).

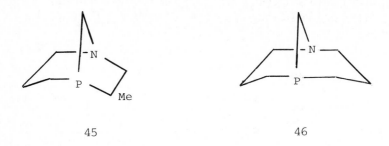

45   46

(iv) 1,4-Azaphosphorines and -azarsorines
Derivatives of 1,4-dihydro-1,4-azaphosphorine
(47) are readily accessible by the cyclisation of
di-(2-oxoalkyl)phosphine oxides with ammonium
acetate. A convenient route to the required phosphine oxides involves ozonolysis of 3-phospholene
oxides, which are readily accessible *via* the
McCormack cyclisation of 1,3-dienes with alkyl-
or aryl-phosphonous dihalides (L.D. Quin, *et al.*,
Synthesis, 1984, 1074).

47   48   49

The reactions of the iminophosphoranes (48) with
dimethyl acetylenedicarboxylate offer a simple
route to the $1,4\lambda^5$-azaphosphorines (49), (49;
R^1 = H, R^2 = *p*-tolyl, m.p. 228-229°.) Nmr
studies of this group of compounds reveal no evidence of aromaticity (J.C. Barlengua, F. Lopez, and
F. Palacios, Chem. Comm., 1985, 1681).
The *o*-aminophenylarsines (50) under cyclisation
to form the benz-[e][1,4]-azarsorin-2-ones (51),
which with lithium aluminium hydride afford the

reduced system (52), (52; R = Me, b.p. 85-87°/ 0.3 mm.) (A. Tzschach and H. Biering, J. organo-metallic Chem., 1977, <u>133</u>, 293).

50   51   52

(f) Diazaphosphorines and related compounds

(i) Benzo-[1,3,2]-diazaphosphorines and related systems
    In a reconsideration of earlier work, it has been shown that the reaction of *o*-aminobenzamide (or the related thioamide) with P_2S_5 in pyridine yields as the initial product, a pyridine complex (hygro-scopic yellow needles, m.p. 170-172°) of the benzo-[1,3,2]-tetrahydrodiazaphosphorine (53), which, on treatment with dimethyl sulphate in the presence of alkali, gives the benzo-[1,3,2]-dihydrodiazaphosphorine (54), m.p. 140-142° (R.M. Acheson, *et al.*, J. Chem. Soc., Perkin II, 1985, 1913). Access to the benzo-[1,3,2]-diazaphosphorine

53   54

system is also afforded by the reactions of anthranilamides with phosphorus trichloride

(G.M. Coppola and R.I. Mansukhani, J. heterocyclic
Chem., 1978, 15, 1169; Coppola, *ibid*, 1979, 16,
897).

Several routes to the benzo-[e][1,3,2]-tetra-
hydrooxazaphosphorine system have been developed.
The reactions of trivalent phosphorus halides with
o-hydroxybenzamides give the phosphorus (III)
systems, (55; m.p. 113-114°)(S. Kobayashi,
Y. Narukawa, and T. Saegusa, Syn. Comm., 1982, 12,
539) (See also, A.K. Kuliev, *et al.*, Zh. obshch.
Khim., 1985, 55, 936). Compounds in which the
phosphorus is in the pentavalent state (56;
m.p. 149°) have been obtained from the reactions
of *o*-hydroxybenzamide with Lawesson's reagent
(S. Scheibye, P. S. Pederson, and Lawesson, Bull.
Soc. chim. Belg., 1978, 87, 229; 299), and also
with the pyridinium dithiophosphoric acid
chloride betaine (M. Meisel and C. Donath,
Phosphorus Sulfur, 1983, 18, 159).

55   56

The reaction of salicylic acid azide with
1-phenyl-1,3,2-dioxaphospholane leads to the ring
system (57) which involves a pentacoordinated
phosphorus atom, cyclisation of the intermediate
iminophosphorane being promoted by the five-
membered ring phosphorus moiety. The corresponding reaction of 3-hydroxyalkanoic acid azides
affords the related tetrahydro-1,3,2-oxazaphos-
phorin-4-ones, (58; R = H or Ph) (P. Pöchlauer,
et al., Chem. Comm., 1985, 1764). The 2,3-
dihydro-4*H*-1,3,2-oxazaphosphorin-4-thione system
(59; R^1, R^2 = H, Me or Ph), (59; R^1 = Ph; R^2 =
H; m.p. 56-60°) is formed in 70-95% yield in the

reactions of primary 3-oxoamides or 3-oxonitriles
with Lawesson's reagent (B.S. Pederson and Lawesson
Tetrahedron, 1979, 35, 2433).

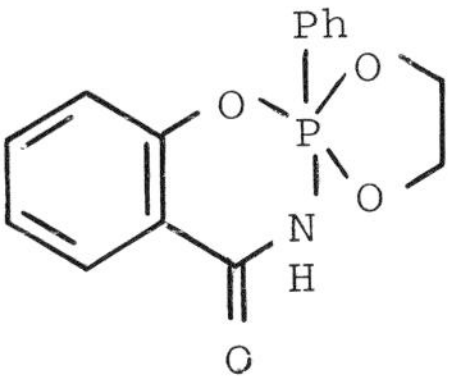

57 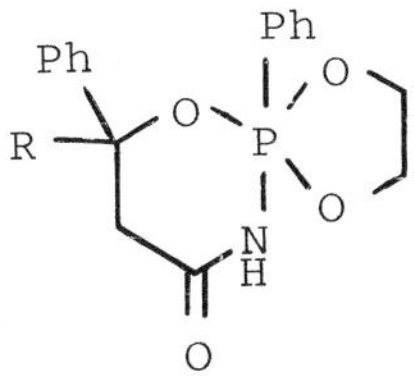   58 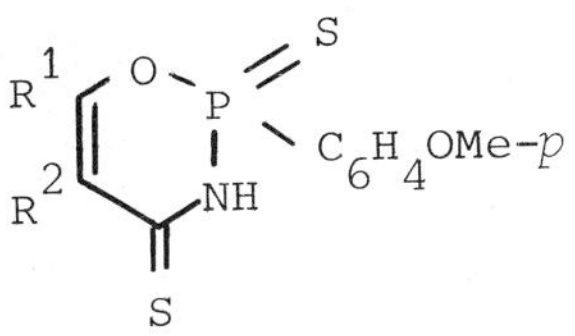   59

Examples of the benz-[3,1,2]-oxazaphosphorine
system (60) have been obtained by the reactions of
anthranilic acid and its derivatives with phosphorus trichloride (A.K. Kuliev, et al., Zh.
obshch. Khim., 1984, 54, 1671; 1985, 55, 457),
and also by the reactions of anthranilic acid and
its derivatives with Lawesson's reagent, giving
compounds such as (61; m.p. 168°; 50%)
(El-Barbary and Lawesson, Tetrahedron, 1981, 37,
2641).

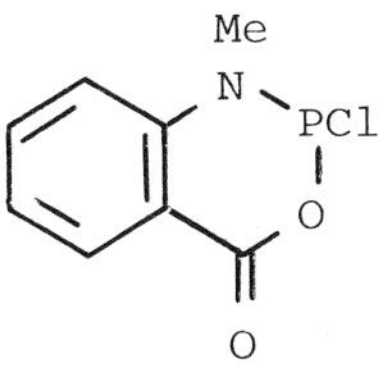

60 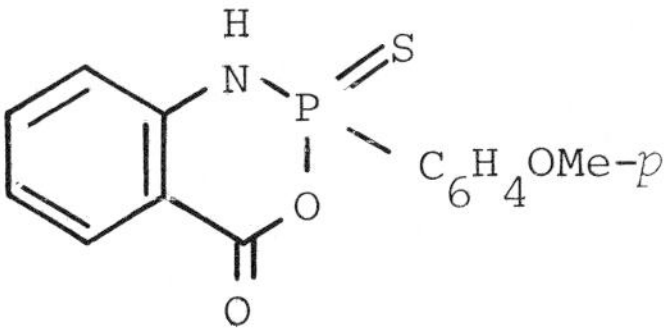   61

(ii) 1,2,3-Diazaphosphorines and related
compounds
The products of cycloaddition of diphenylcarbodiimide to the P=C double bond of 1,2,3-diaza-
phospholes rearrange to form the tetrahydro-
1,2,3λ^3-diazaphosphorine system (62)

(B.A. Arbuzov, E.N. Dianova, and Yu. Yu. Samitov, Dokl. Akad. Nauk SSSR, 1979, 244, 117). The phosphorus atom in this system behaves normally in its reactions with electrophilic reagents (Arbuzov and Dianova, Izv. Akad. Nauk SSSR, Ser. Khim., 1983, 418). A related dihydro system (63)

62   63   64

is claimed to be formed in the cycloaddition reactions of the iminophosphoranes $R_2C=N-N=PPh_3$ with dimethyl acetylenedicarboxylate. However, as yet, no definitive proof of 1,4-addition has been provided. Two alternative isomeric structures arising from 1,2-addition are possible (I. Shahack and Y. Sasson, Isr. J. Chem., 1973, 11, 729). The tetrahydro-1,2,6λ^5-oxazaphosphorine system (64) is formed in the cyclisation reactions of aminoalkenylphosphonium salts (I.V. Megera, L.M. Gertsyuk, and M.I. Shevchuk, Zh. obshch. Khim., 1978, 48, 2030).

   (iii) 1,5,2-Diazaphosphorines and related compounds
   Interest in the synthesis of possible transition state analogues of the nucleoside deaminases has prompted the development of synthetic routes to "phosphapyrimidines" such as (65; R = H, m.p. 220°), and (65; R = Ph, m.p. 180-181°) and also the "phosphapurine" (66; m.p. 199-200°) (P.A. Bartlett, et al., Bio-org. Chem., 1978, 7, 421). The enzyme-inhibiting activity of appropriately functionalised derivatives of (65; R = H) has been studied (G.W. Ashley and Bartlett, Biochem. biophys. res. Comm., 1982, 108, 1467;

J. biol. Chem., 1984, 259, 13621). Benzo-fused
systems derived from the 1,5,2-diazaphosphorine
ring, (67), (a benzo-[2,4,1]-diazaphosphorine),
have been described in the patent literature
(S.F. Lee, U.S. P. 4,433,149; Chem. Abs., 1984,
101, 55327).

65   66   67

(iv) Perhydro-1,3,5-diazaphosphorines
This ring system is easily accessible by the
reactions of alkyl- or aryl-bis(phenylaminomethyl)
-phosphines with formaldehyde, giving, (68,
R = Ph; m.p. 115°) and (68; R = Me, m.p.
90-91°) (G. Märkl and G. Yu. Jin, Tetrahedron
Letters, 1981, 22, 229), and also by the reactions
of phenylbis(hydroxymethyl)phosphine (or
$1,3,5\lambda^3,2$-dioxaphosphaborins) with primary amines
in the presence of formaldehyde (B.A. Arbuzov,
O.A. Erastov, and G.N. Nikonov, Izv. Akad. Nauk
SSSR, Ser. Khim., 1979, 2771; 1980, 952). These
compounds exhibit the expected reactivity due to
the trivalent phosphorus atom, readily forming
phosphine-oxide, -sulphide, and -selenide
derivatives (Arbuzov, *et al.*, *ibid.*, 1980, 721;
1982, 127). Treatment with alkyl halides
results in alkylation at phosphorus only
(Erastov, Nikonov, and Arbuzov, *ibid.*, 1983,
1379). Exchange of substituents of nitrogen has
been observed on treatment with a second primary
amine in benzene solution (Arbuzov, Erastov, and
Nikonov, *ibid.*, 1980, 2417), implying the
existence of complex equilibria involving ring-
opened systems in solution (*idem.*, *ibid.*, 1980,
1438).

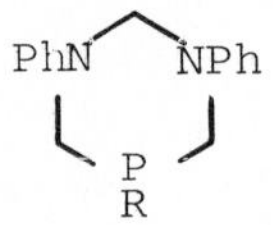

68

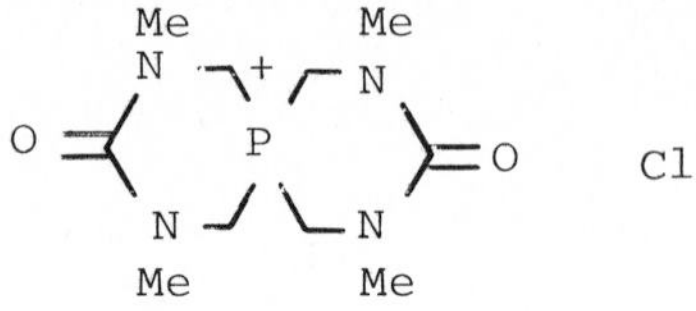

69

The reactions of 1,3-dimethylurea with tetra-kis(hydroxymethyl)phosphonium chloride give rise to both polymeric and heterocyclic products, (69; m.p. 217°) (A.W. Frank, Phosphorus Sulfur, 1981, 10, 147; 207).

(g) Triazaphosphorines and related systems

The 1,3,5,2-triazaphosphorine system is afforded by the reaction of methylthiophosphonyl di-isocyanate with a bis(trimethylsilyl)amino compound, giving compounds of the type (70;, m.p. 85°). The related thiadiazaphosphorine (71, m.p. 86°), is formed in the corresponding reaction with bis(trimethylsilyl)sulphide (W. Roesky and H. Zamankhan, Z. Naturforsch., 1977, 32B, 229). The 1,3,5,2-triazarsenine (72; m.p. 138-140°), is formed in the reactions of trimethylsilyl derivatives of ureas with arsenic

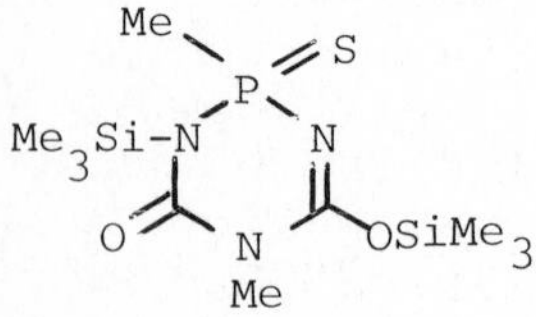

70

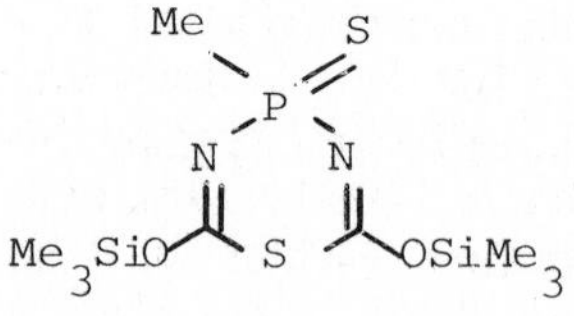

71

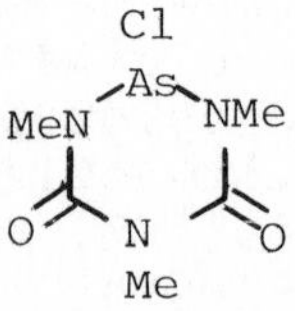

72

trichloride (Roesky and G. Sidiropoulos, *ibid.*, 1978, **33**, 756; W.S. Sheldrick, Zamankhan, and Roesky, Ber., 1980, **113**, 3821).

(h) Diazadiphosphorines

A number of 1,3,2,4-diazadiphosphorines have been prepared by the reactions of alkyl nitriles with hexachlorodiphosphazonium chloride, $[Cl_3P=N-PCl_3]^+Cl^-$. Thus, with acetonitrile, a 1:4 mixture of (73; R = H) and (73; R = Cl) is formed, from which (73; R = H, b.p. 70°/0.1 mm.) has been obtained pure. The related reaction with chloroacetonitrile gives (73; R = Cl, m.p. 53°) and that with phenylacetonitrile gives (73; R = Ph; m.p. 111°) (E. Fluck, E.Schmid, and W. Haubold, Z. anorg. allgem. Chem., 1977, **433**, 229; 1977, **434**, 95; Fluck and Schmid, Phosphorus Sulfur, 1977, **3**, 209).

73

74

Various perhydro-1,4,2,5-diazadiphosphorines are formed in the reactions of azomethines with halogenophosphorus compounds. Thus, the reaction of *N*-phenylbenzaldimine with methylphosphorodichloridite gives (74; R = Cl) and that with ethylene chlorophosphite gives (74; R = CH_2CH_2Cl) (N.K. Maidanovich, S.V. Iksanova, and Yu. G. Gololobov, Zh. obshch. Khim., 1982, **52**, 930; A.M. Kibardin, *et al.*, Izv. Akad. Nauk SSSR., Ser. Khim., 1983, 432).

Chapter 50

SEVEN-MEMBERED RING COMPOUNDS CONTAINING OXYGEN OR SULPHUR IN THE RING

J.T. SHARP

This supplement covers a period of intense activity in the exploration of the chemistry of seven-membered heterocyclic systems, particularly in the development of routes to the unsaturated compounds and the study of their reactions and rearrangements. The flood of patents and papers dealing with the chemistry of the several classes of compounds in this group which have central nervous system activity and with related systems which would hopefully mimic or extend their physiological activity continued through this period. For reasons of space much of this chemistry has had to be omitted, particularly that dealing with functional group transformations where these appeared to be unexceptional, further examples of known reactions unless they revealed new chemistry and, as in the second edition, bridged systems and compounds with annelated heterocyclic rings.

Reviews

(i) *Oxepins.* D.M. Jerina, H. Yagi, and J.W. Daly, Heterocycles, 1973, 1, 267; D.M. Jerina and J.W. Daly, Science, 1974, 185, 573; T.C. Bruice and P.Y. Bruice, Acc. chem. Res., 1976, 9, 378; G.S. Shirwaiker and M.V. Bhatt, Adv. heterocyclic Chem., 1984, 37, 67; D.R. Boyd and J.M. Jerina, Chem. heterocyclic Compd., 1985, 42, 197.

(ii) *Thiepins.* I. Murata and K. Nakasuji, Top. curr. Chem., 1981, 97, 33; D.N. Reinhoudt, Recl:

J.R. Neth. chem. Soc., 1982, 101, 277.

(iii) *Oxepins and Thiepins.* D.R. Boyd in "Comprehensive Heterocyclic Chemistry", Pergamon, 1984, vol. 7, p. 547.

1. *Seven-membered rings containing one oxygen atom*

1.1 *Monocyclic oxepins*

(a) *Fully unsaturated.* (i) *Synthesis.*
Syntheses of substituted derivatives of this system lead to equilibrium mixtures of the oxepin (1) and its valence tautomer the benzene oxide (2). The effect of the nature and position of the substituents on the position of this equilibrium is discussed in recent reviews (see above). In annelated systems the size of a ring fused at the 2,3-position (R^1/R^2) has a marked effect, a seven-membered ring favours (2) while a six-membered ring shifts the equilibrium strongly in favour of (1) (B. Epe, P. Rösner, and W. Tochtermann, Ann., 1980, 1889).

A new route to the system is provided by the decomposition of 4-diazomethyl-4H-pyranes (3) catalysed by μ-allyl palladium chloride (allyl $PdCl)_2$. The diazopyranes (3) are prepared by the reaction of lithiated diazo compounds ($Li^{+-}CRN_2$; R=PO(OMe)$_2$, PO(OMe)Ph, POPh$_2$, CO$_2$Et) with pyrylium salts, and their decomposition gives (1; $R^{1,6}$=Bu^t, R^3=Me, R^4=R) (K-L. Hoffmann and M. Regitz, Tetrahedron Letters, 1983, 24, 5355;

Ber., 1985, <u>118</u>, 3700). Another new route - the consecutive elimination of hydrogen bromide and carbon dioxide from the bromo-β-lactone epoxides (4) - has been used to prepare a number of 3-substituted derivatives (2; $R^{1,3,4,5,6}$=H, R^2=Me, CH_2OCOPh, CH_2CO_2Et) (B. Ganem, G.W. Holbert, L.B. Weiss, and K. Ishizumi, J. Amer. chem. Soc., 1978, <u>100</u>, 6483). The base-induced elimination of two molecules of hydrogen halide from a halogenated cyclohexene epoxide (e.g. 5), a method originated by Vogel (see 2nd. edn.), has been extensively used in several variations, to prepare a wide range of substituted oxepins. These include 1-carboalkoxy and 1-carboxy derivatives (D.R. Boyd and G.A. Berchtold, J. Amer. chem. Soc., 1979, <u>101</u>, 2470; 1978, <u>100</u>, 3958), 1-trimethylsilyl and its 2- and 4-methyl substituted analogues (D.R. Boyd and G.A. Berchtold, J. org. Chem., 1979, <u>44</u>, 468; and with J.E. Van Epp, *ibid*, 1981, <u>46</u>, 1817), 3-carbo-*tert*-butoxy (B.A. Chiasson and G.A. Berchtold, J. org. Chem., 1977, <u>42</u>, 2008), 4-carbo-*tert*-butoxy (G.A. Berchtold, *et al*., J. Amer. chem. Soc., 1974, <u>96</u>, 1193), 4-carboxy *via* its trimethylsilyl ester (J.D. Richardson, *et al*., J. org. Chem., 1974, <u>39</u>, 2088), 3-phenyl (M.P. Servé and D.M. Jerina, *ibid*, 1978, <u>43</u>, 2711), 3- and 4-chloro (H.G. Selander, *et al*., J. Amer. chem. Soc., 1975, <u>97</u>, 4428), 2-, 3- and 4-(β-aminoethyl) (W.H. Rastetter and L.J. Nummy, J. org. Chem., 1980, <u>45</u>, 3149), and a range of 2,6-disubstituted derivatives (1; $R^{1,6}$=CHO, CO_2H, CO_2Me, CN; R^{2-5}=H) (E. Vogel, *et al*., Tetrahedron Letters, 1976, 1167). Many of these compounds and their deuteriated derivatives have been used for the study of their acid or base catalysed ring contraction to give benzene derivatives and as models for the study of the participation of benzene oxides in biochemical processes. Prinzbach's synthesis of substituted oxepins *via* the photoisomerisation of 7-oxanorbornadienes (6) into the oxaquadricyclanes (7) and their subsequent thermal rearrangement has been extended to give the parent compound (1; R^{1-6}=H) *via* a new route to the unsubstituted oxanorbornadiene

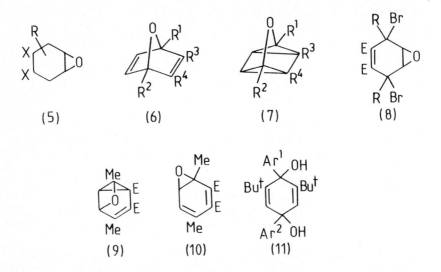

(5)   (6)   (7)   (8)

(9)   (10)   (11)

precursor (H. Prinzbach and H. Babsch, Angew. Chem., internat. Edn., 1975, $\underline{14}$, 753). Substituted derivatives e.g. (1; $R^{1,2,5}$=H, R^6=Ph, $R^{3,4}$=CO$_2$Me; $R^{1,6}$=Ph, $R^{2,5}$=H, $R^{3,4}$=CO$_2$Me) (R.K. Bansal et al., Canad. J. Chem., 1975, $\underline{53}$, 138) and the interesting bridged system (1; $R^{1,6}$=H, R^2/R^5=(CH$_2$)$_6$, $R^{3,4}$=CO$_2$Me) (W. Tochtermann and P. Rösner, Tetrahedron Letters, 1980, $\underline{21}$, 4905; Ber., 1981, $\underline{114}$, 3725) have been prepared by the same sequence. Reactions of the latter give no indication of the existence of an equilibrium with the arene oxide (2). Hydrogenation and epoxidation occur at the 2,3- and the 6,7-double bond and bromination gives (8, R/R=(CH$_2$)$_6$) (P. Rösner, C. Wolff, and W. Tochtermann, Ber., 1982, $\underline{115}$, 1162). The isomerisation of 3-oxaquadricyclanes (7) into oxepins promoted by transition metal catalysts can take a different path to the thermal reaction. Thus thermolysis of (7; $R^{1,2}$=Me, $R^{3,4}$=CO$_2$Me) gives (1; $R^{1,6}$=Me, $R^{3,4}$=CO$_2$Me) but the rhodium catalysed reaction gives (1; $R^{1,4}$=Me, $R^{2,3}$=CO$_2$Me) formed via (9) and (10) (R. Roulet et al., Tetrahedron Letters, 1974, 1479). Full details are available of the synthesis of the highly substituted oxepins (1;

$R^{1,5}=Bu^t$, $R^3=Ar^2$, $R^6=Ar^1$, $R^{2,4}=H$) by the acid catalysed ring expansion of the cyclohexadiene-1,4-diols (11) (A. Rieker, G. Henes, and S. Berger, Ber., 1975, 108, 3700). The oxepin structure of one product of this reaction (1; $R^{1,5}=Bu^t$, $R^3=Ph$, $R^6=p$-tolyl) has been confirmed by X-ray structure analysis (A. Rieker, et al., Ber., 1982, 115, 385).

(ii) *Reactions.* A re-examination of the photolysis of oxepin/benzene oxide in acetone-d_6 at room temperature, using the deuterium labelled substrate (1/2; $R^{2,5}=^2H$, $R^{1,3,4,6}=H$) and monitoring the reaction by nmr spectroscopy, has revealed an oxygen-walk rearrangement to give (12), and eventually a complete conversion into phenol. Irradiation of a thin film at 77K produces an intermediate, stable only at low temperature, which has a strong ir-absorption at 2112 cm^{-1}. This intermediate is thought to be the ketene $CH_2=CHCH=CHCH=C=O$, formed by the reversible ring opening of the keto-tautomer of phenol (13) - itself not detected (D.M. Jerina, et al., J. Amer. chem. Soc., 1974, 96, 5578). Many

(12)   (13)

oxepins react with dienophiles *via* the Diels-Alder reaction of their benzene oxide valence tautomer (2) because of the greater planarity of the diene unit. Oxepin reacts in this way with singlet oxygen to give the highly unstable peroxide (14) which on heating rearranges to the benzene trioxide (15) (C.H. Foster and G.A. Berchtold, J. org. Chem., 1975, 40, 3743). Reaction with the nitroso-compound $Cl_2C=C(Cl)N=O$ gives (16) probably *via* cycloaddition to the N=O group to give an

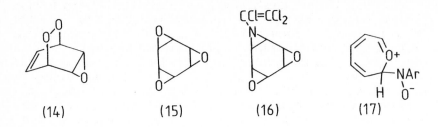

(14)   (15)   (16)   (17)

intermediate of the same type as (14) (E. Francotte, R. Merényi, and H.-G. Viehe, Angew. Chem., internat. Edn., 1978, *17*, 936). The reaction with *p*-chloronitrosobenzene however results in ring cleavage to give the nitrone $OHC(CH=CH)_2CH=N^+(O^-)C_6H_4Cl(p)$, possibly formed *via* ring opening of (17) (cf. (13) Ch. 51) (G. Kresze and W. Dittel, Ann., 1980, 1630). Oxepin itself reacts with azodicarboxylic esters $(RO_2C-N=N-CO_2R)$ *via* the Diels-Alder reaction of its benzene oxide form to give (46) but its 2,7-dimethyl derivative (1; $R^{1,6}$=Me, R^{2-5}=H) reacts *via* the oxepin form only giving (18) as the primary product, most likely because 2,7-substitution strongly favours the oxepin tautomer. Compound (18; $E=CO_2CH_2CCl_3$) rearranges to give (19) in high

(18)   (19)

yield *via* a Claisen rearrangement. Interestingly, Wittig methylenation of the latter, followed by thermal rearrangement, gives the carbocyclic analogue of (18) - O replaced by CH_2 - which after removal of the N-N bridge gives 2,7-dimethylcycloheptatriene (W.H. Rastetter and T.J. Richard, Tetrahedron Letters, 1978, 2999).

In reactions with 4π electron systems oxepins give products involving the participation of both 6π and 2π electrons. Thus the reaction with the cyclopentadienone (20) gives both the *exo* $(6+4)\pi$ adduct (21) and the *endo* $(2+4)\pi$ adduct (22) as primary products. The latter rearranges readily at 80°C to give (23), presumably *via* the *anti-endo* $(4+2)\pi$ adduct analogous to (25) (cf. also the analogous secondary product of addition to 1*H*-azepine, (17) Ch. 51). Compound (22; E replaced by Me) however, decomposes only *via*

(20)   (21)   (22)   (23)

(24)   (25)   (26)

cycloreversion (T. Ban, Y. Wakita, and K. Kanematsu, J. Amer. chem. Soc., 1980, 102, 5415). Rearrangement of the primary *endo* $(2+4)\pi$ adduct of the diazacyclopentadienone (24) is apparently even easier than that of (20), as (25) in equilibrium with *ca* 20% of (26) are the only isolated products (T. Ban and K. Kanematsu, Heterocycles, 1981, 15, 373). Reaction of oxepin with 3,6-dicarbethoxy-1,2,4,5-tetrazine occurs by similar primary

(27)

(28)

processes, followed by loss of nitrogen and
further rearrangements to give, via (2+4)
cycloaddition (27, R=H), and via (6+4)
cycloaddition (28). 3-Cyano-4,6-dicarbethoxy-
1,2,4-triazine reacts similarly. 2,7-Dimethyl-
oxepin, however, reacts only by $(2+4)\pi$ cycloaddition to give (27, R=Me) (R. Dhar, et al.,
Ber., 1983, <u>116</u>, 97). Diazomethane, a 4π system,
reacts only with the double bonds in the benzene
oxide tautomers of various substituted oxepins to
give fused pyrazolines, with the O-atom in (2)
exerting a cis-directing effect (H. Prinzbach, et
al., Ber., 1976, <u>109</u>, 2823).

Polar reactions can occur via the reaction of
either (1) or (2). Both oxepin and its
2,7-dimethyl derivative - which differ in their
reactions with dienophiles (see above) - are
readily ring-opened by peroxy acids or by the
reaction with N-bromosuccinimide in aqueous DMSO
to give 1,6-dioxohexadienes $RCO(CH=CH)_2COR$ (R=H
and Me respectively), apparently via epoxidation
of the 2,3-bond of the oxepin tautomers (S.G.
Davies and G.H. Whitham, J. chem. Soc., Perkin I,
1977, 1346). The reactions of oxepin/benzene
oxide with nucleophiles have been extensively
investigated: methyl-lithium for example adds to
give the cis-alcohol (29) while dimethylmagnesium
gives a mixture of the cis- and the trans-isomer;
alcohols and alkoxides are not very reactive but
give (30; R=OMe, OEt) in good yield; the
hydroperoxide ion similarly gives (30; R=OOH)
which may be reduced to the trans-diol; ammonia
and NH_2^- are unreactive, but more polarizable
nucleophiles like N_3^- amd RS^- add readily to give
(30; R=N_3 and SR) respectively (D.M. Jerina, et

al., J. Amer. chem. Soc., 1974, <u>96</u>, 6929). 4-Carbo-*tert*-butoxyoxepin (1; $R^3=CO_2Bu^t$, $R^{1,2,4,5,6}$ =H) exists predominantly as the oxepin tautomer

(29)   (30)   (31)

but reacts with nucleophiles by attack at the 3-position of the benzene oxide tautomer. Thus the reaction with MeO^- gives (31, $E=CO_2Bu^t$) (G.A. Berchtold, *et al.*, J. Amer. chem. Soc., 1974, <u>96</u>, 1193). A similar attack by trimethylamine catalyses its aromatisation to give *tert*-butyl *m*-hydroxybenzoate (D.M. Johnson and T.C. Bruice, J. Amer. chem. Soc., 1975, <u>97</u>, 6901). The isomeric 3-carbo-*tert*-butoxy derivative, however, which exists mainly in the benzene oxide form does not react readily with either hard or soft nucleophiles (e.g. HO^-, MeO^-, N_3^-, PhS^-), either not reacting or giving a complex mixture of products (B.A. Chiasson and G.A. Berchtold, J. org. Chem., 1977, <u>42</u>, 2008).

The involvement of benzene oxides in normal and aberrant biochemical processes and in particular their aromatisation to phenols has stimulated much work on the mechanisms of these reactions. Space limitation prevents detailed discussion here but this area has been well reviewed (see reviews listed above). Under acid catalysis the 2-carbomethoxy derivative and analogous esters are converted exclusively into salicylic esters by an NIH shift of the ester group, for example (1; $R^1=CO_2R$, $R^{2-6}=H$) → (32) → (33). The 2-carboxy derivative of (1; $R^1=CO_2H$, $R^{2-6}=H$) however gives a mixture of phenol and salicylic acid in proportions dependent on pH (D.R. Boyd and G.A. Berchtold, J. Amer. chem. Soc., 1979, <u>101</u>, 2470).

(32) (33)

(b) *Dihydro-oxepins.*   *Synthesis and Reactions.*
     Thermolysis of butadienyloxiranes, e.g. (34)
leads to 2,3-dihydro-oxepins (36) but only when
the α,β bond has the *Z*-configuration.   The

(34)   (35)   (36)

reaction is considered to proceed *via* an 8π-
electron electrocyclisation of the diene-
conjugated carbonyl ylide intermediate (35, R=H)
followed by successive hydrogen shifts.   3-
Alkenyl-2,3-dihydrofurans are also obtained *via* a
competing 6π-electrocyclisation (W. Eberbach and
U. Trostmann, Ber., 1981, 114, 2979; Tetrahedron
Letters, 1979, 4649). The cyclisation to give a
cyclohexa-fused analogue of (36) takes place in
the conrotatory fashion expected for intermediates
of type (35). The 1,7-cyclisation is disfavoured
in relation to the competing 1,5 (6π electron)
cyclisation when R=Me (35) rather than H (W.
Eberbach, E. Hädicke, and U. Trostmann,
Tetrahedron Letters, 1981, 22, 4953; Ber., 1985,
118, 4035). This parallels a similar observation
in the cyclisation of analogous diene-conjugated
diazo compounds (Ch. 51 section 2.1.1(b)).
Examples of the synthesis of 4,5-dihydro-oxepins
(37) by the Cope rearrangement of the 2,3-divinyl

epoxide (40) have been reported (J.-P. Beny, J.-C. Pommelet, and J. Chuche, Bull. Soc. chim. Fr., 1981, Pt. II, 377). The latter may be generated

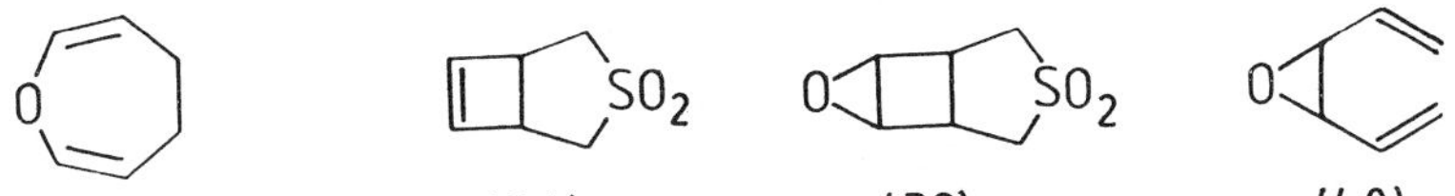

(37)   (38)   (39)   (40)

conveniently *via* epoxidation of the bicyclic sulphone (38) to give (39), followed by thermal extrusion of sulphur dioxide (J.I.G. Cadogan, I. Gosney, *et al.*, Chem. Comm., 1982, 1164). The

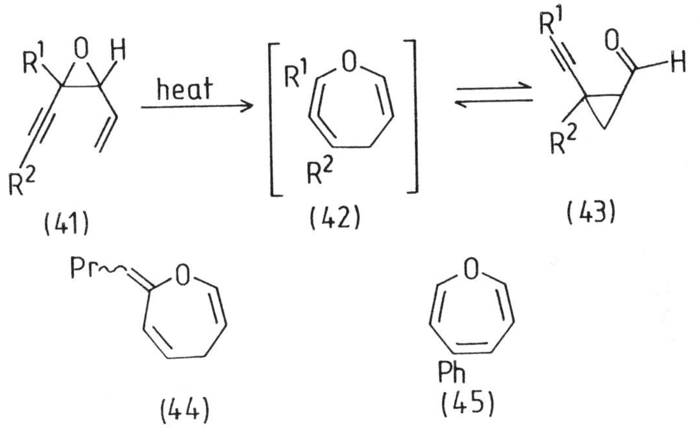

(41)   (42)   (43)

(44)   (45)

thermal rearrangement of the alkynyl analogue (41) has also produced some interesting chemistry. Gas-phase thermolysis gives the cyclopropane-carboxaldehyde (43) *via* the strained intermediate (42) but in the liquid phase both (41) and (43) are converted into oxepins *via* hydrogen shifts in (42). When $R^1=Bu^n$ and $R^2=H$ the dihydro system (44) is obtained, and with $R^1=H$ and $R^2=Ph$ the product is the fully unsaturated compound (45) which reacts further to give 4-hydroxybiphenyl (F. Bourelle-Wargnier, M. Vincent, and J. Chuche, J.

org. Chem., 1980, <u>45</u>, 428; Chem. Comm., 1979, 584). As discussed earlier, oxepin adds to

(46)   (47)   (48)   (49)

azodicarboxylic esters *via* a Diels-Alder reaction of its benzene oxide tautomer to give (46, $R=CO_2CH_2CCl_3$). This reaction provides the first step in a specific route to oxepin-3,4-epoxide. Further epoxidation gives (47) and then deprotection, decarboxylation and loss of nitrogen produces (48, R=H) (W.H. Rastetter, J. Amer. chem. Soc., 1976, <u>98</u>, 6350). Synthesis of the 2,7-dideuterio analogue (48, $R=^2H$) and a kinetic study by nmr-spectroscopy revealed the existence of a degenerate Cope rearrangement (48) ⇄ (49) at ambient temperature (W.H. Rastetter and D.D. Haas, J. Amer. chem. Soc., 1976, <u>98</u>, 6353; Chem. Comm., 1978, 377). The oxepin epoxide has also been generated by the pyrolysis of (50), and its Cope rearrangement shown to be slow on the nmr time scale (H. Klein, and W. Grimme, Angew. Chem., internat. Edn., 1974, <u>13</u>, 672). The dimethyl derivative (49; R=Me), prepared by a similar route, rearranges readily to give (48, R=Me) and thence (51; R=Me) (W.H. Rastetter and T.J. Richard, Tetrahedron Letters, 1978, 2995). All

(50)   (51)   (52)

of these symmetrical oxepin oxides, on treatment
with acid, are converted very rapidly into
4-substituted pyrans of type (51) (*loc. cit.*).
Opening of the epoxide ring occurs on attack by
amine and thiol nucleophiles (W.H. Rastetter, T.
Chancellor, and T.J. Richard, J. org. Chem., 1982,
47, 1509). 2,3-Dihydro-oxepin (52) is deprotonated at the 3 (allylic) position by alkyl-
lithium reagents to give the anion which rapidly
ring-opens to give $^-O(CH{=}CH)_2CH{=}CH_2$ (F.T. Oakes,
Fu-An Yang, and J.F. Sebastion, J. org. Chem.,
1982, 47, 3094).

(c) *Dihydro-oxepinones and diones.* The 3,3-
disubstituted dihydro-oxepin-2-ones (53; $R^1{=}Ph$ or
COR) show interesting photochemistry. Sensitized

photolysis gives (54) as the primary product *via*
1,5-phenyl or acyl shifts. Further reactions of
(54) are strongly dependent on solvents,
temperature, sensitizers and the presence of acid.
When $R^1{=}Ph$ a further sensitized photochemical
transformation gives (56) as the major product
but, in the presence of traces of acid, a rapid
reaction gives the styryl-furanones (57). A
separate synthesis of (54; $R^1{=}Ph$) shows that it
readily rearranges (above 4°C) to give (55) by a

1,5 hydrogen shift and (54; R^1=COR) shows the same reaction. Compounds (55; R^1=Ph, R^2=H or Me) undergo the expected photochemical electrocyclisation reaction of the butadiene moiety to give fused cyclobutenes (N. Hoshi, H. Hagiwara, and H. Uda, Chem. Letters, 1979, 1291; 1295; N. Hoshi and H. Uda, Chem. Comm., 1981, 1163; J. chem. Res. (S), 1985, 70). Compounds of the same type (53; R^1=Ph, Me; R^2=Me), prepared by the reaction of $^1O_2$ with fulvenes, eliminate carbon monoxide on flash vacuum pyrolysis above 220° to give R^1R^2C=CHCH=CHCHO (W. Skorianetz and G. Ohloff, Helv., 1975, 58, 1272).

(58)   (59)   (60)

Ring expansion of the peroxy-ester (58) in the presence of trifluoroacetic anhydride gives the oxepin-4-one derivative (60) via the oxepinium cation (59) (A. Nishinaga, K. Nakamura, and T. Matsuura, Tetrahedron Letters, 1980, 21, 1269).

(61)

A number of muconic acid anhydrides (61; R=H, OMe, Bu^t) have been simply prepared by the reaction of ortho-benzoquinones with MCPBA at 0°C. They are stable at this temperature when protected from moisture but undergo ring opening reactions in methanol to give open chain muconic acid derivatives and γ-lactone esters (T.R. Demmin and M.M. Rogić, J. org. Chem., 1980, <u>45</u>, 1153). Various substituted derivatives of type (61) are obtained by oxidation of phenol derivatives over ruthenium catalysts (Jap. Pat., 82,108,033 (1982), Chem. Abs., 1982, <u>97</u>, 198128s), and catechols by V^{IV} complexes (U. Casellato *et al.*, Inorg. chim. Acta, 1984, <u>84</u>, 101).

(d) *Tetrahydro-oxepins*. Decomposition of the diazonium ion (62) in methanol gives the tetrahydro-oxepin derivatives (63) and (64).

These compounds - remarkable as the first isolated heterocyclic *trans*-cycloheptenes - rearrange slowly to give the *cis*-isomers at room temperature, but can be trapped in the *trans* form as their Diels-Alder adducts (H. Jendralla, Ber., 1982, <u>115</u>, 220). The *trans*-isomer of (65) has been proposed as an intermediate in the photo induced addition of alcohols to (65) giving (66) (Y. Inoue, N. Matsumoto, and T. Hakushi, J. org. Chem., 1981, <u>46</u>, 2267). The 2,3,6,7-tetrahydro derivative (67), together with a 5,6-dihydro-oxin derivative, is produced by the cyclisation of HC≡ $CCH_2O(CH_2)_2CH(CO_2Et)_2$ in the presence of sodium ethoxide (A.T. Bottini, J.G. Maroski, and V. Dev, J. org. Chem., 1973, <u>38</u>, 1767). The tetrahydro-oxepins (68) are produced in varying yields, together with oxetane derivatives, by the photochemical reaction of the vinylcyclopropanes (69) with aryl ketones (PhCOR^2), *via* the diradical $\overset{\bullet}{C}H_2CH_2CH=C(R^1)CH_2O\overset{\bullet}{C}R^2Ph$ (S. Nishida, *et al.*, J.

$$(65) \qquad (66) \qquad (67)$$

$$(68) \qquad (69)$$

Amer. chem. Soc., 1974, <u>96</u>, 6456). The oxacyclo-heptyne (70) is formed by the oxidation of the

$$(70)$$

dihydrazone of 3,3,6,6-tetramethyl-oxepan-4,5-dione, and subsequently trapped by cycloaddition to phenyl azide. It is much less stable than the all carbon- or thia-analogues (A. Krebs and G. Burgdörfer, Tetrahedron Letters, 1973, 2063). The 4,5-dione precursor has also been synthesised by another group (P.Y. Johnson, J. Zitsman, and C.E. Hatch, J. org. Chem., 1973, <u>38</u>, 4087). 2,3,4,5-Tetrahydro-oxepin (65) like its 5- and 6-membered ring analogues, is deprotonated by alkyl-lithium reagents at the α (vinyl) position to give the anion which reacts with acetone to give the expected alcohol. This contrasts with deprotonation of 2,3-dihydro-oxepin (52) which occurs at the allyl position (F.T. Oakes, Fu-An Yang, and J.F. Sebastion, J. org. Chem., 1982, <u>47</u>, 3094).

(e) *Oxepanes*.   A   useful   1,2   carbonyl   trans-
position   has   been   developed   which   transforms
lactones   of   type   (71)   into   β-keto   ethers   (73).
The   process   involves   conversion   of   (71)   into   an
enol   phosphate   (72;   R=OP(O)(OEt)$_2$),   reductive
removal   of   the   phosphate   to   give   (72;   R=H)
followed   by   hydroboration   and   Collins   oxidation.
The   method   has   been   applied   in   the   preparation   of
a   key   intermediate   for   the   synthesis   of
±zoapatanol   (V.V.   Kane,   D.L.   Doyle,   and   P.C.
Ostrowski,   Tetrahedron   Letters,   1980,   21 ,2643;
V.V.   Kane   and   D.L   Doyle,   *ibid,*   1981,   22,   3027).

(71)   (72)   (73)   (74)

The   lactones   (71;   R^1/R^2=   =C(R^3)COR^4)   have   been
prepared   in   high   yield   by   the   MCPBA   oxidation   of
the   fused   furan   derivatives   (74)   *via* rearrangement
of   the   *trans* di-epoxide   (S.B.   Gingerich,   *et al.*,
J.   org.   Chem.,   1981,   46,   2589).   Oxepane   itself
has   been   prepared   in   high   yield   by   the   dehydrative
cyclisation   of   hexan-1,6-diol   using   the   mild
DEAD/TPP   technique   ['oxidation/reduction'
dehydration   using   diethyl   azodicarboxylate   (DEAD)
and   triphenylphosphine   (TPP)]   (J.T.   Carlock   and
M.P.   Mack,   Tetrahedron   Letters,   1978,   5153),   in
lower   yield   using   alumina   (Y.   Inoue,   S.   Deguchi,
and   T.   Hakushi,   Bull.   chem.   Soc.   Japan,   1980,   53,
3031)   and   also   by   the   use   of   *N*-methyl-
N,*N*-di-*tert*-butylcarbodiimidinium   ion   (R.C.   Schnur
and   E.E.   van   Tamelen,   J.   Amer.   chem.   Soc.,   1975,
97,   464).
     The   photolysis   of   oxepane   using   short   wave-
length   radiation   (185   nm)   proceeds   *via* cleavage   of
the   1,2-bond   to   give   the   diradical   ·O(CH$_2$)$_5$CH$_2$·
which   reacts   further   to   give   hexanal,   hex-5-
en-1-ol   and   3'-hydroxypropylcyclopropane   plus
other   products   (H.P.   Schuchmann   and   C.   von

Sonntag, J. Photochem., 1980, <u>13</u>, 347). Oxepane is oxidised by either RuO_4 and sodium periodate (A.B. Smith and R.M. Scarborough, Synth. Comm., 1980, <u>10</u>, 205) or oxygen with a platinum catalyst (K. Heyns and H. Buchholz, Ber., 1976, <u>109</u>, 3707) to give ε-caprolactone and then adipic acid. Hydrogen atom abstraction by *tert*-BuO· occurs selectvely at the 2-position of oxepane and the derived radical undergoes cleavage at 160°C to give $OHC(CH_2)_4\overset{\bullet}{C}H_2$ which recyclises to cyclohexanol (F. Flies, R. Lalande, and B. Maillard, Tetrahedron Letters, 1976, 439). Reaction with *tert*-butylhydroperoxide in the presence of copper salts at a lower temperature however gives 2-*tert*-butylperoxyoxepane (C. Filliatre, *et al.*, Compt. Rend., 1980, <u>291C</u>, 223).

1.2 *1-Benzoxepins*

(a) *Fully unsaturated.* (i) *Synthesis.*
1-Benzoxepin (75; R^{2-5}=H) is formed cleanly by the silver, palladium or rhodium catalysed isomerisation of its valence tautomer (76) (M. Uyegaki, *et al.*, Tetrahedron Letters, 1976, 4473).

(75)   (76)   (77)

Cyclisation of the propargyl anion (77) occurs predominantly *via* pathway (b) to give (75; R^4=Ph, R^5=NHR, $R^{2,3}$=H) together with only 4% of the benzofuran resulting from the alternative pathway (a) (O. Tsuge, K. Ueno, and K. Oe, Chem. Letters, 1981, 135). Full details are available of the synthesis of the 5-amino derivatives (75; $R^{3,4}$=CO$_2$Me, R^5=NR^{6}R^7, R^2=H) *via* the cycloaddition of DMAD to 3-aminobenzofurans to give (78) and

R^6R^7N E / E / O

(78)

their subsequent thermal ring expansion. The last step is much facilitated by the presence of the amino group and also by polar solvents. Acid hydrolysis of these 5-amino derivatives gives the 5-hydroxy compounds (D.N. Reinhoudt and C.G. Kouwenhoven, Rec. Trav. Chim., 1974, 93, 129; J. org. Chem., 1981, 46, 172). Earlier reports of

(79)   (80)   (81)

the preparation of fully unsaturated 1-benzoxepins by the acylation of the diones (79) have been followed by further examples of their alkylation and acylation to give (75; R^2=H, R^4=Ph, R^3 and R^5=MeO or AcO) (H. Hofmann, R. Heidrich, and A. Seubert, Z. Naturforsch., 1983, 38B, 895) and by extensions of the same principle to monocarbonyl derivatives. Thus the anions produced by the deprotonation of (80; R^3=H, Me, CN with R^4=Ph; and $R^{3,4}$=H) are alkylated and acylated to give the appropriate compound (75; R^5=OMe or OAc) (H. Hofmann and P. Hofmann, Ber., 1973, 106, 3571; and with H.-J. Haberstroh, Ann., 1973, 2032; H, Hofmann and M. Djafari, ibid, 1985, 599). The 3-oxo derivatives (81; R^5=H or Cl) similarly give (75; R^3=OMe or OAc) (H. Hofmann, and P. Hofmann, Ann., 1974, 1301). The 5-chloro-4-phenyl derivative (75; $R^{2,3}$=H, R^4=Ph, R^5=Cl) is also prepared by 2,3 dehydrochlorination (Ann., 1973,

2032).

(ii) *Reactions*.   1-Benzoxepins (75; R^2=H) ring-
contract on heating to give naphthols (82) *via*
(83) (D.N. Reinhoudt and C.G. Kouwenhoven, Rec.
Trav. Chim., 1974, *93*, 129;   H. Hofmann and P.
Hofmann, Ber., 1974, *107*, 2259).

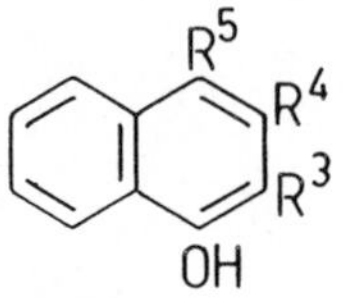

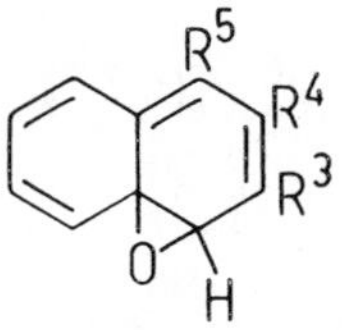

(82)   (83)

(b) *Dihydro-1-benzoxepins*.   Several new routes to
dihydrobenzoxepins involve the ring expansion of
chromones and coumarins.   The 2,3-dihydro-2-one
derivative (84) is prepared by the action of
sodium hydroxide on 3-bromoacetylchromone (U.S.P.,
3,991,082 (1976), Chem. Abs., 1977, *86*, 72475
g).

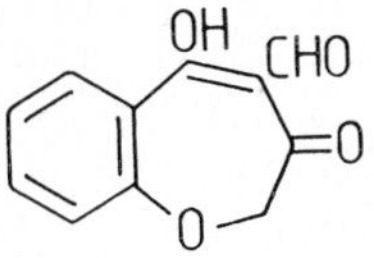

(84) 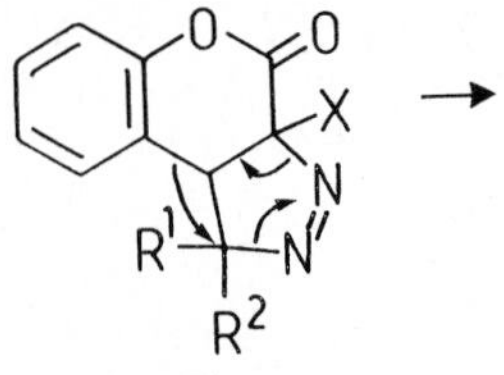   (85) 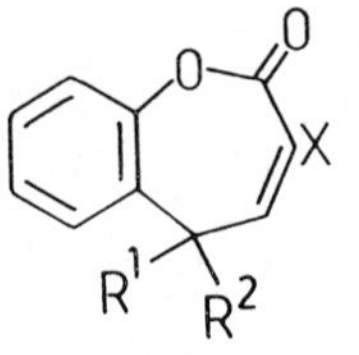   (86)

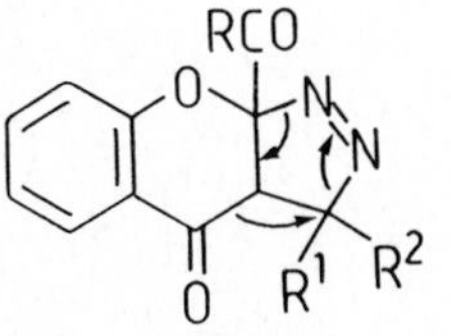

(87) 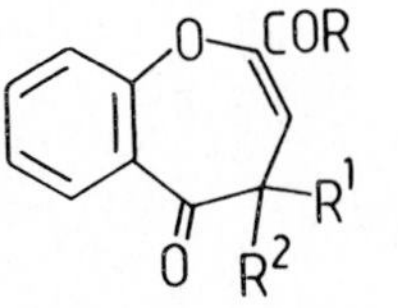   (88)

The adducts (85) of higher diazo-compounds with
coumarins which have an electron withdrawing
substituent (X) at the 3 position undergo ring

expansion to give the 2,5-dihydro-2-ones (86) (F.M. Dean and B.K. Park, J. chem. Soc., Perkin I, 1980, 2937 and references cited). The analogous adducts of 2-acylchromones (87) similarly undergo ring enlargement by carbonyl migration to give the 5-oxo derivatives (88) (F. M. Dean and R.S. Johnson, *ibid*, 1981, 224). The dichlorocarbene

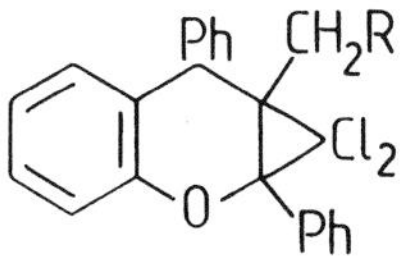

(89) 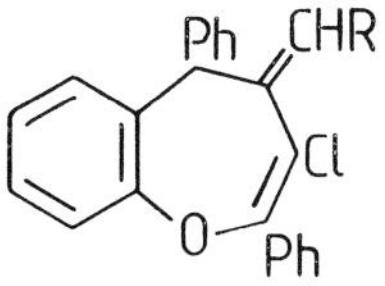 (90)

adducts (89) rearrange on refluxing in DMF to give (90) (A.V. Koblik and K.F. Suzdalev, Zh. org. Khim., 1982, 18, 1778). A similar ring expansion is also involved in the formation of (91) by the reaction of iodomethylene (:CHI) with the chalcone (92) (A.C. Jain, N.K. Nayyar, and B.N. Sharma, Ind. J. Chem., 1984, 23B, 1211). A synthesis of

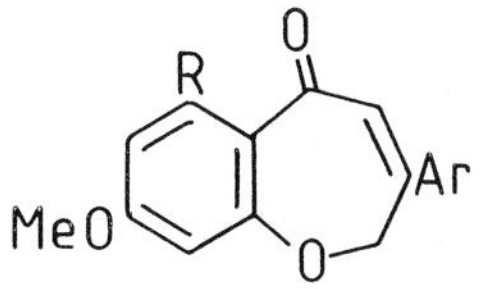

(91) 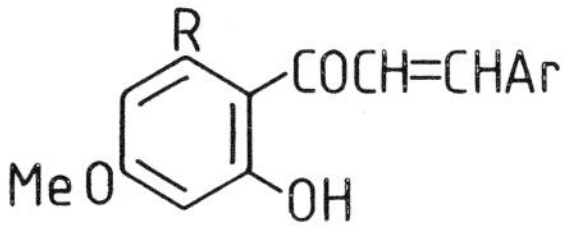 (92)

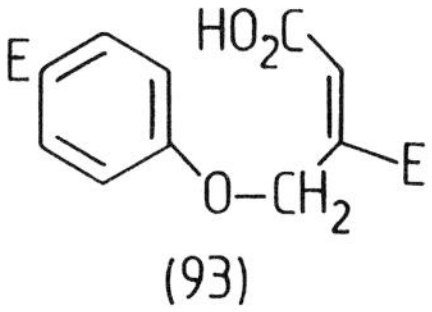

(93) 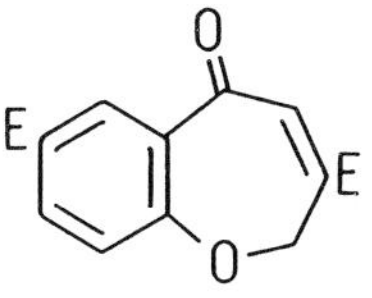 (94) 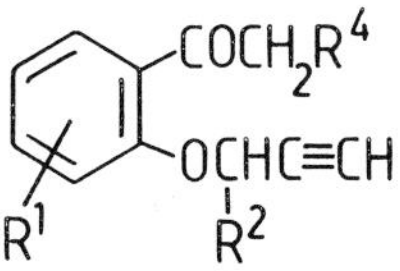 (95)

the benzoxepin-5-one system (94; E=CO_2Me) has been effected to confirm the identity of a related fungal metabolite, the first such compound of natural origin. The intermediate (93; E=CO_2Me) is prepared by the reaction of 4-HO-$C_6H_4CO_2$Me with $MeO_2C(CH_2Br)C=CHCO_2H$, converted into the acid chloride and then cyclised with aluminium trichloride (J.K. Holroyde, A.F. Orr, and V. Thaller, J. chem. Soc., Perkin I, 1978, 1490). The same system is produced by the dimsyl sodium mediated cyclisation of the 2-acylaryl propargyl ethers (95) (M. Jackson-Mülly, J. Zsindely, and H. Schmid, Helv., 1976, 59, 664).

(c) *Tetrahydro-1-benzoxepins.* A straightforward route to compounds unsubstituted in the heterocyclic ring (96) is provided by the butyl-lithium induced cyclisation of o-BrArO$(CH_2)_4$Br. Metal-halogen exchange at -100°C gives the aryl-lithium which, on warming to 25°C, gives (96; R=H or Br) in good yield (C.K. Bradsher and D.C. Reames, J. org. Chem., 1981, 46, 1384). 3-Hydroxybenzoxepins (97; R^1=Me, R^2=OH)

(96)   (97)   (98)

(99)   (100)

are formed in moderate yield, together with 2-hydroxymethylchromans as the major products, by the cyclisation of the epoxides (98) (K.J. Baird and M.F. Grundon, J. chem. Soc., Perkin I, 1980, 1820). However, in a formally similar reaction, treatment of the ketones (99) with $Me_2SO^+CH_2^-$ gives the benzoxepins (97; R^1=Me, Et or Ph; R^2=OH) in high yield. It was suggested that this reaction involves the direct cyclisation of the inter- mediate (100) without formation of the oxirane (98) (P. Bravo, C. Ticozzi, and D. Maggi,

Chem. Comm., 1976, 789; Gazz., 1979, <u>109</u>, 137).
This interpretation has been questioned. It has
been suggested that the intermediate (100) does
lead to the oxiranes (98) and thence to the
oxepins (97; R^2=OH) and it has been shown that
2-hydroxymethyl-chromans can be formed via the
acid catalysed ring contraction of oxepins of this
type (F.M. Dean and M.A. Jones, Tetrahedron
Letters, 1983, <u>24</u>, 2495). The β-dicarbonyl
compound (101), when treated with N-phenyl-
selenophthalimide, cyclises to give (102;
R=CH$_2$SePh), via attack of the oxygen atom of the

(101)   (102)

enolic form on a presumed seleniranium ion intermediate. Reduction gives (102; R=Me) in high
yield (W.P. Jackson, S.V. Ley, and J.A. Morton,
Chem. Comm., 1980, 1028). The 4-oxo derivative
(103; E=CO$_2$Et) is obtained from chroman-4-one by

(103)   (104)

treatment with ethyl lithiodiazoacetate to give
(104; E=CO$_2$Et), followed by acid catalysed ring
expansion (R. Pellicciari, et al., J. chem. Res.
(S), 1979, 142). The 5-oxo system (106) has been
prepared by two routes. Nucleophilic ring
opening of the cyclopropabenzopyran (105) by its
reaction with alcohols or water (ROH) gives (106;
R^2=OR, R^3=H, R^4=NO$_2$) (F. M. Dean and R.S. Johnson,
J. chem. Soc., Perkin I, 1980, 2049). The
photocycloadduct (107) from an alkene and
3-acetoxy-2-acetylbenzo[b]furan, on retro-aldol
ring expansion gives (106; R^2=COMe) in good yield

(105)   (106)   (107)

(J.H.M. Hill and S.T. Reid, Chem. Comm., 1983, 501). The 6,7,8,9-tetrahydro analogue of (106; R^{2-4}=H) is obtained by the cyclisation of (108) (G. L'hommet and P. Maitte, Bull. chem. Soc. Fr., 1976, Pt. 2, 1913). The hydroxy group in (109; E=CO_2Et) promotes the cyclisation to give the

(108)   (109)   (110)

3,5-dione system (110) by deactivating the acetyl group towards nucleophilic attack and so suppressing benzofuran formation (B.K. Wasson, P. Hamel, and C.S. Rooney, J. org. Chem., 1977, 42, 4265).

1.3 *2-Benzoxepins*

(a) *Dihydro-2-benzoxepins.* The 1,7-cyclisation of diene-conjugated carbonyl ylides (35) to give monocyclic oxepins (36) has been extended to give the 1,5-dihydro-2-benzoxepin system (111; $R^{3,4}$=H) by the thermolysis of Z-styryloxiranes (112) (W. Eberbach and U. Trostmann, with B. Burchardt, Tetrahedron Letters, 1979, 4049; Ber., 1985, 118, 4035). Several members of the same system (111; R,R^2=H, R^1=Me; R^3/R^4=H/H, Me/Me, H/Me) have been prepared by the p-toluenesulphonic acid catalysed cyclodehydration of (113). The latter are prepared from 1,2-dihydronaphthalenes by

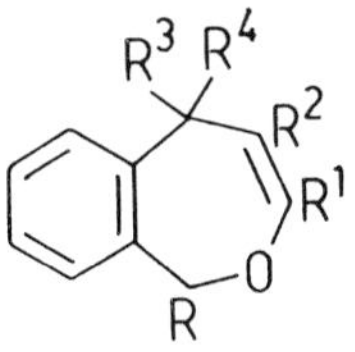

(111) 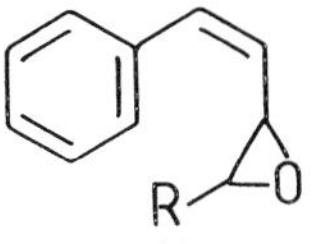 (112) 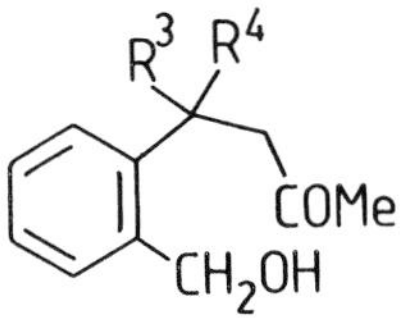 (113)

ozonolysis and selective reduction. The thermolysis of (111; R^1=Me) gives 2-acetylindene and other products depending on the nature of the substituents (F. Collonges and G. Descotes, Bull. Soc. chim. Fr., 1975, 789). The 3-oxo

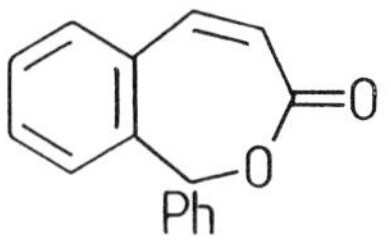

(114) 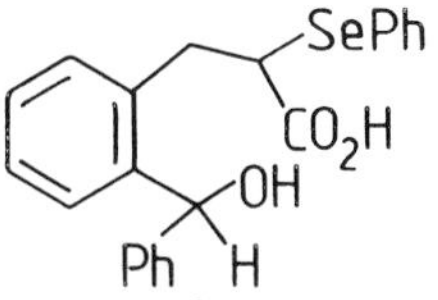 (115)

derivative (114) has been synthesised in six steps from o-aminobenzophenone via lactonisation of (115); its decomposition by F.V.P. gives o-benzoylstyrene and carbon monoxide (cf. the decomposition of (53)) (R.F.C. Brown et al., Austral. J. Chem., 1981, 34, 1467).

(b) *Tetrahydro-2-benzoxepins*. The acetoxy aldehydes (116), prepared by the LTA/I_2 cleavage of tetralols, cyclise in the presence of methanol and sulphuric acid to give the acetals (117) (cf.

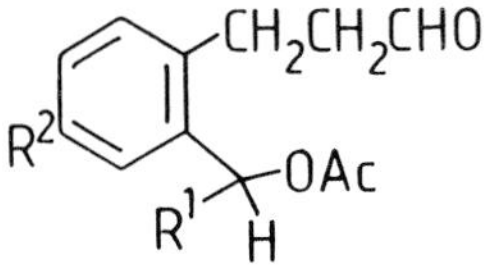

(116) 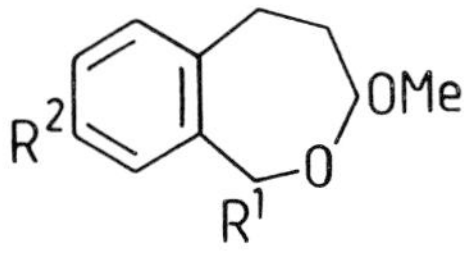 (117)

the formation of the analogous enol ethers (111) from (113)) (D.G. Talekar and A.S. Rao, Synthesis, 1983, 595). 1-Aminoalkyl derivatives of

2-benzoxepins have useful hypotensive and anti-depressant activity. 1-Haloalkyl derivatives (118; R=H or Me, $R^1=(CH_2)_nX$) used in their synthesis are prepared either by the Lewis acid catalysed reaction of the acetals of chloropropionaldehyde or bromoacetaldehyde with $3,4\text{-di-MeOC}_6H_3CH_2CR_2CH_2OH$, or by the dehydration of (119) (R.E. TenBrink and J.M. McCall, J. heterocyclic Chem., 1981, <u>18</u>, 821; see also Ger. Pat., 3,011,916 (1980), Chem. Abs., 1981, <u>94</u>, 121601w; Ger. Pat., 2,846,043 (1979), Chem. Abs., 1979, <u>91</u>, 94447b). 1-Isocyanato derivatives are

(118)　　　　(119)　　　　(120)

prepared by the reaction of 1-bromo-compounds with silver isocyanate, and a variety of other 1-substituted derivatives are formed by the reaction of 1-cyano compounds with electrophiles (H. Boehme and V. Hitzel, Arch. Pharm., 1974, <u>307</u>, 442). The reaction of $o\text{-Br-}C_6H_4(CH_2)_3OH$ with carbon monoxide in the presence of palladium acetate and triphenyl phosphine gives (120) (M. Mori, *et al.*, Heterocycles, 1979, <u>12</u>, 921). The 4-oxo derivative is prepared by the method used for the conversion of (104) into (103) (*loc. cit.*).

1.4 *Dibenzoxepins*

Several new routes provide access to the relatively rare dibenz[b,d]oxepin system. The fully unsaturated compound (121) is produced in 20% yield by the photolysis of phenanthrene-9,10-epoxide and its identity has been confirmed by an independent synthesis in which the 6,7-dihydro compound is formed by the photolysis of $o\text{-Br-}C_6H_4OCH_2CH_2Ph$ and then converted into (121) *via* bromination with NBS and dehydrobromination with DBN (N.E. Brightwell and G.W. Griffin, Chem. Comm., 1973, 37; K. Shudo and T. Okamoto, Chem. Pharm. Bull., 1973, 21, 2809). The tetrahydro-6-one derivative (122; R^1-R^3=H or Me, R/R=O) is

(121) (122) (123)

synthesised in one step by condensation of the substituted phenol $R^1R^2R^3C_6H_2OH$ with ethyl 2-(2'-oxocyclohexyl)acetate. One example, (122, R/R=O, $R^{1,2}$=H, R^3=Me) on treatment with methyl magnesium iodide, ring-opens to give (123) which may be cyclised to give (122; R=Me, $R^{1,2}$=H, R^3=Me) (U. Kraatz, M.N. Samimi, and F. Korte, Synthesis, 1977, 430). Intramolecular cyclisation of the vinyl cation (124) provides the dibenz[b,f]oxepin system (125) in quantitative yield (T. Kitamura, S. Kobayashi, and H. Taniguchi, Chem. Letters, 1984, 547). An extension of carbonyl ylide cyclisation [cf. (35) → (36) and (112) → (111)] gives the dibenz-[c,e]oxepin (126) (W. Eberbach and U. Trostmann, Ber., 1985, 4035).

(124) (125) (126)

2. *Seven-membered rings containing two oxygen atoms*

2.1 *1,2-Dioxepins*

Compounds of this type, containing the O-O

peroxide bond are relatively rare. The saturated monocyclic system (127) has been prepared in rather low yield by the LTA oxidation of hydroperoxides of the type $Ph(R)CH(CH_2)_3CMe_2OOH$ (H. Kropf and H. van Wallis, Synthesis, 1981, 633), and the 3,7-methano-bridged system (128; X=O) in high yield by reaction of the diazepine (128; X=-N=) with hydrogen peroxide (U.S.P., 4,291,051 (1981), Chem. Abs., 1981, 95, 204014d). The ozonide of 2-(4-nitrobenzyl)-3-phenylindenone has been shown by X-ray crystallography to have the 1,4-epoxy-2,3-benzodioxepin-5-one structure (129; $Ar=4-NO_2C_6H_4-$) - the first X-ray crystallographic structure determination of an ozonide (J.L. McAtee Jr., W.O. Milligan et al., Chem. Comm., 1978, 729; J. chem. Soc., Perkin II, 1979, 1703). Similar benzo fused and unfused ozonides e.g. (130) react with methanol in the presence of chlorosulphonic acid to give, among other products the 1,2-dioxepanes (131). Their formation involves heterolytic fission at the peroxide bridge and attack by methanol to give α-methoxyhydroperoxides e.g. $PhCO(CH_2)_3CH(OMe)OOH$ from (130) which cyclises via acetal formation to give (131) (M. Miura, M. Nojima, and S. Kusabayashi, J. chem. Soc., Perkin I, 1980, 2909).

(127)

(128)

(129)

(130)

(131)

2.2 1,3-Dioxepins

(a) *Fully unsaturated.* The parent compound (132) has been prepared (*ca.* 12% yield) by the HMPA induced dehydrobromination of (133). It gives a Diels-Alder adduct in high yield with 4-phenyl-1,2,4-triazoline-3,5-dione at 20°C but is unreactive towards both *N*-phenylmaleimide and benzoquinone at 80°C. Irradiation with uv-light gives (134) (J.F.W. Keana and R.H. Morse, Tetrahedron Letters, 1976, 2113). The 1,3-bridged analogue (135; X=O) is formed, in moderate yield, by the photolysis of (136). Catalytic reduction of (135; X=O) gives the saturated analogue and Wittig methylenation gives (135; X=CH$_2$) which on hydrolysis with aqueous oxalic acid ring-opens to give MeCOC(=CH$_2$)(CH$_2$)$_2$CMe$_2$CO(CH$_2$)$_2$COMe (B. Frei, *et al.*, Helv., 1979, 62, 553).

(132)   (133)   (134)

(135)   (136)

(b) *Dihydro-1,3-dioxepins.* (i) *Synthesis.* The major route to the 4,7-dihydro system (137) is *via* the condensation of (Z)HOCH$_2$CH=CHCH$_2$OH with aldehydes, acetals or ketals. In an improvement

(137)   (138)

to this method a range of 2-substituted derivatives of (137), and their saturated analogues, have been prepared by carrying out the reaction with aldehydes in the presence of the adduct of DMF and dimethyl sulphate ($Me_2N\!=\!\overset{+}{C}H\!=\!OMe$ $MeOSO_3{}^-$) (W. Kantlehner and H.-D. Gutbrod, Synthesis, 1979, 975). Treatment of the substituted diols $(Z)-R^2CH(OH)CH=C(R^3)CH_2OH$ with acetone in the presence of boron trifluoride etherate can give the dioxepin system (137; $4-R^2,6-R^3$, $R=R^1=Me$), or an alkenyl-substituted 1,3-dioxolane, or an $(E)-\alpha,\beta$-unsaturated aldehyde $(R^2CH_2CH=C(R^3)CHO)$ depending on the reaction conditions (G. Zweifel, et al., J. org. Chem., 1985, 2004). The reaction of the unsaturated diol with $ClCH_2CH(OMe)_2$ in the presence of Amberlite 15 gives (137; $R=H$, $R^1=CH_2Cl$) (Jap. Pat., 58,134,087 (1983), Chem. Abs., 1984, 100, 6569x), which by alkaline dehydrobromination gives the 2-methylene derivative (Jap. Pat., 58,154,575 (1983), Chem. Abs., 1984, 100, 85731v). Similar reactions with (138) also give (137) (Russ. Pat., 771,099 (1980), Chem. Abs., 1981, 94, 156980w).

(139)   (140)

Several examples of the fused system (140) have been prepared in moderate yield by a one-pot photochemical reaction between 2-alkoxy-3-bromo-1,4-naphthoquinones (139; R=Br) and 1-aryl-1-trimethylsilyloxyethene ($CH_2=C(Ar)OSiMe_3$); the reaction proceeds *via* the primary formation of (139; R=CH$_2$COAr) and its subsequent photochemical cyclisation (K. Maruyama, *et al.*, Chem. Letters, 1980, 859; Heterocycles, 1981, *16*, 1963)

(ii) *Reactions.* The conversion of the 4,7-dihydro derivatives (137) into the more stable 6,7-dihydro isomers (141) occurs quantitatively in the presence of either ruthenium hydride complexes (H. Suzuki, *et al.*, Tetrahedron Letters, 1980, *21*, 4927), or palladium catalysts (M. Dumic, *et al.*, Chem. Abs., 1985, *102*, 45897p). The cations

(141) (142) (143)

(142), produced by protonation or boron trifluoride treatment of (141), undergo fragmentation and recyclisation to give tetra-hydrofuran-3-aldehydes (143) in preparatively useful yields (H.-D. Scharf and H. Frauenrath, Ber., 1980, *113*, 1472; H. Suzuki, *et al.*, Tetrahedron Letters, 1980, *21*, 4927). The reaction is stereoselective when under kinetic control at low temperature and favours the formation of the *Z*-isomer, e.g. (141; R=H, R^1=Me) gives predominantly (143; R^1=Me, R=H, at -78°C), however at higher temperatures this isomerises to give the more stable *E* isomer (H. Frauenrath, J. Runsink, and H.-D. Scharf, Ber., 1982, *115*, 2728). The vinyl halides (144) produced by dehydro-halogenation of 2-substituted analogues of (133)

(144) (145)

give (145) on treatment with alkoxide ions (R^1O^-), apparently *via* an elimination/addition process (M.V. Prostenik, M. Dumic, and I. Butula, Chem. Abs., 1984, 101, 90895b). Oxidation of (137) with permanganate provides an effective route to the *bis*(carboxymethyl)acetals $RR^1C(OCH_2CO_2K)_2$, of possible value in detergent formulation (M.S. Niewenhuizen, A.P.G. Kieboom, and H. van Bekkum, Synthesis, 1981, 612).

2.3 *1,4-Dioxepins*

(a) *Monocyclic*

Members of the 2,3-dihydro-5H-1,4-dioxepin system have been prepared *via* the ring expansion of the 1,4-dioxine/dichlorocarbene adducts (146), either on heating in xylene to give (147) when R^1=H, or by reaction with nucleophiles (RX^-, X=O or S) to give (148) (W. Schroth and W. Kaufmann, Z. Chem., 1977, 17, 331; N.V. Kuznetsov and I.I. Krasatsev, Chem. Abs., 1977, 86, 16,648). The saturated system (150) is obtained by the trans-acetalisation of (149) on treatment with MeOH/H^+ (A. Espinosa Ubeda, and M.A. Gallo Mezo, An. Quim., 1983, 79C, 210).

(146)   (147)   (148)

(149)   (150)

(b) *Benzodioxepins*

The thermal ring expansion of the dihalogenocarbene adducts of 1,4-benzodioxine

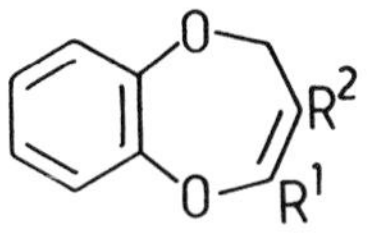

(151)

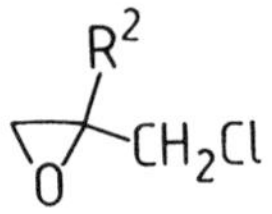

(152) 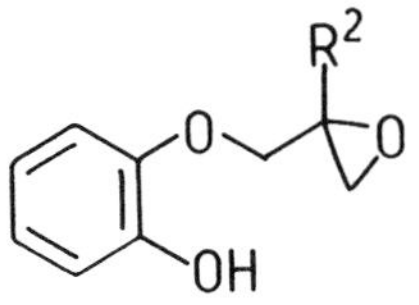   (153) 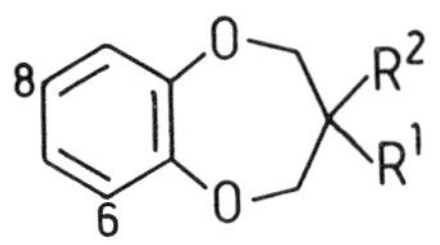   (154)

(cf. 146) gives the dihalogeno 2*H*-1,5-benzo-dioxepins (151; $R^{1,2}$=Cl or Br) in high yield. These compounds may be converted into (151; R^1=H, R^2=Cl or Br) by reduction (G. Guillaumet, G. Coudert, and B. Loubinoux, Angew. Chem., intern. Edn., 1983, 22, 64). The 3-hydroxy-3,4-dihydro-2*H*-1,5-benzoxepins (154; R^1=OH) are obtained in moderate yield, together with 2-hydroxymethyl-1,4-benzodioxane derivatives, by the reaction of catechol with the chloromethylepoxides (152). The primary step is thought to be formation of (153) which undergoes preferential ring opening by attack at the quaternary carbon atom (cf. the conversion of (98) into 1-benzoxepin and chroman derivatives) (A. Salimbeni and E. Menghisi, J. heterocyclic Chem., 1980, 17, 489). The 3,3-dimethyl-6,8-dinitro compound (154; $R^{1,2}$=Me, 6,8-diNO$_2$) is produced by heating the Meisenheimer

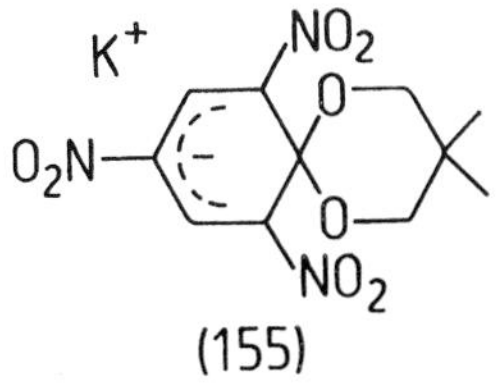

(155)

complex (155) in refluxing DMSO (V.N. Knyazev, V.N. Drozd, and V.M. Minov, Zh. org. Khim., 1978, 14, 105). The 2-oxo derivative (156) has been prepared *via* the oxidative ring opening of chromanone with hydrogen peroxide and perchloric acid to give $o\text{-HOC}_6\text{H}_4\text{O(CH}_2)_2\text{CO}_2\text{H}$, followed by

(156)   (157)   (158)

lactonisation (F. Eiden and C. Schmiz, Arch. Pharm., 1979, 312, 741). The 3-oxo analogue (157), important in perfumery has been prepared by two routes. Either *via* the Dieckmann cyclisation of dimethyl o-phenylenedioxydiacetate $(o\text{-C}_6\text{H}_4\text{(OCH}_2\text{CO}_2\text{Me})_2)$ (U.S.P., 3,799,892 (1974), Chem. Abs., 1974, 81, 3974), or the reaction of catechol with chloroacetonitrile (C.S. Rooney, *et al.*, Canad. J. Chem., 1975, 53, 2279). The reaction of 2-chloroethanol with sodium salicylate gives 2,3-dihydro-5H-1,4-benzodioxepin-5-one (158) (J. Gilbert, D. Rouselle, and P. Rumpf, Bull. Soc. chim. Fr., 1975, 277).

3.   *Seven-membered rings containing one sulphur atom*

3.1 *Monocyclic Thiepins*

(a)   *Fully unsaturated.*   Simple monocyclic thiepins (159) cannot normally be isolated because they readily extrude sulphur to give the related benzene derivative (161) *via* the thianorcaradiene valence tautomer (160). However, the kinetic stability of the thiepin can be increased to a remarkable degree by the introduction of bulky groups at the 2- and 7-positions ($R^{2,7}$ in 159) which inhibit the formation of (160). Thus

(159); $R^2,7=Bu^t$, $R^3,6=H$, $R^4=Me$, $R^5=CO_2Et$) is highly stabilised and requires prolonged heating ($t_{\frac{1}{2}}=7$ h at 130°C) to bring about the extrusion of sulphur whereas the 2,7-di-isopropyl analogue loses sulphur readily even at -70°C. These derivatives were prepared by the palladium catalysed ring expansion of (162; $E=CO_2Et$) and of its 2,6-di-isopropyl analogue (I. Murata *et al.*, J. Amer. chem. Soc., 1979, 101, 5059).

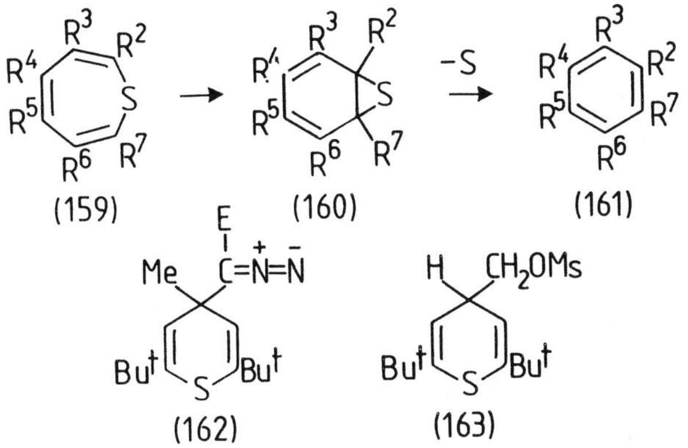

(159) (160) (161)

(162) (163)

Interestingly the simpler derivative (159; $R^2,7=Bu^t$, $R^{3-6}=H$), prepared by the solvolysis of (163), is more stable ($t_{\frac{1}{2}}=195$ h at 130°C) (I. Murata, N. Kasai, *et al.*, Tetrahedron Letters, 1982, 3195), and further stabilisation is provided by a single additional substituent at the 4-position, for example (159; $R^2,7=Bu^t$, $R^3,5,6=H$, $R^4=CO_2Et$ or CH_2OTMS or CH_3) do not extrude sulphur even on prolonged heating at 130°C (I. Murata *et al.*, Chem. Letters, 1982, 1843). The substituted derivatives (165; $R^2/R^6=Me$ or H) also

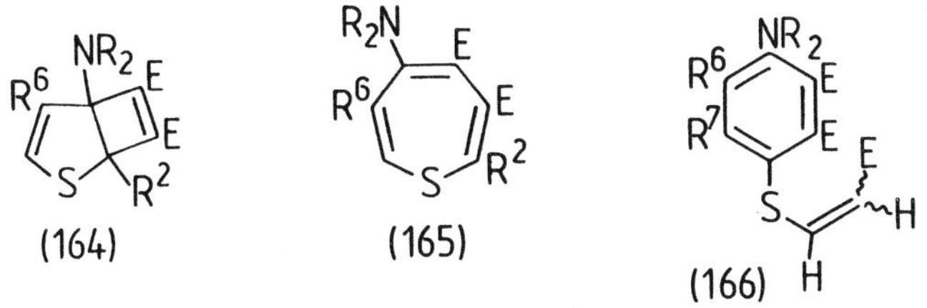

(164) (165) (166)

show some increased stability, but to a much lesser extent, probably arising from push-pull stabilisation and the electron withdrawing effects of the ester groups. These compounds are prepared at -30°C by the cycloaddition of 3-pyrrolidinylthiophenes to DMAD, giving (164), followed by ring expansion, and are stable enough to persist for some time at that temperature before extruding sulphur (D.N. Reinhoudt and C.G. Kouwenhoven, Tetrahedron, 1974, 30, 2093). In this reaction, in some cases the thiepin (or the intermediate 160) reacts with excess DMAD by S-alkylation to give products of the type (166) (D.N. Reinhoudt *et al.*, Tetrahedron Letters, 1979, 1529, J. org. Chem., 1981, 46, 424). The 1,1-dioxides (167) are much more stable and can be

(167)   (168)

prepared by the photochemical reaction of (168) with alkynes PhC≡CR (N. Ishibe, K. Hashimoto, and M. Sunami, J. org. Chem., 1974, 39, 103).

(b) *Dihydrothiepins*. Reinhoudt's preparation of thiepins (above) has been extended to provide good syntheses of the 2,3- and 2,7-dihydro derivatives (169; R^4=NR_2, R^5=CO_2Me, R^6=H or CO_2Me, R^7=H) and (170; R^3=NR_2 or H, $R^{6,7}$=H) respectively, by the use of 4,5- and 2,5-dihydro- 3-aminothiophenes (D.N. Reinhoudt and C.G. Kouwenhoven, Rec. Trav. Chim., 1973, 92, 865).

(169)   (170)   (171)   (172)

The oxide (171) reacts with DMAD in a formally similar reaction to give the sulphoxide (169; $R^{6,7}=CO_2Me$, $R^{4,5}=H$, S=O for S) but in this case not *via* 2+2 cycloaddition (K. Gollnick and S. Fries, Angew. Chem. intern. Edn., 1980, 19, 833). The 2,3-dihydro system has also been prepared by a new ring expansion involving the treatment of (172) with chlorotrimethylsilane and zinc to give (169; R^4=Me, $R^{5,6}$=H, R^7=Bu^t) (I. Murata, K. Yamamoto, and S. Yamazaki, Bull. chem. Soc., Japan, 1982, 55, 3057). 2,3-, and 2,7-, and 4,5-Dihydrothiepins are all produced by the thermolysis of 2,3-divinylthiiranes (173). At 90°C the *cis*-isomer gives the expected 4,5-dihydrothiepin (174; R=H) by a Cope rearrangement (cf. the oxirane analogue, section 1.1(b)) but also the 2,7-dihydrothiepin (170; R^{3-7}=H) by a diradical mechanism. The *trans*-isomer at 100°C gives the latter as the primary product which then rearranges to the more stable 2,3-dihydrothiepin (169; R^{4-7}=H) (M.P. Schneider and M. Schnaithmann, J. Amer. chem. Soc., 1979, 101, 254). Further work on the synthesis of 2,7-dihydrothiepin-1,1-dioxides (175) by the cycloaddition of sulphur dioxide to *cis*-3-hexatrienes

(173)   (174)   (175)

has produced a range of substituted derivatives but generally in low yield because of the inaccessibility of the trienes (W.L. Mock and J.H. McCausland, J. org. Chem., 1976, 41, 242). However, the 4,5-disubstituted derivatives (175; R=H or Ph) are prepared in *ca* 30% yield by the reaction of the complex $(PhC\equiv CR)Co_2(CO)_6$ with $(H_2C=CH)_2SO_2$ (I.U. Khand and P.L. Pauson, Heterocycles, 1978, 11, 59). The unsubstituted analogue of (175) functions as a dienophile in

'inverse' Diels-Alder reactions with tetrazines
(G. Seitz, et al., Arch. Pharm., 1981, 314, 892).
     Homothiopyrylium ions (176) have been prepared
by two routes; the substituted species (176;
$R^{2,7}=Bu^t$, $R^4/R^5=H$ or Me) by the protonation of the
appropriate thiepin (159) using SO_2/FSO_3H at
-78°C, and the unsubstituted analogue (176;
$R^{2,4,5,7}=H$) by treatment of (174, R=OH) with the
same reagent. Their nmr-spectra support the
charge delocalisation shown and when quenched with
methanol they give compounds of the type (174;
R=OMe) (I. Murata, et al., Chem. Comm., 1984, 604;
Angew. Chem., intern. Edn., 1985, 24, 214).

(176)   (177)   (178)

Decomplexation of the thiacycloheptyne complex
(177) yields (178), the first cyclobutadiene
derivative sufficiently stabilised by steric
effects to be isolable at room temperature (A.
Krebs, H. Kimling, and R. Kemper, Ann., 1974,
2074; 1978, 431; 1979, 473).

(c) *Thiepanes.* Thiepane can be obtained in high
yield from the reaction of 1,6-dibromohexane with
sodium sulphide (L. Mandolini, and T. Vontar,
Synth. Comm., 1979, 9, 857; A. Singh, A.
Mehrotra, and S.L. Regen, *ibid*, 1981, 11, 409).

(179)   (180)

3-Oxo-1-phenylthiepinium perchlorate (179) is obtained in high yield by treatment of the diazoketone $PhS(CH_2)_4COCHN_2$ with perchloric acid (W.T. Flowers, et al., J. chem. Soc., Perkin I, 1979, 1119). 2-Methylenethiepin-1-oxide (180) is prepared by the intramolecular cyclisation of the sulphenic acid $HC\equiv C(CH_2)_5SOH$ (R. Bell, et al., ibid, p. 2106).

3.2 1-Benzothiepins

(a) *Fully unsaturated*. 1-Benzothiepins (181) are more stable than the monocyclic species (159) but are still labile enough to require low-temperature preparative and work-up conditions to avoid sulphur extrusion (stability is discussed in more detail below). Several general routes are available, giving access to the parent compound and a range of substituted derivatives. The most straightforward preparation is the rhodium catalysed isomerisation of the valence tautomer (182) (cf. the isomerisation of 76) which has been used to prepare the unsubstituted compound (181; R^{2-5}=H) (I. Murata, T. Tatsuoka, and Y. Sugihara, Angew. Chem., intern. Edn., 1974, 13, 142; Tetrahedron Letters, 1973, 4261; 1974, 199) and the 3-, 4-, and 5-Me; 4,5-di-Me, 4-CO_2Me and 4-CHO derivatives (I. Murata and T. Tatsuoka, Tetrahedron Letters, 1975, 2697). All had half-lives at 47°C in the range 30-100 min. The

(181)   (182)   (183) R^2

second method involves the base induced elimination of hydrogen halides from 2-halogeno-2,3-dihydro-1-benzothiepins of the type (183). The parent system and the 2- and the 4-chloro derivatives have been produced via hydrogen

chloride elimination with potassium *tert*-butoxide
(V.J. Traynelis *et al.*, J. org. Chem., 1973, <u>38</u>,
3979) and the reaction has been extended to give a
range of 5-alkyl- and 5-aryl-4-halogeno-
derivatives *via* DBN induced hydrogen bromide
elimination. The products have half-lives at
30°C in the range 12-29 h (V.J. Traynelis, *et al.*,
J. org. Chem., 1978, <u>43</u>, 3379). The third route
is Reinhoudt's extension of his preparation of
monocyclic thiepins (159) and requires the
reaction of 3-pyrrolidinylbenzo[b]thiophenes with
DMAD to give (181; R^2=H, $R^{3,4}$=CO_2Me, R^5=NR_2). As
before this substitution pattern produces push-
pull enhanced stability. Hydrolysis gives the
5-hydroxy derivatives which are obtained in the
enolic form, thought to be stabilised by hydrogen
bonding to the 4-CO_2Me group (D.N. Reinhoudt and
C.G. Kouwenhoven, Tetrahedron, 1974, <u>30</u>, 2431;
see also B. Lamm and C.-J. Aurell, Acta Chem.
Scand., 1982, <u>36B</u>, 435). A number of substituted

(184)

(185)

derivatives have been prepared by the palladium
catalysed ring expansion of diazo-compounds (cf.
the ring expansion of 162); for example (184,
E=CO_2Et) gives a mixture of the 2- and the 3-
ethoxycarbonyl derivative which have been
separated by low temperature chromatography. The
4-substituted isomer of (184) similarly gives the
4- and the 5-ethoxycarbonyl derivatives (K.
Nishino, K. Nakasuji, and I. Murata, Tetrahedron
Letters, 1978, 3567). The earliest route to this
system involves the alkylation or acylation of
enolised ketones e.g. (185;, n=0) (see 2nd Edn.)
and further examples and developments have been
reported (H. Hofmann, *et al.*, Ber., 1975, <u>108</u>,
3596).

The effects of substitution on stability appear to be complex; theoretical calculations have suggested that electron withdrawing groups will reduce the anti-aromatic character (B.A. Hess, L.J. Schaad, and D.N. Reinhoudt, Tetrahedron, 1977, *33*, 2683). However experimental results (*loc. cit.*) show that for 3-, 4-, or 5-substitution, methyl groups have a stabilising effect while ethoxycarbonyl groups are destabilising - the greatest effect in both cases occurs with 3-substitution. In contrast the 2-ethoxycarbonyl derivative is strongly stabilised.

The 1,1-dioxides of (181) are much more stable than the benzothiepins themselves and can be produced either by MCPBA oxidation or directly from the dioxides of (183) and (185) (*loc. cit.*). The sulphoxides (186) however are much less stable but a number of substituted derivatives (186; R=Me or COMe) have been prepared either by controlled MCPBA oxidation of (181) at low temperature (-20°C) or by alkylation or acylation of (185; n=1) at -120°C. Compound (186; R=COMe) has a half-life of 155 min. at 10°C compared to 180 min. at 59°C for the similarly substituted thiepin (H. Hofmann and H. Gaube, Ber., 1979, *112*, 781). Thermolysis of (181), its dioxide, or (186) leads in each case to a naphthalene by the extrusion of S, SO_2 and SO respectively.

(186)

(187)

1-Benzothiepinium salts (187; R^3=H, OAc, OMe; R^5=OAc, OMe) have been prepared by the S-methylation of the parent thiepins with methyl fluorosulphonate. In contrast to (186) they are more stable than the parent compounds. Thus (187,

$R^{3,5}$=OMe) has a half-life of 68 h. at 80°C compared with 2 h. for the parent, but is less stable than the 1,1-dioxide. Thermolysis gives an α-naphthyl thiomethyl ether (H. Hofmann, A. Molnar, and C. Gottfert, Tetrahedron Letters, 1977, 1985; Ann., 1983, 425). The 5H- and 3H-benzothiepinium ions (188) and (189) are generated by protonation using fluorosulphonic acid. A ^{1}H-nmr spectroscopic study shows that (188) has the benzohomothio- pyrylium ion structure shown while in (189) the charge is localised (I. Murata, S. Yamazaki, and K. Yamamoto, Chem. Letters, 1984, 919).

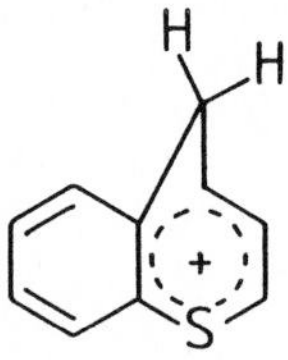

(188) 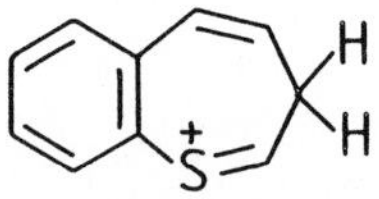   (189)

(b) *Dihydro- and tetrahydro-1-benzothiepins.* The dihydro-derivative (190; E=CO$_2$Et) is prepared by the reaction of ethyl lithiodiazoacetate with thiochroman-4-one to give (191; E=CO$_2$Et) followed by acid catalysed ring expansion. Hydrolysis and decarboxylation gives 2,3,4,5-tetrahydro-1-benzothiepin-4-one. Similar reactions with isothiochroman-4-ones give the 2-benzothiepin analogues (R. Pellicciari and B. Natalini, J. chem. Soc., Perkin I, 1977, 1822). Thiochroman-4-ones can also be converted into thiepins by

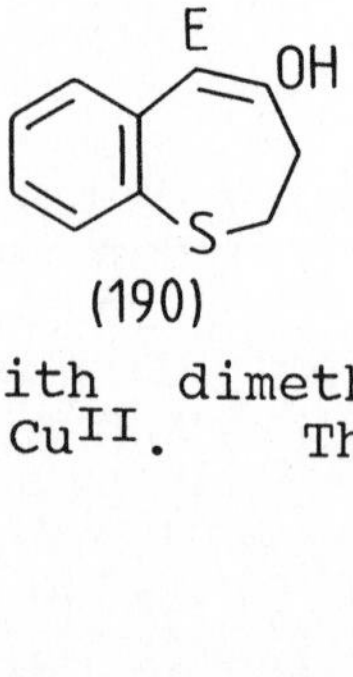

(190) 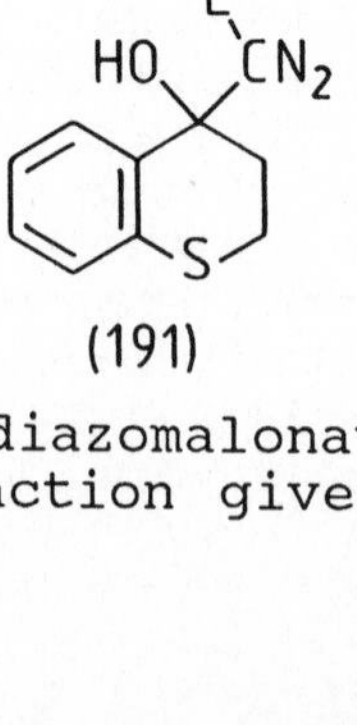   (191)

treatment with dimethyl diazomalonate in the presence of Cu^{II}. The reaction gives a mixture of the thiepinone (192; R=CO$_2$Me) and the methanide (193; E=CO$_2$Me). The latter can be converted into (192; R=CO$_2$Me) by treatment with triethylamine, *via* ring cleavage and subsequent cyclisation by a Michael addition (Y. Tamura, *et al.*, J. chem. Soc., Perkin I, 1981, 2978; Heterocycles, 1981, 15, 875). The photo-adducts of 3-acetoxybenzo[b]thiophene 1,1-dioxide and alkenes (R^{3}CH=CHR^4) undergo retro-aldol cleavage when treated with base to give (192; R=H, SO$_2$ for S) (cf. the cleavage of 107) (N.V. Kirby and S.T. Reid, Chem. Comm., 1980, 150). Extensive work

(192)   (193)   (194)

has been done on the mono- and di-halogenation of (192; R and R^{3,4}=H) and subsequent dehydro-halogenation. The 2,5-dihalogeno derivatives undergo a high-yielding ring contraction to give (194; X=Cl or Br) when treated with triethylamine (V.J. Traynelis, *et al.*, J. org. Chem., 1973, 38, 2623; 2629).

3.3 *3-Benzothiepins*

The fully unsaturated compound (195) (t$_{\frac{1}{2}}$=24 min. at 24°C) is prepared from benzo[c]thio-pyrylium fluoroborate and ethyl lithiodiazoacetate *via* a diazo intermediate analogous to (184). Palladium catalysed ring expansion at -10°C, followed by chromatography at -40°C gives pure (195) which readily extrudes sulphur to give 2-ethoxycarbonylnaphthalene (K. Nakasuzi, *et al.*, Angew. Chem., intern. Edn., 1976, 15, 611). The 2-chloro derivative (196) is formed in moderate yield together with 1-(chloromethyl)isothiochroman

by the reaction of diazomethane with 1-chloroiso-
thiochroman.   Treatment of (196) with triethylamine gives (197), which may be oxidised to the

(195)   (196)   (197)

(198)

sulphone (H. Böhme and F. Ziegler, Ber., 1974,
107, 605).   Böhme has also investigated the
chemistry of 4,5-dihydro-2H-3-benzothiepin-1-one
(198) and its 3,3-dioxide including its conversion
into an enamine;   Beckmann ring expansion to a
thiazocine;   alkylation;   acylation and arylation
at C-2;   and formation of 2-methylene derivatives
(H. Böhme and R. Malcherek, Arch. Pharm., 1979,
312, 648, 653, 714;   1980, 313, 15, 81).

3.4 *Dibenzothiepins*

The dibenzo[b,d]thiepin (199;   R=CO$_2$Et) is
prepared in quantitative yield by a palladium-
catalysed diazo ring-expansion (cf. the
preparation of 195 and decomposition of 184).
This compound shows remarkable thermal stability
with no tendency to extrude sulphur, even at an
elevated temperature.   Hydrolysis and
decarboxylation at 150°C gives (199;   R=H) which
is somewhat less stable and gives phenanthrene on
prolonged heating with triphenyl phosphine in
boiling xylene (I. Murata, *et al.*, Angew. Chem.,
intern. Edn., 1976, 15, 611).

(199)   (200)   (201)

Space limitations do not permit coverage of the extensive work on the functional group chemistry of the dibenzo[b,e]- and [b,f]-thiepins. Compounds of these types have central nervous system activity and in consequence many derivatives have been prepared. The 11-oxo derivative of dibenzo[b,e]thiepin (200) can be deprotonated at C-6 and alkylated at that position provided the operation is carried out at low temperature. At higher temperatures the 6-anion readily rearranges to give, after methylation, (201). The anion formed from the sulphone of (200) does not rearrange (J. Ackrell, J. org. Chem., 1978, 43, 4892). The sulphone (202; R= cyclopropyl), on treatment with sulphuric acid at

(202)   (203)   (204)

-20°C isomerises to (202; R= CH=CHMe) and at +20°C cyclises to give the dihydrodibenzo[b,f]-thiepin dioxide (203) in ca 80% yield (E.V. Pisanova, L.G. Saginova, and Yu. S. Shabarov, Zh. org. Khim., 1982, 18, 322; 1981, 17, 643). During Friedel-Crafts acylation dibenzo[b,f]-thiepins (204) can undergo ring contraction to give acylated thioxanthenes (M. Protiva, *et al.*, Coll. Czech. chem. Comm., 1978, <u>43</u>, 471). Bromine adds to the 10,11 double bond of benzo[b,f]thiepin (204; $R^{1,2}$=H) to give a mixture of the *cis*- and the *trans*-dibromide (*trans* only when under kinetic control), each of these is cleanly converted into the 10-bromo derivative (204; R^1=H, R^2=Br) by dehydrobromination with potassium *tert*-butoxide. Interestingly the bromination of the sulphone in carbon tetrachloride solution gives the *cis* dibromide whereas the *trans*-dibromide is formed in acetic acid solution (N. Nógrádi, W.D. Ollis, and I.O. Sutherland, J. chem. Soc., Perkin I, 1974, 621). The strained cyclic alkyne (204; R^1/R^2=bond) is generated by the thermolysis of the [d] fused 1,2,3-selenadiazole, and may be trapped *in situ* by tetracyclone or other reagents (M. Lorch and H. Meier, Ber., 1981, <u>114</u>, 2382).

4. *Seven-membered rings containing two sulphur atoms*

4.1 *1,2-Dithiepins*

The lead dithiolate Pb[S(CH$_2$)$_5$S] reacts with sulphur in benzene to give 1,2-dithiepane in high yield. This is an example of a general route to cyclic disulphides and is facilitated by easy preparation of the thiolates from the dithiol and lead acetate (R.H. Cragg and A.F. Weston, Tetrahedron Letters, 1973, 655). 1,2-Dithiepane, when treated with an alkyl lithium (RLi) and carbon monoxide at -110°C, undergoes nucleophilic acylation and ring cleavage to give RCOS(CH$_2$)$_5$SLi which with methyl iodide gives RCOS(CH$_2$)$_5$SMe (D. Seyferth and R.C. Hui, Organometallics, 1984, <u>3</u>, 327). 3*H*,7*H*-Naphtho-[1,8-d,e]1,2-dithiepin has been prepared by the oxidation of 1,8-bis(mercaptomethyl)naphthalene for a dynamic nmr-study (J.P. Snyder *et al.*, J. Amer. chem. Soc., 1981, <u>103</u>, 159).

4.2 *1,3-Dithiepins*

2*H*-1,3-Dithiepin (205) has been prepared by the dehydrochlorination of (206; R=Cl). It reacts

(205)   (206)   (207)   (208)

with butyl-lithium to give the deep red anion (207) which when quenched with D_2O, Me_3SiCl, or Ph_2CO gives the expected 2-substituted derivative. The anion (207) is of interest as a potential 10π heteroaromatic system. A [1]H-nmr study indicates some negative charge delocalisation but does not permit any firm conclusion about the ion's aromatic character. However (205) is *ca.* 150x more acidic than its saturated analogue and this has been interpreted as indicating a small (*ca.* 12kJmole^{-1}) resonance stabilisation of (207) (C.L. Semmelhack, I-Ching Chiu, and K.G. Grohmann, Tetrahedron Letters, 1976, 1251; J. Amer. chem. Soc., 1976, <u>98</u>, 2005). In support of this the nmr-spectrum of (208) indicates some anionic character of the dithiepin ring (Y. Sugihara, Y. Fujiyama, and I. Murata, Chem. Letters, 1980, 1427). 4,7-Dihydro-2*H*-1,3-dithiepin (206; R=H) has been prepared by the reaction of methylene bromide with either *cis*-but-2-en-1,4-dithiol ($HSCH_2CH=CHCH_2SH$) or with the analogous *bis*-isothiouronium salt (prepared by reaction of $ClCH_2CH=CHCH_2Cl$ with thiourea) (D.N. Harpp, K. Steliou, and B.T. Friedlander, Org. prep. Proced. Int., 1978, <u>10</u>, 133). This compound reacts with di-iron nonacarbonyl to give the expected 1-substituted iron tetracarbonyl derivative and the dinuclear complex $H_2C(S_2Fe_2CO)_6)$ by the cleavage of two C-S bonds (A. Shaver, *et al.*, J. Amer. chem. Soc., 1979, <u>101</u>, 1313). 2-Substituted 1,3-dithiepanes are produced when the corresponding dioxepane is heated with sulphur at

120-280^{C}C in the presence of calcium carbonate (D.L. Rakhmankulov, Chem. Abs., 1978, **88**, 105326q).

Me—CH$_2$SH / Me—CH$_2$SH **(209)**   R^2—S—R / R^2—S—R^1 **(210)**   —SO$_2$ / R **(211)**

1,5-Dihydro-3H-2,4-benzodithiepins **(210)** are usually prepared by the reaction of dithiols of type **(209)** with 1,1-bis electrophilic reagents. It has been shown that alkynes (RC≡CCOR^2) give **(210**; R^1=CH$_2$COR^2) (V.N. Elokhina et al., Khim. Geterotsikl. Soedin, 1984, 1062; 1985, 203), and the α-chloroether ClCH$_2$CHClOEt gives **(210**; R=H, R^1=CH$_2$Cl, R^2=Me) (I.M. Nasyrov et al., ibid, 1985, 997). Alkylation or dialkylation at C-3 of **(210**; R and/or R^1=H, R^2=Me) can be achieved by butyllithium deprotonation followed by treatment with alkyl halides. The products can be cleaved with CuO/CuCl$_2$ to give RR^{1}CO in high yield (K. Mori, et al., Synthesis, 1975, 720). The anion derived from the 2,4-tetra-oxide of **(210**; R^1=H; R=H, Me, Ar) undergoes ring contraction to give **(211)** (S. Rozen, I. Shalak, and E.D. Bergmann, Int. J. Sulphur Chem., Part A, 1972, **2**, 287). 3-Alkyl and 3-aryl derivatives of **(210)** have been prepared by the reaction of Grignard reagents with **(210**; R=H, R^1=Cl) (H. Gross, I. Keitel, and B. Costisella, J. prakt. Chem., 1978, **320**, 255), and the 3-phosphonate **(210**; R=H, R^1=P(O)(OEt)$_2$) has been used to prepare 3-methylene derivatives (M. Mikolajczyk, et al., Tetrahedron, 1978, **34**, 3081).

4.3 1,4-Dithiepins

5H-1,4-Dithiepin **(212**; R^{1,2,3}=H), a reasonably stable oil, is obtained in ca. 60% yield by the elimination of methanol from (213; R=OMe) using lithium dicyclohexylamide. Deprotonation with butyl-lithium followed by treatment with methyl iodide gives (212; R^1=Me, $R^{2,3}$=H) and not the expected 5-methyl derivative; further methylation gives (212; $R^{1,2}$=Me, R^3=H). It has been concluded that the anion of (212) has no aromatic character (cf. 207) (I. Murata, K. Nakasuji, and Y. Nakajima, Tetrahedron Letters, 1975, 1895).

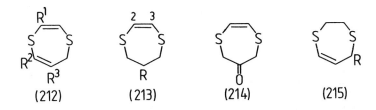

<table>
<tr><td>(212)</td><td>(213)</td><td>(214)</td><td>(215)</td></tr>
</table>

The substituted derivative (212; R^1=H, R^2=Ac, R^3=OH) is produced in moderate yield by the acetylation of (214) (R.R. Schmidt and R. Berrer, Ber., 1976, 109, 2928). 2,3-Dihydro-5H-1,4-dithiepin (215; R=H) is prepared from the saturated analogue of (213; R=OMe) using the same method as for the preparation of (212); in this case methylation does give the 5-methyl derivative (215; R=Me). 6,7-Dihydro-5H-1,4-dithiepins (213) can be prepared either from $^-$SCH=CHS$^-$ on treatment with Br(CH$_2$)$_3$Br, or from $^-$S(CH$_2$)$_3$S$^-$ and ClCH=CHCl (or substituted derivatives in either case; see references above). 1,4-Dithiepin

(216)   (217)   (218)   (219)

derivatives can also be prepared by the ring expansion of 2-substituted 1,3-dithians. The 1,3-dithian-2-ylmethanols (216; R^1=Ar, R^2=H, R^3=H or CO_2Me), on treatment with methanesulphonyl chloride give the 6,7-dihydro-5H-1,4-dithiepins (213; 2-Ar, 3-H or 3-CO_2Me, R=H) (P.M. Weintraub, J. heterocyclic Chem., 1979, 16, 1081). The LTA oxidation of (216; R^1/R^2=$(CH_2)_n$) gives (218; R^1/R^2=O, R^3=$(CH_2)_m COR^2$) (B.M. Trost, K. Hiroi, and L.N. Jungheim, J. org. Chem., 1980, 45, 1839), and the oxidation of (217) gives (218; R^1/R^2=O) (K. Hiroi, S. Sato, and K. Matsuo, Chem. pharm. Bull. Japan, 1980, 28, 558). 1,4-Dithiepane itself (218; R^{1-3}=H) is produced in 60% yield either by the reaction between $HS(CH_2)_2SH$ and $Br(CH_2)_3Br$ under high dilution conditions or that between $HS(CH_2)_3SH$ and $Br(CH_2)_2Br$ (E. Weissflog, Phosphorus Sulphur, 1979, 6, 489). Ethane-1,2-dithiol reacts with acetophenone and formaldehyde to give 6-benzoylthiepane in good yield (V.I. Dronov, et al., Zh. org. Khim., 1979, 15, 1709) and (218; R^1=NHAc, R^2=Me, R^3=H) has been prepared via the reductive ring-opening of (219) to give $AcNH(Me)C=CHS^-$, its subsequent treatment with $Cl(CH_2)_3SH$ to give $AcNH(Me)C=CHS(CH_2)_3SH$, followed by cyclisation (S. Hoff, A.P. Blok, and E. Zwanenburg, Rec. Trav. chim., 1973, 92, 890). A member of the 2H-3,4-dihydro- 1,5-benzodithiepin system has been prepared by rearrangement of a Meissenheimer complex analogous to (155) (R. Cabrino, et al., J. chem. Soc., Perkin I, 1976, 2214), and the 2-oxo derivative by the phosphorus pentoxide induced cyclisation of 3-(2-mercaptophenylthio)propionic acid (G.A. Bistocchi, G. de Méo, and A. Ricci, Compt. Rend., 1977, 284C, 795).

Chapter 51

SEVEN-MEMBERED RING COMPOUNDS CONTAINING NITROGEN
IN THE RING

J.T. SHARP

The contents of this chapter are discussed in the
general introduction at the beginning of chapter
50.

Reviews

(i) *Azepines.* "Azepines" (R.K. Smalley, Comp.
Heterocycl. Chem., 1984, 7, 491). "The 1-, 2-,
and 3-Benzazepines" (S. Kasperek, Adv.
heterocyclic Chem., 1974, 17, 45). "Synthesis of
3-Benzazepines" (K. Orito, Mem. Fac. Eng.,
Hokkaido Univ., 1979, 15, 223, Eng.). "Synthesis
of Benz[d]indeno[1,2-b]azepines" (K. Orito and M.
Itoh, *ibid*, 1979, 15, 235, Eng.). "Synthesis of
Benzazepine Alkaloids" (T. Kametani and K.
Fukumoto, Heterocycles, 1975, 3, 931). "Dibenz-
[b,f]-azepines and related systems" (A. Ledwith,
Chem. Rev., 1974, 74, 101). "Dibenzazepines and
other Tricyclic Azepines" (B. Renfroe and C.
Harrington, Chem. heterocyclic Compd., 1984, 43,
Pt. 1, 1). "Azepine Ring Systems Containing Two
Rings" (G.R. Proctor, Chem. heterocyclic Compd.,
1984, 43, 637).

(ii) *1,2-Diazepines.* "The Chemistry of
1,2-Diazepines" (M. Nastasi, Heterocycles, 1976,
4, 1509). "Pyridinium *N*-Ylides to Diazepines and
Beyond" (J. Streith, Heterocycles, 1977, 6, 2021).
"1,2-Diazepines: A New Vista in Heterocyclic
Chemistry" (V. Snieckus and J. Streith, Acc. chem.
Res., 1981, 14, 348). "Photochemical Synthesis
and Reactions of 1,2-Diazepines" (J. Streith,

Lect. heterocycl. Chem., 1974, 2, 17).
"Chemistry of 1,2-Benzodiazepines and Related
Compounds" (T. Tsuchiya, Yuki Gosei Kagaka
Kyokaishi, 1981, 39, 99, Japan).

(iii) *1,3-Diazepines*. (T. Tsuchiya, J. synth.
org. Chem. Japan, 1983, 41, 641).

(iv) *1,4-Diazepines*. "2,3-Dihydro-1,4-
diazepines" (D. Lloyd, H.P. Cleghorn and D.R.
Marshall, Adv. heterocyclic Chem., 1974, 17, 1;
D. Lloyd and H. McNab, Heterocycles, 1978, 11,
549). "1,5-Benzodiazepines" (D. Lloyd and H.
Cleghorn, Adv. heterocyclic Chem., 1974, 17, 27).
"The Benzodiazepine Story" (L.H. Sternbach, Drug
Res., 1978, 22, 229; J. med. Chem., 1979, 22, 1).
"New Methods for Preparation of 1,4-Benzo-
diazepines and Their Applications" (H. Yamamoto,
Chem. Economy and eng. Rev., 1977, 9, 22).

(v) *General Diazepine*. "Dibenzodiazepines and
other Tricyclic Diazepine Systems" (J.W.H. Watthey
and J. Stanton, Chem. heterocyclic Compd., 1984,
43, Pt. 2, 1). "Seven-membered Rings with Two or
More Heteroatoms" (J.T. Sharp, Comp. heterocycl.
Chem., 1984, 7, 593).

(vi) *Triazepines and Tetrazepines*. "Monocyclic
and Condensed Triazepines and Tetrazepines" (N.P.
Peet, Chem. heterocyclic Compd., 1984, 43, 719).
"Chemistry and Biological Activity of
1,3,4-Benzotriazepines" (P. Richter and O.
Morgenstern, Pharmazie, 1984, 39, 301).

A general review covering advances in the
chemistry of heteroepins, particularly azepines,
diazepines and oxazepines is available (T. Mukai,
T. Kumagai, and Y. Yamashita, Heterocycles, 1981,
15, 1569).

1. *Seven-membered rings containing one nitrogen atom*

1.1 *Monocyclic azepines*

1.1.1. *1H-Azepines.* (a) *Synthesis.*

An improved route to the unsubstituted system (1, R^{1-7}=H) involves treating (1, R^{2-7}=H, R^1=CO$_2$SiMe$_3$) with methanol at -78°C and then warming the mixture to room temperature. The compound is very unstable (it readily tautomerises to give the more stable $3H$ isomer, accompanied by polymeric material), but it can be acylated, sulphonated and silylated at nitrogen (E. Vogel, *et al.*, Angew. Chem., intern. Edn., 1980, 19, 1016).

(1)   (2)

NH-Tautomers of this type can however be stabilised by hydrogen bonding to an α-carbonyl group. Thus the stable 2-acyl and 2-alkoxycarbonyl derivatives (1; R^1=H, R^2=Ac, R^{3-7}=H) and (1; R^1=H, $R^{2,5}$=CO$_2$Me, $R^{3,4,6,7}$=H) are obtained by the Raney nickel detosylation of *N*-tosyl precursors (N.R. Ayyangar, A.K. Purohit, and B.D. Tilak, Chem. Comm., 1981, 399).

(i) *via Ring expansion reactions.* Observations on the well established route to *N*-substituted azepines *via* azide decomposition in aromatic substrates have shown that yields of *N*-sulphonylazepines e.g. (1; R^1=Ts, R^{2-7}=H) and (1; R^1=Ts, $R^{2,5}$=CO$_2$Et, $R^{3,4,6,7}$=H) are much enhanced when the reaction is carried out at high pressure (60-85 atm.) (N.R. Ayyangar, R.B. Bambal, and A.G. Lugade, Chem. Comm., 1981, 790; Heterocycles, 1982, 18, 77). In reactions of this type the nature of the substituents on the azide and the resulting nitrene markedly affect

its reactivity. Thus the N-cyanonitrene
(::N-C(OR)=NCN), like ethoxycarboylnitrene
(::N-CO$_2$Et) reacts readily with benzene to give an
azepine in good yield but the N-methanesulphonyl
analogue (::N-C(OR)=NSO$_2$Me) does not (W. Lwowski
and O. Subba Rao, Tetrahedron Letters, 1980, 21,
727).

A route to the rare perfluoroazepine system (1;
R^1=Ph, R$^{2-7}$=F) is provided by the cycloaddition of
phenyl azide to hexafluoro-Dewar-benzene giving
the adduct (2) which on heating loses nitrogen and
undergoes ring expansion to give the azepine (M.G.
Barlow, et $al.$, J. chem. Soc., Perkin 1, 1982,
2101).

1H-Azepines have been prepared by the expansion
of 6-membered rings carrying an N-substituent.
Thus (3) on treatment with acid gives the cation
(4) which rearranges to give (1; R^1=Ts, R^2=Me,
R^{3,5,7}=Ph, R^{4,6}=H) in 72% yield.

(3) (4) (5) (6)

Similar reactants having an N-acyl group give
bicyclic oxazolines as primary products but these,
on heating, also give azepines (H. Eckhardt, et
$al.$, Angew. Chem., intern. Edn., 1981, 20, 699).
In a formally similar reaction the cyclohexadienyl
acetate (5) reacts with triethylammonium cyanide
to give (6) and thence (1; R^1=SO$_2$Ph, R^2=CN,
R^{3,5,7}=Me, R^{4,6}=H) (H. Perst, et $al.$, Angew.
Chem., intern. Edn., 1985, 24, 875). This
compound - apparently the first 1H-azepine to be
prepared - remained unrecognised for many years
because of structural uncertainty.

(ii) via $Rearrangement$. Further examples of the
3σ-3π-route to 1H-azepines from 7-azanorborna-
dienes (7) have been reported and the influence of

N- and *C*-substitution on the ease and direction of bond cleavage in the intermediate (8) examined (H. Prinzbach, *et al.*, Ber., 1973, <u>106</u>, 3824).

(7)   (8)   (9)

The method has been used to prepare the *N*-tosyl substituted parent molecule (1; R^1=Ts, R^{2-7}=H) (H. Prinzbach and H. Babsch, Heterocycles, 1978, <u>11</u>, 113) and, of particular interest, to generate azepines (1; R^1=Ts, $R^{2,4,5,7}$=H, $R^{3,6}$=Cl) and (1; R^1=Ts, $R^{2,7}$=H, $R^{3,6}$=Cl, $R^{4,5}$=CO$_2$Me) which have been shown by ^{1}H- and ^{13}C- nmr spectroscopy to be in equilibrium with appreciable amounts (1% at 0°C and 10% at -70°C respectively) of the benzeneimine tautomers (9), this is the first direct observation of benzeimine tautomers of 1*H*-azepines (H. Prinzbach, *et al.*, Tetrahedron Letters, 1977, 1355). In systems where the concentration of such tautomers is too low to be detected by ^{1}H-nmr spectroscopy e.g. (1; R^1=SO$_2$Ar, $R^{2,3,6,7}$=H, $R^{4,5}$=CO$_2$Me), reaction with diazomethane occurs predominantly by addition to the double bonds of the benzeneimine tautomer (9) (H. Prinzbach, *et al.*, Ber., 1976, <u>109</u>, 3505). However in analogous compounds (1) where R^1=COCH$_3$ or CO$_2$CH$_3$, addition takes place at the 4,5 bond of the azepine tautomer.

(10)

N-Phenylazepine (1; R^1=Ph, R^{2-7}=H) has been
prepared by the thermal isomerisation of (10) (M.
Christl and H. Leininger, Tetrahedron Letters,
1979, 1553).
(b) *Reactions.*
(i) Cycloaddition. - A great deal of work on the
reactions of 1*H*-azepines concerns their
cycloaddition reactions with a variety of 2π- and
4π-electron reactants. Among the 2π-electron
species it has been shown that nitrosobenzene
reacts with *N*-ethoxycarbonylazepine to give both
the (4+2)π adduct (11) and the (6+2)π adduct (12).
Since formation of the latter by a concerted
supra-supra reaction is forbidden it is suggested
that both may be formed in stepwise reactions *via*
(13) (W.S. Murphy and K.P. Raman, Tetrahedron
Letters, 1980, <u>21</u>, 319).

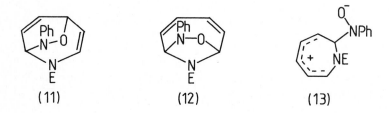

(11)　　　　　　(12)　　　　　　(13)

Reaction of *N*-ethoxycarbonylazepine with singlet
oxygen also gives both (4+2)π and (6+2)π adducts.
The thermal and base induced ring cleavage of the
latter yields $MeO_2CN(CHO)CH=CHCH=CHCHO$ and
$MeO_2CNHCOCH=CHCH=CHCHO$ respectively (T. Kumagai,
et al., Tetrahedron Letters, 1982, <u>23</u>, 873). The
reactions of *N*-ethoxycarbonylazepine with cyclopentadienone derivatives have been extensively
studied by Kanematsu. The major primary product
of reaction with the 4π system of (14,E'=CO_2Et) is
the *endo* (2+4)π adduct (15) together with some of
the *exo* (6+4)π adduct (16).

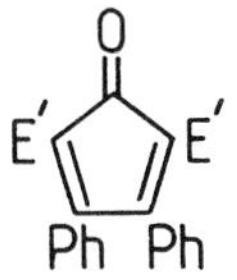

(14)

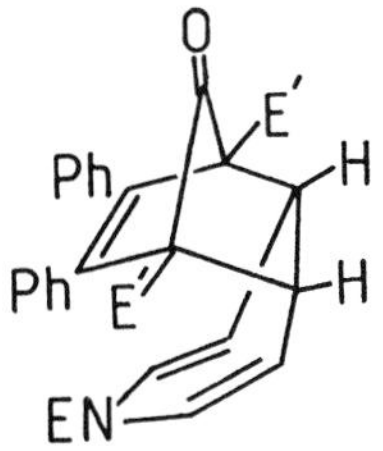

(15)

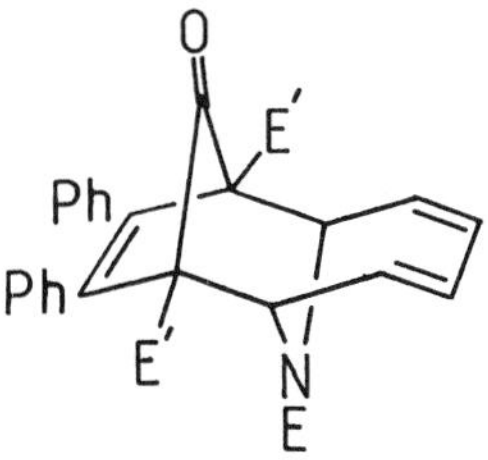

(16)

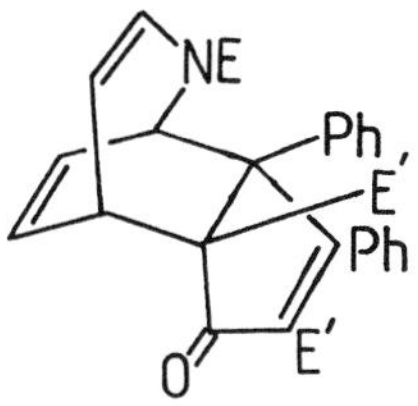

(17)

However the former is thermally unstable and rearranges readily by a Cope process to give the *anti-endo* (4+2)π adduct (17) (K. Harano, *et al.*, J. org. Chem., 1980, 45, 4455; Tetrahedron Letters, 1979, 1599). Phencyclones react in a similar way (M. Yasuda, K. Harano, and K. Kanematsu, J. org. Chem., 1980, 45, 2368; Heterocycles, 1980, 14, 129; and with T. Ban, Tetrahedron Letters, 1978, 4037) and the reaction with 2,5-diphenyl-3,4- diazacyclopentadienone also gives an *anti-endo* (4+2)π adduct as the isolated product (K. Harano, T. Ban, and K. Kanematsu, Heterocycles, 1979, 12, 453). 3,6-Dimethoxycarbonyl-1,2,4,5-tetrazine reacts similarly by a (2+4)π cycloaddition at the 4,5-bond of the azepine and after loss of nitrogen gives (18) – subsequently aromatised with chloranil (G. Seitz, T. Kaempchen, and W. Overheu, Arch. Pharm., 1978, 311, 786). o-Chloranil reacts with the azepine itself to give rather surprisingly the (6+4)π adduct (19) as the major product together with smaller amounts of the (4+2)π adducts arising from reaction at the 4,5- and the 2,3-bond of the azepine. The formation of (19) may be favoured by polar attraction (K. Saito, S. Iida, and T. Mukai, Heterocycles, 1982, 19, 1197).

(18)   (19)   (20)

(ii) *Other reactions.* The perfluoroazepines (1; R^1=CN or CO_2Et or $CONH_2$, R^{2-7}=F) undergo the expected photochemical electrocyclisation of the diene moiety to give 2-azabicyclo[3.2.0]-hepta-3,6- diene systems (M.G. Barlow, *et al.*, J. chem. Soc., Perkin I, 1982, 2105). 1-Ethoxycarbonyl-azepine reacts with dichlorosilanes in the presence of magnesium and HMPA to give the 1:2 adducts (20; R'=R_2SiOH) which may be ring closed to give a siloxane (K. Saito, and K. Takahashi, Heterocycles, 1979, 12, 263).

1.1.2   *2H-Azepines*
(a)   *Synthesis and Structure.* Several routes are available to fully unsaturated 2H-azepines. The reaction of 3-hydro-1-azirines with cyclopentadienones is an established and effective route to 3H-azepines (see later), and it has now been shown that the highly strained trisubstituted azirine (21) will add to a range of monocyclic and fused cyclopentadienones to give 2H-azepines, e.g. (22; R^1/R^2=$(CH_2)_2$, R^3=H, R^{4-7}=Ph) *via* addition to tetraphenylcyclopentadienone   (O.   Tsuge,   T. Ohnishi, and H. Watanabe, Heterocycles, 1981, 12, 2085). The 2H-azepine (24) is produced by the reaction of (23) with pyridine *N*-oxide but on heating it reforms the pyridine ring giving 5-(2-pyridyl)-   3,3,6,6-tetramethylthiepin-4-one

(A. Krebs, Heterocycles, 1979, <u>12</u>, 1153).

(21)

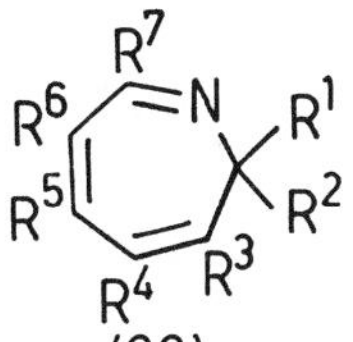

(22)

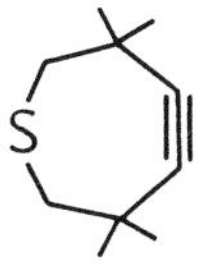

(23)

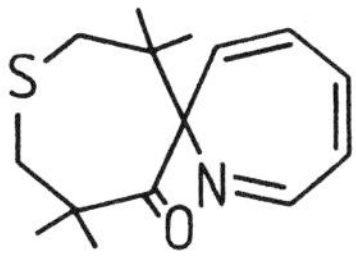

(24)

2*H*-Azepines have also been formed in aryl azide decompositions. Thus (22; $R^1/R^2=-C(CN)_2C(CN)_2-$, $R^3=Ph$, $R^{4-7}=H$) is formed when o-azidobiphenyl is photolysed in the presence of tetracyanoethylene (S. Murata, T. Sugawara, and H. Iwamura, Chem. Comm., 1984, 1198), and the perfluoro dimer (25) is formed by the pyrolysis of perfluorophenyl azide (R.E. Banks and N.D. Venayat, Chem. Comm., 1980, 900). It seems likely that both of these reactions involve intermediates of the type (31).

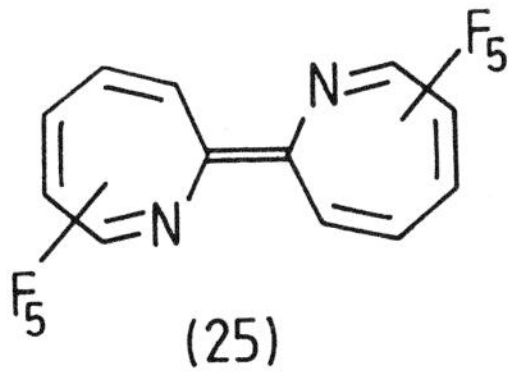

(25)

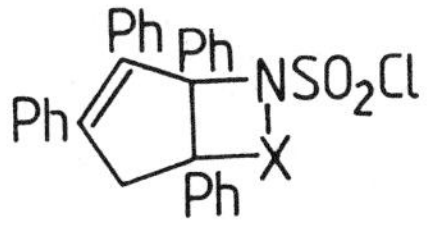

(26)

Tetraphenylcyclopentadiene reacts with chloro-sulphonyl isocyanate to give (26; X=CO) as the primary product which may be reduced to (26;

X=CH$_2$) with lithium aluminium hydride. Ring expansion of both these compounds has been achieved by successive reactions with p-cresol and sodium hydride giving respectively the azepine-2-one (22; R^1/R^2=O, R^{3,5,6,7}=Ph, R^4=H) from (26; X=CO) and the analogous 2,2-dihydroazepine (22; R^{1,2}=H, R^{3,5,6,7}=Ph, R^4=H) from (26, X=CH$_2$) (K. Narasimhan, P.R. Kumar, and T. Selvi, Heterocycles, 1984, 22, 2751). The stability of the latter is unexpected in view of earlier reports on the thermal and base catalysed isomerisation of 2-hydro-2H-azepine species to their 3H-isomers.

Photochemical cycloaddition of alkynes (R^{1}C≡CH) to Δ^2-pyrroline-4,5-dione gives the cycloadducts (27) which undergo thermal ring cleavage to give the azatropolones (28).

(27)   (28)   (29)   (30)

In some cases the adducts (27) cannot be isolated as they undergo photochemical transformation into (29). A route to (28) which avoids this problem is via the base induced expansion of the 6,7-dihydro analogue of (27), to give (30) followed by dehydrogenation (T. Sano, Y. Horiguchi, and Y. Tsuda, Heterocycles, 1978, 9, 731; 1979, 12, 1427; 1981, 16, 355).

1.1.3   3H-*Azepines*
(a)   *Synthesis.* The photolysis of aryl azides in the presence of nucleophiles such as amines (see 2nd edn.) or alcohols (*vide infra*) has been much used as a route to 3H-azepines (34). Investigation of the mechanism of the reaction using matrix techniques has demonstrated the presence of the highly unstable didehydroazepine (31) as a transient intermediate (O.L. Chapman and

J.P. Le Roux, J. Amer. chem. Soc., 1978, <u>100</u>, 282;
I.R. Dunkin *et al.*, J. chem. Soc., Perkin II,
1985, 307).

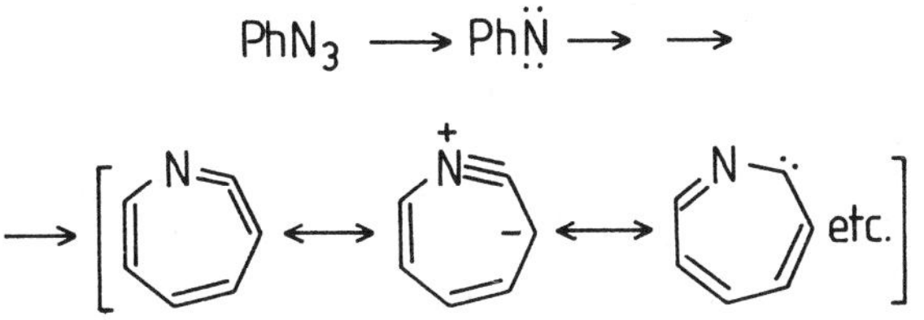

$$PhN_3 \longrightarrow Ph\ddot{N} \rightarrow \rightarrow$$

(31)

(32)

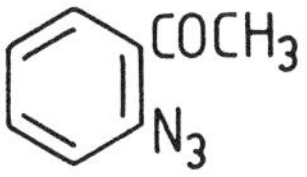

(33)

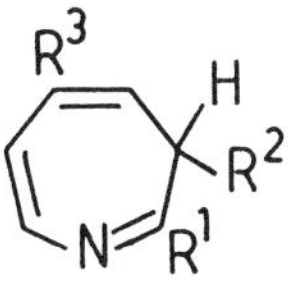

(34)

It seems likely that it is this species which
reacts with nucleophiles to give the isolated
3*H*-azepines (*e.g.* 34). There has been much
discussion on the possible involvement of
benzazirines (*e.g.* 32) and, although no
experimental evidence has been found for their
occurrence in phenyl azide photolysis, the
naphtho-fused analogues have been observed in the
photolysis of naphthyl azides (I.R. Dunkin and
P.C.P. Thomson, Chem. Comm., 1980, 499). Much
work has been done on the photolysis of a range of
aryl azides in the presence of alcohols to provide
a practical synthesis of 2-alkoxy-3*H*- azepines
(*e.g.* 34; R^1=OR, R^2=COCH$_3$, R^3=H), by the
photolysis of (33) in the presence of ROH.
However, good yields are obtained only when the
aromatic nucleus contains an electron withdrawing
group - preferably in the *ortho*-position (R.K.

Smalley and H. Suschitzky - with W.A. Strachan,
Synthesis, 1974, 503; with W.A. Strachan and R.
Purvis, J. chem. Soc., Perkin I, 1978, 191; and
with R. Purvis and M.A. Alkhader, *ibid*, 1984, 249:
T.B. Brown, *et al.*, *ibid.*, 1983, 2485; R.A.
Mustill and A.H. Rees, J. org. Chem., 1983, 48,
5041). Attempts to carry out an intramolecular
attack by OH resulted in an unexpected
intermolecular dimerisation giving a 14-crown-4
analogue (M. Azadi-Ardakani, *et al.*, J. chem.
Soc., Perkin I, 1985, 1121). In a similar
reaction the deoxygenation of nitrobenzene by
tributylphosphine in the presence of a primary or
a secondary alcohol leads to 2-alkoxy-3,3-dihydro-
3*H*-azepines (34; R^1=OR, $R^{2,3}$=H) in good yield.
The alkoxy derivatives are converted into 2-amino-
3*H*-azepines by reaction with ammonium carbonate
(M. Masaki, K. Fukui, and J. Kita, Bull chem. Soc.
Japan, 1977, 50, 2013). Earlier work on the
deoxygenation of nitrobenzenes by phosphines in
the presence of amines has been extended to give a
range of substituted 2-amino-3*H*-azepines (F.R.
Atherton and R.W. Lambert, J. chem. Soc., Perkin
I, 1973, 1079). The thermal decomposition of
N,O-bis(trimethylsilyl)-*N*-phenylhydroxylamine
provides an effective route to phenyl nitrene for
use in this reaction (F.P. Tsui, *et al.*, J. org.
Chem., 1976, 41, 3381). Hassner has published
full details of his synthesis of 3*H*-azepines *via*
the addition of 3-hydro-1-azirines to
cyclopentadienones (see also section b) (A.
Hassner and D.J. Anderson, J. org. chem., 1974,
39, 3070).

(35)   (36)

The 2-ethoxy-3-azatropolones (35; R=H, OH,
OAc, Ph) are obtained by ring opening (36) with

tin(IV) chloride, followed by DDQ dehydrogenation
(T. Sano, *et al.*, Heterocycles, 1981, *16*, 363).
(b) *Reactions.* Phenyl-lithium reacts with
1-alkoxycarbonyl-1*H*-azepines to give a product
shown by X-ray structural analysis to be (34;
$R^{1,3}$=H, R^2=C(OH)Ph_2) (H.J. Linder and B. von
Gross, Ber., 1973, *106*, 1033). Pyrolysis of this
compound gives the parent 3*H*-azepine (34;
$R^{1,2,3}$=H) and benzophenone (G. Schaden, Ber.,
1973, 1038; 2084). 2-Diethylamino-5-phenyl-3*H*-
azepine (34; R^1=NEt$_2$, R^2=H, R^3=Ph) has been deprotonated using lithium di-isopropylamide and the
resulting anion alkylated to give 3-substituted
derivatives in good yield (J.W. Streef and H.C.
van der Plas, Tetrahedron Letters, 1979, 2287).

1.1.4   4H-*Azepines*
(a)   *Synthesis.* Addition of the azirine (37)
to DMAD and subsequent rearrangement gives the 4*H*-
azepine (38) (L. Ghosez, *et al.*, Heterocycles,
1977, *7*, 895). On treatment with Meerwein's
reagent the dihydroazatropolone (30; R^2=CO$_2$Et)
gives the 4*H*-azepine (39).

(37)   (38)   (39)

(b)   *Structure and Reactions.* In general it
appears that 4*H*-azepines (40) are more stable than
their 3-azanorcaradiene tautomers (*e.g.* 41) (A.
Steigel, *et al.*, J. Amer. chem. Soc., 1972, *94*,
2770) but the existence of the equilibrium has
been demonstrated by trapping the latter with
cyclopropenes to give (42) (U. Göckel, *et al.*,
Tetrahedron Letters, 1980, 599). As in related
systems the position of the equilibrium is
affected by the nature of the substituents and it
is notable that (40/41, R=Me) exists entirely in
the bicyclic form. A number of variously
substituted 4*H*-azepines (43), prepared by the

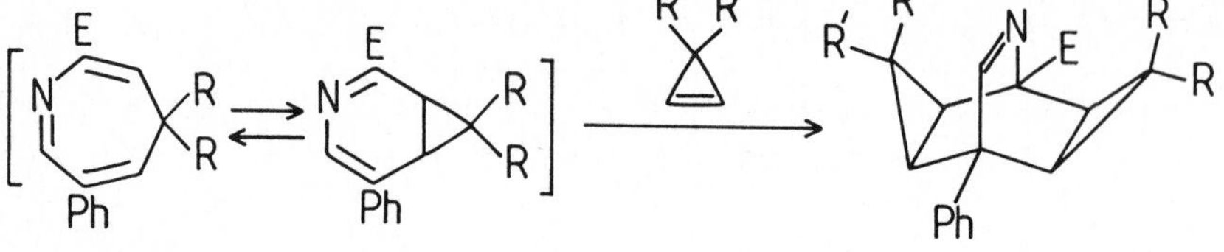

(40)   (41)   (42)

reaction between cyclopropenes and 1,2,4-
triazines, have ^{13}C chemical shifts for C-4
ranging from 24.6 to 52 p.p.m. and for C-7 from
161-171 p.p.m. The energy barriers for ring
inversion range from 7 to 17 kJ mole^{-1}. Those
compounds having two hydrogen atoms on C-4 (43,
$R^{4,5}$=H) undergo thermal [1,5] hydrogen shifts to
give the 3H-isomer (44) and a photochemical
isomerisation into (45) (U. Goeckel, et al.,
Tetrahedron Letters, 1980, 21, 595).

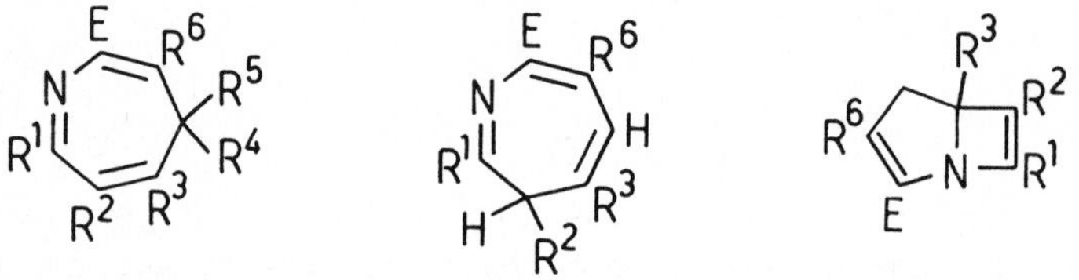

(43)   (44)   (45)

1.1.5   Dihydroazepines
(a)   Synthesis
(i)   1H-2,3-Dihydroazepines. The adducts of
dimethylaminoallene and acetylenic esters (46) are
converted into 1H-2,3-dihydroazepines (50; R^1=Me,
R^2=H, R^3=R, R^4=E) on heating at 120°C. This
remarkable rearrangement is considered to involve
ring opening of (46) to give (47), followed by
proton transfer and cyclisation of the azomethine
ylide (48) (W. Klop and L. Brandsma, Chem. Comm.,
1983, 989). Thermal rearrangement of the
7-azabicyclo[4.1.0]heptene (49) also gives a

1H-2,3-dihydroazepine (50; $R^1=CH_2Ph$, $R^{2,3}=E$, $R^4=H$) (W. Eberbach, J.C. Carré, and H. Fritz, Tetrahedron Letters, 1977, 4385) and *syn-N*-carbomethoxy-2-azatricyclo[4.1.0.0^{3,5}]heptane gives *N*-carbomethoxy-2-dihydroazepine (50; $R^1=CO_2Me$, $R^{2,3,4}=H$) (S.R. Tanny and F.W. Fowler, J. Amer. chem. Soc., 1973, 95, 7320).

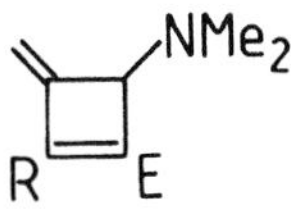

(46) 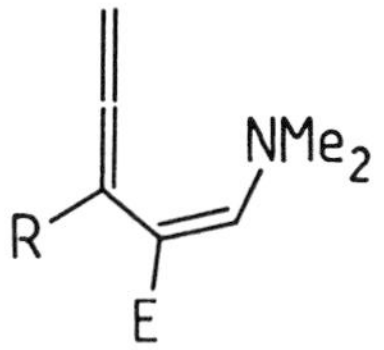   (47) 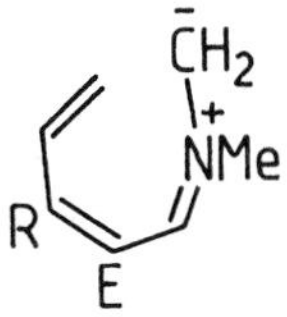   (48)

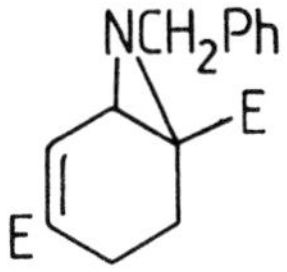

(49) 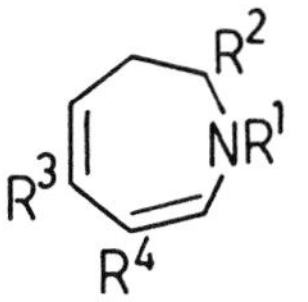   (50)

(ii)   1H-2,5-*Dihydroazepines*.   This system (52) is prepared by a hetero-Cope rearrangement of a compound of the type (51; X=N and Y=CH, or Y=N and X=CH).   The latter can be made by the addition of 2-vinylaziridines to acetylenes. Thus, for example, (51; X=N, Y=CH, $R^{1,2}=CO_2Me$, $R^3=Ph$) rearranges to give (52; $R^1=H$, $R^{2,3}=CO_2Me$, $R^4=Ph$) (A. Hassner, *et al.*, Tetrahedron Letters, 1981, 22, 3691).

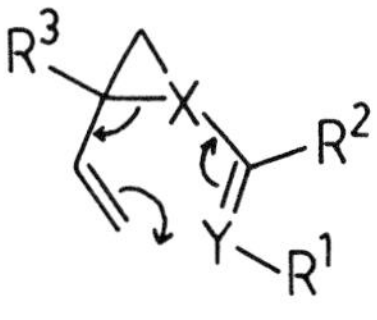

(51) 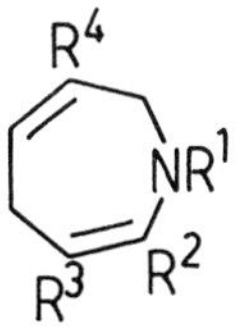   (52)

Some interesting examples of this reaction involving acetylenic phosphonium salts had been reported earlier (M.A. Calcagno and E.E. Schweizer, J. org Chem., 1978, $\underline{43}$, 4207). The cis-iminovinylcyclopropane (51; X=CH, Y=N, R^1=Me, R^2=MeO, R^3=H) equilibrates with the azepine (52; R^1=Me, R^2=MeO, $R^{3,4}$=H) on heating to give a ca. 3:1 excess of the azepine (L.A. Paquette and G.D. Ewing, J. Amer. chem. Soc., 1978, $\underline{100}$, 2908).

(iii)   1H-4,5-*Dihydroazepines*. - Cycloaddition of diphenylnitrile ylide (PhC$\equiv$N$^+$-C$^-$HPh) to 1,2-dicyanocyclobutene and ring opening of the adduct yields the 4,5-dihydroazepine (53; $R^{1,4,5}$=H, $R^{2,7}$=Ph, $R^{3,6}$=CN) (H.-D. Martin, et al., Monatsh. chem., 1983, $\underline{114}$, 1145).

$$R^4 \quad R^3 \quad R^2$$

(53)

This parallels an earlier report (H.-D. Martin, and M. Hekman, Angew. Chem., intern. Edn., 1972, $\underline{11}$, 926) and a later report (I.J. Turchi, C.A. Maryanoff, and A.R. Mastrocola, J. Heterocyclic Chem., 1980, $\underline{17}$, 1593) of the addition of mesoionic oxazolones to give similar products. A number of 4,5-dihydro-1H-azepines (e.g. 53; R^1=H, $R^{2,7}$=Me, $R^{3,6}$=E, R^4=Me, R^5=Ph), of interest as calcium channel blockers, have been prepared utilising the known ring expansion of 4-haloalkyl-1,4-dihydropyridines and related compounds (D.A. Clareman, et al., J. org. Chem., 1984, $\underline{49}$, 3871).

(iv)   1H-2,3-*Dihydroazepin-2-ones*.   The unsubstituted system (54; R^{1-3}=H) has been prepared by the hydrolysis of 2-amino- or 2-butoxy-3H-azepine (M. Masaki, K. Fukui, and J. Kita, Bull. chem. Soc. Japan, 1977, $\underline{50}$, 2013). The decomposition of phenyl azide in the presence of 'naked' acetate or hydroxide anions (potassium salt in the presence of 18-crown-6) gives the

3H-azepine (34; R^1=OH or OAc, $R^{2,3}$=H) which on hydrolysis gives (54;, R^{1-3}=H) (R. Colman, et al., Chem. Ind., 1981, 249). The same compound is also formed when phenyl azide is decomposed in acetic acid - the first such decomposition in an electrophilic reagent. The reaction path is thought to involve addition of acetic acid to (31) to give (34; R^1=OAc, $R^{2,3}$=H), followed by solvolysis (H. Takeuchi and K. Koyama, J. chem. Soc., Perkin I, 1982, 1269).

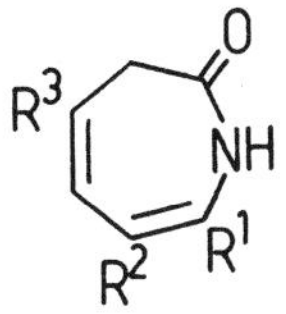

(54) 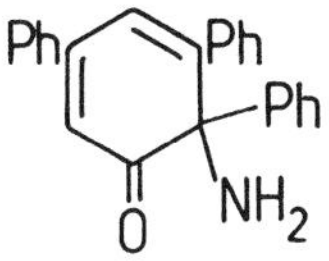   (55)

The triphenyl substituted molecule (54; R^{1-3}=Ph) is produced by the thermally induced ring expansion of the aminocyclohexadienone (55) (H.H. Eckhardt, et al., Tetrahedron Letters, 1979, 4975). It is thought that similar 6-aminocyclo-hexa-2,4-dienones are key intermediates in the formation of azepinones (e.g. 57; R=But) by the attack of a primary amine on the oxepinobenzofuran system (56).

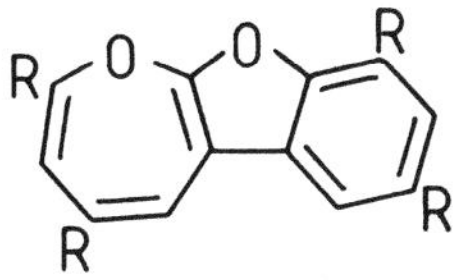

(56) 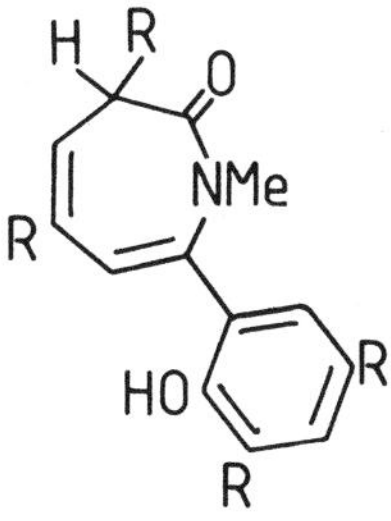   (57)

Oxidation of (57) gives the benzofurano-annelated azepine, and the azepine ring in (57) undergoes the expected photochemical electrocyclisation of its butadiene moiety (H.-D. Becker, *et al.*, Austral. J. Chem., 1979, <u>32</u>, 1931 and reference cited therein).

(v) *Other Dihydroazepinones.* The $1H$-2,5-dihydro-azepin-2-ones (30; R^2=E) are produced by base induced ring opening of the photo-cycloaddition products of alkenes with dioxopyrroline *(loc. cit)*. The $1H$-2,3-dihydroazepin-3-one (59) results from the ring expansion of the pyridone carbanion (58) which is sterically destabilised when R=Me (A.R. Katritzky, J. chem. Soc., Perkin I, 1980, 2851).

(58)

(59)

(b) *Reactions.* The reaction of (50; R^1=CH$_2$Ph, $R^{2,3}$=E, R^4=H) with DMAD follows one of three possible reaction paths (depending strongly on the nature of the solvent) to give either the Diels-Alder adduct (60), the 6-vinyl-2,3-dihydro-azepine (61) or an azacyclononatriene. The reaction in carbon tetrachloride at 100°C gives only (60); that in acetonitrile at 25°C gives only the azacyclononatriene, and that in methanol at 25°C gives only (61). The last two products are thought to be formed *via* the dipolar intermediate (62) (W. Eberbach and J.C. Carré, Tetrahedron Letters, 1980, 1145).

(60)

(61)

(62)

The course of the photochemical rearrangement of (57) also shows strong solvent dependence. In aprotic solvents it gives the expected bicyclic system resulting from electrocyclisation of the butadiene moiety but in methanol solution the reaction proceeds *via* a 1,3-acyl migration to give (63) (H.D. Becker and A.B. Turner, Tetrahedron Letters, 1979, 4871).

(63)   (64)

Irradiation of 4,5-dihydro-2-dimethylamino-3*H*-azepine (64), prepared by selective reduction of the 4,5 double bond of (34; R^1=NMe$_2$, $R^{2,3}$=H, gives 1-dimethylaminocyclopentene and 4,5-tri-methylenepyrimidine as major products (E. Lerner, R.A. Odum, and B. Schmall, Chem. Comm., 1973, 327).

1.1.6   *Tetrahydroazepines*

(a)   *Synthesis*
(i) 2,5,6,7-*Tetrahydro*-1H-*azepines*. - The addition of 2-phenyl-2-vinylaziridine to electron deficient alkenes (CH$_2$=CHX, X=CN, CO$_2$Me, SO$_2$Ph) and cyclisation of the intermediate (65) provides a useful general route to this ring system (66; $R^{1,2,3,5,6}$=H, R^4=Ph, R^7=CN or CO$_2$Me or SO$_2$Ph). A limitation however is that the cyclisation fails when X is an easily enolised group e.*g*. ketone CO or NO$_2$ (A. Hassner, *et al*., Tetrahedron Letters, 1981, 22, 3691; A. Hassner and W. Chaux, *ibid*., 1982, 23, 1989). The 2-vinylpyrrolidine (67) ring-expands on treatment with methyl iodide to give the methiodide of (66; R^1=Me, $R^{2,3,4,6,7}$=H, R^5=Ar) (H.W. Persch, R. Rissmann, and D. Schon, Arch. Pharm, 1982, 315, 749). The pyrrolidine ring in (68) undergoes a remarkable ring expansion

on reaction with pentan-2-4-dione to give (66; R^1=Me, R^2/R^3=CHCOMe, $R^{4,6,7}$=H, R^5=3-indolyl) (D.I. Bishop, I.K. Al-Khawaja, and J.A. Joule, J. chem. Res.(S), 1981, 361).

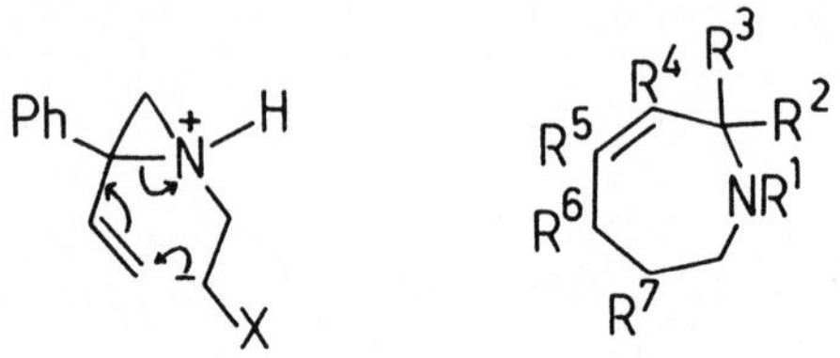

(65)

(66)

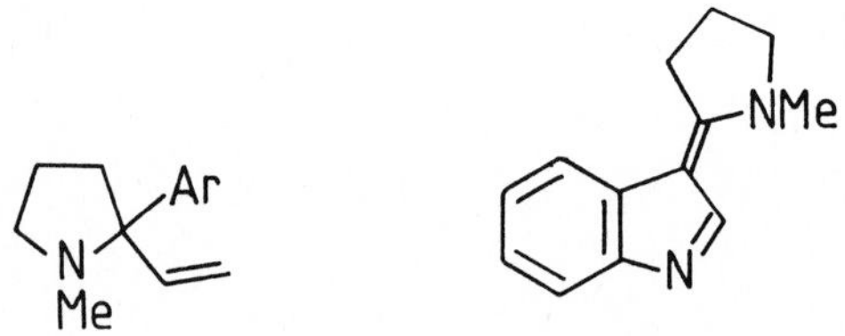

(67)

(68)

(ii) *Tetrahydroazepine-2-ones.* - The base induced cyclisation of the unsaturated *N*-allyl amides (69) provides a good synthetic route to the tetrahydroazepinone system (70), probably *via* cyclisation of a dianion. If the reaction is quenched with methyl iodide rather than water the *N*-methyl analogue of (70) is obtained (M. Bortolussi, R. Bloch, and J.M. Conia, Tetrahedron Letters, 1977, 2289).

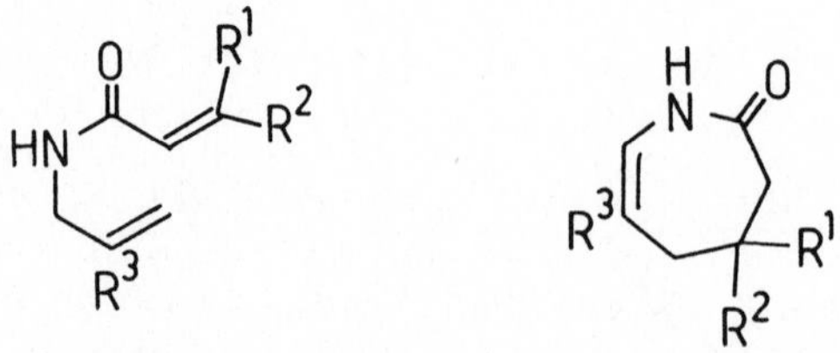

(69)

(70)

N-Methylated analogues of (70) are also produced, but in only moderate yield, by the ring expansion of nitrones (R.H. Prager, K.D. Raner, and A.D. Ward, Austral. J. Chem., 1984, *37*, 381). Further examples have been reported of the preparation of substituted 2,5,6,7-tetrahydro-1*H*-azepine-2-ones *via* Beckmann or Schmidt reaction of cyclohex-2-enone derivatives (G.I. Hutchison, R.H. Prager, and A.D. Ward, Austral. J. Chem., 1980, *33*, 2477). (iii) *Tetrahydroazepindiones*. - The 2,4-dione (72; R=Me or PhCH$_2$ is produced in good yield by the irradiation of (71) (Jap. Pat., 57,146,761 (1982), Chem. Abs., 1983, *98*, 143292f).

(71)　　　　　　　　　(72)

(b) *Reactions*.　The reaction of caprolactam with phosgene at 0°C has been used as a convenient route to the imidoyl chloride (73;　R=Cl) and thence to caprolactim ethers (73;　R=OR') (J. Jurczak, Synthesis, 1983, 382) but it has also been shown that phosgenation at higher temperatures leads to (74) (A.L. Chimishkyan and S.I. Orlov, Chem. Abs., 1982, *97*, 162792q).

(73)　　　　　　　　　(74)

Caprolactim thiomethyl ether (73;　R=SMe) reacts with a range of activated methylene compounds in the presence of triethylamine to give 2-methylene-perhydroazepines (G. L'hommet, M.G. Richaud, and P. Maitte, J. heterocyclic Chem., 1982, *19*, 431).

Caprolactim ethers (73; R=OR') similarly react
with Meldrum's acid to give (75) which on
pyrolysis eliminates carbon dioxide and acetone to
give the ketene (76). The latter reacts with
alcohols, thiols and amines to give (77; X=O,S,N)
(J.P. Célérier, G. L'hommet, and P. Maitte,
Tetrahedron Letters, 1981, $\underline{22}$, 963; and with E.
Deloisy, J. org. Chem., 1979, $\underline{44}$, 3089). The
crystal structure of the ester has been determined
(J.P. Célérier and G. L'hommet, J. org. Chem.,
1982, $\underline{19}$, 481).

(75)   (76)   (77)

1.1.7 Perhydroazepines

(a)   Synthesis of Perhydroazepines
(i) Reduction of azepines.   N-Benzyl-ϵ-
caprolactam is effectively reduced by sodium
borohydride and titanium tetrachloride to give the
saturated azepine (Jap. Pat. 80,162,746 (1980),
Chem Abs., 1981, $\underline{95}$, 62023e). Caprolactam itself
is reduced in virtually quantitative yield by gas
phase hydrogenation at 240-270°C in the presence
of a Group I or II metal catalyst which has been
previously treated with aliphatic hydrocarbons
(Ger. Pat., 2,752,307 (1979), Chem. Abs., 1979,
$\underline{91}$, 56848d).
(ii) Reductive cyclisation.   Perhydroazepine
itself is prepared in very high yield by the
catalytic (Ni,Co) deamination of $NH_2(CH_2)_6NH_2$ at
low temperature and pressure in an inert solvent
(Ger. Pat., 2,532,871 (1976), Chem. Abs., 1976,
$\underline{84}$, 180097c). The same conversion is also
achieved using $RuCl_2(Ph_3P)_3$ as the catalyst in an
inert solvent at 180°C (Bui-The-Khai,

C. Conciliano, and G. Porzi, J. org. Chem., 1981, _46_, 1759). The ruthenium catalyst $RuH_2(Ph_3P)_4$ is highly selective for the activation of alcohols and has been used to effect the ring closure of $HO(CH_2)_6NH_2$ (S.I. Murahashi, K. Kondo, and T. Hakata, Tetrahedron Letters, 1982, _23_, 229). Various _N_-substituted perhydroazepines have also been made in 60-70% yield by the Raney nickel catalysed cyclisation of $RNH(CH_2)_6OH$ (D. Barbry and B. Hasiak, Synth. Comm., 1982, _12_, 733).

(iii) _Other methods._ 1-Methyl-3,6-diethoxy-carbonylazepane has been synthesised by the reaction of $Me_3SiO(OEt)C=CH(CH_2)_2CH=C(OEt)OSiMe_3$ with methylamine (S. Miyazawa, _et al._, Chem. Letters, 1984, 785). The Beckmann rearrangement of cyclohexanone oxime methanesulphonate induced by Grignard reagents provides direct access to α-alkyl- and α,α-dialkyl-perhydroazepines (K. Hattori, K. Maruoka, and H. Yamamoto, Tetrahedron Letters, 1982, _23_, 3395).

(b) _Synthesis of Perhydroazepinones._
A new route to lactams which gives ϵ-caprolactam in 96% yield involves reaction of a 1-nitroso- or 1-nitro-1-halogenocycloalkene (e.g. 78) with tri-phenylphosphine or via the single step reaction of a cycloalkanone oxime with halogen and triphenylphosphine. The intermediate (79) rearranges to give the imidoyl chloride (73, R=Cl) which gives caprolactam on hydrolysis (I. Sakai, N. Kawabe, and M. Ohno, Bull. chem. Soc. Japan, 1979, _52_, 3381).

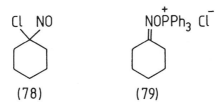

(78)   (79)

3-Aminocaprolactam (81; $R^{1,2}$=H) has been prepared from lysine using an efficient general method for the synthesis of the lactam ring from an ω-aminocarboxylic acid (H. Ogura and K. Takeda,

Heterocycles, 1981, _15_, 467). The protected azepinone ring system of Cobactin T (81; $R^1=OCH_2Ph$, $R^2=BOC$) has been synthesised by cyclisation of the α-N-protected O-alkyl hydroxamate of ϵ-hydroxynorleucine (80) using triphenylphosphine and diethyl azodicarboxylate (P.J. Maurer and M.J. Miller, J. org Chem., 1981, _46_, 2835).

(80)   (81)

1-Acyl-3-piperidones (e.g. 82) undergo ring expansion to give (83) and its 3-methoxy-carbonyl-4-oxo- isomer on reaction with ethyl diazoacetate in the presence of BF_3 (P. Krogsgaard-Larsen and H. Hjeds, Acta chem. Scand., 1976, _B30_, 884).

(82)   (83)   (84)   (85)

Photochemical cycloaddition of enolised 2,3-dioxopyrrolidines to alkenes gives adducts of type (84) which readily undergo base catalysed ring expansion to give the azepine-2,3-diones (85) (S.T. Reid and D. DeSilva, Tetrahedron Letters, 1983, _24_, 1949).

(c) *Reactions*. Two ring contractions have been reported involving the rearrangement of azepinols under acylating conditions *via* transannular attack. Thus the *trans* diol (86) in acetic acid/acetic anhydride undergoes a fundamental reorganisation to give (87) whereas the *cis*-isomer is cleanly converted into the diacetate (P.Y. Johnson and J. Lisak, Tetrahedron Letters, 1975, 3801). The 4-hydroxy-4-phenyl derivative (88) on treatment with acid anhydrides ($(RCO)_2O$) rearranges to give (89) (M.A. Iorio and M. Miraglia, Chem. Abs., 1973, _79_, 78518r).

OH OH (86) AcO ...OAc ...H NAc (87) Ph OH N Me (88) MeCHCH_2O_2CR Ph NMe (89)

2-(Nitromethylene)perhydroazepine rearranges and dimerises under acid conditions to give 2,5-bis-(5-aminopentyl)pyrazine (Brit. Pat., 1,525,672 (1979), Chem. Abs., 1979, _91_, 91667r). N-Alkylation of caprolactam can be carried out under very mild conditions using a phase-transfer method (M. Takahata, T. Hashizame and T. Yamazaki, Heterocycles, 1979, _12_ 1449). 1-Methyl-3-ethylcaprolactam has been methylated at the 3-position *via* deprotonation with a lithium dialkylamide and treatment with methyl iodide (T. Cuvigny, *et al.*, Canad. J. Chem., 1979, _57_, 1201).

1.2 *Bicyclic compounds containing an azepine ring*

1.2.1   *Azepines fused to four-membered rings*
The cyclobuteno]3,4-d]-4,5-dihydroazepine (90; R=H) is prepared by the reaction of (90; R=CO_2Et) with either butyl-lithium or potassium *tert*-butoxide. Its lithium salt reacts readily at N with MeI, ClCONMe_2, ClCO_2Me and ClCOMe. Reduction of (90; R=CO_2Et) with lithium aluminium hydride gives the amino alcohol (90; R=CH_2OH). The nmr chemical shifts of the protons at the γ- and δ-positions have been correlated with the

nature of R.   (A.G. Anastassiou and R.L. Elliot, Chem. Comm., 1973, 601).   The urea (90; R=CONMe$_2$) is thermally unstable and readily rearranges to give (91).   However in general compounds of this type undergo thermal ring opening of the cyclobutene moiety to give azonine intermediates which either collapse to give cyclobuteno[b]azepines and indoles, or can be trapped by reaction with dienes.

(90)   (91)   (92)

Thus, decomposition of (90; R=Ac) in the presence of α-pyrone, followed by loss of carbon dioxide and aromatisation with chloranil affords the 4,5-benzazonine system (92) (A.G. Anastassiou, et al., J. org. Chem., 1973, 38, 1959;   Tetrahedron Letters, 1973, 3805).

1.2.2   *Azepines fused to five-membered rings - Aza-azulenes*

(a) 3a-*Aza-azulenes* and 3a-*Aza-azulenones*. - The dihydropyrroloazepinone (94) is prepared by the base-catalysed cyclisation of the oxo-aldehyde (93) and is dehydrogenated by DDQ or Pd/C to give (95).   Reduction of (94) using borohydride and subsequent dehydration gives (96) - the first preparation of the unsubstituted system (G. Jones and P.M. Radley, J. chem. Soc., Perkin I, 1982, 1123).

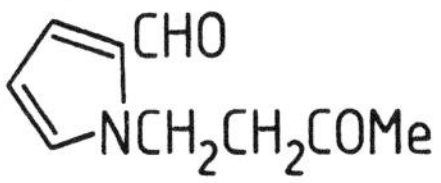

(93)

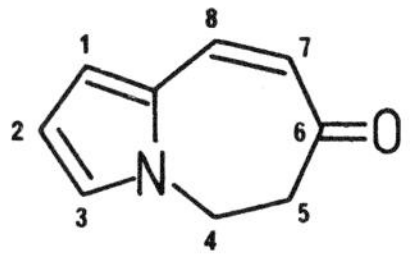

(94)

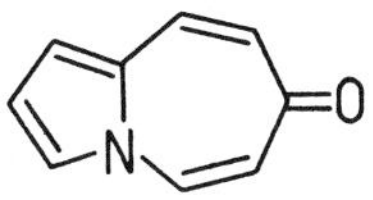

(95)

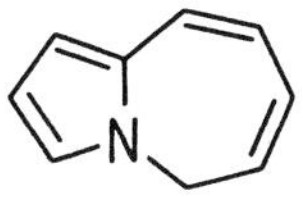

(96)

The reaction of the dehydropyrrolizidine (97) with DMAD gives (98) and (99) in 36 and 20% yield respectively (S. Miyano, *et al.*, Heterocycles, 1983, 20, 2197).

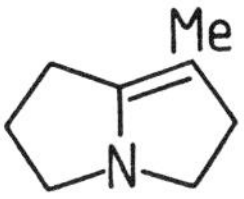

(97)

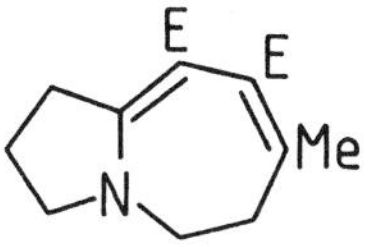

(98)

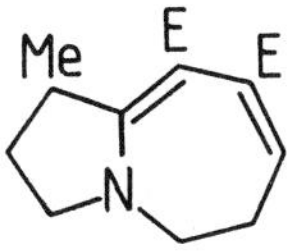

(99)

The cyano-substituted system (100; R=CN) and thence the ester is prepared by the sodium hydride mediated cyclisation of 1-(4-iodobutyl)-2-(cyanomethyl)-pyrrole (J.M. Muchowski, *et al.*, Canad. J. Chem., 1982, 60, 2295). The 3a-aza-azulene-4-one system (102; R[1]=H or Me or Ph) is synthesised by the base induced cyclisation of (101; R=CH=CHC(R')=CHCO$_2$R[2]). The 5,6,7,8-tetrahydro analogue is prepared in a similar way (W. Flitsch, B. Müter, and U. Wolf, Ber., 1973, 106, 1993).

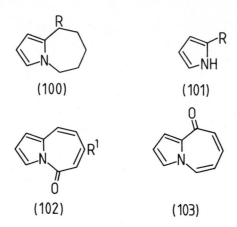

(100)

(101)

(102)

(103)

Olefination of (102; R^1=H) gives the 4-methylene derivatives in some cases but reactions with acylmethylene-triphenylphosphoranes proceed *via* a Michael addition to the electron-deficient diene system followed by cyclisation and lead eventually to 1,4-cyclohexadiene derivatives (W. Flitsch and E.R. Gesing, Tetrahedron Letters, 1979, 4529). The 8-oxo analogue (103) is prepared by heating (101; R=COCH=CHCH=CHNMe$_2$) at 400°C. Nmr studies on (102) and (103) indicate a limited π electron delocalisation which increases on protonation at the exocyclic double bond (W. Flitsch, F. Kappenberg, and H. Schmitt, Ber., 1978, 111, 2407). Benzo fused 3a-aza-azepines have been prepared by the thermolysis of azido di- and tri-arylmethanes (R.N. Carde and G. Jones with P.C. Hayes and C.J. Cliff, J. chem. Soc., Perkin I, 1981, 1132; and with W.H. McKinley and C. Price, *ibid.*, 1978, 1211).

(b) *4-Aza-azulenes*. The first preparation of the parent system (105) involves the spray pyrolysis of the azidoformate (104) with loss of nitrogen, carbon dioxide and hydrogen chloride. The azulene is a liquid which is not very stable in air and undergoes an apparently autocatalytic decomposition. However it is more stable when dissolved in organic solvents or in acids (O. Meth-Cohn and C. Moore, Chem. Comm., 1983, 1246).

(104)   (105)   (106)

Photochemical decomposition of 4-azidoindane in the presence of diethylamine gives 5-diethylamino-1,2,3,6-tetrahydro-4-aza-azulene - another example of the aryl azide ring expansion route to 3H-azepine rings (R.N. Carde and G. Jones, J. chem. Soc., Perkin I, 1975, 519). The octahydro derivatives (106) are prepared by the condensation of 2-(4-aminobutyl)cyclopentanones, and, on autoxidation, undergo ring contraction to give 4-quinolizidone derivatives (D. Schumann and A. Naumann, Ber., 1982,115, 1626).

(c) 5-*Aza-azulenes*. The fulvenoid chloroformamidinium chloride (107) derivative reacts with ammonia to give the stable 4-(dimethylamino)-5-aza-azulene (108) as red crystals (K. Hafner and H.-P. Krimmer, Angew. Chem., intern. Edn., 1980, 19, 199).

(107)   (108)   (109)

2-*tert*-Butyl-7,8-di-ethoxycarbonyl-4,6-dimethyl-amino-5-aza-azulene has been prepared by the addition of DMAD to the 2-azapentalene (109) at 0°C and subsequent ring expansion (K. Hafner, H.-G. Klas, and M.C. Böhm, Tetrahedron Letters, 1980, 21, 41), and the same process has been used to trap a transient intermediate of a type similar to (109) (W. Friedrichsen, W.-D. Schröer, and T. Debaerdemaeker, Ann., 1981, 491).

(d) *6-Aza-azulenes*. The first derivative of 6-aza-azulene (111) has been prepared from (110) *via* bromination/dehydrobromination. The compound is a crystalline blue solid (m.p. 45-46°C) which shows [1]H-nmr chemical shifts and electronic spectra similar to those of azulene itself (M. Kimura and S. Tai, Chem. Comm., 1980, 974).

(110)   (111)

An analogous indeno-azepine having a fully conjugated 14π electron system has also been prepared (M. Kimura, *et al.*, Bull. chem. Soc. Japan, 1980, 53, 3232; 1979, 52, 1437). The 6-chlorosulphonyl-1,2,3,4,5,7,8-heptahydro-7-one derivative has been prepared by the reaction of chlorosulphonyl isocyanate with the spiro-vinylcyclopropane 4-methylene-*spiro*[2,4]-heptane (S. Sarel, A. Felzenstein, and J. Yovell, Tetrahedron Letters, 1976, 451).

1.2.3   *1-Benzazepines*
(a)   *Fully unsaturated - Synthesis*.   A major route to 1*H*-1-benzazepines (113) is provided by the thermal and photochemical ring opening of 2a,7b-dihydro-3*H*-cyclobut[b]indoles (112) − usually prepared by the cycloaddition of indole to an alkyne (P.D. Davies and D.C. Neckers, J. org. Chem., 1980, 45, 456, 462). The thermolysis is relatively easy when ring cleavage is assisted by electron donating R[1] and electron withdrawing R[2] groups. However, in the absence of such groups the temperature required (250-300°C) can be reduced by 100-160°C by the presence of silver ions (M. Ikeda, *et al.*, J. chem. Soc., Perkin I, 1982, 741; Tetrahedron Letters, 1980, 3403).

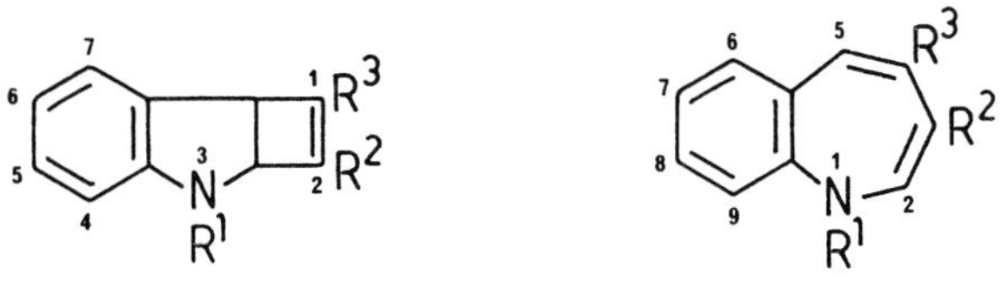

(112)   (113)

The cycloaddition of DMAD with 2-substituted quinoline-1-oxides (114) gives the 3*H*-1-benzazepine system (117; R=Ph or Me) *via* (115) and (116) (Y. Ishiguro, *et al.*, Heterocycles, 1983, 20, 1545).

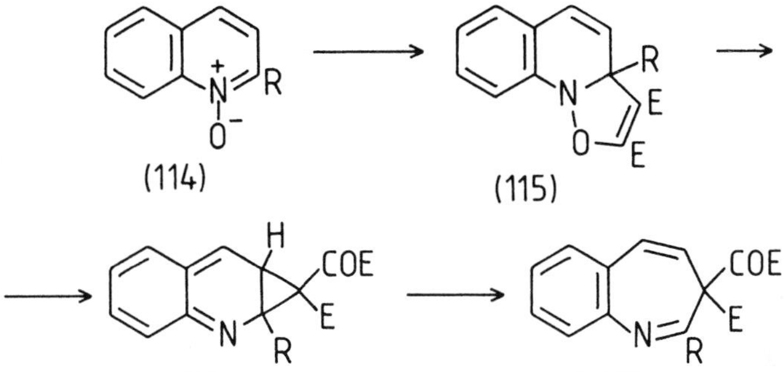

(114)   (115)

(116)   (117)

(b) *Dihydro*-1-benzazepines. - *Synthesis.* The reaction of a 1-methyl-2,3-dialkyl-1,2-dihydro-quinoline (118) with excess ethyl azidoformate (N_3CO_2Et) provides a useful route to the iminoazepines (120; R^1=Me or Et, R^2=CO_2Et). The intermediates (119) in these reactions, also produced by the action of alkali on 1,2,3-trialkylquinolinium chloride, react similarly with phenyl or benzoyl azides to give (120; R^2=Ph or PhCO) (Y. Sato, H. Kojima and H. Shirai, J. org. Chem., 1976, 41, 195; 3325).

(118)   (119)   (120)

(121)   (122)   (123)

The 1H-2,5-dihydro-1-benzazepine system (121) is
formed by the acid catalysed cyclisation of
PhN(R)CHR^{1}C(OH)R^{2}C(Me)=CHR^3 (J. Bonnic, et al.,
Compt. Rend., 1974, 278C, 1461). The reaction of
a 1-alkyl-1-hydroxytetralin (122) with hydrazoic
acid and sulphuric acid gives a 2-substituted-4,5-
dihydro-3H-1-benzazepine (123; R^1=Me or CH$_2$Ph)
(G. Adam, J. Andrieux, and M. Plat, Tetrahedron,
1982, 38, 2403). The photochemical ring opening
of the 3a,7a-dihydro analogue of (112; R^1=CO$_2$Et,
R^{2,3}=H) at -78°C gives the expected 1-ethoxy-
carbonyl-5a,9a-dihydro-1-benzazepine (A.G.
Anastassiou, et al., J. org. Chem., 1978, 43,
315). There is a further report on the Schmidt
ring expansion of substituted 1,4-naphthaquinones
to give 2,5-dihydro-3-hydroxy-1-benzazepine-2,5-
diones (G.R. Birchall and A.H. Rees, Canad. J.
Chem., 1974, 52, 610) and also a report of
increased yields of 2,3-dihydro-1H-1-
benzazepin-2-ones from the reactions of 4-oxo-
carboxylic acids with aromatic amines when methoxy
groups are used to activate the aromatic ring. (V.
Candeloro and J.H. Bowie, Austral. J. Chem., 1978,
31, 203).
(c) Tetrahydro-1-benzazepines. - (i) Synthesis.
Oxazoline activation of adjacent methoxy groups to
nucleophilic displacement has been utilised in the

synthesis of benzazepines (**125**; R=2-(4,4-dimethyloxazolinyl) and the analogous benzoxepins from (**124**; X=NH$_2$ or OH respectively). The oxazoline group can then be converted into other functional groups such as carboxyl by the usual reactions (A.I. Meyers, M. Reuman, and R.A. Gabel, J. org. Chem., 1981, 46, 783).

(124) (125)

The ortho-chloroaniline derivative o-ClC$_6$H$_4$N(Me)CH$_2$CH$_2$CH$_2$CH=CH$_2$ with methylmagnesium bromide in the presence of a catalytic amount of NiCl$_2$(Ph$_3$P)$_2$ gives 5-methylene-1-methyl-1-benzazepine in good yield (M. Mori, S. Kudo, and Y. Ban, J. chem. Soc., Perkin I, 1979, 771).

A new route to 1-benzazepin-2-ones is provided by the reaction of o-lithiomethylphenylisocyanide (**126**) with α,β-unsaturated esters. Thus MeCH=CHCO$_2$Me gives (**127**; R^{1,2}=H, R^3=Me) via 1,4-addition and subsequent hydrolysis and cyclisation (Y. Ito, et al., Chem. Letters,1980, 487).

(126) (127)

The substituted derivative (**127**; R^1=CH$_2$CO$_2$H, R^2=NHCH(CO$_2$H)CH$_2$CH$_2$Ph) has been shown to be a potent inhibitor of angiotensin converting enzyme and a range of similar compounds has been

synthesised *via* condensation of 1-ethoxycarbonyl-
methyl-2,3,4,5-tetrahydro-1*H*-1-benzazepin-2,3-dione
with L-aminoesters and reduction of the resulting
3-imine (J.L. Stanton, *et al.*, J. med. Chem.,
1985, <u>28</u>, 1511; 1603).   The 1-benzazepin-4-one
(129)   is   the   major   product   resulting   from
rearrangement of (128) - the adduct of allene and
diphenylnitrone (J.J. Tufariello, Sk. A. Ali, and
H.O. Klingele, J. org. Chem., 1979, <u>44</u>, 4213).

(128)   (129)   (130)

A   number   of   3-substituted   3,4-dihydro-1*H*-1-
benzazepin-2,5-diones (130) have been prepared by
the thermal cyclisation of *N*-protected kynurenine
derivatives   (D.E.   Rivett,   and   F.H.C.   Stewart,
Austral. J. Chem., 1978, <u>31</u>, 439).
(ii)   *Reactions*.   The   tetrahydro   1-tosyl-2-
benzazepin-5-ones   (131)   have   been   efficiently
detosylated using either H_2SO_4/CH_2CO_2H or sodium
in   liquid   ammonia   and   the   products   N-alkylated
with several reagents (G.R. Proctor, *et al.*, J.
chem. Soc., Perkin I, 1974, 1828;   1973, 1984).
The   sodium   salts   of   the   β-ketoesters   (131;
$R=CO_2Me$) react with DMAD to give benzazonin-5-ones
(G.R. Proctor, *et al.*, J. chem. Soc., Perkin I,
1980, 1251).

(131)

1.2.4   2-Benzazepines

(a)   *Fully unsaturated.*   There are three effective new routes to the 1H-2-benzazepine system (132).   The 3-methoxy derivative (132; $R^{1,3,4,5}$=H, R^2=OMe) is produced in high yield by the photolysis of 2-azidonaphthalene in the presence of a high concentration of potassium methoxide in methanol.   The parallel reaction of 2-azidoanthracene gives 3-methoxy-1H-naphtho-[2,3-c]-azepine in almost quantitative yield. The use of amines as nucleophiles in this reaction is less successful but the 3-amino derivatives can be prepared from the methoxy compounds by heating them with amines (J. Rigaudy, C. Igier, and J. Barcelo, Tetrahedron Letters, 1975, 3845).   The phosphorus oxychloride induced cyclisation of the unsaturated amides (133) gives (132; $R^{1,3}$=H, R^2=NR_2 or CH_3, R^4=Ar) in good yield.

(132)   (133)

The products (132, R^2=NR_2) may be hydrolysed to give the 2,3-dihydro-1H-2-benzazepin-3-one system and the imine bond of (132; R^2=Me) may be reduced by sodium borohydride (G. Gast, J. Schmutz, and D. Sorg, Helv., 1977, 60, 1644).   The cyclisation of diene-conjugated nitrile ylides also provides a good route to 1H-1-benzazepines.   Species such as (134), generated by the action of base on imidoyl chlorides, cyclise at room temperature to give (135) which on thermal ring expansion gives (132; $R^{1,4,5}$=H, R^2=Ph, R^3=Me or Ph) (K.R. Motion, I.R. Robertson and J.T. Sharp, Chem. Comm., 1984, 1531).

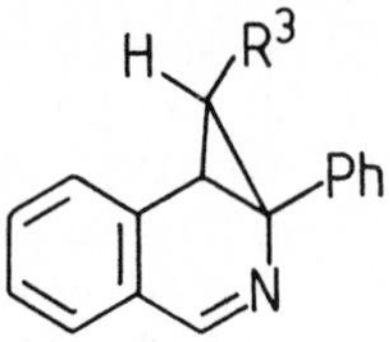

(135)

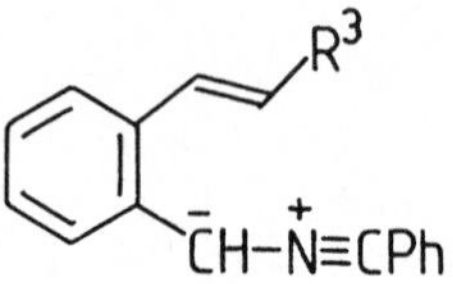

(134)

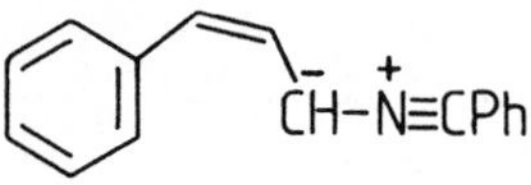

(136)

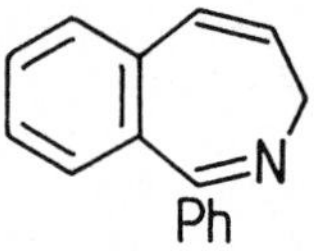

(137)

The nitrile ylide (136), generated by azirine photolysis, similarly gives the 3*H*-2-benzazepine system (137) and similar results are obtained using the naphthyl analogue (A. Padwa, J. Smolanoff, and A. Tremper, J. Amer. chem. Soc., 1975, 97, 4682). The reaction of 1-azirines with 1,3-diphenyl-2-indene and with phencyclone gives 3*H*-2-benzazepines and the analogous phenanthro fused azepines respectively. The latter (138) under the action of heat or strong base undergoes a hydrogen shift converting the 2*H*-azepine ring into the more stable 3*H*-isomer (139).

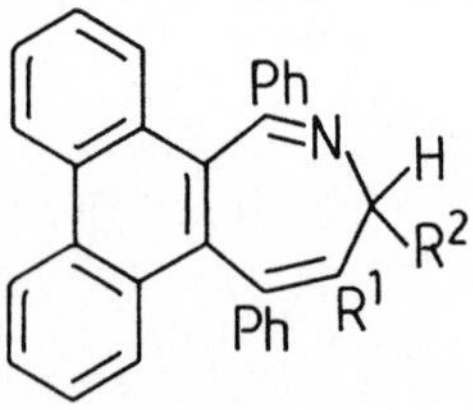

(138)

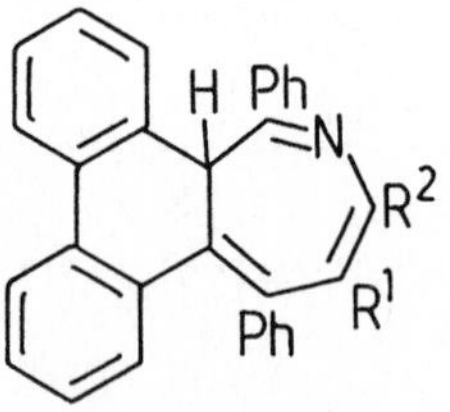

(139)

This reaction does not occur in compounds of type (137) because of the higher resonance energy of benzene (A. Hassner and D.J. Anderson, J. org. Chem., 1974, 39, 3070). A 3*H*-2-benzazepine is a minor product in the photolysis of an iso-quinolinium *N*-ylide (A. Lablache-Conbier and G. Surpateaunu, Tetrahedron Letters, 1976, 3081).

(b) *Dihydro-2-benzazepines*. A good synthetic route to the 3,4-dihydro-5*H*-2-benzazepine-5-one system (141) involves the 2-acetylenic benzophenone (140; R=NH$_2$) as a key intermediate. Its phthalimido precursor (140; R=phth.) is prepared by coupling the 2-iodoketone with propargylphthalimide in the presence of copper(I) iodide and a palladium catalyst. After removal of the phthaloyl group, hydration of the alkyne and concomitant cyclisation gives (141). The intermediate (140; R=NH$_2$) also serves as a precursor for the 2-benzazepine-4-one system (143) *via* reaction with pyrrolidine to give the enamine (142) followed by hydrolysis (E.J. Trybulski, *et al*., J. org. Chem., 1982, 47, 2441).

(140)

(141)

(142)

(143)

The naphthalenone (144) (D. Berney and K. Schuh, Helv., 1980, 63, 924) and the related 3-hydroxy-tetralone (145) (J. Henin, J. Likforman, and J. Gardent, Bull. Soc. chim. France, 1981, 456) both cyclise under basic conditions to give the 2,5-dihydro-2-methyl-1*H*-2-benzazepin-1-one system (146; R^1=CH$_2$Ph or -CH$_2$CH$_2$CONHMe).

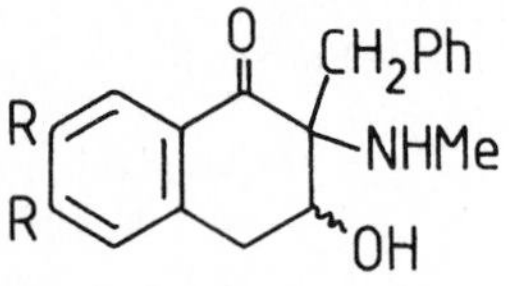

(144)

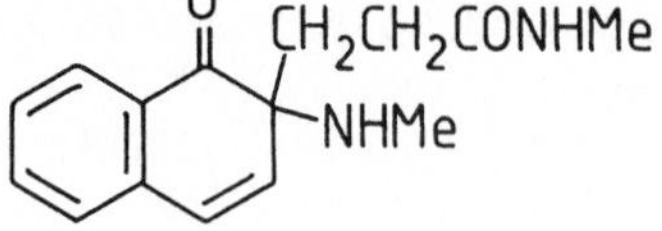

(145)

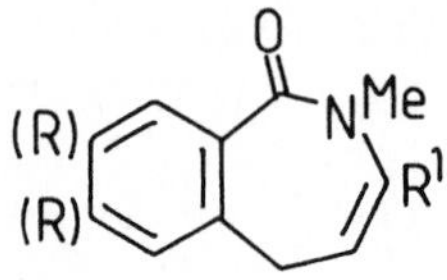

(146)

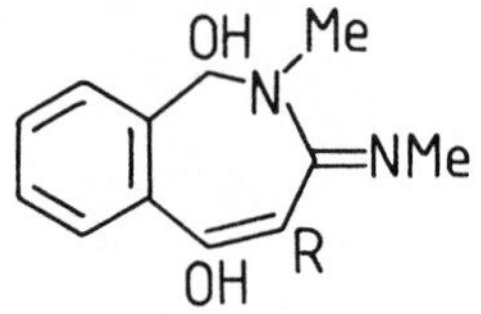

(147)

A similar ring expansion of a 2-methylamino-
naphthaquinone in the presence of methylamine
hydrochloride gives (147) (E.A. Korkhova, V.S.
Kuznetsov, and A.V. El'tsov, Zh. org. Khim., 1975,
11, 2140).

The dibenzylpropynylamine $(PhCH_2)_2NCH_2C\equiv CPh$
cyclises in the presence of triflic acid to give
an almost quantitative conversion into the $2H$-1,3-
dihydro-2-benzazepine (148; $R^{1,2,4,5}$=H, R^3=CH_2Ph,
R^6=Ph) (H. Takayama, T. Suzuki, and T. Nomato,
Chem. Letters, 1978, 865).

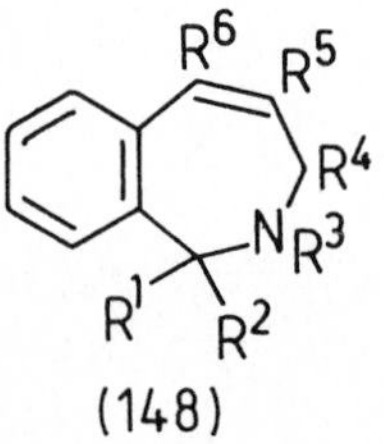

(148)

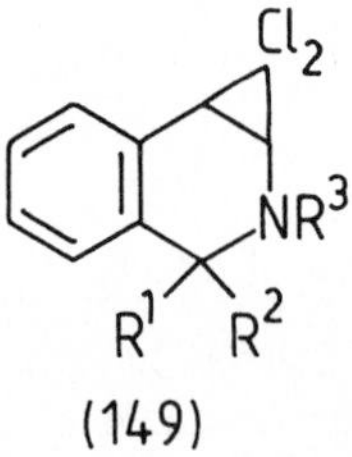

(149)

The same ring system (148; R^4=OR, R^5=Cl, R^6=H),
and its 1-oxo analogue (R^1/R^2=O), are produced by
the reaction of an alcohol (ROH) on the
dichlorcarbene adduct (149) of 1,2-dihydroiso-
quinolines and isoquinoline-1-ones respectively.
In the case where R^1/R^2=O and R^3=Me, thermolysis

and reduction of the primary product with lithium borohydride gives (148; R^1/R^2=O, R^3=Me, $R^{4,6}$=H, R^5=Cl) (C.D. Perchonock, et al., J. org. Chem., 1980, $\underline{45}$, 1950; H.P. Soetens and H.K. Pandit, Heterocycles, 1978, $\underline{11}$, 75; Rec. Trav. chim., 1980, $\underline{99}$, 271). The benzazepinone (148; R^1/R^2=O, $R^{3,4,5}$=H, R^6=Ph) is also produced by treatment of (150; $R^{1,4,5}$=H, R^2=OH, R^3=Ph) with sulphuric acid (K. Ackerman, D.E. Horning and J.M. Muchowski, Canad. J. Chem., 1972, $\underline{50}$, 3886).

1,3-Dihydro-2H-2-benzazepin-1,3-dione is prepared by hydrolysis of 3-chloro-1H-2-benzazepin-1-one. This dione and its N-methyl analogue give 'head to tail' dimers of the 4,5-bond on irradiation (M.S. Puar and B.R. Vogt, Tetrahedron, 1978, $\underline{34}$, 2887).

(c) *Tetrahydro-2-benzazepines*. Three useful new routes have been devised to the 2-benzazepin-1-one system (150). Cyclisation of the isocyanates (151; R^5=MeO, R^4=M,OMe) by treatment with phosphorus oxychloride and tin(IV) chloride give (150; $R^{1,2,3}$=H, R^4=H,OMe, R^5=MeO) (T. Fushimi, *et al.*, Heterocycles, 1979, $\underline{12}$, 1311).

(150) (151) (152)

The o-bromoamides o-$BrC_6H_4(CH_2)_3NHR$ react with carbon monoxide in the presence of $Pd(OAc)_2/PPh_3$ catalyst to give (150; R^1=R, R^{2-5}=H) (H. Mori. K. Chiba, and Y. Ban, J. org. Chem., 1978, $\underline{43}$, 1684), and the amino-lactones (152; R^1=H or Me) undergo spontaneous rearrangement in the absence of solvent to give (150; R^1=H or Me, R^2=OH, R^3=Ph, $R^{4,5}$=H) (K. Ackerman, D.E. Horning, and J.M. Muchowski, Canad. J. Chem., 1972, $\underline{50}$, 3886). The 2-benzazepin-3-one system (154; R=H, OMe) is produced by the acid catalysed cyclisation of the amido-ketones (153).

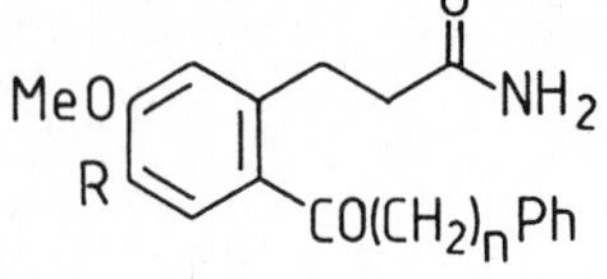

(153)

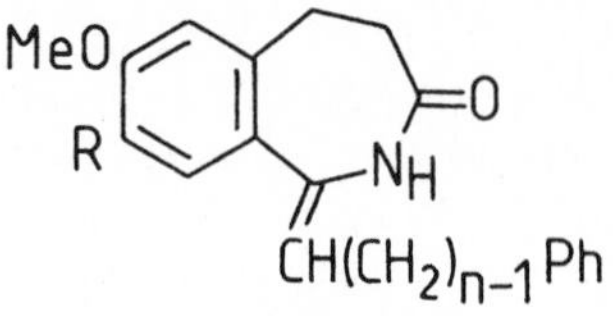

(154)

The products may be hydrogenated to give the 1-aralkyl derivatives and reduced with LAH to give the tetrahydro compounds (D. Berney and K. Schuh, Helv., 1975, _58_, 2228).

The photolytic ring expansion of *N*-alkyl-phthalimides (155) gives the 4-substituted 1,5-diones (157; $R^{1,2}$=H, R^3=R) *via* diradical intermediates of type (156) (Y. Kanaoka, *et al.*, Tetrahedron Letters, 1973, 1193; see also references 1-14 cited in J. Amer. chem. Soc., 1977, _99_, 7063, and Heterocycles, 1978, _10_, 165).

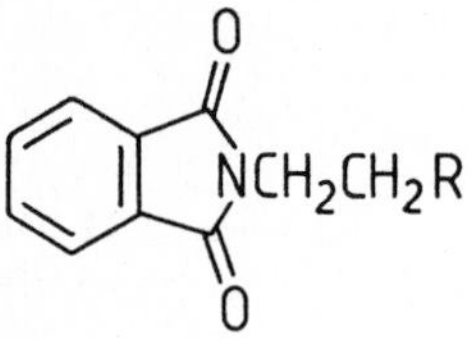

(155)

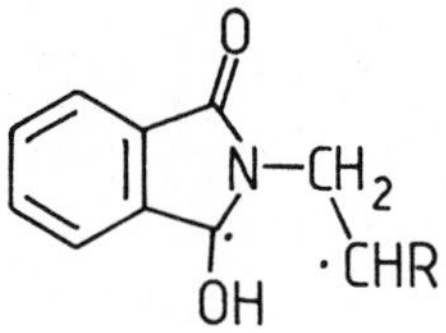

(156)

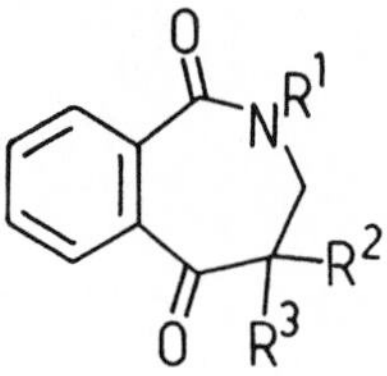

(157)

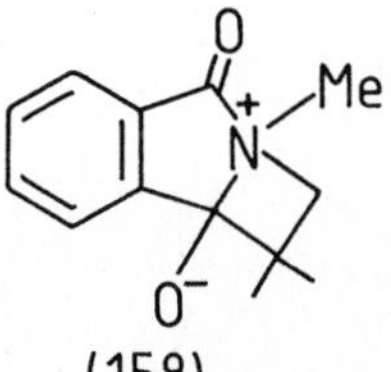

(158)

As an alternative to side-chain involvement it has been found that the photolysis of *N*-alkyl-phthalimides in the presence of alkenes and dienes also gives rise to 2-benzazepin-1,5-diones (e.*g.* 157; $R^{1,2,3}$=Me) from *N*-methylphthalimide and 2-methylpropene, probably *via* intermediates such as (158), but the reaction only gives good yields for alkenes with an ionisation potential of > 9 eV (P.H. Mazzocchi and M.J. Bowen, with N.K. Narain, J. Amer. chem. Soc., 1977, 99, 7063; with S. Minamikawa, J. org. Chem., 1978, 43, 3079).

1.2.5 3-Benzazepines
(a) *Fully unsaturated.* The $1H$-3-benzazepines (160; R^1=Me or Ph, R^2=CO$_2$Et, $R^{3,4}$=H) are formed in 20 and 35% yields respectively, together with other products, in the thermolysis of the azidocinnamates (159; R^1=CH$_2$CH=CH$_2$ or CH=CHPh) (D.M.B. Hickey, C.J. Moody, and C.W. Rees, Chem. Comm., 1982, 1419). The imino-ether (160; R^1=OEt, R^2=H, R^3=Ph, R^4=OMe) is prepared by the reaction of (163) with triethyloxonium fluoroborate (D. Berney and K. Schuh, Helv., 1981, 64, 373). On reaction with amines it gives the 2-amino analogues (R^1=NR$_2$)

(159)   (160)

3-Methylsulphonyl-3H-3-benzazepine is prepared in low yield by the action of the methylsulphinyl-methyl carbanion (CH$_3$SOCH$_2$$^-$) on isoquinoline N-oxide (I. Takeuchi, *et al.*, Chem. Letters, 1976, 519). A new class of isoquinoline alkaloids, the fully unsaturated indeno[2,1-a]-3- benzazepines have been discovered and several structures previously thought to be spirobenzyliso-quinolines have been reclassified (M. Shamma, *et al.*, Tetrahedron Letters, 1981, 22, 3127, 3131, 3135, 3139 and 3143). Transformations of spiroisoquinolines into indeno[2,1-a]-3-benzazepines had been reported previously (T. Kametani, J. chem. Soc., Perkin I, 1975, 2028).
(b) *Dihydro-3-benzazepines.* In a new synthetic approach to the system, the 1,2-dihydro-3H-3-

benzazepine (162) has been prepared by the acid catalysed cyclisation of either the epoxide (161; R,R=O) or the diol (161; R=OH) derived from laudanosine (T. Kametani, J. org. Chem., 1976, 41, 2988).

$$CH_2CH_2N(Me)CO_2CH_2Ph$$

(161)

(162)

A 4,5-dihydro-3H-3-benzazepine has been isolated as one of the products obtained from the treatment of adrenaline with strong hydrochloric acid (J.E. Forrest, R.A. Heacock, and T.P. Forrest. Chem. Ind., 1974, 498). The 3-benzazapine-2-one (163) is prepared in good yield by the DDQ oxidation of (164; R^1=H, R^2=Ph, R^3=MeO) (D. Berney and K. Schuh, Helv., 1981, 64, 373).

(163)

(164)

(c)   *Tetrahydro-3-benzazepines.*   The cyclisation of β-hydroxyethylamines of the type $ArCH_2CH_2NHCH_2CH(OH)R$ is a well established route to 3-benzazepines; such precursors can be conveniently prepared by reduction of 2-substituted 2-oxazolines with diborane (L.N. Pridgen, L.B. Killmer, and R.L. Webb, J. org. chem., 1982, 47, 1985). A ring closure of the same type has been used to prepare the 3-benza-zepin-2-one (164; $R^{1,4}$=H, R^2=Ph, R^3=MeO) by the PPA induced cyclisation of $3,4$-diMeOC$_6$H$_3$CH$_2$CONHCH$_2$CH(OH)Ph (D. Berney and K. Schuh, Helv., 1981, 14, 373). Similar benzazepinones have been prepared by intramolecular amidoalkylation e.g. (164; R^1=NHCO$_2$Me, $R^{2,3,4}$=H) from PhCH$_2$CH$_2$NHCOCH(NHCO$_2$Me)$_2$ (D. Ben Ishai, N. Peled, and I. Sataty, Tetrahedron Letters, 1980, 1085). The same system (164; $R^{1,2,3}$=H, R^4=MeO) can also result from a remarkable photocyclisation reaction of the *N*-chloroacetyl derivative of *m*-methoxy-phenylethylamine (*m*-MeO-C$_6$H$_4$CH$_2$CH$_2$NHCOCH$_2$Cl), the course of the reaction is however strongly dependent on the nature of the substituents and the solvent (O. Yonemitsu *et al.*, Tetrahedron Letters, 1974, 1169; Heterocycles, 1976, 4, 1095, 1371; Chem. pharm Bull., 1975, 2584).

The reduction of quaternary ammonium salts of 1-aroyl-dihydroisoquinolines is an established route to tetrahydro-3-benzazepines; further work has been done on the mechanism of this reaction (C. Rély and J. Gardent, Bull. Soc. chim. France, 1977, 893) and several new methods involving isoquinoline ring expansion have emerged. Thus the 1-(α-hydroxylbenzyl) compounds (165; R=H or OMe, Ar=3,4-diMeOC$_6$H$_3$) on heating in formic acid are converted into (166; R^1=Ar, R^2=H, R^3=Me, R^4=MeO, R^5=H or MeO) (R.M. McMahon, C.W. Thornber, and S. Ruchirawat, J. chem. Soc., Perkin I, 1982, 2163).

(165)

(166)

(167)

(168)

The quaternary salts (167; R=H or Me, R^1=H, Me or Ph), on treatment with potassium *tert*-butoxide, rearrange to give (166; R^1=H, Me or Ph, R^2=p-$NO_2C_6H_4$, R^3=Me, R^4=H or MeO, R^5=MeO) (S. Smith, Jr., V. Elango, and M. Shamma, J. org. Chem., 1984, *49*, 581), and (168) reacts with Me_2SCHCO_2Et to give, after borohydride reduction, (166; R^1=H, R^2=CO_2Et, R^3=CH_2Ar, $R^{4,5}$=MeO) (H. Bierñugel, H.P. Soetens, and U.K. Pandit, Heterocycles, 1977, *7*, 37).

Alkylation of 1,2,4,5-tetrahydro-3-methyl-3*H*-3-benzazepin-2-one at the 1-position has been effected using sodium hydride in THF/DMF and various alkyl halides (K. Orito and T. Matsuzaki, Tetrahedron, 1980, *36*, 1017).

1.3 *Azepines with two annelated rings*

1.3.1 *Dibenzazepines*
(a) Dibenz[b,d]azepines. The tetrahydrodibenz-[b,d]azepinone (169; R=H or Et) is obtained in high yield by the photochemical cyclisation of (170) (I. Iida, Y. Yuasa, and C. Kibayashi, Tetrahedron Letters, 1978, 3817).

(169)   (170)

Unsubstituted 5,7-dihydro-6*H*-dibenz[b,d]-azepin-6-one (171) is formed by the trifluoromethane-sulphonic acid catalysed cyclisation of the phenylhydroxylamine derivative PhN(OH)COCH$_2$Ph (Y. Endo, *et al.*, Heterocycles, 1977, *8*, 367), and the analogous tetrahydro system (172) by the thermal cyclisation of the amino-ester (173) (M.N. Samini, U. Kraatze, and F. Korte, Heterocycles, 1976, 5, 73).

(171)   (172)   (173)

(b)   *Dibenz[b,e]azepines*.   Some   11-substituted 5,6-dihydro-11*H*-dibenz[b,e]azepin-6-ones   (174; R^1=H, R^2=CH(CO$_2$Et)$_2$,   CMe(CO$_2$Et)$_2$,   CMe$_2$NO$_2$)   have been synthesised *via* the Beckmann rearrangement of the corresponding anthrone oximes with phosphorus pentachloride   whereas   the   oxime   tosylates   are resistant to rearrangement (R.H. Williams and H.R. Snyder, J. org. Chem., 1973, 38, 809).

(174)   (175)

Oxidation of *N*-benzylmorphanthridene (175) by ruthenium oxide is selective for the ring benzyl group and gives (174; R^1=CH$_2$Ph, R^2=H) in good yield (G. Bettoni, *et al.*, Tetrahedron, 1981, 37, 4159).   The compound (176;   X=OTs) undergoes an interesting rearrangement to (177) on treatment with   methylamine.   The   reaction   goes   *via* transamidation in (176;   X=MeNH).   Reversal with concurrent reduction of the carbonyl group is induced by treatment with lithium aluminium hydride (R. Süss, Helv., 1979, 62, 1103).

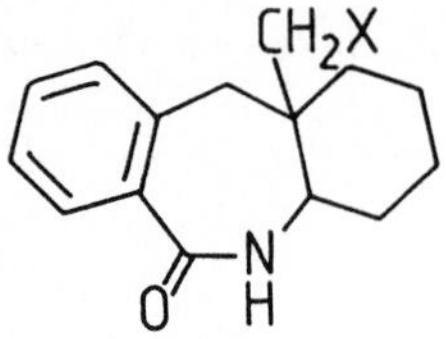

(176)

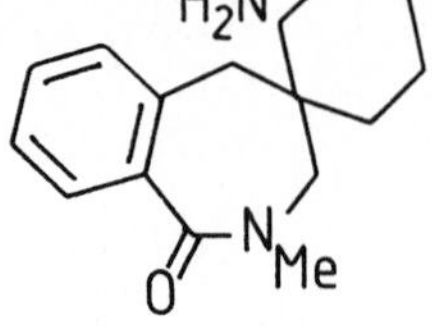

(177)

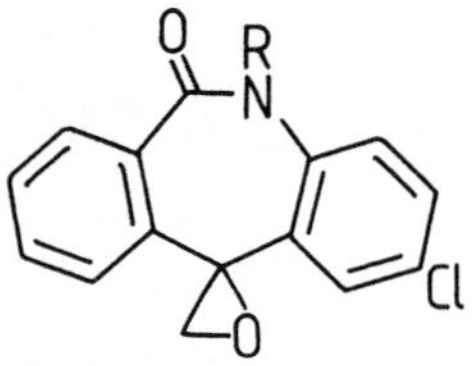

(178)

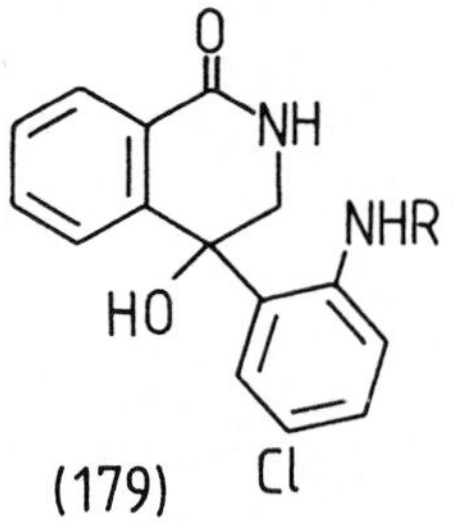

(179)

Several rearrangements of morphanthridenedione derivatives, such as (178) to (179) on treatment with ammonia, proceed *via* transannular attack on the lactam carbonyl group by substituents attached to the 11-position (G.M. Coppola, G.E. Hardtmann, and R.I. Mansukhani, J. org. Chem., 1975, 40, 3602).

(c) *Dibenzo*[b,f]*azepines*. The cyclisation of 2,2'-diformyldiphenylamine using hydrazine and acetic acid provides a new route to 5*H*-dibenz-[b,f]azepine (180) (A.K. Sinha, P.K. Agarwal, and S. Nizamuddin, Ind. J. Chem., 1982, 21B, 237).

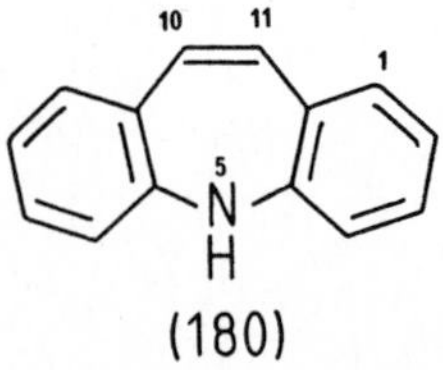

(180)

The 10-amino-10,11-dihydro-5-methyl derivative of (180) is obtained in high yield by the cyclisation of 2-MeC$_6$H$_4$NMeC$_6$H$_4$-2CN in the presence of lithium or sodium amide, followed by hydrogenation (Ger. Pat., 2,345,972 (1974), Chem. Abs., 1974, 81, 3785t).

N-Alkylation of (180) and its 10,11-dihydro

analogue can be carried out under mild conditions using phase-transfer catalysis (I. Gozlan, *et al.*, J. heterocyclic Chem., 1982, <u>19</u>, 1569; E. Hannig, R. Peck, and C. Dressler, Pharmazie, 1979, <u>34</u>, 670). The reaction of (180) and its 10,11-dihydro analogue with 2-3 moles of *n*-butyl-lithium gives the 4,5-di-lithiated derivative which reacts with various electrophilic reagents to give 4-substituted products (T. Dahlgren, *et al.*, J. heterocyclic Chem., 1983, <u>20</u>, 341). The 10,11 diarylethene type double bond in (180) and analogous oxepins can be reduced conveniently and in high yield using magnesium in ethanol (J.A. Profitt, and H.H. Ong, J. org. Chem., 1979, <u>44</u>, 3972). Irradiation of a range of *N*-acyl, -aroyl, -carbamoyl and -ethoxycarbonyl derivatives of (180) has been found to give $(2+2)\pi$ dimers of the 10,11-double bond in good yield (L.J. Kricka, M.C. Lambert, and A. Ledwith, J. chem. Soc., Perkin I, 1974, 52). Friedel-Crafts acylation of the *N*-methyl derivative of (180) gives the 2,8-diacylated product but for *N*-acyl derivatives the reaction occurs only at the 10(11) position (L.J. Kricka and A. Ledwith, J. chem. Soc., Perkin I, 1973, 859).

(d) *Dibenz[c,e]azepines*. The parent molecule (181; $R^{1,2}$=H) has been prepared by the base-induced elimination of HX from (183; R^1=H, X=CH$_3$SO$_2$ or NO$_2$). The precursors are synthesised by the reaction of 2,2-bis(bromomethyl)biphenyl with for example methanesulphonamide. It reacts at nitrogen with alkyl halides and acid anhydrides to give stable salts (R. Kreher and W. Gerhardt, Ann., 1981, 240; Angew. Chem., intern. Edn., 1975, <u>14</u>, 265).

(181)   (182)   (183)

(184)   (185)

Several ingenious routes to (181; R²=Ar), capable of producing compounds with various substitution patterns, involve as the final step, the reaction of the benzophenone derivatives (182; X=H or NMe₂) with N-bromosuccinimide or cyanogen bromide to give the benzylic bromides which are cyclised with ammonia (H.W. Gschwend and A. Hamdan, J. org. Chem., 1982, 47, 3652; with J.J. Fitt, S.K. Boyer, and H.H. Haider, ibid., 1982, 47, 3658). Several other 5-substituted derivatives (181; R¹=H, R²=CN or CO₂H) and their N-oxides have been prepared (R. Kreher and H. Morgenstern, Chem. Ztg., 1983, 107, 70). The N-hydroxy compounds (183; R¹=H or Ph, X=OH) are prepared by oxidation of (183; X=CH₂CH₂Ph) to give the N-oxide followed by a Cope elimination. Oxidation of (183; R¹=H or Ph, X=OH) with hydrogen peroxide gives the cyclic nitrones (184; R¹=H or Ph). Reaction of (184; R¹=H) with phenylmagnesium bromide gives (183; R¹=Ph, X=OH) (R. Kreher and H. Pawelczyk, Z. Naturforsch., 1974, 29B, 425). The nitrone (184; R¹=H) is also produced by treatment of

(183; X=OCH$_2$Ph) with hydrogen peroxide (R. Kreher and H. Morgenstern, Ber., 1982, 115, 2679). Addition of carbon disulphide to the nitrone (184; R^1=H) gives the thiolactam (185) which on alkylation at sulphur gives the thiolactim ether (R. Kreher and H. Morgenstern, Z. Chem., 1982, 22, 258; Tetrahedron Letters, 1980, 2141). The 6-mesyl-6,7-dihydro-5H-dibenz[c,e]azepine (183; X=SO$_2$Me, R^1=H) and various derivatives with substituents (Me, OMe, Cl) in one aryl ring have been prepared by Pschorr cyclisations (J.L. Huppatz, Austral. J. Chem., 1973, 26, 1307).

2   *Seven-membered rings containing two nitrogen atoms*

2.1   *1,2-Diazepines*

2.1.1   *Monocyclic 1,2-diazepines*
(a)   *1H-1,2-Diazepines*. (i) *Synthesis*. The photochemical ring expansion of pyridine *N*-imides (186) provides the most versatile entry to this ring system (188) (see 2nd edn.) and has continued to generate interesting chemistry including work on the influence of substituents on the course of the reaction. This whole area, up to *ca.* 1980 has been thoroughly reviewed (J. Streith, Heterocycles, 1977, 6, 2021; V. Snieckus and J. Streith, Acc. chem. Res., 1981, 14, 348).

(186)   (187)   (188)

(189)   (190)

1-Methyl-3,5,7-triaryl-1H-1,2-diazepines have been

prepared by the reaction of thiapyrylium salts with methylhydrazine, and by methylation of the analogous triaryl-4H-diazepines (200; $R^{1,4,6}$=Ar, $R^{2,3,5}$=H). Unfortunately the former reaction seems to be of limited scope as the use of phenylhydrazine gives only pyrazolines (D.J. Harris, Canad. J. Chem., 1974, 52, 2798). The acylation of 4H-1,2-diazepines (200; $R^{1,4,6}$=Ar, $R^{2,3,5}$=H) provides a good route to a range of 1-acyl-3,5,7-triaryl-1,2-diazepines (188; $R^{1,3,5}$=Ar, $R^{2,4}$=H, X=acyl) (D.J. Harris, et al., Synthesis, 1975, 603). The acylation of the iron tricarbonyl complex of (188; Y=H), followed by decomplexation gives access to a range of N substituted 1H tautomers not available by direct synthesis (D.J. Harris and V. Snieckus, Chem. Comm., 1976, 844).

(ii) *Reactions.* The chemistry of the 1H-tautomers has been explored extensively. Full details have been published of the photochemical ring closure of the butadiene moiety in (188) (J.-P. Luttringer, N. Pérol, and J. Streith, Tetrahedron, 1975, 31, 2435) and it has been shown that for the 4- and 6-chloro-1-benzoyl derivatives (188; Y=COPh, R^2 or R^4=Cl, $R^{1,3,5}$=H) that the photochemical 1,7 shift of the benzoyl group is a minor reaction compared to the formation of the bicyclic isomer (M. Martigneaux, C. Strehler, and J. Streith, Tetrahedron Letters, 1983, 24, 3327). Further work on the thermolysis of these systems has revealed a new reaction path which has provided the first synthesis of fully unsaturated 1,3-diazepines (190) (30-70%). The path is favoured only when electron donating substituents are present in the 4- and/or 6-position (e.g. 188; R^2 or R^4=MeO, Y=COR). These favour N-N cleavage in the first intermediate (187) leading to a walk rearrangement to give (189) and subsequent ring expansion (J.A. Moore, et al., J. org. Chem., 1979, 44, 2683; T. Tsuchiya, Chem. pharm. Bull. Japan, 1981, 29, 3688, 3696).

[1]H Nmr studies on protonation have shown that the 1H compound (188; Y=Me, $R^{1,3,5}$=Ar, $R^{2,4}$=H) and the 4H compound (200; $R^{1,4,6}$=Ar, $R^{2,3,5}$=H) are converted into (191; R=Me and H

respectively). Vigorous acid hydrolysis leads to ring cleavage in both and the formation of pyrazole and pyridine derivatives (D.J. Harris, et al., Canad. J. Chem., 1974, 52, 2805).

(191)

The selective reduction of the C=C and C=N bonds in (188) and its iron tricarbonyl complex has been studied (J. Streith and B. Willig, Bull. Soc. chim. France, 1973, 2847;   T. Tsuchiya and V. Snieckus, Canad. J. Chem., 1975, 519).

Cycloaddition reactions of the compounds (188) have received much attention.   They give Diels-Alder reactions with 4-phenyl-1,2,4-triazoline-3,5-dione (G. Kiehl, J. Streith and G. Taurand, Tetrahedron, 1974, 30, 2851), singlet oxygen (T. Tsuchiya, H. Arai, H. Hasegawa, and H. Igeta, Chem. pharm. Bull. Japan, 1977, 25, 2749), and undergo acid catalysed Diels-Alder dimerisation (B. Willig and J. Streith, Tetrahedron Letters, 1973, 4167).   They also react as a 2π-component, via addition to the 4,5-bond, with diazoalkanes (G. Kiehl, J. Streith and G. Taurand, loc. cit., J.R. Frost and J. Streith, J. chem. Soc., Perkin I, 1978, 1297); the TOSMIC anion (D. Harris, S. Syren, and J. Streith, Tetrahedron Letters, 1978, 4093); cyclopentadienones (Chem. Letters, 1975, 423); chloranil, 2-pyrones and tetrachlorocyclopentadiene (K. Saito, S. Iida, and T. Mukai, Heterocycles, 1982, 19, 1197;   Bull. chem. Soc. Japan, 1984, 57, 3483);   and with tetrazines (G. Seitz, T. Kämpchen, and W. Overheu, Arch. Pharm., 1978, 311, 786).   However, nitrile oxides (J. Streith, G. Wolff, and G. Fritz, Tetrahedron, 1977, 33, 1349) and ketenes add first to the imine bond in stepwise reactions (J. Streith and G.

Wolff, Heterocycles, 1976, 5, 471; J. Streith *et al.*, Ann., 1983, 1361, 1374, 1393). The thermal and photochemical reactions of the azetido-diazepines resulting from ketene addition have been studied (J. Streith and T. Tschamber, with G. Kiehl, Ann., 1983, 2141; and with H. Strub, Canad. J. Chem., 1984, 62, 2440). 1*H*-1,2-Diazepines (188) form iron tricarbonyl complexes as 4π donors; these compounds are useful in synthesis (V. Snieckus and J. Streith, Acc. chem. Res., 1981, 14, 348).

(b) 3H-1,2-*Diazepines.*   (i) *Synthesis.*   This system (193/194) can be prepared by two routes. The first and more versatile synthesis involves the 1,7-electrocyclisation of diene-conjugated diazo compounds (192).   In some cases, for example when R^5=Ar, the ring closure giving (193) is followed by a [1,5] sigmatropic hydrogen shift to give (194).   Such shifts are rapid at room temperature and so the ratio of (193) to (194) obtained reflects their relative thermodynamic stability.   Cyclisation only takes place if (192) has a *cis* hydrogen atom at the diene terminus, other groups, for example Me or Ph, obstruct the cyclisation transition state and the diazo compound then reacts *via* a 1,5-electrocyclisation to give pyrazoles (I.R. Robertson and J.T. Sharp, Tetrahedron, 1984, 3095 and 3113).

(192)   (193)   (194)

(195)

The second method involves the elimination of
p-toluenesulphinic acid from the dihydro-1,2-
diazepines (195; R=Ts) by reaction with sodium
ethoxide (C.D. Anderson, J.T. Sharp, and R.S.
Strathdee, J. chem. Soc., Perkin I, 1979, 2209;
C.B. Argo and J.T. Sharp, *ibid.*, 1984, 1581).
(ii) *Reactions.* 3H-1,2-Diazepines undergo
photochemical electrocyclisation of the
diazabutadiene moiety to give for example (196)
from (193) (C.D. Anderson and J.T. Sharp, J. chem.
Soc., Perkin I, 1980, 1230). They show a rapid
thermal [1,5] hydrogen shift which interconverts
the two isomers (193/194) even at room temperature
or below (C.D. Anderson, J.T. Sharp, and R.S.
Strathdee, J. chem. Soc., Perkin I, 1979, 2209;
C.B. Argo and J.T. Sharp, *ibid.*, 1984, 1581). At
higher temperature they undergo ring contraction
to give 3-vinyl-3H-pyrazoles (e.g. 197) from
(193/194; $R^{1,3,5}$=Me, $R^{2,4}$=H), which rearrange
further *via* migration of the vinyl group and
hydrogen (C.D. Anderson, J.T. Sharp, and R.S.
Strathdee, J. chem. Soc., Perkin I, 1979, 2730;
I.R. Robertson and J.T. Sharp, Tetrahedron, 1984,
<u>40</u>, 3095).

(196)   (197)

Oxidation with MCPBA gave the 2-oxide which in one
case has been found to rearrange to give a 3-
alkenyl-3H-pyrazole-2-oxide at room temperature
(C.B. Argo, I.R. Robertson, and J.T. Sharp, J.
chem. Soc., Perkin I, 1984, 2611). 3H-1,2-
Diazepines form dinuclear iron hexacarbonyl
complexes involving only the azo-group, in
contrast to the 1H- isomers (188) which give diene
complexes (C.B. Argo and J.T. Sharp, J. chem.
Soc., Perkin I, 1984, 1581).

(c) 4H-1,2-*Diazepines*.   (i) *Synthesis*.   The
thermal   or   photochemical   rearrangement   of   3,4-
diazanorcaradienes   (198)   provides   an   effective
route to this system (200)   Both proceed *via* a
walk   rearrangement   to   give   (199)   followed   by
valence   isomerisation   (H.D.   Fühlgruber   and   J.
Sauer,   Tetrahedron   Letters,   1977,   4393,   4398;
H.E.   Zimmerman   and   W.   Eberbach,   J.   Amer.   chem.
Soc.,   1973,   <u>95</u>,   3970;   H.   Kolbinger,   G.
Reissenweber,   and   J.   Sauer,   Tetrahedron   Letters,
1976, 4321).

(198)   (199)   (200)

The   thermal   rearrangement   product   of   (198;
$R^{1,3,4,5,6}$=Ph, R^2=H),   previously   thought   to   be   a
5H-1,2-diazepine   or   a   diazetopyrrole   (see   2nd
edn.) has been shown by X-ray structural analysis
to be (200; $R^{1,3,4,5,6}$=Ph, R^2=H) (J.N. Brown,
R.L.   Towns,   and   L.M.   Trefonas,   J.   Amer.   chem.
Soc.,   1970,   <u>92</u>,   7436),   presumably   formed   by   the
mechanism shown.   The major route to this system
(200), the reaction of hydrazine with pyrylium and
thiapyrylium   salts,   has   been   further   investigated
(D.J.   Harris,   *et al.*,   Canad.   J.   Chem.,   1974,   <u>52</u>,
2798;   with T. Tschamber,   *ibid.*,   1980,   <u>58</u>,   494).
(ii)   *Reactions*.   4H-1,2-Diazepines   (200)
undergo   photochemical   cyclisation   of   the
azabutadiene unit to give for example (201) from
(200; $R^{1,4}$=Ph, $R^{2,3}$=H)   (D.J.   Harris,   *et al.*,
Canad.   J.   Chem.,   1977,   <u>55</u>,   56),   and   the   products
of such reactions (e.*g.* 201) can undergo a further
photo-induced di-π-methane   rearrangement   to   give
1,4-diazepines   (202)   (G.   Reissenweber   and   J.
Sauer, Tetrahedron Letters, 1977, 4389).

(201)   (202)   (203)   (204)

The reaction path followed in the thermolysis of these 4H- tautomers is, as in the 1H analogues, dependent on the nature and position of the ring substituents. The first step is a valence tautomerisation (e.g. 200 → 199), followed by either a C → C walk rearrangement, for example when $R^{4,5}$=H, $R^{2,3}$=Me, $R^{1,6}$=Ph, to give (198) which rearranges further to give 4-isopropyl-3,6-diphenylpyridazine (H.E. Zimmerman and W. Eberbach, J. Amer. chem. Soc., 1973, 95, 3970); or by a C → N walk, for example when $R^{1,4,6}$=Ph, $R^{2,3,5}$=H, to give (203) and thence (200; $R^{1,4,5}$=Ph, $R^{2,3,6}$=H) (V. Snieckus et al., Canad. J. Chem., 1977, 55, 56).

Protonation and acylation of the 4H tautomers have been referred to above (section 2.1.1(a) (ii)). The triaryl derivatives (200; $R^{1,4,6}$=Ar, $R^{2,3,5}$=H) are deprotonated by LDA in the presence of TMEDA and the derived anions react with a variety of electrophilic reagents to give 4-substituted derivatives (L. Bemi, M.T. Thomas, and V. Snieckus, Synthesis, 1979, 130).

(d) Dihydro-1,2-diazepines. (i) Synthesis. $\alpha\beta$,$\gamma\delta$-Unsaturated ketones (e.g. $R^1COC(R^2)=C(R^3)C(R^4)=CHR^5$) react with various substituted hydrazines ($RNHNH_2$) under acid conditions to give 2H-3,4-dihydro-1,2-diazepines (195) (P.N. Anderson and J.T. Sharp, with C.D. Anderson, J. chem. Soc., Perkin I, 1979, 1640; with C.B. Argo, ibid., 1981, 2761). Most of the other new routes to dihydro-1,2-diazepines involve rearrangements of other heterocyclic systems. The photochemical ring expansion of the 2-methoxypyridinium N-imide (186; Y=COPh, R^1=MeO, $R^{2,4}$=H) gives (188;

Y=COPh, R^1=OMe) regiospecifically, and its subsequent reaction with trimethylsilyl chloride gives 1-benzoyl-1,2-dihydro-1,2-diazepin-3-one (T. Kiguchi, *et al.*, J. org. Chem., 1980, *45*, 5095). Photochemical rearrangement of the 2,3-diazabicyclo[4.1.0]heptenone (204) gives 5,6-diphenyl-1,5-dihydro-1,2-diazepin-4-one in a reaction resembling the conversion (198 → 200) (A. Nabeya, K. Kurita, and J.A. Moore, J. org. Chem., 1973, *38*, 2954). Acetylation of (204) gives (188; Y=Ac, $R^{2,3}$=Ph, R^4=OAc, $R^{1,5}$=H) which on hydrolysis is converted into 5,6-diphenyl-2,3-dhydro-1,2-diazepin-4-one. An earlier route to 4,5-dihydro-1,2-diazepines *via* the ring expansion of pyrazolines produced by adding diazo compounds to cyclobutenes, has now been extended to give a good general synthesis of 4,5-dihydro-1*H*-1,2-diazepin-4-ones *via* the use of cyclobutenones (H.-D. Martin *et al.*, Tetrahedron Letters, 1983, *24*, 5469). The thermolysis of (205) leads to the extrusion of $(CF_3)_2C=CH_2$ and the formation of (206; R=Me or Ph) (K. Burger, H. Schickaneder, and C. Zettl, Ann., 1982, 1741).

(205)

(206)

(207)

(208)

The pyrolytic valence tautomerisation of (207) gives (208 (H.-D. Martin, H. Hockstetter, and A. Steigel, Tetrahedron Letters, 1984, *25*, 297).
(ii) *Reactions.* The azadiene moiety in 2,4-dihydro-3*H*-1,2-diazepin-3-ones undergoes photo-

chemical cyclisation (*cf.* the formation of 201 from 200) (J.J. Koenig and C.-G. Wermuth, Compt. Rend., 1977, 284C, 245). Acylation studies have been carried out on 2,3-dihydro-1-acyl-1,2-diazepines and their iron tricarbonyl complexes (T. Tsuchiya and V. Snieckus, Canad. J. Chem., 1975, 53, 519), and on 2,3- and 1,5-dihydro-1,2-diazepin-4-ones (J.A. Moore, *et al.*, J. org. Chem., 1973, 38, 2939).

(e) *Tetrahydro-1,2-diazepines.* The tetrahydro system (209; R^1=Ar) has been obtained by reaction of δ-halogeno aryl ketones (ArCO(CH$_2$)$_4$Cl) with substituted hdyrazines (RNHNH$_2$) (J.J. Koenig and C.-G. Wermuth, Tetrahedron Letters, 1973, 603) and by LAH reduction of the 3-oxo derivatives (210). The cyclodehydration of 5-ketoacid hydrazones has been used to prepare 1,2-diazepin-3-ones (210) for example from $R^1COCHR^2CHR^3CHR^4CO_2H$ and RNHNH$_2$ (C.-G. Wermuth and J.J. Koenig, Angew. chem., intern. Edn., 1972, 11, 152). Similarly the hydrazone PhNHN=C(CN)(CH$_2$)$_3$CO$_2$H, generated by the reaction of 2-cyanocyclopentanone with phenyl diazonium chloride, gives (210; R^1=CN, R=Ph, $R^{2,3,4}$=H) (T.A.N. Trinh, M. Lamant, and H. Quiniou, Compt. Rend., 1984, 299C, 929).

(209)   (210)

Many halogeno derivatives of tetrahydro-1,2-diazepines have been prepared (M. Nastasi, Heterocycles, 1976, 4, 1509; C.-G. Wermuth, Compt. Rend., 1978, 286C, 671), usually with the objective of subsequent dehydrohalogenation. For example the reaction of (209; R=COPh, R^1=Ph) with NBS followed by treatment with pyridine gives (209; 4,5-dehydro), further bromination takes place at the 6-position but this compound cannot be dehydrobrominated to give the fully unsaturated system (J.J. Koenig and C.-G. Wermuth, Tetrahedron Letters, 1973, 603). Dehydrobromination is often accompanied by ring contraction to give diazanorcaradienes (J.J. Koenig and C.-G. Wermuth, Tetrahedron, 1974, $\underline{30}$, 501; Heterocycles, 1975, $\underline{3}$, 15).

2.1.2 1,2-*Benzodiazepines*
(a) 1H-1,2-*Benzodiazepines*. (i) *Synthesis*. This system (211) can be synthesised *via* the cyclisation of diene-conjugated nitrilimines (212; R^1 or R^2=H).

(211) (212) (213)

When the reaction is carried out at 80°C the diazepines are obtained in good yield, but at room temperature the products are cyclopropa[c] cinnolines (213) (L. Garanti and G. Zecchi, J. chem. Soc., Perkin I, 1977, 2092; with L. Chiodini, Synthesis, 1978, 603; and G. Testori, *ibid.*, 1979, 380; with L. Bruché, J. chem. Soc., Perkin I, 1983, 539; with A. Alemagna and E. Licandro, Tetrahedron, 1984, $\underline{40}$, 2165). The latter (213), when R^1 or R^2=H, are converted into the benzodiazepines (211) on heating (A. Padwa and S. Nahm, J. org. Chem., 1981, $\underline{46}$, 1402). 1H-1,2-Benzodiazepines (211) have also been obtained by the photochemical ring expansion of quinoline N-imides (*cf.* the analogous conversion 186 → 188) (T. Tsuchiya, J. Kurita, and V. Snieckus, J. org. Chem., 1977, $\underline{42}$, 1856). This method has been extended to the synthesis of 1H-1,2-diazepines fused to other carbocyclic and heterocyclic rings (T. Tsuchiya, with H. Sashida, Chem. pharm. Bull. Japan, 1981, $\underline{29}$, 1747; with J. Kurita, *ibid*, 1979, $\underline{27}$, 2183; with M. Enkaku and H. Sawanishi, Heterocycles, 1979, $\underline{12}$, 1471). 3,4-Disubstituted

1*H*-1,2-benzodiazepines (211) have also been prepared by the addition of alkynes to diazo-tropylidene followed by a series of rearrangements (K. Saito, Chem. Letters, 1983, 463).
(ii) *Reactions.* The products obtained on acylation of 3-methyl-1*H*-1,2-benzodiazepines (211; R^3=Me, $R^{1/2,4}$=H) depend on the presence of substituents in the benzene ring. Acylation in benzene takes place at N-2 to give the 2-acyl-3-methylene derivatives as the major products, plus some quinoline *N*-imides. However, in cases where the diazepine has an electron donating group (Me or MeO) at the 7-position the intermediate 2-acyl salt also rearranges by a minor reaction path to give 3*H*-1,3-benzodiazepines (*e.g.* 240). Analogous thienodiazepines give the same rearrangement as the major reaction when benzene is used as solvent, but in pyridine acylation occurs at N-1 (J. Kurita, M. Enkaku, and T. Tsuchiya, Chem. pharm. Bull. Japan, 1983, <u>31</u>, 3684). The oxidation of 1*H*-1,2-benzodiazepines (211) with LTA provides the first example of a 5*H*-1,2-benzodiazepine (214); these compounds react with either acetic acid or methanol *via* 1,4-addition to give (215; R^1=H, X=H, R^4=OAc, R^2=OMe or OAc) (T. Tsuchiya and J. Kurita, Chem. pharm. Bull. Japan, 1980, <u>28</u>, 1842).

(214)   (215)

The unsubstituted compound and the 5-methyl derivative (211; $R^{1/2,3}$=H, R^4=H or Me) on reduction with LAH or sodium borohydride give the 2,3-dihydro derivatives (T. Tsuchiya, J. Kurita, and V. Snieckus, J. org. chem., 1977, <u>42</u>, 1856).
(b) 3H-1,2-*Benzodiazepines.* (i) *Synthesis.* Two routes to this system (216) are now available.

$$R^3 \quad R^2$$

(216)

The first, but of limited application, is *via* the 1,7-electrocyclisation of diene-conjugated diazo-compounds with aromatic $\gamma\delta$ unsaturation (192; $R^4/R^5 =(CH=CH)_2$). In cases where the competing 1,5-cyclisation is disfavoured, *e.g.* when R^1/R^2 or $R^2/R^3 =(CH_2)_3$, this reaction gives good yields of $3H$-1,2-benzodiazepines (216) with an annelated cyclopentane ring (J.T. Sharp with R.H. Findlay, and P.B. Thorogood, J. chem. Soc., Perkin I, 1975, 102; with K.L.M. Stanley, J. Dingwall, and T.W. Naisby, *ibid.*, 1979, 1433; with T.K. Miller, H.R. Sood, and E. Stefaniuk, J. chem. Soc., Perkin II, 1984, 823). This route also gives access to analogous thieno-1,2-diazepines (T. K. Miller and J.T. Sharp, J. chem. Soc., Perkin I, 1984, 223). The $3H$ tautomers (216) can also be obtained from the $1H$ tautomers (211) *via* reduction of the imine bond with LAH and dehydrogenation of the resulting NHNH moiety to give the azo group of (216) (T. Tsuchiya and J. Kurita, Chem. pharm. Bull. Japan, 1978, <u>26</u>, 1890, 1896; with H. Sawanishi, *ibid.*, 1979, <u>27</u>, 2188).
(ii)   *Reactions*. $3H$-1,2-Benzodiazepines (216) undergo ring contraction to give 3-vinylindazoles under both thermal and photochemical conditions, this parallels the change 193 → 197, (thermal reactions of the monocyclic analogues are discussed in section 2.1.1 (b)(ii)) (J.T. Sharp, *et al.*, J. chem. Soc., Chem. Comm., 1974, 532; T. Tsuchiya and J. Kurita, Chem. pharm. Bull. Japan, 1978, <u>26</u>, 1890; 1979, <u>27</u>, 2528; T. Tsuchiya, M. Enkaku, and H. Sawanishi, Heterocycles, 1979, <u>12</u>, 1471). They are less stable than the $1H$ tautomers (211) and isomerise to give them in the presence of acid or base (J.T. Sharp, *et al.*, J.

chem. Soc., Perkin I, 1975, 102; T. Tsuchiya and J. Kurita, Chem. pharm. Bull. Japan, 1978, 26, 1890). Oxidation with MCPBA gives both the 1- and the 2-oxide which are useful in the synthesis of 3-substituted 1,2-benzodiazepines *via* reactions with acids or bases (T. Tsuchiya and J. Kurita, Chem. pharm. Bull. Japan, 1978, 26, 1896).

(c) *Dihydro-* and *tetrahydro-1,2-benzodiazepines.* The cyclopropa[c]cinnolines (213) undergo acid catalysed rearrangement to the 4,5-dihydro derivatives (215; X=H) (L Garanti and G. Zecchi, J. chem. Soc., Perkin I, 1979, 1195). If the cyclisation of (212) is carried out in the presence of nucleophiles X$^-$ (*e.g.* CN$^-$, OH$^-$, N$_3^-$) the 5-substituted compounds (215) are obtained (L. Bruché, *et al.*, J. chem. Soc., Perkin I, 1982, 2041 and references cited therein).

2.1.3 2,3-*Benzodiazepines*

(a) 1H-2,3-*Benzodiazepines.* (i) *Synthesis.* The cyclisation of diene-conjugated diazo-compounds having $\alpha\beta$ aromatic unsaturation (192; R^2/R^3=(CH=CH)$_2$) provides a highly effective route to this system (217) (J.T. Sharp, with A.A. Reid, H.R. Sood, and P.B. Thorogood, J. chem. Soc., Perkin I, 1973, 2543; with D.P. Munro, *ibid*, 1984, 849) and to analogous thieno fused diazepines (D.P. Munro and J.T. Sharp, *ibid.*, 1980, 1718).

(217)   (218)   (219)

(ii) *Reactions.* These compounds (217) undergo photochemical cyclisation of the diazabutadiene unit in the same manner as the monocyclic 3H-1,2-diazepines (193 → 196) (A.A. Reid, H.R. Sood, and J.T. Sharp, J. chem. Soc., Perkin I,

1976, 362; D.P. Munro and J.T. Sharp, *ibid.*, 1980, 1718). Their reaction with acid can follow two pathways, either to give the more stable 5*H* isomer (218) or *via* extrusion of nitrogen to give products derived from the carbocation (219) (J.T. Sharp, unpublished work). They rearrange readily in the presence of base to give the more stable 5*H* isomers (218) (J.T. Sharp *et al.*, J. chem. Soc., Perkin I, 1973, 2543), and they react with MCPBA to give the 2-oxide only (C.B. Argo, I.R. Robertson, and J.T. Sharp, J. chem. Soc., Perkin I, 1984, 2611). Their reaction with di-iron nonacarbonyl gives dinuclear iron hexacarbonyl complexes of the azo group (C.B. Argo and J.T. Sharp, *ibid.*, 1984, 1581).

(b) 3H-2,3-*Benzodiazepines*. The only examples of this tautomer are the 3-acyl-3*H*-derivatives (220) prepared by the acylation of some 5*H*-2,3-benzo-diazepines (see section 2.2.2(c)(ii)).

(c) 5H-2,3-*Benzodiazepines*. (i) *Synthesis*. These compounds (218) are the most thermo-dynamically stable tautomers of the 2,3-benzodiazepine system. The most straightforward preparative route is by the base mediated isomerisation of the 1*H* tautomers (see above). Some examples (218; $R^{4,6}$=H, R^1 and R^5=H or Me) have been prepared by the photochemical ring expansion of isoquinolinium *N*-imides in alkaline solution (T. Tsuchiya, M. Enkaku and J. Kurita, Chem. pharm. Bull. Japan, 1982, 30, 3764). The central nervous system active compound Tofisopam (218; R^1=3,4-diMeOC$_6$H$_3$, R^4=Et, R^5=Me, R^6=MeO) and related compounds have been prepared by the reaction of benzopyrylium salts with hydrazine - an extension of the route to monocyclic 4*H*-1,2-diazepines (J. Korosi and T. Láng, Ber., 1974, 107, 3883; U.S.P., 4,322,346 (1982), Chem. Abs., 1982, 97, 23830n).

(ii) *Reactions*. The pathway followed on acylation of (218) is dependent on the nature of the substituents, reagent, solvent, and reaction conditions. In some cases acylation occurs at N-3 to give 3-acyl-3*H*-2,3-benzodiazepines (220) (T. Tsuchiya, J. Kurita and M. Enkaku, Chem. pharm. Bull. Japan, 1982, 30, 3764). This parallels the acylation of monocyclic analogues (200). Tofisopam (218; R^1=Ar, R^4=Et, R^5=Me, R^6=MeO) can follow this path but may also lose a proton from the 4-methyl group to give the 3-acyl-4-methylene derivative (J. Korosi, *et al.*, Ber., 1984, <u>117</u>, 1476).

(220)   (221)

The 4-phenyl derivative (218; $R^{1,4,6}$=H, R^5=Ph) is however more nucleophilic at N-2 and reacts with acid anhydrides, followed by an alcohol or thiol (RXH) to give (221) (D.P. Munro and J.T. Sharp, J. chem. Soc., Perkin I, 1984, 1133), or with acid chlorides to give either dehydrochlorinated dimers or isoquinoline *N*-imides, depending on conditions (K.R. Motion, D.P. Munro, J.T. Sharp, and M.D. Walkinshaw, J. chem. Soc., Perkin I, 1984, 2027). Tofisopam (218; R^1=Ar, R^5=Me, R^4=Et, R^6=MeO) forms stable salts *via* protonation at N-3 and also reacts with methyl iodide at the same position (A. Neszmelyi, T. Lang, and J. Korosi, Acta chim. Hung., 1983, <u>114</u>, 293).

(d) *Dihydro-2,3-benzodiazepines.* 3,5-Dihydro-4*H*-2,3-benzodiazepine-4-ones have been prepared *via* the cyclodehydration of 2-acetylphenylacetic acid phenyl hydrazones using DCCI (A. Sotiriadis, P. Catsoulacos, and D. Theodoropoulos, J. heterocycl. Chem., 1974, <u>11</u>, 401) and the 3-aryl-4,5-dihydro compounds (222) by treatment of (223) with base and then MSH (T. Tamura, *et al.*, Synthesis, 1973, 159; J. Streith and C. Fizet, Tetrahedron Letters, 1977, 3297).

(222)   (223)

2.1.4 Dibenzo-1,2-diazepines

A number of dibenzo[c,f]1,2-diazepines and related heterocyclic fused systems have been prepared by the thermal elimination of HX from precursors of the type (224; X=F or NO$_2$) (A.C. Alty, et al., Chem. Comm., 1984, 832; Tetrahedron Letters, 1985, 26, 1345).

(224)

2.2   1,3-Diazepines

2.2.1   Monocyclic 1,3-diazepines

The fully unsaturated 1H- (190) and 5H- (225) tautomers have been prepared. A number of examples of the 1H system (190) have been obtained by the thermolysis of certain 1H-1,2-diazepines (188), or their photoproducts, as discussed in section 2.1.1(a)(ii). These compounds are very susceptible to hydrolytic ring opening, thus (190: R=OEt, R^2=Me, R^4=H) gives EtO$_2$CNHCH=CHC(Me)= CHNHCHO in high yield. Irradiation results in electrocyclisation of the butadiene moiety as in 1H-1,2-diazepines. A more highly substituted example, the 1-methyl-2-amino-5,6-diethoxy-carbonyl derivative has been prepared in low yield by the reaction between DMAD and 2-amino-1-methylimidazole (F. Troxler, et al., Helv., 1974, 57, 750).

(225)   (226)   (227)

The $5H-$ system (225; R^1 and $R^2=H$ or Me) results from the photolysis of pyridine-3-azides (226) in the presence of methoxide ion (cf. 2-benzazepine synthesis, section 1.2.4(a)).   Benzoylation of (225) induces ring contraction to give 5-benzoyl-amino-2-methoxypyridines (T. Tsuchiya, et al., Chem. pharm. Bull. Japan, 1984, 32, 4694).   In this connection it is also of interest to note that the carbodi-imide (227) is generated by the thermolysis of 2-azidopyridine as an intermediate stable below $-70^{\circ}C$ but detectable by ir-spectroscopy (1975 cm^{-1}).   Similar benzo fused species have also been generated (C. Wentrup, and H.-W. Winter, J. Amer. chem. Soc., 1980, 102, 6159).

(228)   (229)   (230)   (231)

The saturated 2,4- and 2,5-diones (228) and (229, $R^1/R^2=O$) have been prepared by cyclisation reactions.   N-Chlorination of the diamides RCH(CH$_2$CONH$_2$)$_2$ followed by triethylamine induced cyclisation gives (228; R=H, Me or Ph) (A. Brandner, Synthesis, 1982, 973).   Reaction of the di-isocyanate OCNCH$_2$C(-XCH$_2$CH$_2$X-)CH$_2$CH$_2$NCO (X=O or S) with water, followed by deprotection of the carbonyl group gives (229; $R^1/R^2=O$).   Reduction with borohydride converts the latter into (229;

R^1=H, R^2=OH). An alternative route to the latter involves cyclisation of the *cis*-diamine $H_2NCH_2CH=CHCH_2NH_2$ with carbonyl sulphide to give (230), followed by hydroboration and reaction with hydrogen peroxide. The reaction of (230) with *N*-(phenylselenyl)-phthalimide followed by MCPBA gives (231) (V.E. Marquez, *et al.*, J. org. Chem., 1980, 45, 485, 5225; J. med. Chem., 1980, 23, 713). The reaction of 4-chloromethyl- pyrimidine derivatives (*e.g.* 232) with nucleophilic reagents provides a good route to the more highly substituted 1,3-diazepin-2-ones (233; X=RO, CN, CH(CO$_2$Et)$_2$) (E. Bullock, *et al.*, Canad. J. Chem., 1977, 55, 1895; J. Ashby and D. Griffiths, J. chem. Soc., Perkin I, 1975, 657). The halocarbene adducts of 1,3-disubstituted uracils (*e.g.* 234) when heated in the presence of alcohols (R^1OH) give (235) (H.P. Thiellier, G.J. Koomen, and U.K. Pandit, Heterocycles, 1975, 3, 707; Tetrahedron, 1977, 33, 2603). 2,3,4,5-Tetrahydro-1*H*-1,3-diazepin-2,4,5-triones are produced by the reaction of 4-benzoyl-5-phenyl-2,3-dihydro-2,3-furandione with excess aryl isocyanates (G. Kollenz, *et al.*, Ber., 1976, 109, 2503). The reaction of urea with ClCH$_2$COCH$_2$CO$_2$Et gives the saturated 2,4,6-trione but the reaction appears to be of limited range as *N,N'*-dimethyl-urea only gives an imidazole (R.J. Breckenridge and C.J. Suckling, J. chem. Res. (S), 1982, 166).

(232)

(233)

(234)

(235)

2.2.2   1,3-*Benzodiazepines*

Photolysis of isoquinoline- and quinoline-*N*-imides (236) and (237) provides a route to the fully unsaturated 1*H*- (238; X=CO$_2$R) and 3*H*- (239; X = CO$_2$R) systems respectively. In all cases this reaction path is only followed when R and/or R^1 in (237) and R^2 in (236) are electron donating groups. The reaction follows a similar route to that shown for the formation of the monocyclic system (190) in section 2.1.1(a)(ii) (T. Tsuchiya, M. Enkaku and S. Okajima, Chem. pharm. Bull. Japan, 1980, 28, 2602; with J. Kurita, *ibid*, 1982, 30, 3757).

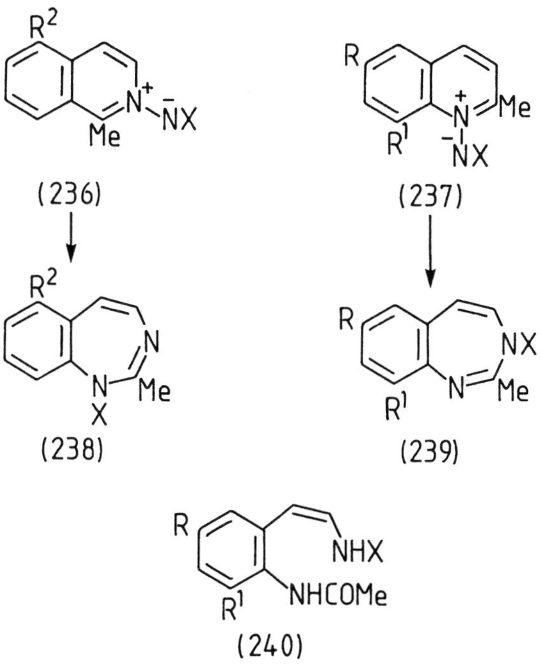

(236)

(237)

(238)

(239)

(240)

1,3-Diazepines fused to a heterocyclic ring have been prepared in a similar way (T. Tsuchiya and M. Enkaku, with S. Okajima, Chem. pharm. Bull. Japan, 1981, 29, 3173; with H. Sawanishi and T. Hirai, *ibid*, 1539). These systems are very susceptible to hydrolysis, giving for example (240) from (239). The photochemistry of these compounds has also been studied. The first *N*- unsubstituted derivatives (239; R^1=H, X=H, R=H or MeO) have been prepared by the reaction of the N-benzyloxycarbonyl compounds (239; R=MeO, R^1=H, X=CO_2CH_2Ph) and (238; R^2=H, X=CO_2CH_2Ph) with trimethylsilyl iodide at room temperature (cf. the similar preparation of 1H-azepine, section 1.1.1(a)). The products are not very stable as free bases at room temperature but their isolation suggests that the 3H-form is the most stable tautomer, contrasting with the 1,2- and the 2,3-benzodiazepine systems (J. Kurita, M. Enkaku, and T. Tsuchiya, Heterocycles, 1983, 20, 2173)

(241)   (242)

3H-1,3-Benzodiazepines (241) have also been prepared via the addition of 1,3-oxazol-5-ones (Munchones) to 2-phenylbenzazete. Thermolysis of (241) at 76°C gives (242, R=Ar^2C=NMe), which on acid hydrolysis yields 2-Ar^1-3-phenylindole. The latter is also produced by direct acid hydrolysis of (241) (P.W. Manley, C.W. Rees, and R.C. Storr, Chem. Comm., 1983, 1007). 3-Acyl-3H-1,3-benzo-diazepines and thieno-fused analogues have also been produced by the acylation of fused 1,2-diazepines (see section 2.1.2(a)(ii)). 3H-1,3-Benzodiazepines, both unsaturated (243; X/Y=CH=CH) and 4,5-dihydro- (243; X/Y = CH_2CH_2) have been prepared by the cyclodehydration of compounds of the type (244) (F.M.F. Chen and T.P. Forrest, Canad. J. Chem., 1973, 51, 881). The reaction of the 1,4-diamines (245; R^1=H, R^2= alkyl) with ortho-esters R^3C(OR)$_3$ provides a general route to the 4,5-dihydro compounds (246, R^2=alkyl). The 1H system (247) has been prepared from (245; R^1=Me, R^2=COR^4) by dehydrative cyclisation (L.L. Martin, L.L. Setescak et al., J. med. Chem., 1982, 25, 340, 346; 1984, 27, 401).

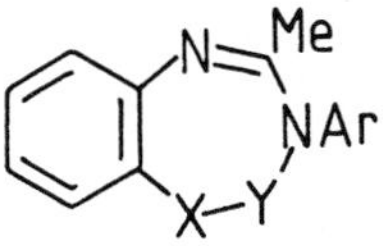

(243)

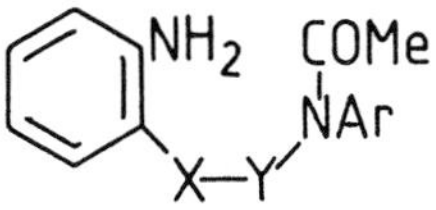

(244)

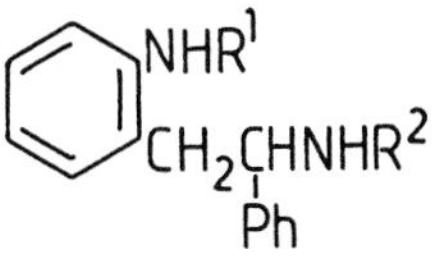

(245)

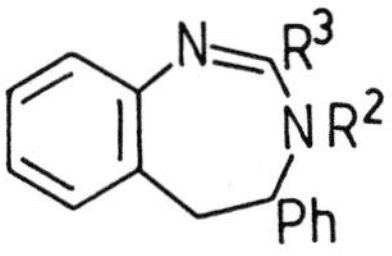

(246)

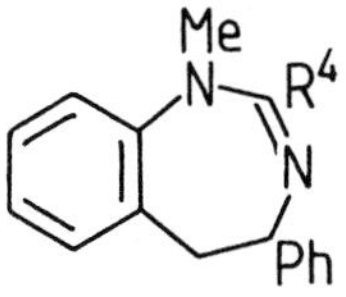

(247)

N-Ethoxycarbonylthioamides $(ArC(=S)NHCO_2Et)$ react with 1,4-diamines to give 1,3-diazepines having a cyclic amidine linkage $(-N=C(Ar)NH-)$ (E.P. Papadopoulos and B. George, J. org. Chem., 1977, **42**, 2530).

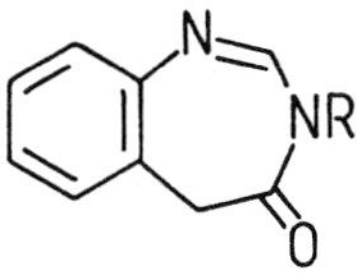

(248)

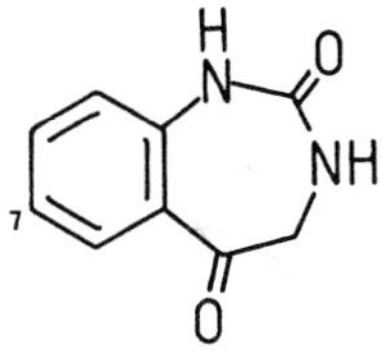

(249)

The 1,3-benzodiazepin-4-one system (248; $R=n\text{-}C_4H_9$, cyclohexyl) is produced by the copper(I) oxide catalysed cyclisation of the isocyanide $o\text{-}CNC_6H_4CH_2CONHR$, but a competing route to indoles is favoured when the R group is bulky ($R=Bu^t$, Ph) (Y. Ito, K. Kobayashi, and T. Saegusa, Tetrahedron Letters, 1979, 1039). The 1,4-diamine $o\text{-}NH_2C_6H_4COCH_2NH_2$ reacts with N,N'-carbonyldiimidazole to give the dione (249) (J. B. Taylor and W.R. Tully, J. chem. Soc., Perkin I, 1976, 1331). The chemistry of this compound has been studied: electrophilic substitution takes place at the 7-position, the 5-oxo group shows typical ketonic reactivity to Grignard reagents and to borohydride reduction, and reaction with Meerwein's reagent gives the expected alkylation at the 2-oxo group.

2.2.3   2,4-Benzodiazepines

The fully unsaturated 1*H* system (250) has been prepared by the photolysis of isoquinoline azide (251) in the presence of methoxide ion (H. Sawanishi and T. Tsuchiya, Heterocycles, 1984, <u>22</u>, 2725).

MeO   N$_3$ O

(250)   (251)   (252)

Treatment of (250) with acid gives 1-amino-isoindolenines, photolysis gives 1-methoxyiso-indolenine, borohydride reduction occurs preferentially at the 2,3-bond, and LTA oxidation yields 1,2-dihydro-3-oxo-derivatives. The dihydro system (252) is prepared by the cyclisation of o-PhCOC$_6$H$_4$CONHCH$_2$Cl with aqueous ammonia, or, better, by a longer route involving its reaction with sodium azide and subsequent reduction to give o-PhCOC$_6$H$_4$CONHCH$_2$NH$_2$ (U. Golik, J. heterocylic Chem., 1975, <u>12</u>, 903). The same compound is also formed by the reaction of 2-(aminomethyl)-phthalimide - which is equivalent to the inaccessible 1,1-diaminomethane - with o-PhCOC$_6$H$_4$COCl (R.F. Lauer and G. Zenchoff, J. heterocyclic Chem., 1979, <u>16</u>, 339). A useful route to either monocyclic or fused 1,3-diazepines is provided by the reaction of N,N'-diaryl-acetamidines (e.g. PhN=C(Me)NMPh) with 1,4-bis-electrophilic reagents such as diacyl halides, thus its reaction with phthaloyl chloride gave (253).

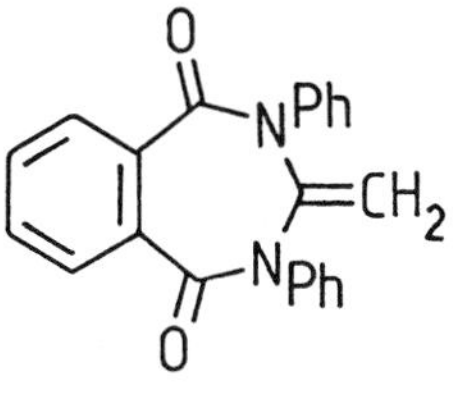

(253) 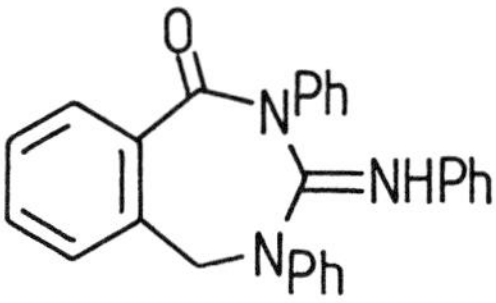   (254)

3-Phenylimino analogues (e.g. 254) can be similarly prepared by the reaction of 1,2,3-triphenylguanidine with *o*-chloromethyl-benzoyl chloride. This compound, when treated with phosphoric acid undergoes an interesting rearrangement to a dibenz[b,e]azepinone derivative (H.W. Heine *et al.*, J. org. Chem., 1979, 44, 3843).

2.3   1,4-*Diazepines*

2.3.1   *Monocyclic 1,4-diazepines*

(i) *Fully unsaturated systems* – *Synthesis and reactions*
Several routes are available to fully unsaturated 6*H*-1,4-diazepines (255). The rather unstable compounds (255, R^1=OMe) are produced from various substituted 4-pyridyl azides either by photolysis in the presence of methoxide ion (T. Tsuchiya et al., Chem. pharm. Bull. Japan, 1984, 32, 4694) (*cf.* sections 1.2.4(a) and 2.2(b)), or by thermolysis in methanol (Y. Ohba *et al.*, Heterocycles, 1985, 23, 287).

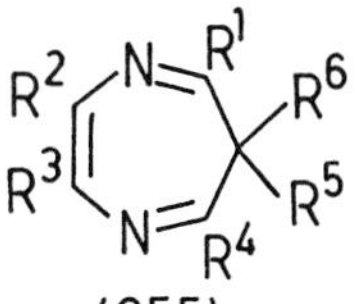

(255) 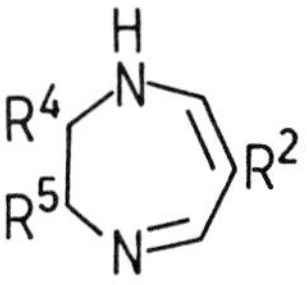   (256) 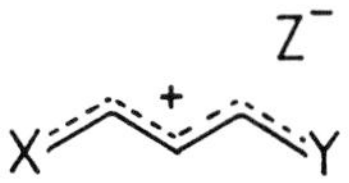   (257)

Benzoylation of (255, R^1=OMe, R^{2-5}=H) gives 1-benzoyl-5-methoxy-1H-1,4-diazepine. The compounds (255; $R^{1,4}$=Me or R^1=Me, R^4=Ph or $R^{1,4}$=Ph, with $R^{2,3}$=CN, $R^{5,6}$=H) are prepared by the condensation of diaminomaleonitrile with pentan-2,4-dione. Its ring inversion has been studied by ^{1}H-n.m.r. spectroscopy and it is reduced by sodium borohydride to give the 1,4,5,7-tetrahydro derivative. The related compound (255; R^1=NMe$_2$, $R^{2,3}$=CN, $R^{4,5,6}$=H) is prepared by DDQ oxidation of (258) (R.W. Begland et al., J. org. Chem., 1974, 39, 2341; Y. Ohtsuka ibid., 1976, 41, 629). Another route to a 6H-1,4-diazepine is given in section 2.1.1(c)(ii). The 6-one and -thione (255; R^{1-4}=Ph, R^5/R^6=O or S) result from air oxidation of (265) (K.N. Mehrotra and G. Singh, Synthesis, 1980, 1001).

(ii) *Dihydro-1,4-diazepines.* *Synthesis*
Most work has been done on the 2,3-dihydro systems (256) and their salts, and the major preparative route continues to be the reaction of a 1,2-diamine with a 1,3-bis electrophilic reagent. 1,3-Dicarbonyl compounds have been commonly used but their derivatives such as diazapentadienium (257, X=Y=NRPh) vinamidinium (257, X=Y=NMe$_2$) and azaoxapentadienium (257; X=RO, Y=NH$_2$) salts offer advantages in allowing the preparation of compounds not otherwise readily obtainable. These methods and their applications and limitations have been fully discussed (D. Lloyd and H. McNab, Heterocycles, 1978, 11, 549; with D.R. Marshall, J. chem. Soc., Perkin I, 1978, 1453; D. Lloyd, A. McCann and D.R. Marshall, J. chem. Res. (S), 1981, 356). 5-Amino derivatives have been prepared from ethylenediamine by reaction with isoxazolium perchlorates (U.S.P., 4,096,140 (1978), Chem. Abs., 1978, 89, 163608f) and the 2-methyl compound by reaction with 1-dialkylamino-but-1-en-3-ynes (E.A. Sokolova, I.A. Maretina, and A.A. Petrov, Zh. org. Khim., 1983, 19, 1541). Further work has been reported on the Cope rearrangement of the imines of diaminocyclopropanes (H. Quast, H.M. Seidenspinner, and J. Stawitz, Ann., 1983, 1207). The 1H-2,3-dihydro-1,4-diazepin-2-one system has been prepared by reaction of diacetylene, aqueous

formaldehyde and ammonia (U.S.S.R. Pat., 740, 772 (1980), Chem. Abs. 1980, <u>93</u>, 150288j).

A range of 6,7-dihydro-1*H*-1,4-diazepines has been produced by the reaction of diamino-maleonitrile with ketones, ketals, α, β-unsaturated-ketones or -amides, e.g. (258) by reaction with *N*,*N*-dimethylacrylamide (R.W. Begland *et al.*, J. org. Chem. 1974, <u>39</u>, 2431).

(258)   (259)   (260)

The thermal rearrangement of (259) gives the 1*H*-4,5-dihydro system (260) (T. Tsuchiya, *et al.*, Chem. Comm., 1983, 941).

Reactions   The dihydrodiazepinium cations (261) typically react with electrophilic reagents by substitution at the 6-position but 6-halogeno derivatives (261, R^2=halogen) - particularly the bromo or iodo compounds - also undergo protiodehalogenation giving (261; R^2=H) in strong acids *via* nucleophilic attack on halogen in the dication (262).

(261)   (262)

Deuteriation studies have shown that in (261; R^{1-3}=H) the 1,-, 4- and 6- protons exchange rapidly in acid but the 5- and 7- protons do not (H. McNab and D. Lloyd, Heterocycles, 1978, <u>11</u>, 549). Electrophilic substitution of the 6-phenyl derivative (261, $R^{1,3}$=H, R^2=Ph) has been studied and - surprisingly at first sight - takes place in the *para* position of the arene ring (D. Lloyd, K.S. Tucker and D.R. Marshall, J. chem. Soc., Perkin I, 1981, 726).

The 6-halogeno derivatives (256; R^2=halogen) readily undergo nucleophilic replacement of the halogen by an alkoxide ion but in many cases nucleophilic protiodehalogenation also occurs to give (256, R^2=H). This mode of reaction becomes more favoured along the series Cl, Br, I and apparently involves nucleophilic attack on halogen in the $6H$ bis-imine tautomer of (256). Protiodehalogenation of the cation (261, R^2=Br) also occurs on reaction with triphenylphosphine, again *via* nucleophilic attack on halogen to give the 'onium anion' (263a) which then abstracts a proton from solvent. MNDO calculations indicate that the alternative formulation (263b) is energetically preferred over the apparently more likely (263a) in which the resonance stabilisation of the vinamidinium π-system is preserved (D. Lloyd *et al.*, Angew. Chem., intern. Edn., 1981, 20, 190; A.F. Cuthbertson, C. Glidewell and D. Lloyd, J. chem. Res. (S), 1982, 80).

(263a)   (263b)

Electrochemical reduction of the diazepinum salts (261) has also been studied (D. Lloyd, *et al.*, Chem. Comm., 1978, 499; Bull. Soc. chim. Belg., 1979, 113; J. chem. Soc., Perkin II, 1980, 668 and 1441; 1982, 801). The thermal decomposition pathways of some substituted 2,3-dihydro-1,4-diazepines (256) show a surprising dependency on the nature of the substituents. For example (256; $R^{4,5}$=Ph, R^2=H) loses ammonia at 150°C to give 2,3-diphenylpyridine but (256; $R^{2,4,5}$=Ph) lost

C_7H_8 to give 2,5-diphenylpyrimidine (H. Quast, H.M. Seidenspinner and J. Stawitz, Ann., 1983, 1230).

(iii) *Tetrahydro-1,4-diazepines*. The reaction between 1,2-diamines and conjugated enones, known as a route to 2,3,6,7-tetrahydro-1H-1,4-diazepines, has been studied in depth by ^{13}C-nmr spectroscopy. In the absence of solvent the diazepine is often formed as the sole or major product but can be lost during work-up by transformation into macrocyclic compounds (D. Lloyd, W. Scheibelein and K. Hideg, J. chem. Res. (S), 1981, 62). The dianion of benzil dibenzylimine, $(PhCH_2N=C(Ph))_2$, reacts with chloroformic esters and with carbon disulphide to give (264, X=O) and (264, X=S) respectively (K.N. Mehrotra and G. Singh, Synthesis, 1980, 1001). Malonimides react with 3-dimethylamino-2,2-dimethyl-2H-azirine to give the diazepinediones (265).

(264)   (266)   (265)

Reduction with sodium borohydride removes the dimethylamino group and reduces the imine bond (B. Scholl, J.H. Bieri, and H. Heimgartner, Helv., 1978, 61, 3050).

iv) *Perhydro-1,4-diazepines*. The reaction of 1,3-diamines $(Me_2C(NH_2)CH_2C(Me)NHR^3)$ with ketones (R^1R^2CO), chloroform and sodium hydroxide under phase-transfer conditions provides a new route to perhydro-1,4-diazepin-2-ones (266) *via* attack of the diamine on a 2,2-dichloro-oxirane intermediate (P. Son and J.T. Lai, J. org. Chem., 1981, 46, 323). A number of 5-oxo derivatives have been prepared by the reductive cyclisation of δ-cyano esters (M. Majchrzak, A Kotelko and R. Guryn, Acta

Pol. Pharm., 1975, $\underline{32}$, 421; Pol. J. Chem., 1978, $\underline{52}$, 1023).

2.3.2   1,4-*Benzodiazepines*
(i) *Fully unsaturated. Synthesis.*
The 1-methyl-1H compound (267; R^1=Me, $R^{2,3,8}$=H, R^5=Ar, R^7=Cl) has been prepared from the 2-one (273) by two routes: (i) LAH reduction to the 2-ol and dehydration; (ii) in lower yield *via* conversion into the 2-thione, S-alkylation to give (267; R^2=S-alkyl) and reductive desulphurisation with Raney nickel. Other reactions of the thioalkyl derivatives have also been studied (R.I. Fryer *et al*, J. org. Chem., 1973, $\underline{10}$, 473). The 5-amino-1H isomers (267; $R^{1,2,3,7}$=H, R^5=NHR, R^8=Cl) are prepared by the photolysis of 7-chloro-4-quinolyl azide in the presence of amines, and the analogous 5-methoxy derivative (and its hydrolysis product the 5-oxo derivative) is formed by photolysis in the presence of methoxide ion (H. Suschitzky *et al.*, Chem. Comm., 1978, 806; J. chem. Soc. Perkin I, 1982, 431).

(267)   (268)

The 2-unsubstituted 3H isomer (268; $R^{2,3}$=H, R^5=Ph, R^7=Cl) is prepared by oxidation of its 1,2-dihydro precursor with manganese dioxide (D.L. Coffen *et al.*, J. org. Chem., 1974, $\underline{39}$, 167), and (268; R^2=NH$_2$, R^3=H, R^5=Ph) is prepared by acid or base catalysed cyclisation of the amino-nitrile O-NH$_2$ArC(Ph)=NCH$_2$CN (K. Meguro *et al.*, Chem. Abs., 1974, $\underline{80}$, 37082f).

$R^2-\underset{Cl}{\overset{R^3}{\triangle}}N$

(269)

R^7 ... $\overset{N=R^2}{\underset{H \overset{|}{R^5} N}{}}R^3$

(270)

R^7 ... $\overset{N=R^2}{\underset{Ar \overset{|}{H} N-OH}{}}$

(271)

5*H*-Isomers have been prepared only recently. The reactions of 2-aminobenzylamines with chloroazirines (269) provide a number of examples (270; R^5=H, Me or Ph, $R^{2,3}$=Me), while reaction with 1,2-diketones (R^2COCOR^3) is successful only for (270; R^5=H, $R^{2,3}$=Ph). Ring inversion has been studied by nmr-spectroscopy and, interestingly, isomerisation to give either the 1*H*- or 3*H*- isomer is not observed (K.R. Randles and R.C. Storr, Chem. Comm., 1984, 1485). One earlier example of this system (270; R^5=Ph, R^2=MeO, R^3=H, R^7=Cl) has been obtained by the dehydration of the cyclic hydroxylamine (271) with phenyl isocyanate. This compound however *does* isomerise to give the 3*H* tautomer when refluxed in *N,N'*-dimethylpiperazine (P. Nedenskov and M. Mandrup, Acta Chem. Scand., 1977, 701).

Reactions. In the 3*H*-isomer (268; $R^{2,3}$=H, R^5=Ph, R^7=Cl) the 1,2-imine bond is more reactive towards nucleophilic attack than the 4,5- bond, and selective addition of a number of nucleophiles (e.g. alcohols, thiols, hydrogen cyanide, and piperidine) has been effected. Oxidation of the hydrogen cyanide adduct gives (268; R^2=CN, R^5=Ph, R^7=Cl) (D.L. Coffen *et al.*, J. org. Chem., 1974, 39, 167). The 4-oxides of the 2-amino derivatives (268; R^2=NHMe, $R^{3,5}$=H, R^7=Cl) and those of the analogous 2-ones (273; $R^{3,5}$=H) add arylmagnesium halides to give cyclic hydroxylamines (e.g. 271) which can be dehydrated or oxidised to give the 3*H*- or the 5*H*- isomer (P. Nedenskov and M. Mandrup, Acta Chem. Scand., 1977, 31, 701). Nitrosation of the 4-oxide of (268; R^2=NHMe, R^3=H, R^5=Ph, R^7=Cl) gives the compound (268; R^2=N(NO)Me) which can react with a variety of nucleophiles (RO^-, HO^-, RS^-, NH_2NH_2, $RNHNH_2$, $RONH_2$, R_2CH^-) to give a 2-substituted derivative (A. Walser, *et*

al., J. heterocyclic Chem., 1974, <u>11</u>, 153).
3*H*-2-Amino derivaties (268; $R^2=NH_2$, $R^3=H$, $R^5=Ph$,
$R^7=Cl$) are converted into their 1-oxides by MCPBA.
These oxides undergo two interesting
rearrangements. Treatment with acetic anhydride
gives the 2-acetamido-3-oxo derivative, and
reaction with phosgene in the presence of
imidazole gives the 3-imidazolyl derivative (H.
Natsugari, K. Meguro, and Y. Kuwada, Chem. pharm.
Bull. Japan, 1979, <u>27</u>, 2608). Functionalisation
of the 3-position is also achieved by the
acylation of (268; $R^2=RNHOH$, $R^3=H$, $R^5=Ph$, $R^7=Cl$);
at low temperature acylation occurs at the
hydroxylamine oxygen atom but at 100°C
rearrangement gives (268; $R^2=NHR$, $R^3=OCOMe$, $R^5=Ph$,
$R^7=Cl$) (H.G. Schecker and G. Zinner, Arch. Pharm.,
1980, <u>313</u>, 926).
(ii) *Dihydro-1,4-benzodiazepines. Synthesis.*
Two routes to the 2,3-dihydro-1*H*-1,4-benzo-
diazepine system (272) have been reported,
involving the oxidation of 1-(2-aminoethyl)
indoles (F. Gatta and S. Chiavarelli, Farmaco Ed.
Sci., 1977, <u>32</u>, 33) and (K. Ishizumi, *et al.*,
Chem. pharm. Bull. Japan, 1972, <u>20</u>, 1628).

(272)

The compounds (272; $R^1=Me$, $R^2=H$, $R^5=Ph$ and $R^3=OEt$
or $-CH_2Br$) are prepared by treatment of
quinazolinium salts or their 4-oxo analogues with
diazomethane (Y. Yamada, T. Oine and I. Inoue,
Bull. chem. Soc. Japan, 1974, <u>47</u>, 343). The ring
expansion of a 2-chloromethylquinazoline 3-oxide
with the anion of nitromethane gives the 4-oxide
of a 2-nitromethylene derivative (272; $R^1=H$,
$R^2/H=CHNO_2$, $R^3=H$ or Me, $R^5=Ar$) (R.I. Fryer *et al.*,
J. heterocyclic Chem., 1976, <u>13</u>, 433). Compounds
of type (272) are also obtained by the reaction of

2-aminobenzophenones with 2-oxazolidone (Patent, Chem. Abs., 1976, $\underline{84}$, 59593h) and 2-bromoethyl-amine (Patents, Chem. Abs., 1978, $\underline{88}$, 190917s; 1978, $\underline{88}$, 152677f). The 1,2-ethanedi-iminium salts $(R^8{}_2N^+{=}CHCH{=}N^+R^8{}_2)$ react with 2-anilino-phenylmethanimine $(o{-}NHRArC(Ar'){=}NH)$ to give the 2,3-diamino derivatives (272; $R^1{=}R$, $R^{2,3}{=}R^8$, $R^5{=}Ar^1$). In cases where R=H, the products easily eliminate the amino group at C-2 to give the 3-amino-3H derivatives (P. Giacconi, E. Rossi and R. Stradi, Synthesis, 1982, 789). A number of substituted benzylidene - ethylenediamines $(o{-}BrArN(R^1)CH_2CH_2N{=}CHAr^1)$ have been cyclised using phenyl-lithium to give (272; $R^{2,3}{=}H$, $R^5{=}Ar^1$) (Patent, Chem. Abs., 1975, $\underline{82}$, 57751t). A 4,5-dihydro-1H-1,4-benzodiazepine has been prepared by the base catalysed cyclisation of the imine 2,5-diMeC$_6$H$_3$N=C(Ar)-C(CN)=NBu^t (J.A. Deyrup and J.C. Gill, Tetrahedron Letters, 1973, 4845).
(iii) *Dihydro-1,4-benzodiazepinones.*
The high yielding routes to the 2,3-dihydro-1H-1,4-benzodiazepin-2-one system (273) *via* the oxidation of 2-aminomethylindoles and by the oxidation or ozonolysis of N-phthalimidoacetyl-indole and subsequent reaction with hydrazine have been discussed in a review which also illustrates their applications to the synthesis of a number of CNS-active compounds (H. Yamamoto, Chemical Economy and Engineering Review, 1977, $\underline{9}$, 22).

(273)   (274)

Another route to (273) from indoles involves the ozonolysis of 2-azidomethyl-3-phenylindole to give (274; X=N$_3$, $R^{1,7}{=}H$, $R^5{=}Ph$) which is then cyclised by reaction with triphenylphosphine *via* the iminophosphorane intermediate (274; X=N=PPh$_3$) (Y. Tamura *et al.*, Chem. pharm. Bull. Japan, 1978, $\underline{26}$,

2874). Similar cyclisations of iminophosphoranes prepared by other routes have been described (J. Ackrell *et al.*, Canad. J. Chem., 1979, 57, 2696; Ger. Pat., 2,504,937 (1975), Chem. Abs., 1976, 84, 44186d). A number of 5-substituted examples of (273) have been prepared *via* reaction of *N*-chloroacetyl isatins with hexamine to give (273; $R^{1,3}$=H, R^5=CO_2R) and subsequent further transformations of the ester group (M. Ogata and H. Matsumoto, Chem. Ind., 1976, 1067). The conversion of 2-(2-chloroacetamido)benzophenones (274; X=Cl, R^5=Ar) (273) by amination and ring closure is a well known route to such compounds. The use of hexamine in this synthesis as the source of N-4 has been discussed in a general review (N. Blazevic *et al.*, Synthesis, 1979, 161) and it has been reported that dinitroso-1,3,5,7-tetra-azabicyclo[3.3.1]nonane is an excellent alternative to hexamine (M. Japelj *et al*, Vestn. Slov. Kem. Drus., 1979, 26, 215). A study has been carried out, for reactions of this type, of the effect of substituents and reagents on the amination (using ammonia, ammonium carbonate and hexamine), and ring closure steps and on the formation of by-products (G.M. Clarke *et al.*, J. chem. Res. (S), 1980, 398, 399, 400). A specific *ortho*-hydroxybenzylation of *sec*-anilines *via* reaction with boron trichloride and aromatic aldehydes provides a useful route to a range of substituted benzhydrols (*o*-NHRArC(Ar^1)HOH) as precursors for (273) *via* known routes (T. Sugasawa, *et al.*, J. heterocycl. Chem., 1979, 16, 445). Diazepam (273; R^7=Cl, R^1=Me, R^3=H, R^5=Ph) has been synthesised in an efficient route from 7-chloro-1-methyl-3,4-dihydro-1*H*-benzazepin-2,5-dione (M. Gates, J. org. Chem., 1980, 45, 1645) and by heating *N*-(chloroacetyl)-*p*-chloro-*N*-methylaniline (*p*-ClC_6H_4N(Me)$COCH_2Cl$) with benzonitrile and titanium tetrachloride (Patent, Chem. Abs., 1973, 78, 58480h). Various 1-substituted derivatives of (273) have been made by the reaction of the appropriate substituted 2-aminobenzophenone with 2,5-oxazolidinedione (Patent, Chem. Abs., 1977, 87, 168104g). 3-Amino derivatives (273; R^3=NMe_2, R^5=Ph) are obtained by treatment of the N,N-dimethylhydrazone of (274, X=Cl) with base, and, in a related reaction, by a Stevens type rearrangement of 1,4,5-benzotriazocinium salts (H. Natsugari, K. Meguro and Y. Kuwada, Chem. pharm. Bull. Japan., 1979, <u>27</u>, 2084). 4,5-Dihydro-3H-1,4-benzodiazepin-3-one is obtained by photolysis of 3-azidoquinoline in the presence of methoxide (see section 2.3.2(i)), and various di- and tetrahydro-1,4-benzodiazepin-3-ones have been prepared by the reaction of Grignard or lithium reagents with N-acylated anthranilonitriles (o-CNC$_6$H$_3$NHCOCR^{1}R^{2}Br), most likely via aziridone intermediates (J. Bergman, A. Brynolf, and B. Elman, Heterocycles, 1983, <u>20</u>, 2141; with K.W. Tornroos, B. Karlsson, and P.E. Werner, p. 2145).

<u>Reactions</u> O-Phosphorylation of 1-unsubstituted 2-oxo compounds (273; R^1=H) provides a mild method of chemically activating the amide function. Thus (268; R^2=OP(O)(morpholinyl)$_2$, R^3=H, R^5=Ar, R^7=Cl), produced by phosphorylation of (273), reacts readily with nucleophiles to give 2-substituted derivatives (R.Y.Ning et al., J. org. Chem., 1976, <u>41</u>, 2720, 2724). 1,4-Benzodiazepin-2-ones (273; R^3=H) can be hydroxylated at the 3-position in a two-step process involving first the synthesis of the 3-acetate either by reaction with lead tetra-acetate or by treatment with NBS and sodium acetate, and then with sodium methoxide, giving for example (273; R^1=H, R^3=OH, R^5=Ar, R^7=Cl) (T. Kovac, et al., J. heterocyclic Chem., 1979, <u>16</u>, 1449). This provides an alternative to the Polonovski route. Aerial oxidation also provides a convenient route to 3-hydroxy derivatives (M. Gall, B. Kamdar, and R.J. Collins, J. med. Chem., 1978, <u>21</u>, 1290) and both 3-hydroxy and 3-alkoxy derivatives (273; R^1=H or Me, R^3=OH or OR, R^5=Ph, R^7=Cl) are produced by treatment of the oxirane (275) with iron(III) chloride or iron(II) sulphate in alcohols or aqueous THF (R.Y. Ning, Wen Yean Chen, and L.H Sternbach, J. org. Chem., 1973, <u>38</u>, 4206).

(275)

A route to 3-amino derivatives (273; R^1=H, R^3=NR$_2$, R^5=Ph, R^7=Cl) is provided by treating oxazepam, the 3-hydroxy analogue, with the appropriate 2-amino-4,5-dihydro-1,3,2-dioxaphosphole (F. Gatta, M.R. del Giudice, and G. Settimj, Synthesis, 1979, 718; and with C. Pandolfi, Farmaco Ed. Sci., 1982, 37, 343). A direct route to 3-amino derivatives has also been reported (Z.I. Zhilina et al., Khim. Geterotsikl Soedin., 1979, 545). The preparation of derivatives of (273) with functionalisation at the 3 position often involves either conversion of R^3=OH into R^3=Cl by reaction with thionyl chloride and subsequent displacement, or is achieved via deprotonation at C-3. Thus reaction of (273; R^1=H or Me, R^3=Cl, R^5=Ar, R^7=Cl) with trialkyl phosphites gives the phosphonates (273; R^3=PO(OR)$_2$). After C-3 deprotonation the phosphonate carbanion is treated with an aldehyde to give a 3-methylene derivative or with oxygen to give the 2,3-dione (J.H. Sellstedt, J. org. Chem., 1975, 40, 1508). Several other 3- substituted diazepam derivatives are prepared by LDA deprotonation of (273; R^1=Me, R^3=H, R^5=Ph R^7=Cl) and reaction with an alkyl iodide, a carbonyl compound or an ester (B.E. Reitter, Y.P. Sachdeva, and J.A. Wolfe, J. org. Chem., 1981, 46, 3945; see also V. Sunjic et al., Helv., 1981, 64, 1145). Formylation at C-3 has been achieved either by the reaction of the anion with ethyl formate or by hydrolysis of the C-3 dimethyl aminomethylene (=CHNMe$_2$) derivative (Ger. Pat. 2,304,095 (1974), Chem. Abs., 1974, 81, 136197z). The latter is obtained by the reaction of (273) with DMF diethyl acetal (Ger. Pat., 2,234,150 (1974), Chem. Abs.,

1974, <u>80</u>, 108589q). 3-Fluorodiazepam (273; R^1=Me, R^3=F, R^5=Ph, R^7=Cl) has been made by LDA deprotonation and treatment with ClO_3F (Ger. Pat., 2,460,360 (1974); 2,542,251 (1977), Chem. Abs., 1977, <u>87</u>, 85063s).

(iv) *Tetrahydro-1,4-benzodiazepines.*
A versatile route to a range of 1,4-benzo-diazepin-5-ones (276), substituted in the aromatic ring and at N-1 is provided by the reaction of methyl anthranilates (o-MeO$_2$CArNHR) with aziridine (C. Corral, R. Madronero, and S. Vega, J. heterocyclic Chem., 1977, <u>14</u>, 99).

(276)   (277)

A number of 3-substituted derivatives of (276; R^1=H; R^4=Ph; R^3=MeO, AcO, H, SPh, SEt) are obtained by the reaction of a nucleophile with the azirinoquinazoline intermediate (277) generated by the reaction of 2-chloromethyl-3-aryl-4-oxo-1,2,3,4-tetrahydroquinazolines with sodium ethoxide (Y. Yamada, T. Oine, and I. Inoue, Chem. pharm. Bull. Japan, 1974, <u>22</u>, 601). Other examples of this system prepared by new routes are (276; R^1=Ar, $R^{3,4}$=H) prepared by the reductive cyclisation of o-MeO$_2$CC$_6$H$_4$N(Ar)CH$_2$CN (A. Baur and K.H. Weber, Z. Naturforsch., 1974, <u>29</u> B, 670) and (276; R^1=Me, $R^{3,4}$=H) prepared by the sodium hydride induced cyclisation of o-H$_2$NCOC$_6$H$_4$N(Me)CH$_2$CH$_2$Cl (U.S. Patent, 3,717,654 (1973), Chem. Abs., 1973, <u>78</u>, 136352j). A range of 2,5-diones (278; R^1, R^3, R^4 = H and/or Me, R^7 = H or Cl) has been prepared in good yield by the reaction of an isatoic anhydride with an α-amino-ester using improved methodology (Dong Han Kim, J. heterocyclic Chem., 1975, <u>12</u>, 1323). These products, when R^1 and R^4 = H, rearrange when

(278)   (279)

treated with acetic anhydride in pyridine to give
for   example   6-chloro-2-methyl-4H-3,1-
benzoxazin-4-ones (Dong Han Kim, J. heterocyclic
Chem., 1976, <u>13</u>, 1165). The 1-methyl-7-chloro
derivative (278; R^1 = Me, $R^{3,4}$=H, R^7=Cl) is
prepared by a similar route (M. Gates, J. org.
Chem., 1980, <u>45</u>, 1645), and the 4-acetyl analogue
(278; R^1=Me, R^4=Ac, R^3=H, R^7=Cl) is obtained by
the palladium catalysed carbonylation of
o-BrArN(Me)COCH$_2$NHAc (H. Ishikura, et al., J. org.
Chem., 1982, <u>47</u>, 2456). 4-Substituted
1,4-benzodiazepin-3,5-diones (279) have been
synthesised via two routes, in each of which the
final step is the DCCI ring closure of
o-RHNCOC$_6$H$_4$NHCH$_2$CO$_2$H (G. Stavropoulos and D.
Theodoropoulos, J. heterocyclic Chem., 1977, <u>14</u>,
1139).

2.3.3 1,5-$Benzodiazepines$.
The major route to this ring system is via the
reaction of o-phenylenediamine with various
1,3-bis electrophilic reagents; some examples are
given below.

(280)   (281)   (282)

3,3-Dimercapto-2-propen-1-ones ((HS)$_2$C=CHCOR) give the 2-thiones (280) (D. Nardi et al., J. heterocyclic Chem., 1973, 10, 815; 1975, 12, 825; Farmaco Ed. Sci., 1975, 30, 727), and the analogous α-oxoketenethioacetals ((MeS)$_2$C=CHCOR) give (281; R^2=SMe, R^3=H, R^4=R) which are readily oxidised to give the 1H-2,3-dihydro-2-one (G. Kobayashi et al., Heterocycles, 1980, 14, 7). The reaction of o-phenylenediamine itself with alkynones (R^{4}C≡CCOPh) gives (281; R^2=Ph, R^3=H) (W. Ried and E. König, Ann., 1972, 755, 24: S.P. Korshunov et al., Khim. Geterotsikl. Soedin, 1973, 1421), and N-substituted o-phenylenediamines give (282; R^1=Me or Ph, R^2=Ph, R^4=H) (W. Ried and R. Teubner, Ann., 1978, 741; R.L. Amey and N.D. Heindel, Org. Prep. Proced. Int., 1976, 8, 306). Acetylenic imidates (ArC≡CC(OEt)=NH) similarly give 2-amino derivatives (281; R^2=Ar, R^3=H, R^4=NH$_2$) (P.C. Unangst and P.L. Southwick, J. heterocyclic Chem., 1973, 10, 399; 1981, 18, 1257). Phosgeneiminium salts (Me$_2$NCCl=C(F)CCl=N$^+$Me$_2$Cl$^-$) give (281; R^{2,4}=NMe$_2$, R^3=F) (J. Gorrissen and H.G. Viehe, Bull. Soc. chim. Belg., 1978, 87, 391). 2-Phosphoniovin-amidinium salts give (282; R^{1,2,4}=H, R^3=P$^+$Ph$_3$ClO$_4^-$) (R. Gompper, E. Kujath and H-U. Wagner, Angew. Chem., intern. Edn., 1982, 21, 543). The allenic diester EtO$_2$CCH=C=CHCO$_2$Et gives the 4-ethoxycarbonylmethylene-2,3,4,5-tetrahydro-2-one derivative (J. Ackroyd and F. Scheinmann, J. chem. Res.(S), 1982, 89). Cyclopenta[b] fused 1,5-benzodiazepines have been produced by the addition of o-benzoquinone di-imines to fulvenes (W. Friedrichsen and H-G. Oeser, Tetrahedron Letters, 1974, 4373; Ann., 1978, 1161).

3. Seven-membered rings containing three nitrogen atoms

3.1 Monocyclic triazepines

3.1.1 1,2,4-Triazepines (i) Fully unsaturated Examples of the 2H (286), 4H (290) and 6H (287) systems have been prepared. The 2H isomers (286; R^1=Ar, CO$_2$Et; R^2=H, Ph; R^3=H, alkyl, Ph) are the major products of the cycloaddition of 1- azirines (284) to 1,2,4,5-tetrazines (283).

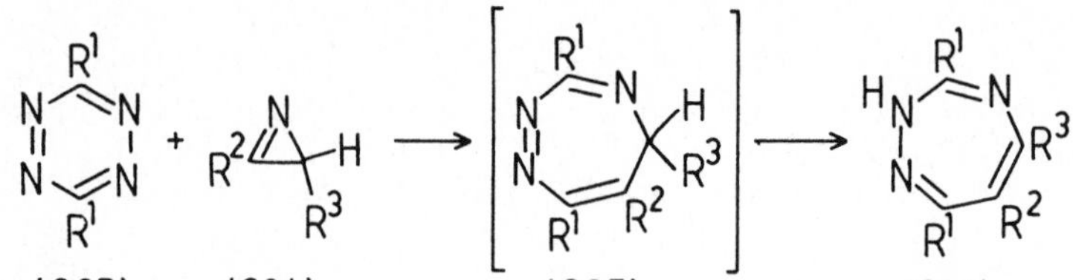

(283)   (284)   (285)   (286)

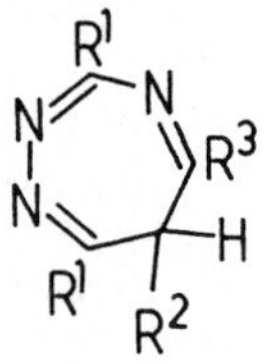

(287) 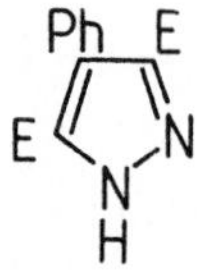   (288)

The primary product (285), which has not been isolated, rearranges by a [1,5]-hydrogen shift to give (286), and in some cases a further shift gives the 6*H*-isomer (287) (D.J. Anderson and A. Hassner, Chem. Comm., 1974, 45; G.C. Johnson and R.H. Levin, Tetrahedron Letters, 1974, 2303; V. Nair, J. heterocycl. Chem., 1975, 12, 183; M. Takahashi, N. Suzuki and Y. Igari, Bull. chem. Soc. Japan, 1975, 48, 2605). The 2*H*- system is rather unstable and can decompose to give a variety of products, depending on the substituents. The most common path leads to pyrazoles (e.g. 288) together with benzonitrile (from 286; R^1=CO$_2$Me, R^{2,3}=Ph), but pyrimidines and 1,2,4-triazoles have also been obtained.

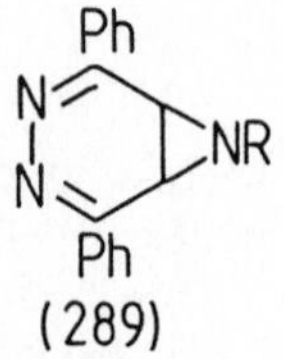

(289) 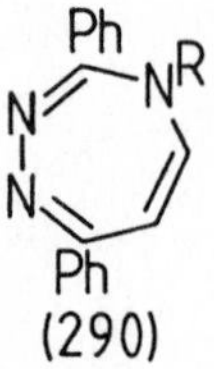   (290) 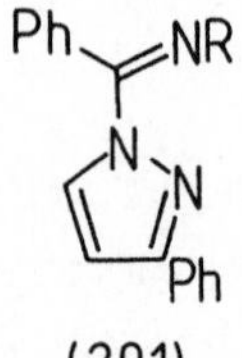   (291)

The 4H- system (290; R=PhCH$_2$ or C$_6$H$_{11}$) has been prepared from the triazanorcaradienes (289) *via* a photochemical walk and ring expansion (I. Saito, A. Yazaki and T. Matsuura, Tetrahedron Letters, 1976, 2459) (cf. section 2.1.1(b)(ii)). The photolysis of (290; R=C$_6$H$_{11}$) gives 1-cyclohexyl-2,3-diphenylpyrrole in quantitative yield. The products of thermolysis depend strongly on the nature of R; thus (290; R=CH$_2$Ph) gives 2,4-diphenylpyrimidine while 1-cyclohexyl-4-phenylimidazole is obtained when R=C$_6$H$_{11}$. The presence of water induces hydrolytic ring opening and recyclisation to give (291) (I. Saito, A. Yazaki and T. Matsuura, Tetrahedron Letters, 1976, 4753).

(ii) *Tetrahydro-1,2,4-triazepines*
The reaction of β-aryl isocyanates and isothiocyanates (RCOCH$_2$C(Me)$_2$N=C=X, X=O or S) with hydrazines (R^{1}NHNH$_2$, R^1=H or alkyl) leads to the 3-one and 3-thione systems respectively (292, R^{2,3}=Me) (C. Zigeuner, A. Fuchsgruber and F. Wede, Monatsh., 1975, 106, 1495; R. Neidlein and W-D. Ober, *ibid.*, 1976, 107, 1251; R. Lantzsch and D. Arlt, Synthesis, 1977, 756). In a similar process the α,β-unsaturated ketone PhCH=CHCOPh reacts with thiocyanic acid to give PhCOCH$_2$CH(Ph)NCS which is subsequently cyclised to give (292; X=S, R=Me, R^{1,2}=Ph, R^3=H) by reaction with methyl hydrazine (P. Richter and K. Steiner, Chem. Abs., 1985, 103, 104934u).

(292)

Alkylation of (292; R=H or Me, R^{2,3}=Me, X=S) gives the 2H-6,7-dihydro-3-thiomethyl ethers. Further examples have been reported of the formation of 2,3,4,6-tetrahydro-3-thiono-5H-1,2,4-triazepin-5-ones by the reaction of a thiosemicarbazide

$(NH_2N(R)CSNH_2)$ with a β-keto ester or a malonic acid derivative (A. Hasnaoui, J-P. Lavergne, and Ph. Viallefont, J. heterocyclic Chem., 1978, 15, 71; Rec. Trav. chim., 1980, 99, 301).

3.1.2 1,2,5-*Triazepines*. (1) *Dihydro*.

The 6,7-dihydro-1H system (293) has been synthesised by the reaction of 1-(2-aminoethyl)-1-methylhydrazine $(H_2NCH_2CH_2N(Me)NH_2)$ with a benzil $((RCO)_2)$. Sodium borohydride selectively reduces the 4,5 bond (D.L. Trepanier, S. Wang and C.E. Moppett, Chem. Comm., 1973, 642).

(293)   (294)   (295)

Oxidation of (293) with MCPBA gives initially the 2-oxide which rearranges to produce the bridged compound (294). Acylation of (294) at low temperature gives the expected O-acyl derivative but at 100°C this rearranges to a triazabicyclo-[3.2.0]octadiene. Halogenation of the 5,6-dihydro-4H derivative (295) induces ring contraction to give 1-benzyl-4-phenylimidazole plus benzonitrile, probably *via* a triazanor-caradiene intermediate (O. Tsuge and K. Kamata, Heterocycles, 1975, 3, 547).

(ii) *Tetrahydro*. The 4,5-dihydro analogue of (293) is prepared by the reaction of the α-haloketone PhCOCHBrPh with methylhydrazine (D.L. Trepanier *loc. cit.*), and the analogous 6-one (296) by the reaction of an α-anilinoacetophenone hydrazone $(Ar^2NHCH_2C(Ar)=NNHAr^1)$ with chloroacetyl chloride to give $ClCH_2CON(Ar^2)CH_2C(Ar)=NNHAr^1$, conversion into the iodide by the Finkelstein reaction, and cyclisation (H. Gnichtel and W.I. Salem, Ann., 1982, 729).

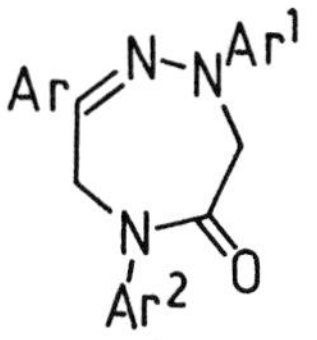

(296) 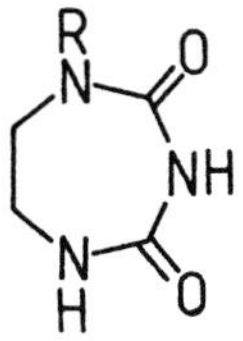   (297)

3.1.3. 1,3,5-*Triazepines*

A number of 2,4-diones (297) have been prepared by
the reaction of 1,2-diamines ($RNHCH_2CH_2NH_2$) with
urea (Pol. Pat., 84, 508 (1977), Chem. Abs., 1977,
<u>86</u>, 43755t). An earlier report that the
thiocarbamate ($EtO_2CNHC(S)OEt$) reacts with
ethylenediamine to give a triazepine has been
corrected; the product is an imidazole derivative
(N.P. Peet and S. Sunder, Ind. J. Chem., 1978,
<u>16B</u>, 207).

3.2 *Triazepines fused to carbocyclic rings*

3.2.1 1,2,3-*Triazepines*

Only two structures containing the 1,2,3-
triazepine ring have been reported. The
cyclobuta[d] fused system (299) is obtained as a
secondary product after chromatography of the
pyrolysis products of (298) (L.A. Paquette and
R.J. Maluska, J. Amer. chem. Soc., 1972, <u>94</u>, 534).

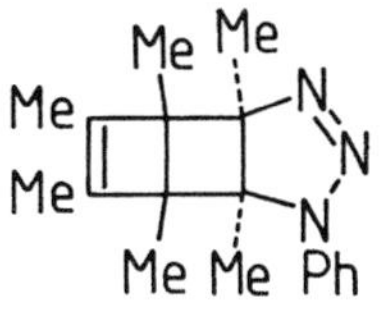

(298) 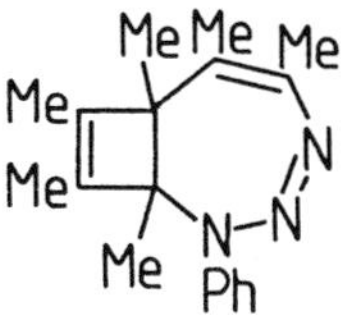   (299)

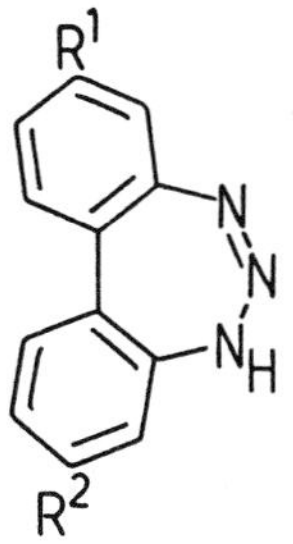

(300)

Tetrazotization of 2,2'-diaminobiphenyl and treatment with aqueous ammonia gives the $5H$-dibenzo[d,f] system (300; R^1 = H, Me, MeO; R^2 = H, MeO). These compounds can be alkylated at nitrogen but decompose on heating to give either the benzo[c]cinnoline N-imide at 80°C or at higher temperatures the carbazole and nitrogen (C.W. Rees, *et al.*, J. chem. Soc., Perkin I, 1974, 1248).

3.2.2   *Fused 1,2,4-triazepines*
(i) 1,2,4-*Benzotriazepines*
There are only two reports of the synthesis of this system. The fully unsaturated $1H$ compounds (301; R^1 = RS, alkyl, aryl) is prepared in good yield by the reaction of a hydrazidoyl chloride of the type PhN(Ar)N=C(Cl)Ph with a nitrile R^1CN in the presence of aluminium trichloride. The reaction is considered to involve formation and cyclisation of the nitrilium ion PhN(Ar)N=C(Ph)N=C$^+$R^1. The compounds undergo ring contraction in acid to give 1-phenyl-3R^1-indazoles but are stable to alkali (S. Conde, C. Corral and R. Madronero, Tetrahedron, 1974, $\underline{30}$, 195).

(301)   (302)

The 4,5-dihydro-5-ones (302) are synthesised by cyclisation of the diazonium salt of an amine of the type o-NH$_2$C$_6$H$_4$CON(R)CH(CO$_2$Et)$_2$ followed by successive treatment with base and acid to give (302, R^1=CO$_2$H) which on decarboxylation gives (302, R^1=H). These compounds, like the analogous phenylhydrazones, are unreactive to alkylation at N-1, but (302; R^1=H, R = Me) gives a quaternary salt on reaction with methyl fluorosulphonate *via* methylation at N-2 (M. Bianchi *et al.*, Eur. J.

med. Chem. chim. Ther., 1977, 12, 263; M. Bianchi,
E. Hausermann and S. Rossi, J. heterocyclic Chem.,
1979, 16, 1411).
(ii) 1,3,4-*Benzotriazepines*. (a) *Dihydro*
The reaction of 2-aminobenzoyl hydrazides (303)
with ortho esters $(R^1C(OR)_3$ provides a
straightforward and effective route to the 5-one
systems (304) and (305) in cases where R^3 in (303)
is not hydrogen. The hydrazides are readily
obtained by the reaction of the appropriate
isatoic anhydride with the hydrazine (R^3NHNH_2).
When R^2=H in (303) the reaction gives the 3*H*
system (304) and when R^2=Me or Ph it gives the 1*H*
system (305) (S. Sunder, N.P. Peet and D.L.
Trepanier, J. org. Chem., 1976, 41, 2732; R.W.
Leiby and N.D. Heindel, *ibid*. 2736; N.P. Peet and
S. Sunder, J. heterocyclic Chem., 1976, 13, 967).
Cyclisation of (303; R^2=H, Me) with cyanogen
bromide gives the 2-amino derivative (305; R^1=NH_2)
(R.W. Leiby and N.D. Heindel, Synth. Comm., 1976,
6, 295; J. pharm. Sci., 1977, 66, 605). Reactions
of similar type involving unsubstituted
2-aminobenzoylhydrazide (303; $R^{2,3}$=H) are however
much more complex and can give all or any of
several possible products, depending on the
reactants, solvents and reaction conditions.

(303)   (304)   (305)

Small to moderate yields of the benzotriazepines
(e.g. 304; R^3=H, R^1=H, Me, Ph) can be obtained
(J.Y. Merour, J. heterocyclic Chem., 1982, 19,
1425; N.P. Peet and S. Sunder, *ibid*., 1984, 21,
1807; P. Scheiner *et al.*, *ibid*., 1984, 21, 1817;
R.W. Leiby, *ibid*., 1984, 21, 1825; N.P. Peet,
Synthesis, 1984, 1065) and it appears that these
may be the kinetically favoured products but they

react further and rearrange to give quinazoline derivatives and 1,3,4-oxadiazoles. This whole area of chemistry seems to have been a minefield for investigators and is littered with the remains of triazepine structures proposed but then "exploded" by other workers. Most of these are noted in the last reference of the set given above. This paper refutes a claim that the reaction of isatoic anhydride with benzoyl-hydrazine in acetic acid and p-toluenesulphonic acid gives (304; R^3=H, R^1=Ph) (Ch.K. Reddy, P.S.N. Reddy and C.V. Ratnam, Synthesis, 1983, 842). The product is in fact 3-amino-2-phenyl-4(3H)-quinazolinone but it *is* formed *via* the benzotriazepine as an intermediate. The triazepinones (305) also undergo ring cleavage under basic conditions to give quinazolinones (R.W. Leiby and N.D. Heindel, J. org. Chem., 1977, 42, 161). 2,3-Dihydro-1H-1,3,4-benzotriazepin-2-thiones (306) have been prepared by the reaction of 2-benzoylphenyl isocyanate with a hydrazine (cf. the formation of the monocyclic analogue (292) above) (P. Richter and K. Gerish; Pharmazie, 1979, 34, 847), and their hydrolysis by acid and by alkali has been studied (P. Richter et al., Pharmazie, 1983, 38, 269; 1984, 39, 505).

(306)   (307)

The 1H-4,5-dihydro system (307; R = CO_2Me) has been made by the thermal cyclisation of [(2,6-dimethylphenyl)-imino]phenylazoacetate (2,6-diMeC$_6$H$_3$N=C(CO$_2$Me)N=NPh) (cf. the 1,4-benzodiazepine synthesis of Deyrup and Gill, section 2.3.2 (ii)), hydrolysis gives (307; R=H) (R. Fusco and F. Sannicolo, Tetrahedron Letters, 1982, 23, 1829). An extension of the much used photo-

chemical ring expansion of heterocyclic N-imides
to quinazoline N-imides also gives dihydro-1,3,4-
benzotriazepines (J. Fetter *et al.*, Tetrahedron
Letters, 1975, 2775). An indeno fused 1,2,4-tri-
azepin-3-one has been prepared by reaction of a
1,3,4-oxadiazepine with phenyl isocyanate (K. Oe
and O. Tsuge, Heterocycles, 1976, **4**, 989)

(b) *Tetrahydro.* The irradiation of 2-amino-
benzoylhydrazide in ethanol provides an easy route
to (308; $R^{1,2,3,5}$ = H, R^4 = Me) (I. Ninomiya and
O. Yamamoto, Heterocycles, 1976, **4**, 475), and a
more heavily substituted example (308; R^1 = H,
$R^{2,5}$ = Ac, R^3 = CH_2CO_2Me, R^4 = CO_2Me) has been
proposed as the product of acetylation of the
adduct (o-$NH_2C_6H_4CONHN=C(CO_2Me)CH_2CO_2Me$) of
2-aminobenzoylhydrazide and DMAD (M.D. Nair, Ind.
J. Chem., 1973, **11**, 109). Many compounds have
appeared in the literature in the guise of the
2,5-diones (309) but authentic examples of such
compounds have been made *via* the reaction of
methylhydrazine with an o-substituted phenyl
isocyanate (310).

(308)   (309)   (310)

The ester (310, X=OMe) reacts to give (309; R^1=H,
R^2=Me) but the acid chloride (310, X=Cl) shows
reverse regioselectivity and gives (309; R^1 = Me,
R^2 = H) (N.P. Peet and S. Sunder, J. org. Chem.,
1975, **40**, 1909). The latter has also been
prepared by the reaction of (303; R^3 = Me, R^2 = H)
with chloroformic esters. Alkylation of these
diones has been studied (N.P. Peet and S. Sunder,
J. org. Chem., 1977, **42**, 2551).

3.2.3 *Fused 1,2,5-triazepines.* 3H-Benzo[c]-
[1,2,5]-triazepino[1,2-a]cinnolines (311; R/R =

2,2'-biphenyl, $R^{1,3}$ = CO_2R, R^2 = H or CO_2R) have been prepared by the cycloaddition of extended azomethine imines to acetylenes ($R^3C{\equiv}CR^3$) (C.W. Rees *et al*., J. chem. Soc., Perkin 1, 1975, 556).

(311)

3.2.4. 1,3,5-*Benzotriazepines*.

1,3,5-Benzotriazepin-2,4-diones (312) are yet another group of triazepines whose history is complicated by structural misassignments. It has been shown (N.P. Peet and S. Sunder, Ind. J. Chem., 1978, **16B**, 207) that earlier reports of their synthesis by the acid catalysed cyclisation of 1,2-phenylenebisureas (o-RHNCONHC$_6$H$_4$NHCONHR) are wrong and the products are benzimidazole derivatives.

(312)   (313)

The reported formation of related systems by the reaction of o-phenylenediamine with ethyl carbethoxythiocarbamate (EtOCNHC(S)OEt) has also been shown to be wrong and again the product is a benzimidazole derivative. An 'authentic' preparation of (312, R = Ph) has been achieved by the reaction of N-phenyl-o-phenylenediamine (o-PhNHC$_6$H$_4$NH$_2$) with phenyl isocyanate to give o-PhNHC$_6$H$_4$NHCONHPh followed by cyclisation with phosgene. Interestingly, the application of the reagents in the reverse order gives the isomeric benzimidazolin-2-one (313) (F. Weigert, J. org. Chem., 1973, <u>38</u>, 1316)

4. *Seven-membered rings containing four nitrogen atoms*

4.1 *1,2,3,4-Tetrazepines*

The only example of this system (314) has been prepared by the oxidation of the dihydrazine $H_2NN(Me)-(CH_2)_3-N(Me)NH_2)$ with potassium ferricyanide. Thermal decomposition gives (315) (C.J. Michejda, S.R. Koepke and D.H. Campbell, J. Amer. chem. Soc., 1978, <u>100</u>, 5978).

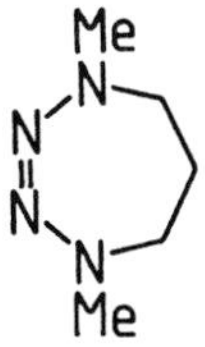

(314)

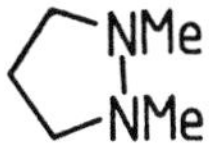

(315)

4.2. *1,2,4,5,-Tetrazepines*

The reaction of 1,2-dihydrazines (eg. $R^1NHN(Me)CH_2CH_2N(Me)NHR^1$) with aldehydes gives (316; R^1 = H or alkyl, R^2 = H or Me) (G.S. Gol'din *et al.*, Zhur. org. Khim., 1974, <u>10</u>, 2218).

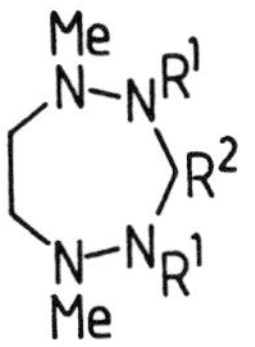

(316)

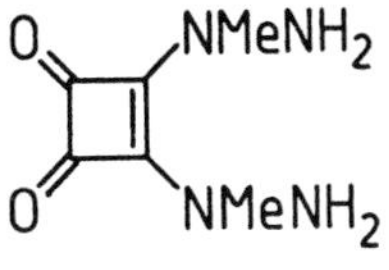

(317)

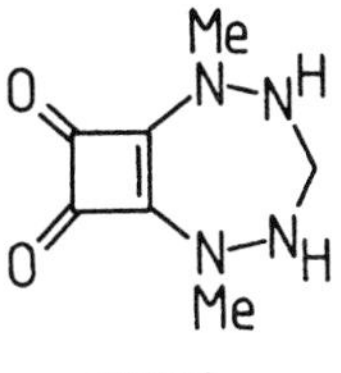

(318)

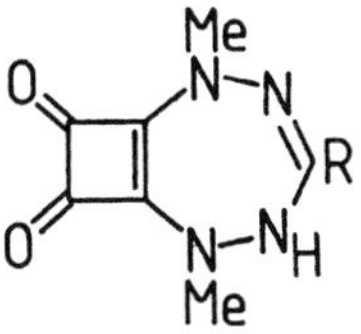

(319)

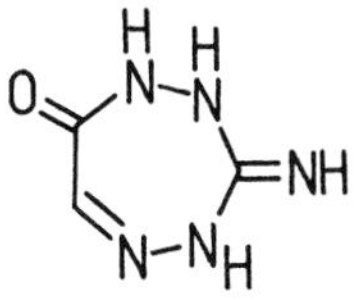

(320)

Similarly (317) reacts with formaldehyde to give (318) and with orthoformic esters to give (319) (G. Seitz and H. Morck, Synthesis, 1973, 355). The iminotetrazepinone (320) has been synthesised by reaction of diaminoguanidine ($H_2NNHC(=NH)NHNH_2$) and chloral (Cl_3CCHO). Compounds of this type have anti-inflammatory activity (D. Bierowska-Charytonowicz and M. Konieczny, Chem. Abs., 1974, 81, 3903e).

Chapter 52

COMPOUNDS WITH LARGER HETEROCYCLIC RINGS:

SEVEN-MEMBERED RING COMPOUNDS WITH TWO OR MORE

DIFFERENT ELEMENTS IN THE RING

C.D. GABBUTT and J.D. HEPWORTH

Interest in this field has continued unabated over the last ten years. Particular attention has been directed towards novel polyheteroatom ring systems. In addition, the parent member of several previously investigated heterocycles has been obtained for the first time. On the other hand many derivatives of the more familiar systems have been prepared using both established and new methodologies; this chapter highlights only the latter.

An excellent review of this area of heterocyclic chemistry has appeared (J.T. Sharp, in 'Comprehensive Heterocyclic Chemistry' ed. A.R. Katritzky and C.W. Rees, Pergamon, Oxford, 1984, vol. 8, p.593).

1. *Rings containing oxygen and nitrogen*

(i) *1,2-Oxazepines*

Since the publication of the 2nd edition, many novel 1,2-oxazepines have been prepared. Of particular interest are the 1,2-oxazepinium salts which appear to be the first examples of cationic seven-membered rings possessing 8π electrons. 4-Aryl-2,6-diphenylpyrylium perchlorates readily condense with nitrones in boiling dimethylformamide to give the oxazepinium salts in good yields. The initial step probably proceeds via a 1,3-dipolar cycloaddition to the pyrylium ring, giving the bicyclic intermediate (1) from which aryl-aldehyde elimination affords (2). Use of a *N*-arylhydroxylamine in place of the nitrone component is also successful, but a *N*-alkylhydroxylamine gives only an intractable product (E.V.A. Zvezdina *et al.*, Chem. heterocyclic Compd., 1982, 340).

The oxazepine (3), isolated in low yield, is amongst the products formed when *trans*-3-[3-(2-nitrophenyl)-2-propenylidene]-2,4-pentanedione is condensed with hydroxylamine in acetonitrile (T. Kurihara *et al.*, Chem. pharm. Bull., 1982, 30, 2742).

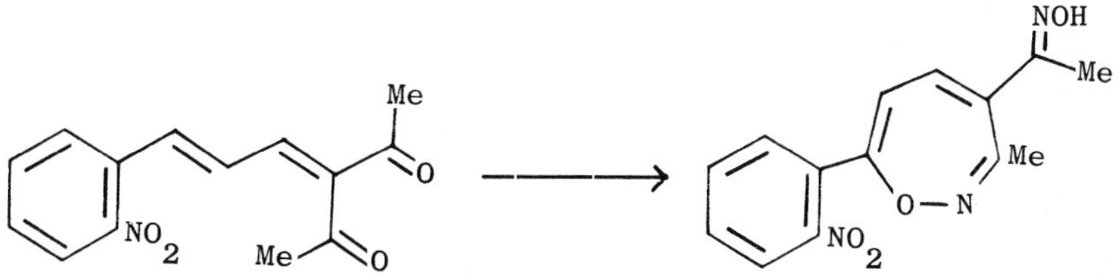

(3)

Silver tetrafluoroborate catalyses the cyclisation of β-allenic oximes to 4,7-dihydro-1,2-oxazepines (4). The ^{1}H nmr spectrum of each of the products exhibits a singlet at ca. δ1.2 for the *gem*-dimethyl group, suggesting that the ring undergoes a rapid conformational change (J. Grimaldi and A. Cormons, Tetrahedron Letters, 1985, 825).

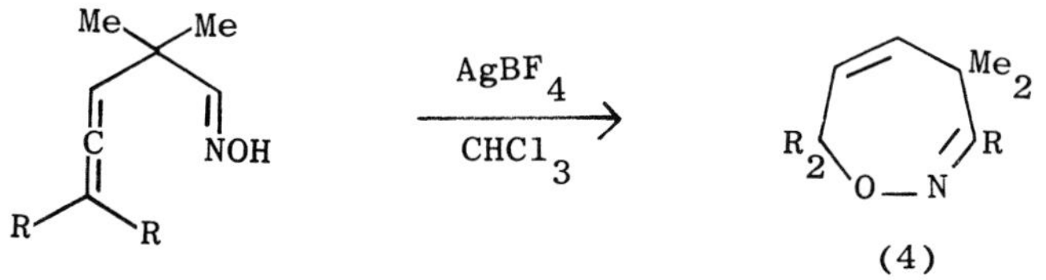

(4)

R = H, Me or Et

In an extension of earlier work (see 2nd Edn. Vol.IVK, p.284), the thermolysis of 1-methyl-2-phenyl-1,2,3,6-tetrahydropyridine-1-oxide results in the formation of 2,3,4,7-tetrahydro-1,2-oxazepines *via* a [1,2] Meisenheimer rearrangement (H. Sashida and T. Tsuchiya, Chem. pharm. Bull., 1984, <u>32</u>, 4117).

Oximation of 6-chlorohexan-2-one in the presence of potassium carbonate occurs with concomitant halide elimination to give 3-methyl-4,5,6,7-tetrahydro-1,2-oxazepine, b.p. 53-57°C/0.6 mm. (S. Shatzmiller and E. Shalom, Ann., 1983, 897).

Methanolysis of *N*-(4,5-epoxypentyloxy)phthalimide under basic conditions affords a mixture of the easily separable oxazepine (5) and the ring-contracted isomer (6). The products are derived by intramolecular attack of the N-atom on the oxirane ring. Both modes of attack appear equally

facile, since (5) and (6) are formed in equal amounts.
Hydrolysis of (5) affords 4-hydroxyhexa-hydro-1,2-oxazepine,
m.p. 98-99°C, in high yield (N. Amlaiky, *et al.*, J. org.
Chem., 1982, 47, 517; Synthesis 1982, 426).

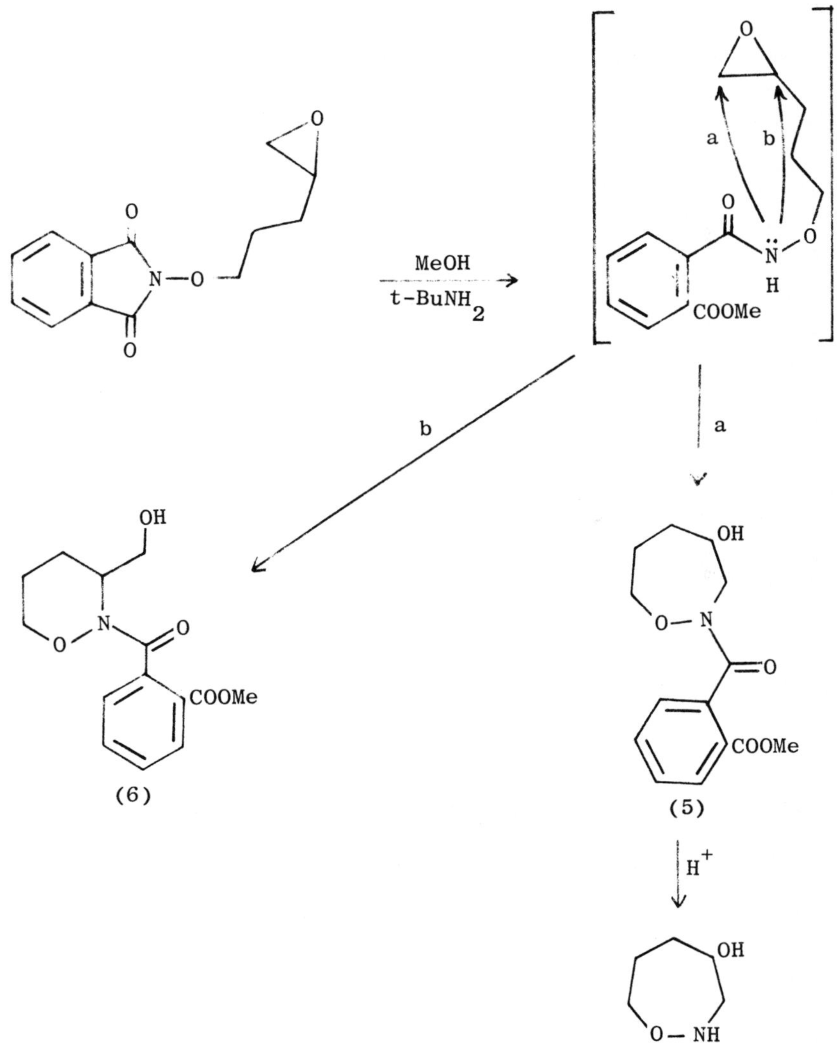

(6)

(5)

Silver tetrafluoroborate promotes heterolysis of the N–Cl bond in O-3-phenylpropyl-N-chlorohydroxamate to give an acyl nitrenium ion which cyclises to 1-benzoyl-2,1-benzoxazepine (7), apparently the first example of this ring system (S.A. Glover *et al.*, J. chem. Soc. Perkin I, 1984, 2255).

(7)

Direct melt pyrolysis at 150°C or thermolysis in refluxing acetonitrile of either of the isomeric tetrahydro-isoquinoline N-oxides (8) provides an efficient entry to the 2,3-benzoxazepine system. Homolytic cleavage of the Cl–N bond generates a stabilised 'double benzylic' radical (9) and intramolecular coupling with the pendant nitroxyl function affords the ring expanded product (J.B. Bremner *et al.*, Austral. J. Chem., 1980, 33, 833).

(8) → (9)

The readily available 3-pyrrolidinoisocoumarin reacts with 1,3-dipoles. The expected fused pyrazole results from reaction with a nitrile imine, but with benzonitrile oxide an alternative reaction supervenes to give a 1,5-dipolar intermediate which rearranges to afford the 2,3-benzoxazepin-(3H)-one (10). *N*-Benzylideneaniline-*N*-oxide reacts to give (11) by an analogous pathway (G.V. Boyd and R.L. Monteil, J. chem. Soc. Perkin I, 1980, 846).

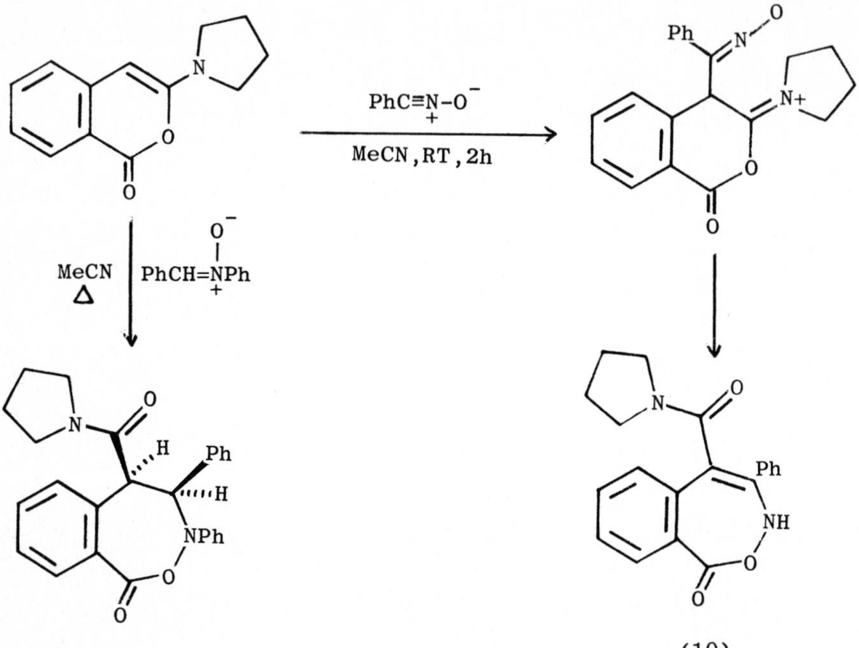

(11)　　　　　　　　　　　　　　　　　(10)

Irradiation (> 300 nm) of 9-chloro- or 9-cyano-acridine-*N*-oxide in benzene affords the corresponding 11-substituted dibenzo[c,f][1,2]oxazepine.　　The ring expansion proceeds through an oxaziridine intermediate (S. Yamada and C. Kaneko, Tetrahedron, 1979, 35, 1273).

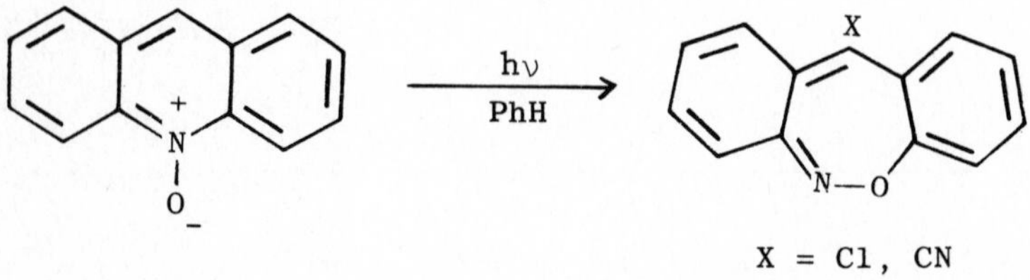

X = Cl, CN

These dibenzo[c,f][1,2]oxazepines are particularly labile and undergo a variety of transformations under mild conditions (Scheme 1). Some of these transformations have been rationalised by invoking participation of the oxaziridine valence tautomer.

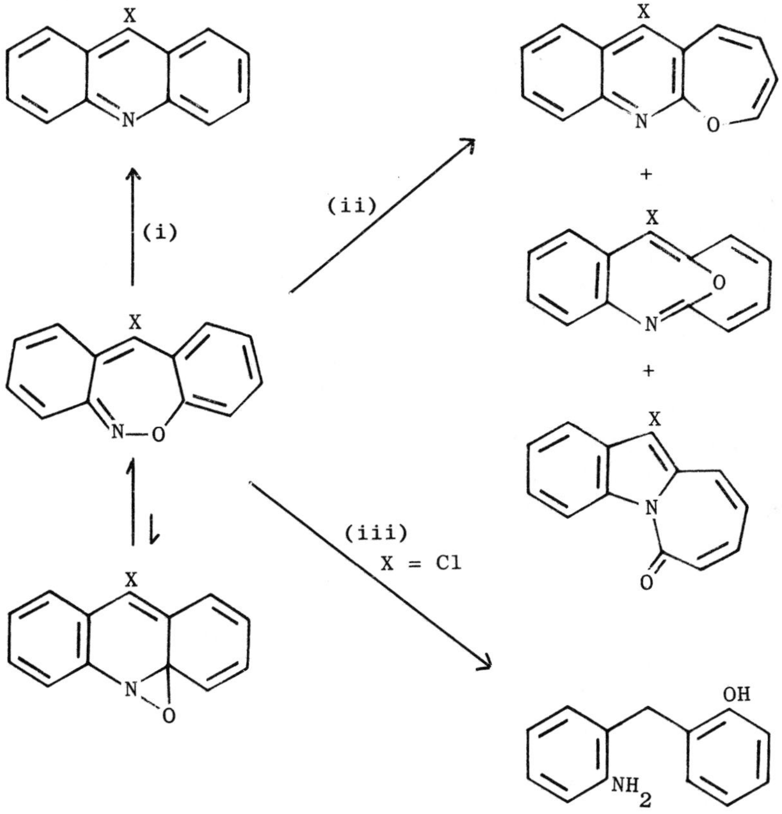

X = Cl, CN

Reagents: (i) PPh$_3$; (ii) silica gel; (iii) LiAlH$_4$

Scheme 1

236

Whereas the 5-membered (12a) and the 6-membered (12b) cyclic imino-ether afford a variety of products on photolysis resulting from fission of the N-O bond, 3-phenyltetrahydro-1,2-oxazepine (13) is inert under these conditions. This stability is attributed to photochemical equilibration between the *syn* (13a) and the *anti* (13b) isomer (T. Mukai *et al.*, J. Photochem., 1981, __17__, 365).

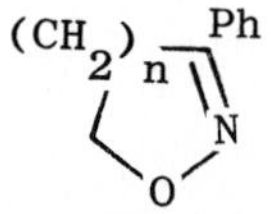

(12)
a, n = 1; b, n = 2

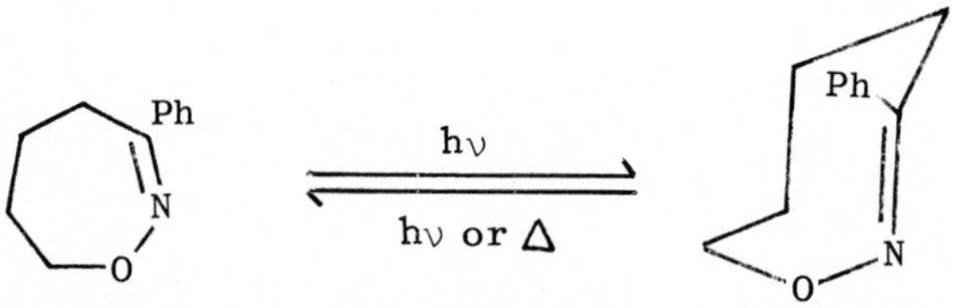

syn

(13a)

anti

(13b)

(ii) *1,3-Oxazepines*

Azapyrylium (1,3-oxazinium) salts are readily available and are useful precursors of a variety of heterocycles (R.R. Schmidt, Synthesis, 1972, 333). 2,4,6-Triaryl-1,3-oxazinium perchlorates undergo nucleophilic attack at C-6 and in the case of attack by the anion derived from ethyl diazoacetate, the adduct spontaneously eliminates nitrogen.   Concomitant ring expansion provides a neat route to 1,3-oxazepines (14), which show a low field nmr absorption for H-5 at about δ 8.1.   (R.R. Schmidt and G. Berger, Synthesis, 1974, 187).

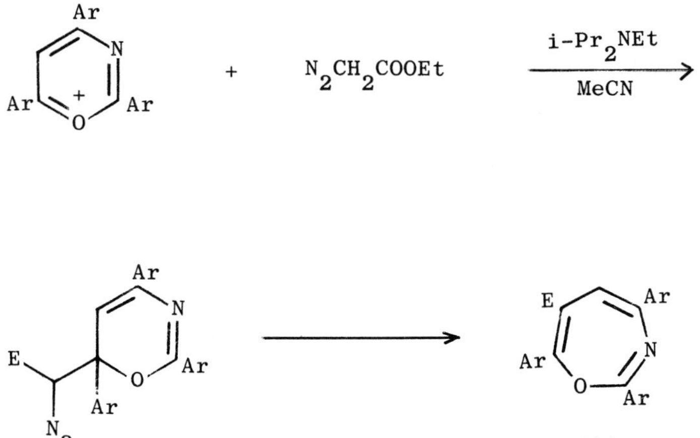

E = COOEt

A related synthesis of 1,3-oxazepines utilises tetra-substituted pyrylium salts which on treatment with sodium azide in acetonitrile afford the highly unstable 2-azido-2H-pyrans (15). Decomposition, by loss of nitrogen from their valence tautomers, gives a keto-vinyl azirine intermediate from which the oxazepine (16) is derived by opening of the azirine ring and a subsequent dipolar electrocyclisation. The products are generally obtained in high yields, particularly when tetra-aryl pyrylium salts are employed. Interestingly, 2,3,5,6-tetraphenylpyrylium perchlorate under these conditions affords an excellent yield of the intermediate *(Z)*-ketovinylazirine, which on heating rearranges to 2,4,6,7-tetraphenyl-1,3-oxazepine (P.L. Desbene and J.-C. Cherton, Tetrahedron, 1984, **40**, 3559).

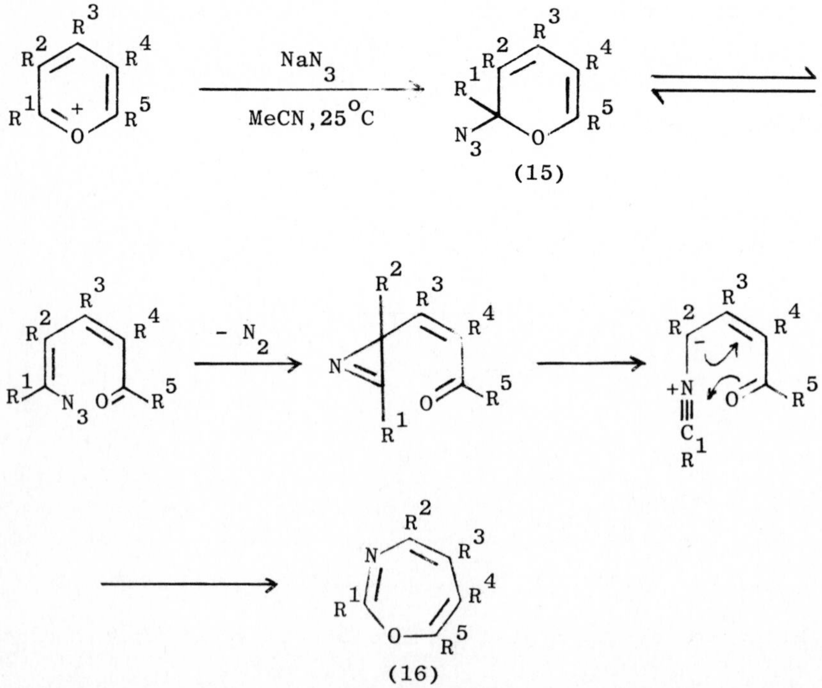

(15)

(16)

The above azirine -> oxazepine rearrangement may also be accomplished by the action of $Mo(CO)_6$ in THF, but phenylpyrroles and 2,3,5,6-tetraphenylpyridine are also formed. These products result from decomposition of a complexed nitrene species, whereas the oxazepine arises by a purely thermal process (F.D. Bellamy, Tetrahedron Letters, 1978, 4577; T. Kobayashi and M. Nitta, Bull. chem. Soc. Japan., 1985, 58, 1057).

Lithiation of benzyl isocyanide followed by treatment with oxetane affords an alcohol which may be cyclised in boiling benzene in the presence of copper(I) oxide to 4-phenyl-4,5,6,7-tetrahydro-1,3-oxazepine (17), b.p. $125°C/0.1$ mm (U. Schollkopf et $al.$, Ann., 1976, 2105).

(17)

The ^{1}H nmr-spectrum of (17) indicates the presence of two conformers in a ratio of 5:1. The preferred conformation is that in which the phenyl group is equatorially disposed. Rapid interconversion between the conformers is observed at 100°C. This also holds true for 4-(2-methoxyethyl)-4,5,6,7-tetrahydro-1,3-oxazepine (U. Schöllkopf *et al.*, Ann., 1976, 2105).

Irradiation of isoquinoline-*N*-oxide in benzene affords 1,3-benzoxazepine. The product is only stable in dilute solution and any attempt to isolate it results in the formation of polymeric material. The structure of the benzoxazepine has been established spectroscopically and confirmed by the nature of the ring opened product obtained by the addition of piperidine (A. Albini, G.F. Bettinetti and G. Minoli, Tetrahedron Letters, 1979, 3761).

Irradiation of 1-methoxyisoquinoline-*N*-oxide or 3-methoxyisoquinoline-*N*-oxide under identical conditions afford the 1,3-benzoxazepines (18) and (19) respectively, as oils of sufficient stability to permit structure determination.

(18)   (19)

Irradiation of 4-methoxyisoquinoline-*N*-oxide gives a mixture of 4-methoxyisoquinoline and 4-methoxy-1(2H)isoquinolone (A. Albini *et al.*, J. chem. Soc. Perkin I, 1980, 2738).

A particularly facile preparation of substituted 1,3-benzoxazepines is based on the 2-azidopyran decomposition reaction described earlier (page 237). Addition of sodium azide to a 1-benzopyrylium perchlorate in acetonitrile at 0°C affords either a 2-azido-2H-chromene (20) or a 4-azido-4H-chromene (21) depending on the substitution pattern of the starting material.

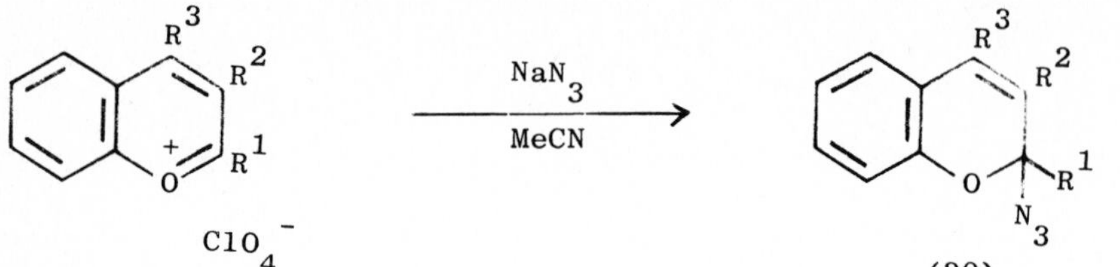

R^1 = Aryl R^2 = R^3 ≠ H
R^1 = t-Bu R^2 = R^3 = H

or

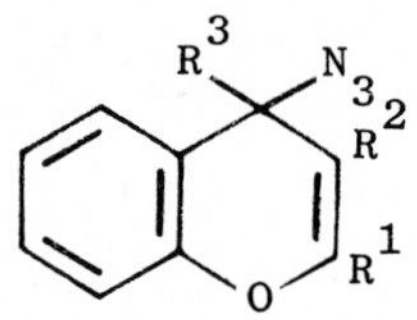

(21)

R^1 = Aryl R^2 = R^3 = H

These azido compounds are stable isolable materials, which in refluxing xylene afford 1,3-benzoxazepines in high yield.

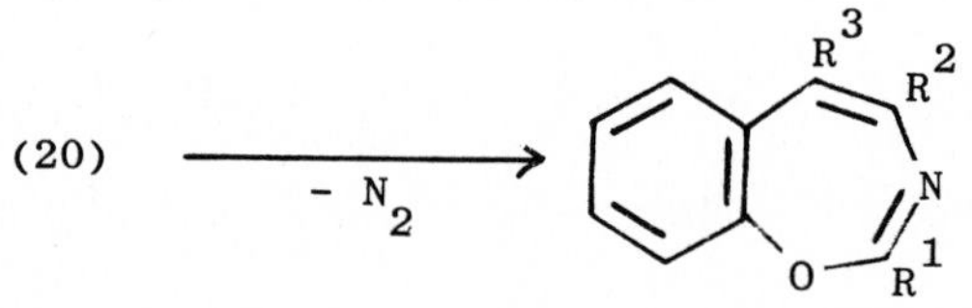

Surprisingly, the formation of 1,5-benzoxazepines from (21) is not observed. Decomposition of the 4-azidochromenes proceeds via rearrangement to the isomeric 2-azido compound leading ultimately to the 1,3-benzoxazepine. For example, 4-azido-2-phenyl-4H-chromene affords 2-phenyl-1,3-benzoxazepine, m.p. 56-67°C, in 70% yield.

Tetracyanoethylene (TCNE) rapidly adds to the 4-methylenehomochromene (22) affording the homo-aromatic zwitterion (23) which readily cyclises to the benzoxepin (24), the latter undergoing a Cope rearrangement to the 1,3-benzoxazepine (25). The reaction proceeds rapidly at room temperature (H. Yamaoka *et al.*, Chem. Letters, 1979, 523)

Treatment of the 3,4-dihydro-2*H*-1,3-benzoxazin-2-one (26) with hydrochloric acid in boiling dioxane results in a novel ring expansion to the 1,3-benzoxazepine (27). Initial ring opening is followed by attack of the carbamate function on the protonated nitro group and dehydration completes the reaction sequence (A.R. Katritzky *et al.*, Heterocycles, 1984, **22**, 1155).

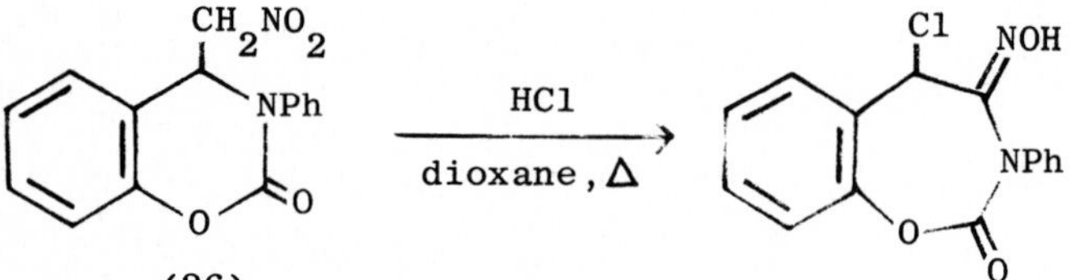

(26)   (27)

Cyclisation of the amide (28) to 8-chloro-2-(3,4-dimethoxybenzyl)-4,5,6,7-tetrahydro-1,3-benzoxazepine is initiated by treatment with phosphoryl chloride in acetonitrile (H. Iida *et al.*, Chem. pharm. Bull., 1978, **26**, 3229).

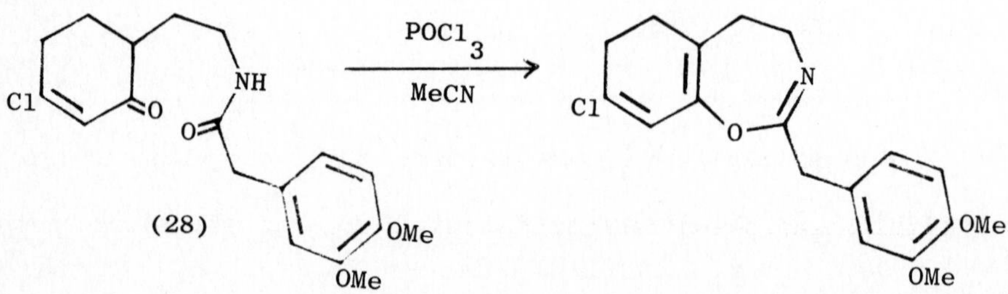

(28)

2-Phenyl-3,1-benzoxazepines result from heating 1-azido-1-phenylisochromenes in xylene. The iminopyrans (29) also obtained are formed by aryl migration in the nitrene intermediate (J.-P. Le Roux, P.L. Desbene and J.-C. Cherton, J. heterocyclic Chem., 1981, 18, 847).

(29)

Irradiation of quinoline-N-oxides has been used to prepare a variety of substituted 3,1-benzoxazepines (A. Albini *et al.*, J. chem. Soc. Perkin I, 1980, 2738; R.W. Irvine, J.C. Summers and W.C. Taylor, Austral. J. Chem., 1983, 36, 1419). Photo-isomerisation of quinoline-N-oxide affords 3,1-benzoxazepine as a pale yellow oil, b.p. 80°C/0.2 mm, in *ca.* 50% yield. (A. Albini, G.F. Bettinetti and G. Minoli, Org. Synth., 1983, 61, 98).

o-(β-Hydroxyalkyl)phenylisocyanides, readily available from *o*-lithiomethylphenylisocyanide and either an aldehyde or a ketone, cyclise to 4,4-disubstituted-4,5-dihydro-3,1-benzoxazepines when refluxed in benzene containing a catalytic amount of copper(I) oxide (Y. Ito, K. Kobayashi and T. Saegusa, Tetrahedron Letters, 1978, 2087).

The phenethyl alcohol (30) is oxidised by manganese dioxide in dichloromethane to 6-nitro-1-methyl-1,2,4,5-tetrahydro-3,1-benzoxazepine in excellent yield. Formation of the corresponding phenylacetaldehyde derivative is not observed. It is proposed that the heterocyclic product results from initial formation of a carbinol amine (F. Kienzle, Tetrahedron Letters, 1983, 2213).

1-Phenyl-1,5-dihydro-2,4-benzoxazepin-3(4H)-ones are obtained by treatment of the benzyl alcohols (31) with phosgene in the presence of sodium carbonate (L. Fontanella and L. Mariani, Farmaco Ed. Sci., 1975, 30, 773).

(31)

The photochemical rearrangement of 6-cyanophenanthridine-5-oxide yields dibenzo[d,f][1,3]oxazepin-6-carbonitrile, although the solvent can influence the outcome of the reaction (C. Kaneko *et al.*, Chem. pharm. Bull., 1978, 26, 2508). Evidence for the participation of an oxaziridine intermediate has been obtained (K. Tokumura *et al.*, J. Amer. chem. Soc., 1980, 102, 5463).

Whilst the irradiation of pyridine-, quinoline-, isoquinoline- and acridine-N-oxides provides a convenient synthesis of 1,3-oxazepines and their benzologues, numerous by-products often result from rearrangement or fragmentation of the initially formed oxazepine. Decomposition of 1,3-oxazepines may also be initiated thermally or by treatment with acid or base.

The thermal and photochemical reactions of 1,3-oxazepines have been discussed in some detail (F. Bellamy and J. Streith, Heterocycles, 1976, 4, 1391; T. Mukai, T. Kumagai and O. Seshimoto, Pure appl. Chem., 1977, 49, 287). For example, 2-phenyl-1,3-oxazepine when heated in benzene undergoes ring contraction to 3-hydroxy-2-phenylpyridine in quantitative yield (T. Tezuka et al., Chem. Comm., 1974, 373). The reaction proceeds through the oxazanorcaradiene valence tautomer.

A solution of boron trifluoride etherate in dry benzene readily promotes ring contraction of 2,4,6,7-tetraphenyl-1,3-oxazepine to the pyridones (31) and (32). The products result from ring cleavage and concomitant phenyl migration within the BF_3-coordinated valence tautomer (T. Kobayashi and M. Nitta, Heterocycles, 1985, 23, 1399). The use of moist benzene at ambient temperature results in the rapid formation of numerous products.

Aryl substituted 1,3-oxazepines undergo a [4+2] cycloaddition with the potent dienophile 4-phenyl-1,2,4-triazoline-3,5-dione (PTAD). The cycloaddition occurs across the C-4 – C-7 diene system of the heterocyclic ring. The adducts (33) are readily hydrolysable to the hydroxyamides (34). Indeed, the cycloaddition with 2-phenyl-1,3-oxazepine affords (34, R=H) directly. Tetra- and penta-phenyl-1,3-oxazepines appear to be unreactive towards PTAD (T. Toda *et al.*, Heterocycles, 1978, **11**, 331).

R = H or Ph

Irradiation of 2-cyano- or 2-phenyl-3,1-benzoxazepine in acetonitrile or methanol results in ring contraction to the corresponding indole-3-carboxaldehyde. The same product is formed by prolonged irradiation of the corresponding quinoline-N-oxide, thus precluding the necessity to isolate the benzoxazepine.

The isolation of methyl 3-acetyl-2-phenyl-3H-indole-3-carboxylate (36) from the irradiation of (35) supports the view that ring contraction proceeds by $[\pi 2s + \sigma 2s]$ pericyclic reaction affording a 3H-indole which rearranges to the product (C. Kaneko and R. Kitamura, Heterocycles, 1977, 6, 111).

Heating the ester (35) in methanol leads to a mixture of the 3H-indole and the indole (37) (C. Kaneko *et al.*, Chem. pharm. Bull., 1980, **28**, 1157).

Irradiation of 4-carboxyquinoline-*N*-oxides affords indole-3-carboxaldehydes directly; neither the derived benzoxazepine nor 3H-indole could be isolated. This contrasts with the behaviour of the corresponding methyl esters (38, R=Me, Ph) which, on irradiation in acetone or acetonitrile, afford isolable methyl 3,1-benzoxazepine-5-carboxylates. The latter undergo photochemical decarbonylation to methyl indole-3-carboxylates (R. Kitamura *et al.*, Tetrahedron Letters, 1977, 2911).

The addition of methanol to methyl 2-phenyl-3,1-benzoxazepine-5-carboxylate (39) occurs both photochemically and thermally. The former method affords *(Z)*-methyl 3-methoxy-2-(2-benzamidophenyl)acrylate (40) as the major product, whereas the *(E)*-isomer (41) predominates under thermal conditions. A mechanistic rationale to account for the differing stereoselectivity in these reactions has been advanced (M. Somei *et al.*, Chem. Comm., 1977, 899; C. Kaneko, Chem. pharm. Bull., 1982, 30, 74).

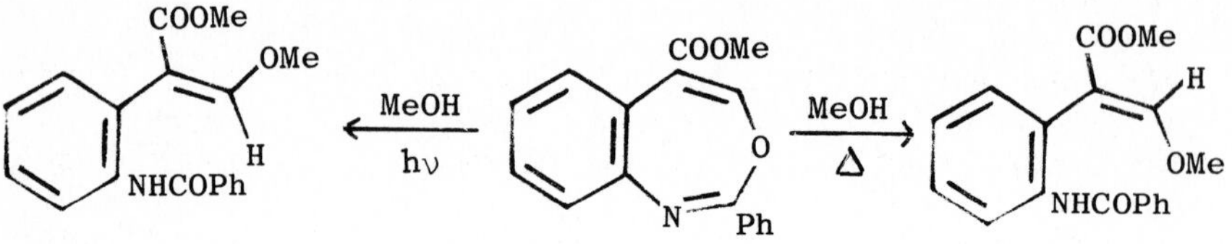

254

(40)   (39)   (41)

Nucleophilic attack of ethanol on methyl 2-cyano-3,1-benzoxazepine-6-carboxylate occurs at C-2 with subsequent ring cleavage and lactonisation giving the isocoumarin (42) in quantitative yield. Subsequent hydrolysis affords 5-amino-isocoumarin. The indoline (43) is obtained by treatment of methyl 2-cyano-3,1-benzoxazepine-6-carboxylate with aqueous methylamine (C. Kaneko, A. Yamamoto and M. Hashiba, Chem. pharm. Bull., 1979, 27, 946).

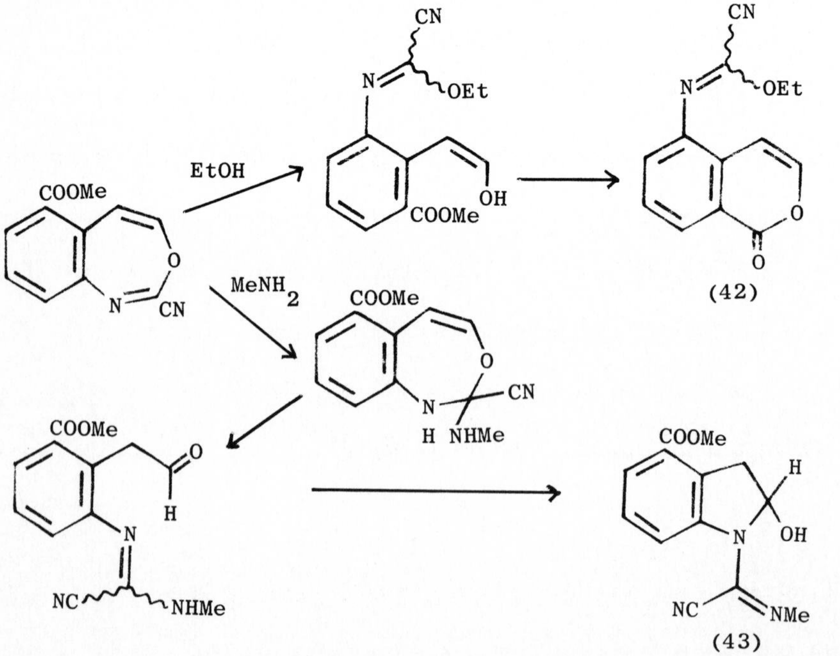

(42)

(43)

(iii)   1,4-Oxazepines

Fully unsaturated 1,4-oxazepines have been obtained for the first time.   Treatment of a substituted pyridine with phenyl magnesium bromide in the presence of benzyl chloroformate and subsequent irradiation affords a 2-azabicyclo[2.2.0]hex-5-ene (44) (F.W. Fowler, J. org. Chem., 1972, 37, 1321). The latter is converted by a three-step sequence of reactions into a 3-aza-7-oxatricyclo[4.1.0.0^{2,5}]hept-3-ene which upon brief irradiation in acetonitrile undergo valence isomerisation to give the corresponding 1,4-oxazepine (45) in excellent yield   (J. Kurita, K. Iwata and T. Tsuchiya, Chem. Comm., 1986, 1188).

Bn = CH$_2$Ph

Reagents: (i) MCPBA; (ii) H$_2$, Pd-C; (iii) t-BuOCl, DBU

The electrophilic oxazolinium salt (46) readily condenses with the carbanion derived from ethyl cyanoacetate. Ring opening of the initially formed oxazolidine under the basic reaction conditions and subsequent lactonisation affords 6-cyano-4-methyl-5-phenyl-2,3,4,7-tetrahydro-1,4-oxazepine-7-one in high yield (M. Dreme *et al.*, Tetrahedron Letters., 1982, 73; Tetrahedron, 1984, **40**, 349).

The Michael addition of *erythro*-1,2-diphenyl-2-phenylaminoethanol to dimethyl acetylenedicarboxylate (DMAD) in methanol at ambient temperature gives the *cis*-adduct (47) together with the 1,4-oxazepine-7-one (48). Exclusive formation of (47) is observed when the condensation is performed in refluxing benzene. Irradiation, or treatment with acetic acid in benzene, effects ring closure of (47) to produce (48), together with the derived hemi-acetal (49). *threo*-1,2-Diphenyl-2-phenylaminoethanol affords the *trans* adduct with DMAD. Subsequent ring closure to (48) is effected photochemically (O. Tsuge, K. Oe and T. Ohnishi, Heterocycles, 1982, **19**, 1609).

E = COOMe

2,4,5-Trimethoxybenzamide   and   2,4'-dibromoacetophenone
condense to give a complex mixture of products, a major
component of which is the 4,5-dihydro-1,4-benzoxazepine (50)
(F. Sanchez-Viesca and M.R. Gomez, Rev. Latinoamer. Quim.,
1982, 13, 67).

(50)

Ring expansion of chroman-4-ones by means of a Schmidt or
Beckmann reaction provides a convenient route to 2,3,4,5-
tetrahydro-1,4-benzoxazepinones   as   a   result   of   alkyl
migration.   Many examples of these reactions have been
reported (I.M. Lockhart in 'Chromenes, Chromanones and
Chromones' ed. G.P. Ellis, Wiley, New York, 1977, p.314).
The   oxazepines   (51),   which   exhibit   anti-inflammatory
activity   are   obtained   by   a   Schmidt   reaction   on   the
corresponding chromanone (D.R. Shridhar *et al.*, Indian J.
Chem., 1979, 17B, 155).

(51)

The formation of 1,4-benzoxazepines by ring closure of phenoxyethyl amides in a Bischler-Napieralsky type reaction has been exploited extensively (A. Waefelaer, *et al.*, Bull. Soc. chim. Belg., 1976, **85**, 787; R.C. Effland *et al.*, J. heterocyclic Chem., 1982, **19**, 537). The reaction can be used to prepare 5-aryl-2,3-dihydro-1,4-benzoxazepines. For example, benzoylation of 2-(3-methoxyphenoxy)ethylamine followed by treatment with phosphorus oxychloride in acetonitrile effects regioselective cyclisation to 8-methoxy-5-phenyl-2,3-dihydro-1,4-benzoxazepine. Methylation and reduction affords the tetrahydro compound (52) (J.B. Bremner, E.J. Browne and I.W.K. Gunawardana, Austral. J. Chem., 1984, **37**, 129).

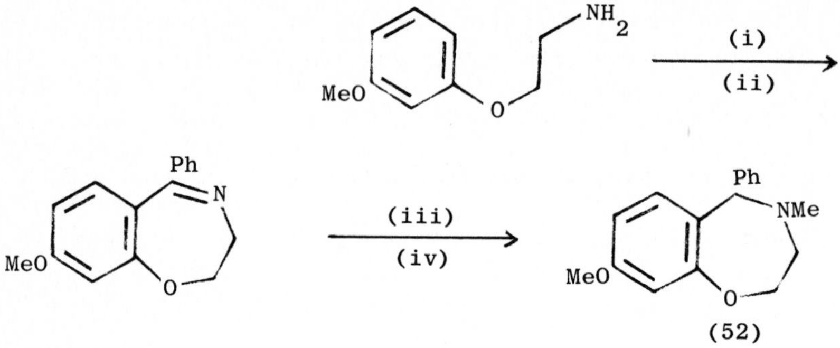

Reagents: (i) PhCOCl, pyridine; (ii) POCl$_3$, MeCN;
(iii) MeI; (iv) NaBH$_4$

A related approach to 5-aryl-2,3-dihydro-1,4-benzoxazepines involves cyclodehydration of 2-(aminoethoxy)benzophenones (54) which occurs readily in boiling pyridine. The starting material is available from a 2-hydroxybenzophenone as illustrated below; μ-oxo-bis(chlorotriphenylbismuth) promotes selective oxidation of the carbinol (53) (*idem, ibid.*).

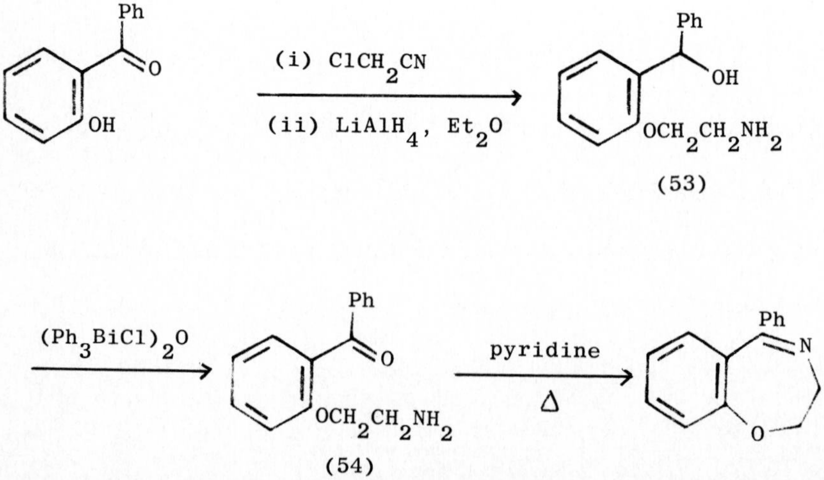

Amino acids readily participate in the Mannich reaction with phenols. Thus *N*-(3,5-dimethyl-2-hydroxybenzyl)sarcosine (55) cyclises with thionyl chloride to 4,7,9-trimethyl-2,3,4,5-tetrahydro-1,4-benzoxazepine-2-one (J.H. Short and C.W. Ours, J. heterocyclic Chem., 1975, 12, 869).

Hydrogenation of the ketone (56) induces cyclisation to the 4,1-benzoxazepine; the succinimide moiety is readily cleaved giving 2-aminomethyl-1,2,3,5-tetrahydro-4,1-benzoxazepine (C. Banzatti *et al.*, J. heterocyclic Chem., 1983, 20, 139).

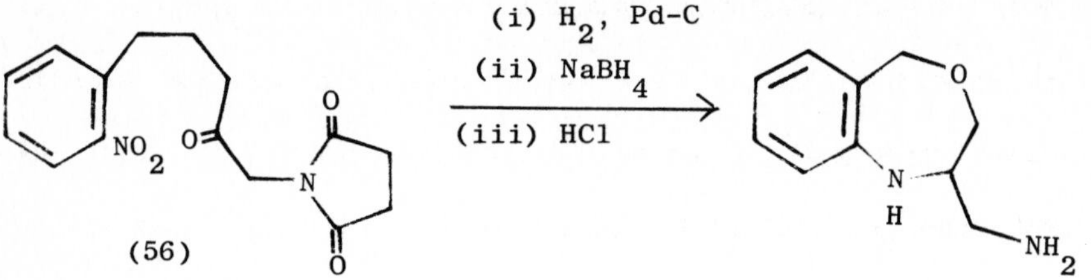

(i) H_2, Pd-C

(ii) $NaBH_4$

(iii) HCl

(56)

The dicarboxylic acid (57) may be cyclodehydrated to the 4,1-benzoxazepine–3,5–dione (58) by treatment with acetic anhydride at 70°C. At elevated temperatures, an alternative pathway supervenes leading to exclusive formation of (59) (E. Tighineanu, F. Chiraleu and D. Raileanu, Tetrahedron, 1980, 36, 1385).

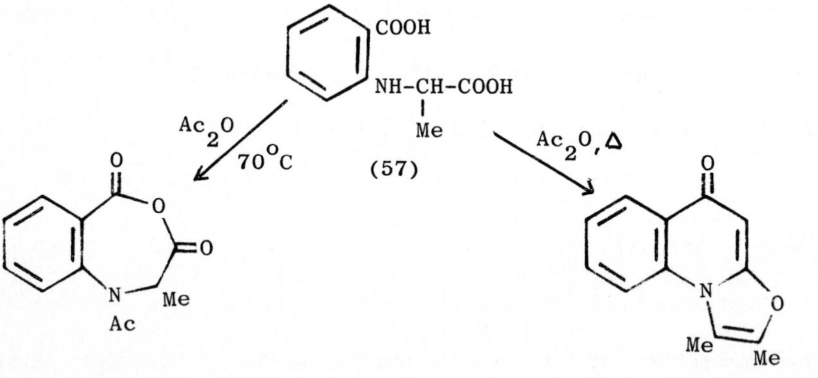

Ac_2O

70°C

(57)

Ac_2O, Δ

(58)

(59)

The broad spectrum of biological activity exhibited by the dibenzo[1,4]oxazepines has ensured that synthetic work in this area has continued unabated.

Many dibenzo[b,f][1,4]oxazepines have been prepared; the major synthetic routes still depend upon formation of the seven membered ring by cyclisation of either an *ortho* substituted diphenyl ether or an *ortho* substituted benzanilide. In contrast, the dibenzo[b,e][1,4]oxazepine system has received much less attention and is accessible by cyclisation of substituted benzyl phenyl ethers or diphenylamine derivatives. The synthesis, reactivity and biological activity of dibenzoxazepines have been reviewed (K. Nagarajan, Stud. org. Chem. (Amsterdam), 1979, 3 (New Trends Heterocyclic Chem.), 317). The syntheses of dibenzo[b,f][1,4]oxazepine and its mono–aza and di–aza derivatives and 1,4–benzoxazepine with a fused thiophene ring have been the subject of a separate review (C.O. Okafor, Heterocycles, 1977, 7, 391).

Ring closure of the anthranilic acid (60) to 7–chlorodibenzo[b,e][1,4]oxazepine–11–(5H)–one is readily accomplished under mild conditions by treatment with cyanogen bromide in the presence of an organic base. The product results from ring closure *via* an intermediate acyl cyanate (D.H. Kim, J. heterocyclic Chem., 1981, 18, 855).

(60)

Treatment of 2-aminodiphenyl ether with two equivalents of n-butyllithium affords the C,N-dilithio derivative which may be quenched with any one of a variety of electrophiles to afford a dibenzo[b,f][1,4]oxazepine (N.S. Narasimhan and P.S. Chandrachood, Synthesis, 1979, 589; N.R. Shete, Indian J. Chem., 1982, 21B, 581).

The synthesis of the eight monomethoxydibenzo[b,f][1,4]-oxazepin-11(10H)-ones has been accomplished by cyclodehydration of the appropriate methoxy 2-amino-2'-carboxydiphenylether. The lactams are readily demethylated to the corresponding hydroxy compound by treatment with sodium ethanethiolate in DMF (K. Brewster *et al.*, J. chem. Soc. Perkin I, 1976, 1286).

The imine (61) undergoes hydrolysis with concomitant cyclodehydration to 2,3-dihydro-8H-anthra[9,1-e,f][1,4] oxazepin-8-one when subjected to chromatography on silica gel (A.P. Krapcho and K.J. Shaw, J. org. Chem., 1983, **48**, 3341).

(61)

Syntheses of 1,5-benzoxazepines almost invariably utilise 2-aminophenol or a derivative. For example, *N*-propylidene-and *N*-benzylidene-2-hydroxyaniline undergo cyclocondensation with carbon suboxide to afford the novel 5-alkylidene-4-oxido-2,5-dihydro-1,5-benzoxazepinium betaines (62) (L. Bonsignore *et al.*, Synthesis, 1982, 945).

$$\text{(benzene ring with } N=CHR, OH) + C_3O_2 \xrightarrow[0^\circ C]{Et_2O} \text{(product)}$$

(62)

2-Aminophenol and carbon suboxide condenses to give 1,5-benzoxazepine-2,4-(3H,5H)-dione in low yield. ^{1}H nmr-spectroscopy indicates that in DMSO solution the dione and enol tautomers are present in the ratio of 1:1 (L. Bonsignore *et al.*, J. heterocyclic Chem., 1982, **19**, 1241).

$$\text{(aminophenol)} + C_3O_2 \longrightarrow \text{(product)}$$

$$\rightleftharpoons \text{(enol form)} \rightleftharpoons \text{(keto form)}$$

Perfluoro-2-methyl-2-pentene reacts with 2-aminophenol in a triethylamine-ether mixture to give 4-fluoro-2-(pentafluoroethyl)-3-(trifluoromethyl)-1,5-benzoxazepine. The reaction is initiated by attack of the phenoxide ion at the less substituted olefinic carbon followed by expulsion of fluoride, generating a highly reactive terminal perfluoro-olefin which cyclises to the product.

However, when perfluoro-2,4-dimethyl-3-heptene reacts with 2-aminophenol in DMF, the sole product is the benzoxazepino-benzoxazepine (63). The initially formed 1,5-benzoxazepine could not be isolated, being readily attacked by a second molecule of 2-aminophenol (M. Maruta *et al.*, J. fluorine Chem., 1980, 16, 75).

(63)

Dimethyl allene-1,3-dicarboxylate condenses with 2-aminophenol giving 4-methoxycarbonylmethylene-2,3,4,5-tetrahydro-1,5-benzoxazepin-2-one (64) in low yield (J. Ackroyd and F. Scheinmann, J. chem. Res., (M), 1982), 1012).

(64)

Sodium 2-nitrophenoxide in acetonitrile effects the stereospecific *cis* ring opening of methyl *trans* 3-(4-methylphenyl)glycidate to the *threo* ester (65). The latter on catalytic hydrogenation and hydrolysis gives the corresponding amino-acid which is cyclodehydrated to the *cis*-1,5-benzoxazepinone by a mixture of dicyclohexylcarbodiimide (DCC) and 1-hydroxybenzotriazole (BtOH).

Stereospecific *trans* ring-cleavage to the *erythro* ester occurs when ethanol is substituted for acetonitrile. The *erythro* ester gives the *trans* 1,5-benzoxazepinone (T. Hashiyama *et al.*, Chem. pharm. Bull., 1985, 33, 634).

(65)

Reagents: (i) H$_2$,Pd-C; (ii) OH$^-$; (iii) DCC-BtOH

Heating a 2-unsubstituted 5-phenyl-1,4-oxazepine results in ring contraction with the formation of a 2-hydroxypyridine, which probably arises by isomerisation of an oxazanorcaradiene intermediate.

On the other hand, 2,7-dimethyl-5-phenyl-1,4-oxazepine undergoes isomerisation to the 1,3-oxazepine (66) when heated. Again participation of an oxazanorcaradiene is invoked enabling the oxygen atom to 'walk' around the ring. Acid hydrolysis of (66) gives the pyrrole (67) (J. Kurita *et al.*, Chem. Comm., 1986, 1188).

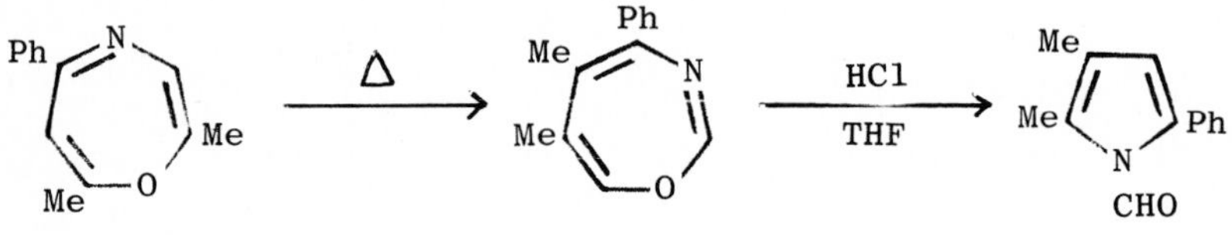

(66)   (67)

Attempted reduction of the amide group in 7-chloro-1,5-dihydro-1-methyl-5,5-diphenyl-4,1-benzoxazepin-2(3H)-one with lithium aluminium hydride results in ring contraction to (68). A mechanism involving initial enamine formation has been proposed (J. Szmuszkovicz *et al.*, J. org. Chem., 1986, **51**, 5001).

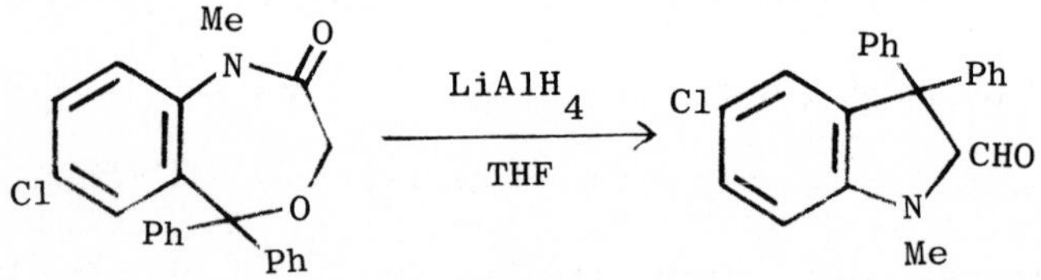

(68)

There are few reports concerning the reactivity of other benzologues of 1,4-oxazepines, but a number of references to the reactivity of the azomethine unit in dibenzo[b,f][1,4]oxazepine have appeared.

The dibenzoxazepines (69) are reductively methylated by sodium borohydride in acetic acid in a simple one-pot operation. The reaction appears to be general and is compatible with a variety of substituents at C-11 (T.C. McKenzie, Synthesis, 1983, 288).

(69)

Dibenzo[b,f][1,4]oxazepine reacts with sodium dichloroisocyanurate under both homogeneous (aqueous ethanol) and heterogeneous (chloroform-water) conditions. The latter conditions result in the formation of nuclear chlorinated derivatives; thus the 7-chloro-, 9-chloro-, and 7,9-dichloro- derivatives are obtained. Under homogeneous conditions, oxidation of the azomethine unit occurs. The major product is salicylaldehyde, but smaller amounts of the ring contracted products (70) and (71) are also formed (I.W. Lawson et al., J. chem. Soc. Perkin I, 1979, 2642).

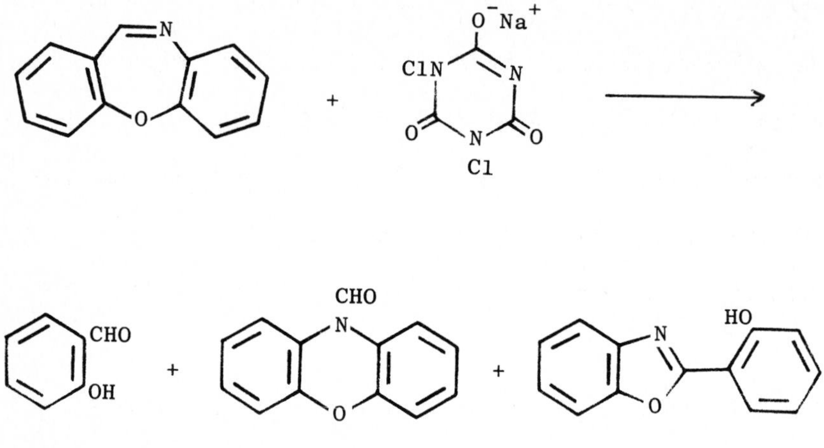

(70)   (71)

Oxidation using peracetic acid follows a similar course producing (70) and (71) together with dibenzo[b,f][1,4]oxazepin-11(10H)-one, once more implicating an oxaziridine intermediate (K. Brewster *et al.*, J. chem. Soc. Perkin I, 1976, 1291).

The structures of the novel 6-substituted-4-methyl-5-phenyl-2,3,4,7-tetrahydro-1,4-oxazepin-7-ones (72), obtained from the oxazolinium salt (46), have been confirmed by X-ray crystallography (M. Dreme *et al.*, Acta Cryst., 1983, C39, 1127).

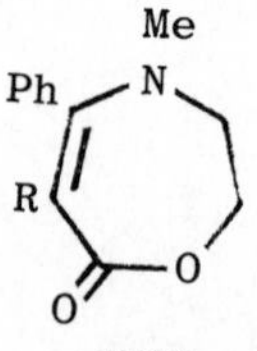

R = CN or COOEt

(72)

Analysis of the 270 MHz ^{1}H nmr spectrum of *cis*-2-ethyl-7-
methylhexahydro-1,4-oxazepines and that of the *trans*-isomer
allows an unambiguous assignment of configuration. The C-7
methine proton absorbs at δ 3.78 in the *trans* isomer, whilst
in the *cis* isomer this proton resonates at δ 3.94 (S.
Hernestam, J. heterocyclic Chem., 1983, 20, 1681).

The fragmentation pathways in the mass spectrum of 2,3,4,5-
tetrahydro-1,5-benzoxazepine have been elucidated (H.
Budzikiewicz and U. Lenz, Org. mass Spec., 1985, 10, 992).

The X-ray crystal structure of loxapine,
dibenzo[b,f][1,4]oxazepine,(73), as the succinate monohydrate,
has been determined. The seven membered ring adopts a
boat conformation, whilst the piperazine ring is in the
chair form. The dihedral angle between the planes of the
benzene rings is 121° (J.P. Fillers and S.W. Hawkinson, Acta
Cryst., 1982, B38, 3041). Loxapine has been used in the
treatment of schizophrenia.

(73)

(b) Oxadiazepines

A number of new syntheses of oxadiazepines have been developed since the publication of the 2nd edition. However of the nine possible isomeric oxadiazepines there are no examples of 1,2,3- or 1,2,7-oxadiazepine ring systems. Fully unsaturated monocyclic systems in this series are uncommon.

(i) 1,2,4-Oxadiazepines

4,7-Dihydro-5-phenyl-1,2,4-oxadiazepin-3(2H)-ones result from the base catalysed condensation of alkylidene acetophenones with N-hydroxyurea. Deprotonation of the intermediate β-ureidoxy ketone initiates cyclodehydration to the product via a 7-exo-trig ring closure (R. Jacquier et al., Tetrahedron Letters, 1975, 2979; C. Lassalvy, C. Petrus and F. Petrus, Gazz., 1981, 111, 273).

Treatment of these compounds with sodium hydroxide and either methyl iodide or dimethyl sulphate affords exclusively the O-methylated product (e.g. 74).

(74)

The dihydro-1,2,4-oxadiazepin-3(2H)-ones undergo a novel ring contraction to 2-hydroxypyrimidines on treatment with phosphorus oxychloride in pyridine. A mechanism involving co-ordination of the phosphorus reagent to the ring oxygen of the imidol tautomer with subsequent 1,7-bond cleavage and aromatisation has been proposed. This transformation may also be effected thermally (C. Lassalvy *et al.*, Gazz., 1981, 111, 273).

(ii)   1,2,6-Oxadiazepines

4,5,6,7-Tetrahydro-1,2,6-oxadiazepines are readily obtained from *syn-(E)-* ω -(benzylamino)propiophenone oximes.   The reaction with phosgene affords the 7-oxo compounds (75) as the predominant product (H. Gnichtel and K. Hirte, Ber., 1975, <u>108</u>, 3387), whilst condensation with formaldehyde provides (76) (H. Gnichtel *et al.*, Ber., 1980, <u>113</u>, 3373).

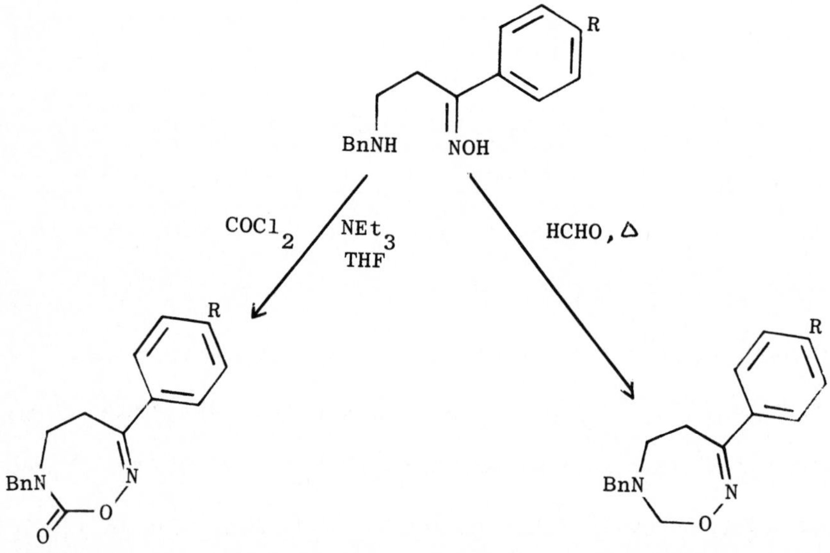

Bn = CH$_2$Ph

Similarly, the carbonylation of the aminobenzophenone oxime (77) affords 7-chloro-5-phenyl-3,1,4-benzoxadiazepin-2(1H)-one (T.S. Sulkowski and S.J. Childress, J. org. Chem., 1962, <u>47</u>, 4424).   Prolonged irradiation of the 3,1,4-

benzoxadiazepinone in solution results in elimination of carbon dioxide and ring contraction to the indazole (78) (K.-H. Pfoertner and J. Foricher, Helv., 1982, 65, 798).

(iii) *1,3,4-Oxadiazepines*

A particularly facile synthesis of 2,5,6-trisubstituted 6,7-dihydro-1,3,4-oxadiazepines utilises the readily available 1-aryl-3-hydroxy-2-methylpropanones (79). These hydroxy ketones condense with hydrazine and an alkyl carboxylic acid in a one-pot operation to afford the 2-alkyl-5-aryl-1,3,4-oxadiazepines (80). The corresponding 2-aryl compounds are accessible by acylation of (79), cyclisation to the product being accomplished by heating the intermediate ester with hydrazine and benzoic acid in a separate step (G. Cignarella *et al.*, Synthesis, 1984, 342).

Alkyl hydrazones of diacetone alcohol when heated with an aliphatic aldehyde in vacuo undergo cyclo-condensation to 2,3-dialkyl-5,7,7-trimethyl-2,3,6,7-tetrahydro-1,3,4-oxadiazepines (81) (G. Yu. Gadziev and G.I. Alekperov, Zh. org. Khim., 1977, 13, 1563).

By a similar process, salicylaldehyde butyl hydrazone and benzaldehyde condense to give 3-butyl-2-phenyl-2,3-dihydro-

1,3,4-benzoxadiazepine (82), b.p. 84–85°C/0.007 mm. (G. Yu. Gadzhiev and G.I. Alekperov, Khim. Geterotsikl. Soedin., 1982, 1319).

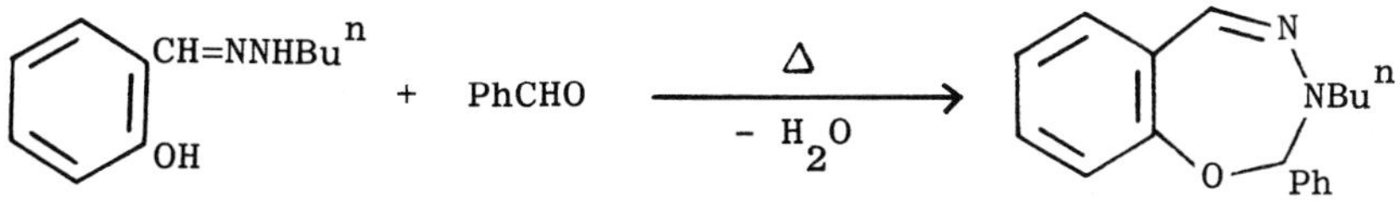

(82)

The *N*-acyl derivatives of salicylaldehyde phenylhydrazone cyclise on treatment with perchloric acid to 2-substituted-3-phenyl-1,3,4-benzoxadiazepinium perchlorates (V.I. Fomenko and A.I. Fomenko, Chem. Abstr., 1982, 96, 199645).

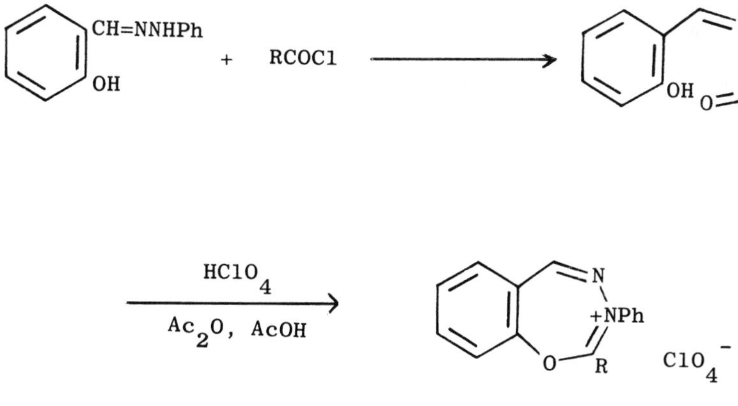

R = Me or Ph

(iv) 1,3,5-Oxadiazepines

Flash vacuum pyrolysis of 4-phenyl-1,2,3-benzotriazine affords the highly reactive 2-phenylbenzazet as the major product, which may be intercepted by a 1,3-dipole. For example, reaction with 4-methylbenzonitrile oxide affords the unstable cycloadduct (83) which through a complex rearrangement sequence affords 2-phenyl-4-(4-methylphenyl)-1,3,5-benzoxadiazepine (84). The latter may also be obtained by irradiation of 2-(4-methylphenyl)-4-phenylquinazoline-3-oxide (C.W. Rees *et al.*, Chem. Comm., 1975, 740).

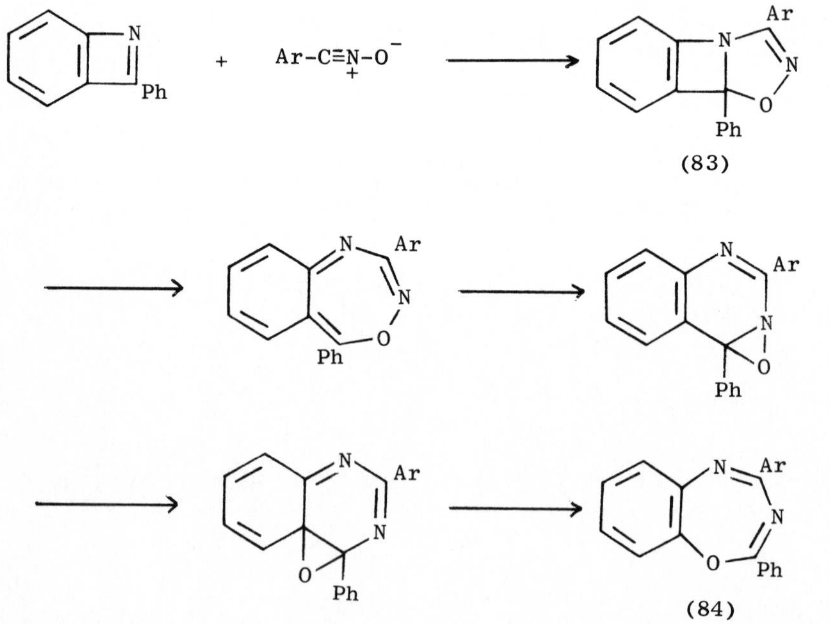

(83)

(84)

The X-ray crystal structure of (84) reveals that the molecule adopts a boat conformation in which there appears to be a significant degree of bond localisation in the heterocyclic ring (A.F. Cameron and A.A. Freer, Acta Cryst., 1976, B32, 1561).

Base catalysed condensation of the immonium salt (85), obtained from 1,1-dimethylthiourea, with 2-aminophenol affords 2-dimethylamino-4-methylthio-1,3,5-benzoxadiazepine, m.p. 89-90°C (M. Yokoyama, K. Arai and T. Imamoto, J. chem. Soc., Perkin I, 1982, 1059).

(85)

(v)   1,3,6-Oxadiazepines

The first fully unsaturated monocyclic 1,3,6-oxadiazepine has been prepared. Oxygenation of the 2,5-dimethoxypyrazine derivative (86) with singlet oxygen gives the corresponding endoperoxide which is ring cleaved with triphenylphosphine to give the oxadiazanorcaradiene (87) which undergoes spontaneous valence isomerisation to the 2,5-dimethoxy-1,3,6-oxadiazepine (88).

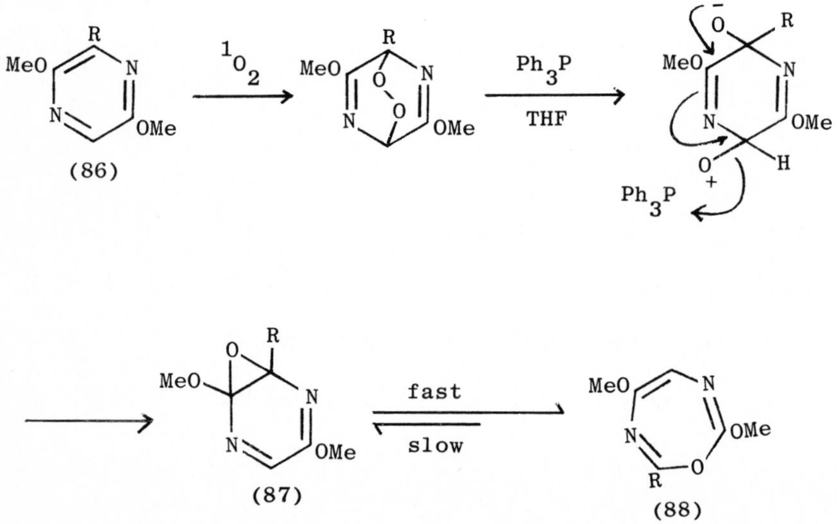

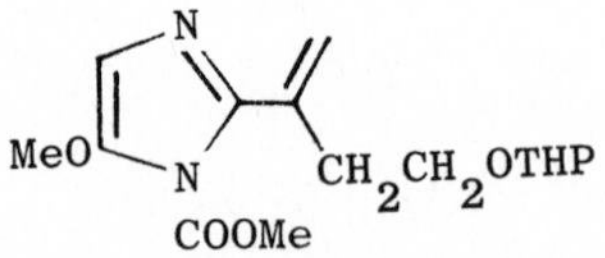

$$R = H_2C=C(CH_2CH_2OTHP)$$

THP = 2-Tetrahydropyranyl

The ^{13}C nmr-spectrum of the oxadiazepine (88), an unstable yellow oil which rapidly polymerises, exhibits ring carbon signals at δ 107, 142, 144 and 151. In solution in THF, (88) slowly converts into the imidazole (89) *via* the valence isomer (87).

Attempts to effect similar transformation of the endoperoxides from other alkyl and dialkyl substituted dimethoxy pyrazines results only in formation of imidazole derivatives (A.J. O'Connell, C.J. Peck and P.G. Sammes, Chem. Comm., 1983, 399).

Further examples of the photolytic rearrangement of quinoxaline-N-oxides to 3,1,5-benzoxadiazepines have been reported. Thus, either 2-methyl or 3-methyl quinoxaline-1-oxides afford 2-methyl-3,1,5-benzoxadiazepine, m.p. 53-55°C, in good yield. Similarly, irradiation of (90) affords 2,4-dimethyl-3,1,5-benzoxadiazepine, m.p. 71-73°C. Ring cleavage products result from irradiation of the quinoxaline-1-oxide or its 3-methoxy derivative (A. Albini, R. Colombi and G. Minoli, J. chem. Soc., Perkin I, 1978, 924).

(90)

These benzoxadiazepines are highly moisture sensitive, being readily converted into N,N'-diacyl-1,2-phenylenediamines which may exist in equilibrium with their cyclic tautomers.

Irradiation of a 2,3-diarylquinoxaline-1,4-dioxide results in formation of a 2,4-diaryl-3,1,5-benzoxadiazepine-1-oxide, albeit in low yield (A.A. Jarrar, J. heterocyclic Chem., 1978, 15, 177).

(vi)   1,4,5-Oxadiazepines

A number of 4,5-diacyl derivatives of perhydro-1,4,5-oxadiazepine have been obtained by condensation of 1,2-diacylhydrazines with bis(2-chloroethyl)ether (K. Krakowiak and B. Kofelko, Acta Pol. Pharm., 1975, **32**, 673).

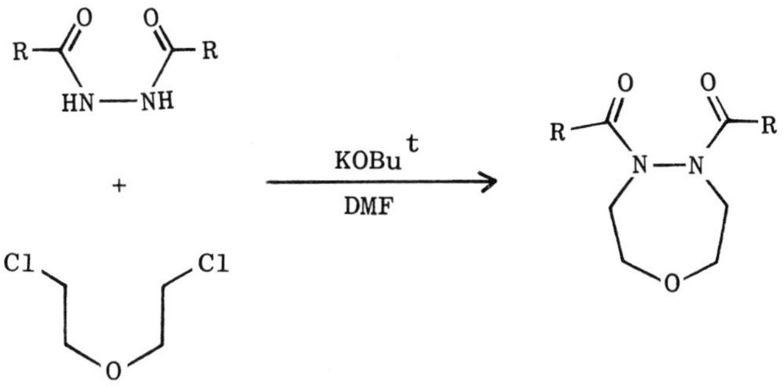

(c)   Oxatriazepines

The cyclo-condensation of benzoyl isothiocyanate with a ketone hydrazone affords a 2,3,4,5-tetrahydro-1,3,4,6-oxatriazepine-5-thione (G.J. Durant, J. chem. Soc., C, 1967, 952). This reaction is apparently the only means of obtaining this ring system. A mechanistic appraisal of this reaction has provided evidence for the involvement of betaine intermediates.

The thione (91), obtained from cyclohexanone phenylhydrazone, undergoes acid hydrolysis to the 1,2,4-triazolinethione (92) (I. Yamamoto *et al.*, J. chem. Soc., Perkin I, 1976, 2243).

PhCON=C=S   +   [cyclohexanone phenylhydrazone structure, NNHPh] $\xrightarrow[\text{room temp.}]{\text{PhH}}$

[structure (91)] $\xrightarrow{H^+}$ [structure (92)]

(91)   (92)

(d) *Dioxazepines*

The oxime ether (93) on treatment with base undergoes ring closure to the 6,7-dihydro-1,5,2-dioxazepin-4-one (94). Stereo-selective reduction of the *(S)* (+)-isomer affords *(R)* (+)-phenylglycine in 6% enantiomeric excess (J.I. Perez, J.M. Juarez and A.B. Rovira, Ann. Quim., 1979, 75, 958).

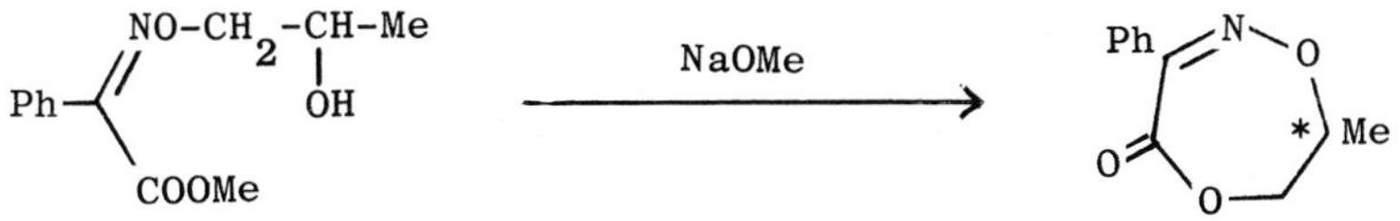

NO-CH$_2$-CH-Me
|
OH

Ph

COOMe

(93)

NaOMe →

(94)

* (S) (+)

$\xrightarrow{\text{Al/Hg}^{2+}}$ Ph-CH-COOH
|
NH$_2$

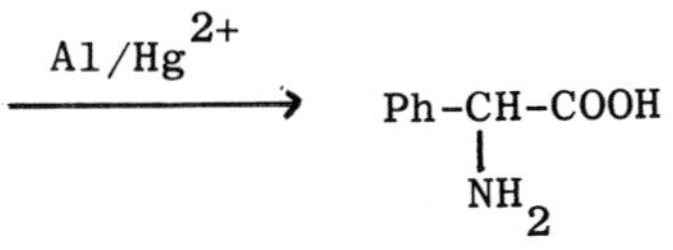

Primary amines condense with 1,2-diols and paraformaldehyde to give perhydro-1,5,3-dioxazepines. For example, isopropylamine and ethylene glycol give (95) b.p. 73°C/13 mm. The fused system (96) is obtained from cyclohexylamine and *trans*-cyclopentane-1,2-diol (H. Kapnang and G. Charles, Tetrahedron Letters, 1980, 2949).

$(CH_3)_2CHNH_2$ + $HOCH_2CH_2OH$ + $(CH_2O)_n$ $\xrightarrow[\Delta]{PhH}$

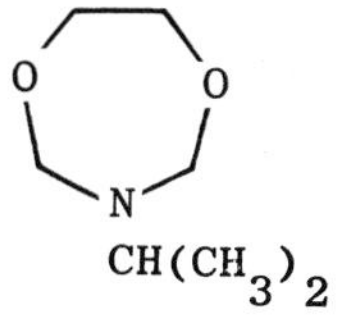

CH(CH$_3$)$_2$

(95)

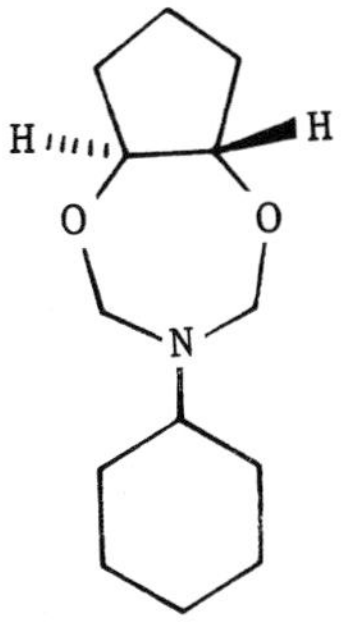

(96)

This system is readily cleaved by a Grignard reagent to afford an amine (H. Kapnang and G. Charles, Tetrahedron Letters, 1983, 1597).

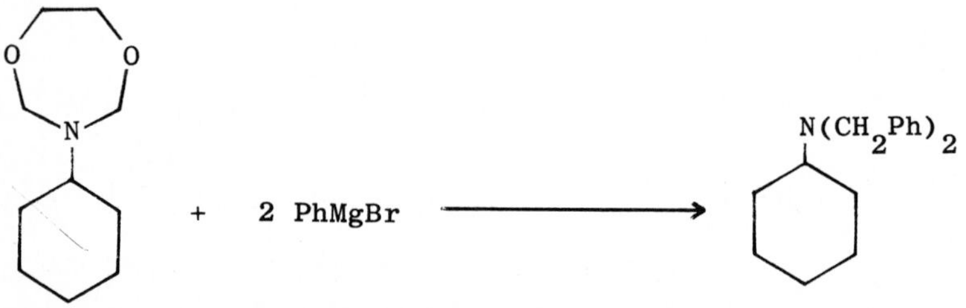

+ 2 PhMgBr ⟶

(e) *Dioxadiazepines*

The *(Z,Z)*-1,2-diketone dioximes (97) are alkylated by dichloromethane under strongly basic conditions to give a dialkyl 1,6,2,5-dioxadiazepine; dicyclohexyl derivative b.p. 135-136°C/1 mm; diisopropyl analogue, b.p. 60-61°C/1 mm.

The stereochemistry of the oxime (97) affects the outcome of the reaction, the *(E,E)-* or *(Z,E)-* isomer affording only a macrocyclic product (T. Hosokawa, T. Ohta and S.I. Murahashi, Chem. Comm., 1982, 7).

$$R \overset{N}{\underset{R}{\diagdown}} \overset{OH}{\underset{N}{\diagup}} \quad + \quad CH_2Cl_2 \quad \xrightarrow[\text{THF, reflux}]{\text{t-BuOK, 18-crown-6}} \quad R \overset{N-O}{\underset{R}{\diagdown}} \overset{O}{\underset{N-O}{\diagup}}$$

(97)

2. Rings containing oxygen, sulphur and selenium or tellurium

(a) Oxathiepins

(i) 1,2-Oxathiepins

The addition of dichlorocarbene to 3,4-dimethyl-2,5-dihydrothiophene dioxide affords the adduct (98). Reduction by lithium aluminium hydride gives a mixture of the *exo* - and the *endo-* 6-chloro-3-thiabicyclo[3.1.0]hexane 3,3-dioxide. The two anions derived from these adducts fragment by different pathways. The *endo*-isomer affords the thiopyran (99) exclusively, whilst the *exo*-isomer furnishes only the oxathiepin-2-oxide (100) (Y. Gaoni, J. org. Chem., 1981, **46**, 4502).

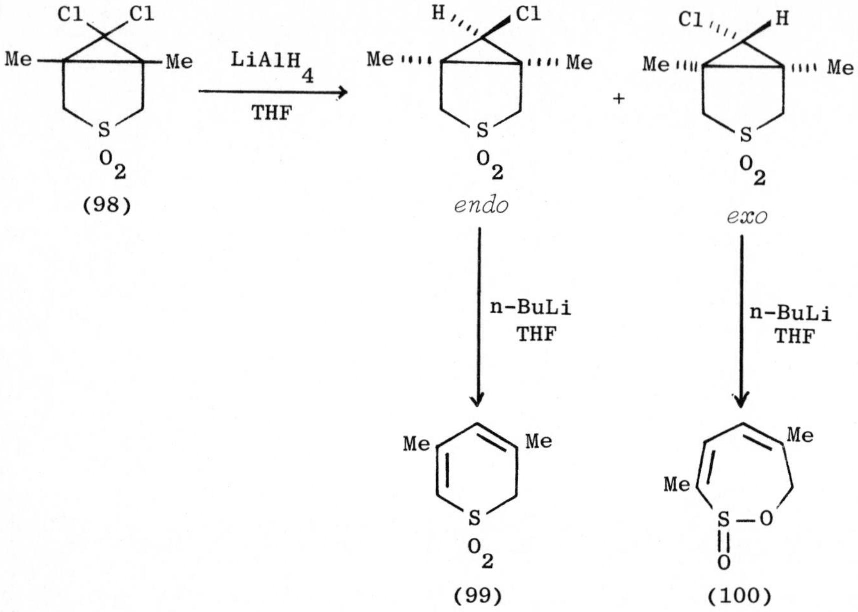

(98) → endo + exo

(99)   (100)

Exposure of the sultone (101) to peroxytrifluoroacetic acid effects oxidative ring expansion to 3,3-diphenyl-4,5,6,7-tetrahydro-1,2-oxathiepin-4(3H)-one-2,2-dioxide (T. Durst *et al.*, Canad. J. Chem., 1978, **57**, 258).

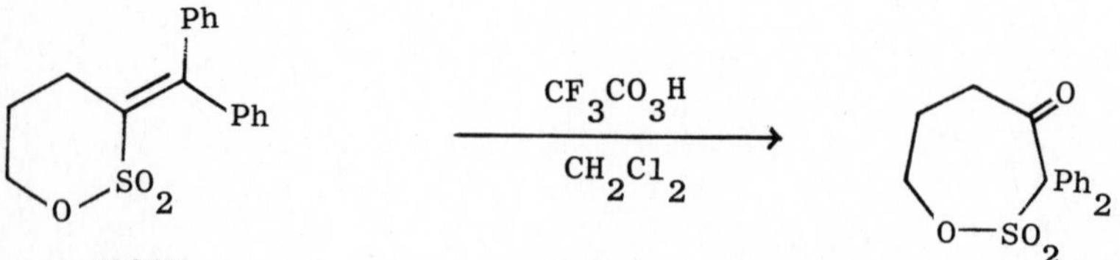

(101)

(ii)   1,3-Oxathiepins

1-Diethylamino-1-propyne adds rapidly to the electrophilic 2,2-bis(trifluoromethyl)-1,3-oxathiolan-5-one with ring expansion to yield the 4,5-dihydro-1,3-oxathiepin-5-one (102) (K. Burger *et al.*, Tetrahedron Letters, 1975, 3223).

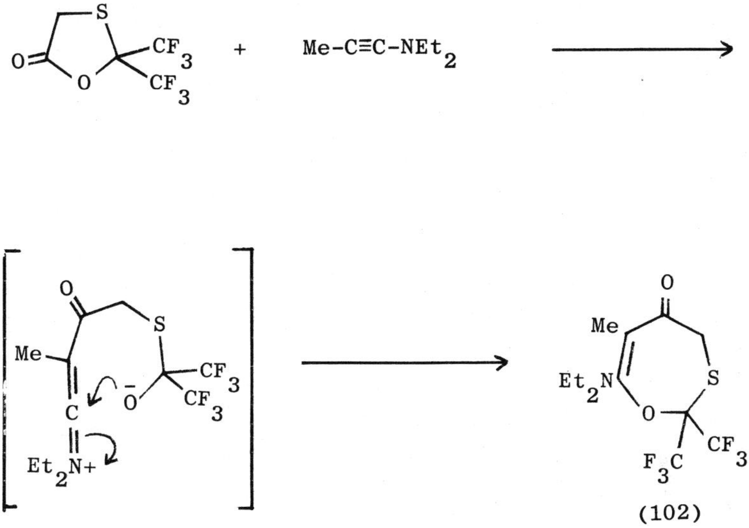

(102)

On heating with di-tert-butyl peroxide, 2-propyl-1,3-oxathiepane undergoes free radical isomerisation to the thio ester (103) (V.V. Zorin *et al.*, Zh. org. Khim., 1984, **20**, 387).

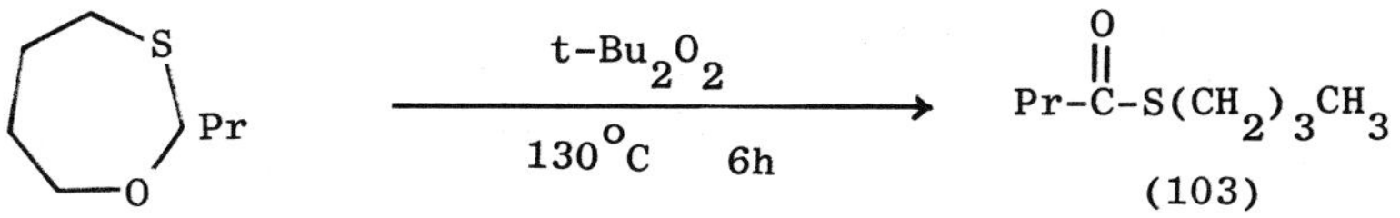

$$Pr-\overset{O}{\underset{\|}{C}}-S(CH_2)_3CH_3$$

(103)

(iii) 1,4-Oxathiepins

The acetylenic thiol (104) cyclises under alkaline conditions to 7H-2,3-dihydro-1,4-oxathiepin, b.p. 82-83°C/18 mm, by intramolecular nucleophilic addition of thiolate to the triple bond. A competing reaction pathway results in formation of the 1,4-oxathiin (105) (C. Dupuy and J.-M. Surzur, Bull. Soc. chim., 1980, 353).

(104)   (105)

Heating of the thiol ester (106) in dilute solution gives a small amount of the expected ten-membered lactone. The major product is 2,3-dihydro-5-(6-hydroxyheptyl)-1,4-oxathiepin-7-one (107), resulting from cleavage of the 1,3-oxathiolane ring by the pendant hydroxy-alkyl function. Subsequent elimination of thiolate affords a spirocyclic intermediate, which fragments on deprotonation (T. Ishida and K. Wada, J. chem. Soc., Perkin I, 1979, 323).

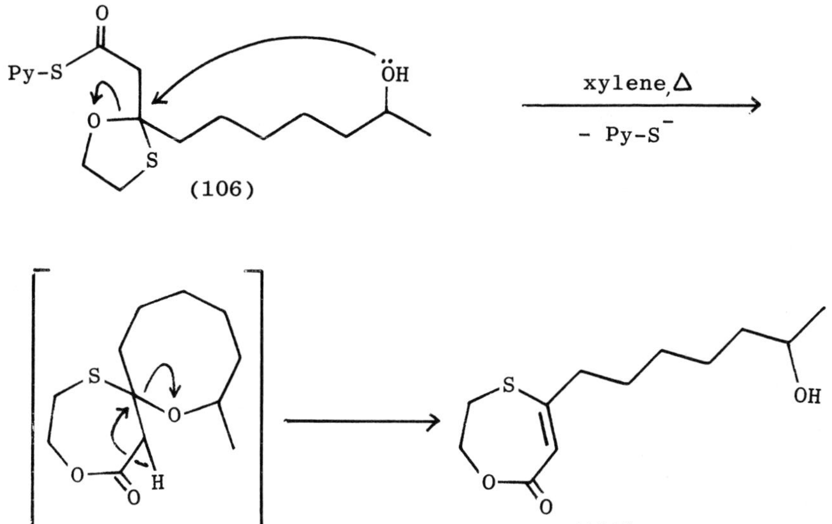

(106)

xylene, Δ

− Py−S⁻

(107)

Py = 2-Pyridyl

Free radical addition of thioglycollic acid to allyl chloride gives 3-chloropropylthioacetic acid, which on treatment with anhydrous potassium fluoride in glacial acetic acid cyclises to 1,4-oxathiepan-2-one, b.p. 86°C/0.06 mm. (D.I. Davies *et al*., J. chem. Soc., Perkin I, 1977, 2476).

$$ClCH_2CH=CH_2 + HSCH_2COOH \xrightarrow[\text{sealed tube 6h}]{70 - 80°C}$$

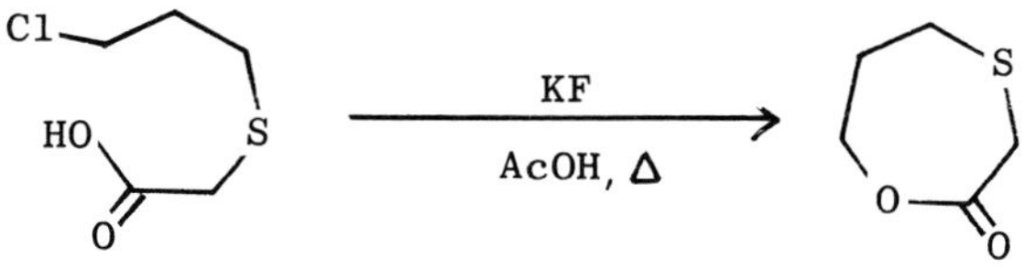

Cyclisation of the ester (108) to ethyl 2,3-dihydro-5H 1,4-benzoxathiepine-5-carboxylate may be accomplished by an intramolecular Friedel-Crafts reaction of the α-chlorosulphide.

(108)

Similarly, the sulphoxide (109) on treatment with acid cyclises to the naphthoxathiepin via an α-acyl carbenium ion intermediate (H. Ishibashi *et al.*, J. heterocyclic Chem., 1986, <u>23</u>, 1163).

(109)

Cyclo-condensation of carbon suboxide with 2-hydroxythiophenol affords $3H$-1,5-benzoxathiepin-2,4-dione, m.p. 130°C. The ^{1}H nmr-spectrum exhibits a singlet at δ 3.76, corresponding to the C-3 protons and an absorption at 1720 cm^{-1} (carbonyl) is observed in its ir-spectrum. The dione is not enolised in solution or in the solid state (L. Bonsignor et al., J. heterocyclic Chem., 1982, 19, 1241).

3-Hydroxy-3,4-dihydro-2H-1,5-benzoxathiepins are obtained by base catalysed condensation of 2-hydroxythiophenol with epichlorohydrin derivatives (S. Cabiddu et al., Phosphorus Sulfur, 1983, 14, 151).

Treatment of 1-(3-hydroxypropylthio)-2,4,6-trinitrobenzene with base effects cyclisation to the benzoxathiepin (110). In addition a small amount of the isomeric compound (111), resulting from a Smiles rearrangement, is obtained (V.N. Kayazev, V.N. Drozd and V.M. Minov, Zh. org. Khim., 1978, 14, 105).

(110)   (111)

Base catalysed ring closure of the diphenyl sulphide (112) affords 6H-dibenzo[b,e][1,4]oxathiepin as an oil. Bromination with NBS gives the 2-bromo compound, m.p. 66–

68°C.   Similarly,   cyclodehydration   of   2-carboxy-2'-
hydroxydiphenyl sulphide gives the lactone (113) m.p. 132.5-
133.5°C.   These compounds and their derivatives are of
interest   since   they   exhibit   neuro-   and   psycho-tropic
activity (K. Sindelar *et al.*, Coll. Czech. chem. Comm.,
1982, **47**, 1367, 3114).

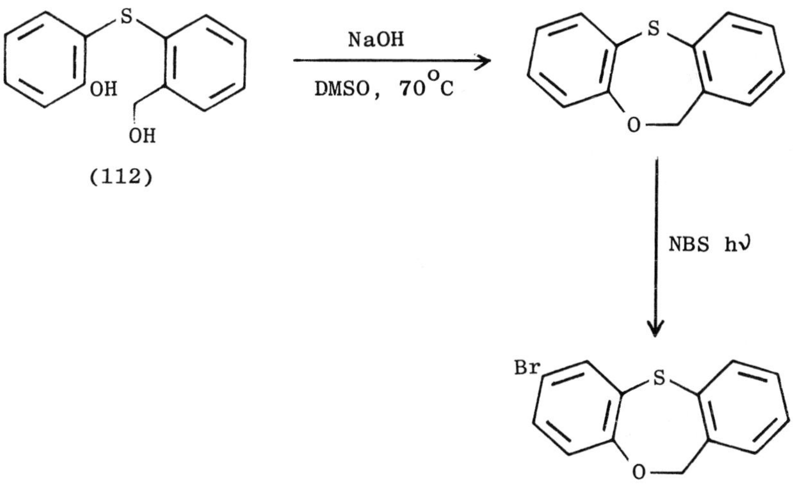

(112)

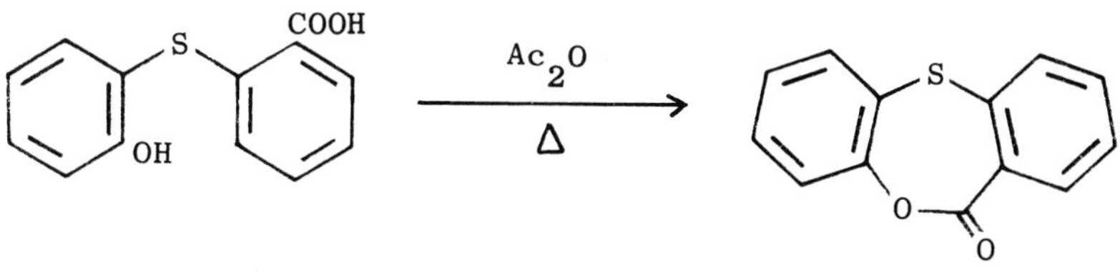

NaOH

DMSO, 70°C

NBS hν

Ac$_2$O

Δ

(113)

2-Hydroxythiophenol may be alkylated with 2-bromobenzyl bromide under mild conditions and the product cyclised in-situ to 11H-dibenzo[b,f][1,4]oxathiepin, m.p. 49–50°C. Lithiation provides access to the carboxylic acid (114), m.p. 150.5–151.5°C (K. Sindelar, Coll. Czech. chem. Comm., 1982, **47**, 967).

(114)

Reagents: (i) DMF, K_2CO_3, 1h, 25°C; (ii) Add Cu,Δ, 6h; (iii) n-BuLi, Et_2O; (iv) CO_2

(b)   Oxadithiepins

Oxidative cyclisation of bis(2-mercaptoethyl)ether to 1,4,5-

oxadithiepane is effected by the gradual addition of a

solution of iodine and the dithiol in chloroform to a

solution of triethylamine in the same solvent, thus ensuring

that cyclisation occurs under high-dilution conditions (M.H.

Goodrow and W.K. Musker, Synthesis, 1981, 457).

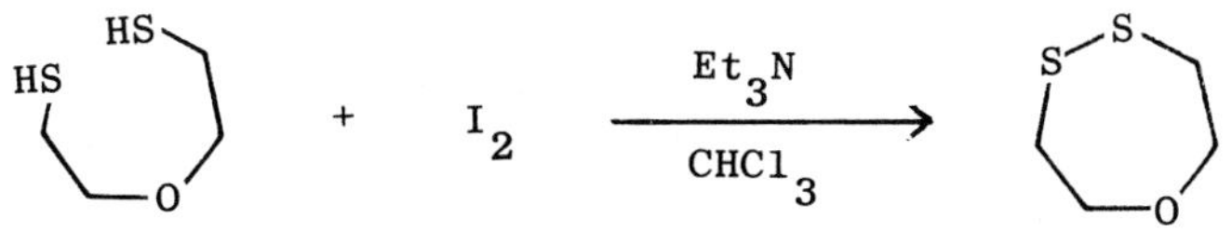

(c)   Oxaselenepins

The oxyselenation of diallyl ether produces a mixture of the

isomers   3-methoxymethyl-6-methoxy-1,4-oxaselenepane   and

(115) in a ratio of 1:4.   Equal amounts of the *cis* and the

trans isomer of each product have been shown to be present

by ^{13}C-nmr-spectroscopy (A. Toshimitsu, S. Hemura and M.

Okano, J. chem. Soc., Perkin I, 1979, 1206).

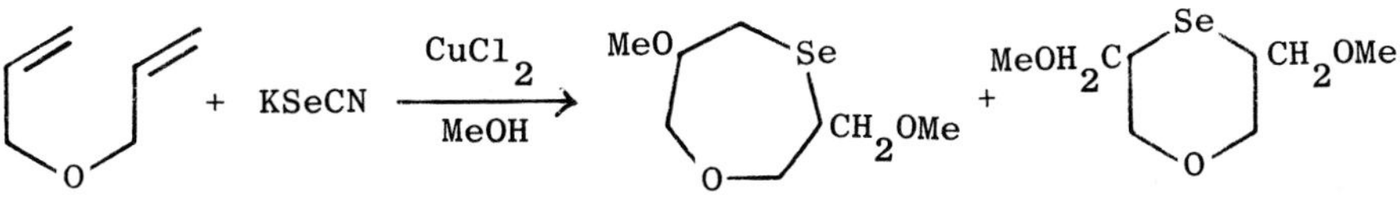

(115)

(d) Dioxathiepins

The most frequently encountered members of this group are the dioxathiepin–S–oxides and S,S–dioxides, which are cyclic sulphites and sulphates, respectively. Consequently these systems are readily accessible by well known methods from 1,2–diols and a 'sulphur transfer' reagent such as thionyl chloride. The chemistry of sulphites and sulphates has been reviewed (K.K. Andersen, in 'Comprehensive Organic Chemistry', ed. D.H.R. Barton and W.D. Ollis, Pergamon, Oxford, 1979, vol. 3, p.373).

Photochemical addition of sulphur dioxide to hexafluorobutadiene affords perfluoro–4,7–dihydro–1,3,2–dioxathiepin, b.p. 80–82°C. Although of reasonable stability when purified, traces of moisture or even aromatic compounds bring about ring cleavage with the formation of difluoromaleyl difluoride and difumaryl difluoride. Heating a mixture of hexafluorobutadiene and sulphur trioxide in a sealed tube gives many products including perfluoro–4,7–dihydro–1,3,2–dioxathiepin–2,2–dioxide (N.B. Kaz'mina *et al.*, Izv. Akad. Nauk SSR, 1978, 163; 1979, 106).

Treatment of 2-hydroxyphenethyl alcohol with the 'sulphinyl transfer' reagent N,N'-thionyldiimidazole affords 4,5-dihydro-1,3,2-benzodioxathiepin-2-oxide, m.p. 78°C (M. Ogata et al., Chem. Ind., 1980, 85).

Im =

N-(2,4,6-Trichlorophenyl)dichlorosulphilimine condenses with the diol (116) giving the 2,4,3-benzodioxathiepin (117), having the novel cyclic imido sulphite function (C. Picard, L. Cazaux and P. Tisnes, Phosphorus Sulfur, 1981, <u>10</u>, 35).

Similarly, the reaction of perfluoroethylimino sulphur difluoride with 2,2'-biphenol affords the 2-iminodibenzo[d,f][1,3,2] dioxathiepin (118) (M.M. Mohammadi and J.M. Shreeve, J. fluorine Chem., 1981, <u>18</u>, 357).

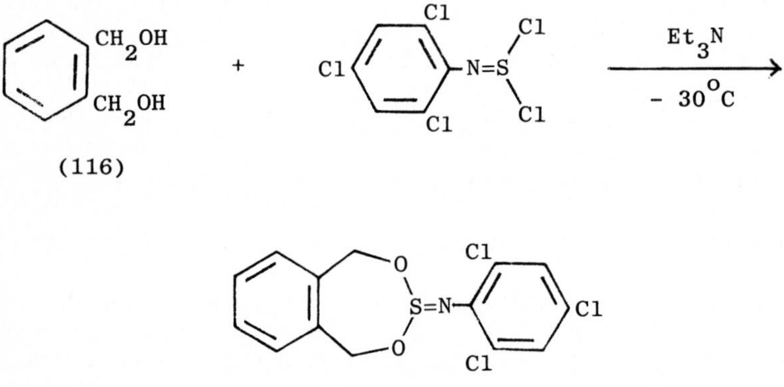

(116)

(117)

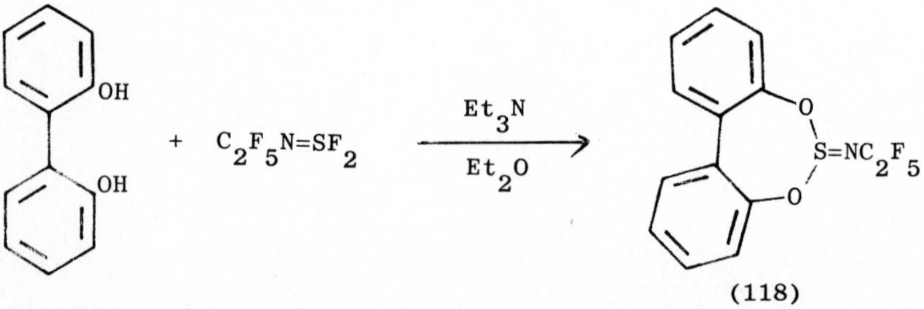

(118)

The electron impact mass spectrum of (117) exhibits an intense peak corresponding to an $[M-SO_2H]^+$ ion, resulting from rearrangement of the molecular ion (G. Puzo *et al.*, Org. mass Spec., 1981, 16, 405).

The conformational preferences of a number of compounds possessing the 1,3,2-dioxathiepin-2-oxide ring have been investigated in considerable detail, principally by nmr-spectroscopy.

From variable temperature ^{1}H- and ^{13}C-nmr studies of 4,7-dihydro-1,3,2-dioxathiepin-2-oxide (119, R=H) it has been concluded that the ring exists predominantly in a twist-boat conformation at low temperature (- 160°C) in CHF_2Cl. However, at room temperature in CCl_4 a small amount of the chair conformer, in which the S=O function is axially disposed, is present. The benzologue (120, R=H) exists in $CHFCl_2$ as a 1:1 mixture of the twist-boat- and the chair-conformer; in less polar solvents the proportion of the latter increases (H. Faucher et al., Tetrahedron, 1981, 37, 689).

(119)   (120)

Conformational analysis of the diastereoisomers of 4,7-dimethyl-1,3,2-dioxathiepane-2-oxide by [1]H-nmr-spectroscopy involving lanthanide induced shift studies, has established that the twist-chair form appears to be the most stable conformer. A similar study of the diastereoisomers of the corresponding dioxathiepin (119, R=Me) shows that these compounds equilibrate between chair and twist-boat conformations (A. Celso *et al.*, Org. Mag. Res., 1983, <u>21</u>, 711).

For (120, R=Me) in which the methyl group is *cis* to the S=O bond, the twist-boat conformer in which the methyl group adopts an equatorial orientation is favoured. In the *trans*-isomer, the chair conformer (axial S=O) predominates (A. Lachapelle *et al.*, Canad. J. Chem., 1983, <u>61</u>, 2244). Whereas 1,1-dimethyl-1,5-dihydro-2,4,3-benzodioxathiepin-2-oxide exists predominantly in the twist-boat conformation, the 6-methyl-1,5-dihydro-compound adopts this structure exclusively (A. Lachapelle and M. St.-Jacques, *ibid.*, 1985, <u>61</u>, 2185).

The fragmentation pathways in the mass spectrum of 1,3,2-dioxathiepan-2-oxide have been investigated (E.J. Lloyd and Q.N. Porter, Austral. J. Chem., 1977, <u>30</u>, 569).

(e) Dioxaselenepins

The 1,3,2-dioxaselenepane (121) results from the reaction of
cis-4-octene-1,8-dial with selenium dioxide (V.N. Odinokov
et al., Zh. org. Khim., 1978, 14, 1617).

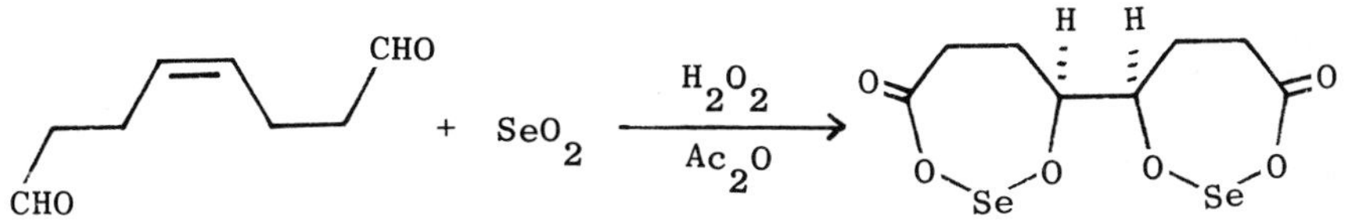

(121)

Cyclo-condensation of selenium dioxide with (122) gives
6,7,8-trimethoxy-1,5-dihydro-2,4,3-benzodioxaselenepine-2-
oxide (S.M. Al-Mousawi, R.J.S. Green and J.F.W. McOmie,
Bull. Soc. chim. Belg., 1979, 88, 883).

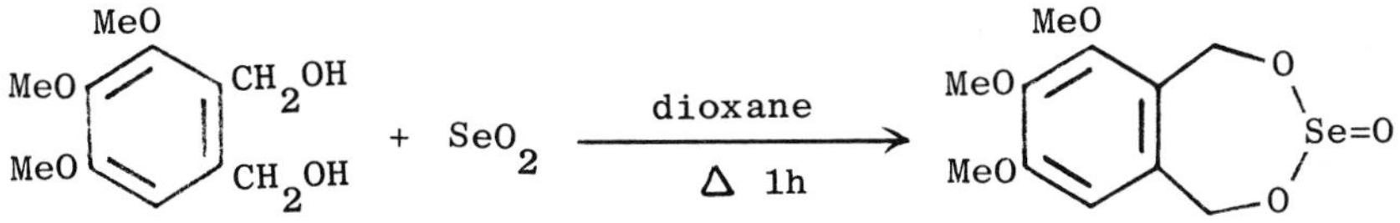

(122)

(f) Dioxatellurepins

3,3-Diphenyl-1,5-dihydro-2,4,3-benzodioxatellurepin-1,5-
dione, m.p. 181-185°C, is readily obtained from phthalic
anhydride and diphenyl telluroxide (S. Tamagaki, I. Hatanaka
and S. Kozuka, Bull. chem. Soc. Japan, 1977, 50, 2051).

(g) Thiaselenepins and Thiatellurepins

A high yielding route to the novel heterocycles 6,7-dihydro-5H-1,4-thiaselenepin (123, X=Se), b.p. 83°C/15 mm, and 6,7-dihydro-5H-1,4-thiatellurepin (123, X=Te), b.p. 133°C/15 mm, has been reported. Treatment of the readily available (3-chloropropylthio)trimethylsilylacetylene with either lithium selenide or telluride in liquid ammonia followed by addition of absolute ethanol results in formation of the heterocyclic products. It is likely that desilylation occurs prior to ring closure (R.S. Sukai et al., Rec. Trav. chim., 1981, 100, 368).

(123)   (124)

Application of the above procedure to phenylacetylene affords the 3-phenyl compounds (124, X=Se or Te).

3. *Rings containing nitrogen and sulphur or selenium*

(a) *Thiazepines*

(i) *1,2-Thiazepines*

Although no developments have been made in the chemistry of
simple 1,2-thiazepines, several new synthetic routes to 1,2-
benzothiazepines have been described.

Pyrolysis of 3-phenylpropanesulphonyl azide in Freon 113
affords   1H-2,3,4,5-tetrahydro-2,1-benzothiazepine-1,1-
dioxide (125), m.p. 173-174.5°C. Hydrogen abstraction and
solvent insertion are competing processes in hydrocarbon
solvents.   The method appears to be of general utility,
although concomitant 1,2-shifts of substituents in the
phenyl ring have been observed. The structure of (125) has
been confirmed by its unequivocal synthesis from 3-(2-
nitrophenyl)-1-propanesulphonyl chloride (R.A. Abramovitch
et al., J.Org.Chem., 1984, <u>49</u>, 3115).

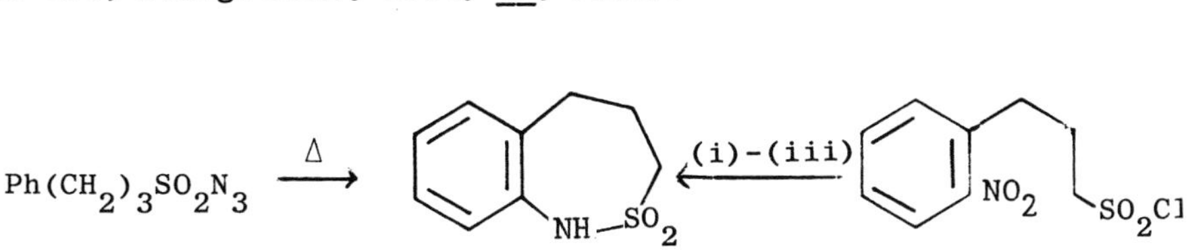

$Ph(CH_2)_3SO_2N_3 \xrightarrow{\Delta}$ (125) $\xleftarrow{(i)-(iii)}$

(125)

Reagents: (i) NaOH; (ii) H_2, Pd-C; (iii) PCl_5, CH_3COCl

Intramolecular sulphonyl-amidomethylation of 2-phenylethane-sulphonamide occurs on treatment with trioxane in methane-sulphonic acid containing trifluoroacetic acid to yield 1,2,4,5-tetrahydro-3,2-benzothiazepine-3,3-dioxide (126), m.p. 160-162°C (O.O. Orazi and R.A. Corral, Chem. Comm., 1976, 470).

(126)

Reagents: CH_3SO_3H, trioxane

Cyclisation of the sulphonamido carboxylic acid (127) can be achieved by boiling the derived acid chloride in xylene (P. Catsoulacos and Ch. Camoutsis, J. heterocyclic Chem., 1976, 13, 1309). ^{13}C-nmr data for the resulting 1,2-benzothiazepin-3-one (128), m.p. 205-206°C, are given below (Fig. 1) (C.I. Stassinopoulou, P. Catsoulakos and Ch. Camoutsis, Org. mag. Res., 1983, 21, 187). The crystal structure for the related N-(4-bromophenyl)-4,5-dihydro-7,8-dimethoxy-1,2-benzothiazepin-3-one-1,1-dioxide has been described (N.C. Panagiotopoulos, S.E. Filippakis and P. Catsoulakos, Cryst. Struct. Comm., 1980, 9, 313).

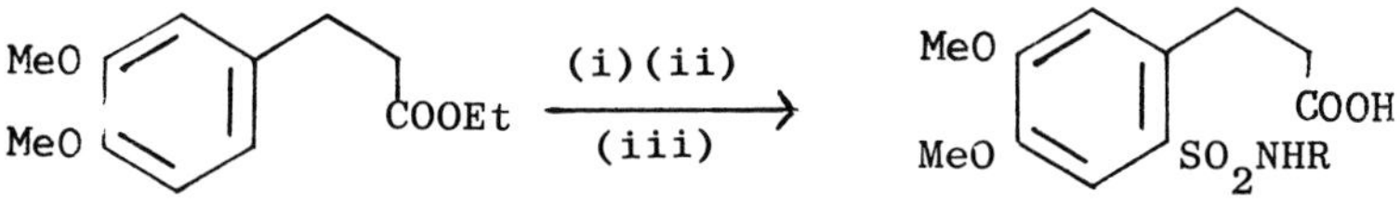

(127)

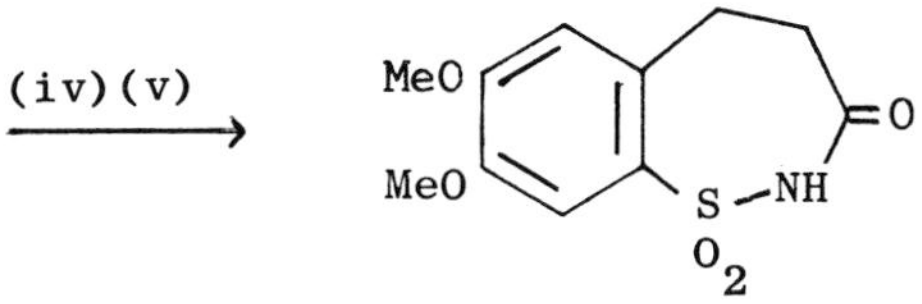

(128)

Reagents: (i) $ClSO_3H$; (ii) $ArNH_2$, C_6H_6; (iii) KOH, MeOH; (iv) PCl_5, C_6H_6; (v) Δ, xylene

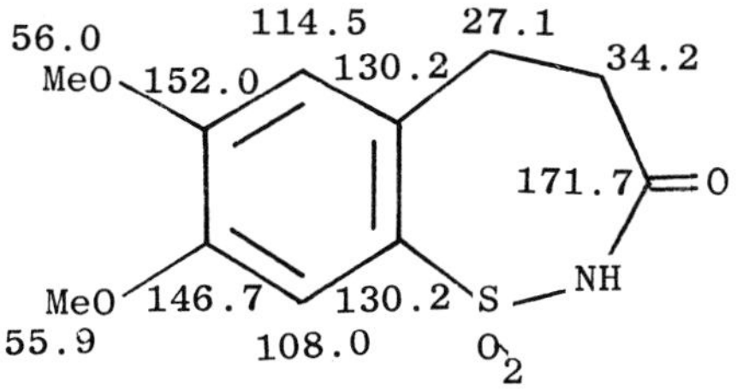

Fig. 1 ^{13}C chemical shifts (δ) for (128)

1,2-Benzothiazepines may be obtained by methods involving the ring expansion of other heterocycles. Both thiochroman-4-ones (129) and benzo[b]thiophenones (130) afford tetrahydro-1,2-benzothiazepin-5-ones on treatment with chloramine-T. The intermediate N-tosylsulphimine cyclises through an intramolecular Michael addition (Y. Tamura et $al.$, J. chem. Soc. Perkin I, 1981, 1037). Oxidation of the benzothiazepines by m-chloroperbenzoic acid yields a mixture of the 1-oxide and 1,1-dioxide.

The photochemical rearrangement of 2-azidothiochroman-4-one-1,1-dioxide leads to the dihydrobenzothiazepinone in good yield using acetophenone as solvent, implying a triplet nitrene intermediate. The analogous 1-oxide does not show parallel behaviour (I.W.J. Still and T.S. Leong, Canad. J. Chem., 1979, 58, 369).

2-Phenylbenzo[b]thiophenones also yield 1,2-benzothiazepinones on amination with O-mesitylsulphonylhydroxylamine, presumably via a sulphenamide. On tosylation and peracid oxidation of the 4-methyl-2,3-dihydrobenzothiazepinone, a mixture of the sulphinamide (131) and sulphonamide (132) results (Y. Tamura et al., J. chem. Soc. Perkin I, 1980, 2830).

Reagent: O-mesitylsulphonylhydroxylamine

(131)   (132)

3-Methyl-1,2-benzisothiazole-1,1-dioxide (133) undergoes a [2+2] cycloaddition with 1-diethylaminopropyne in acetonitrile and the initial adduct rearranges to 3-diethylamino-4,5-dimethyl-1,2-benzothiazepine-1,1-dioxide (134). Various 3-substituents in the benzisothiazole component are compatible with the process. The seven-membered ring system is remarkably stable in acidic and in basic conditions and towards oxidation, reduction, acylation and alkylation, although lithium aluminium hydride cleaves the hetero ring. The 5-ethoxy derivative may be oxidised to the 5-ketone from which an ene-hydrazine has been prepared (R.A. Abramovitch *et al.*, Heterocycles, 1976, **5**, 95). [1]H nmr-spectroscopic data suggest that in the compounds studied the heterocyclic ring in benzothiazepine dioxide is not aromatic, the chemical shift of H-4 and that of H-5 corresponding closely to the values observed for the vinyl protons in cinnamic acid. The coupling (12.7 Hz) between these two protons is also larger than expected for an aromatic system (R.A. Abramovitch, B. Mavunkel and J.R. Stowers, Chem. Comm., 1983, 520).

(133)   (134)

Reagents: $MeC{\equiv}CNEt_2$, MeCN

The dibenzo[c,f][1,2]thiazepine system (135) is formed in low yield during the thermolysis of 2-azidosulphonylbenzophenones in Freon 113 (R.A. Abramovitch *et al.*, J. org. Chem., 1978, **43**, 1218). The 6-methyl derivative of (135; R=R'=H) results from the Lewis acid catalysed cyclisation of the sulphonamide (136).

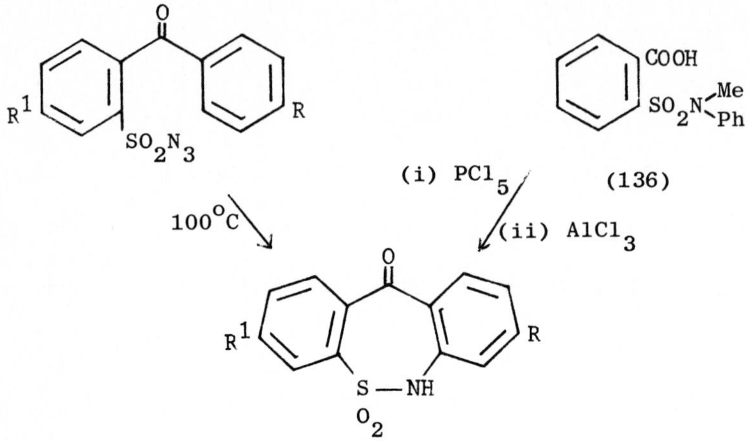

(135)

(ii) 1,3-Thiazepines

Hexahydro-1,3-thiazepin-4-ones (137) result from the reaction of formaldehyde with secondary amides possessing a γ-mercapto group. Reduction furnishes the corresponding 1,3-thiazepine (T. Kametani *et al.*, Heterocycles, 1978, **9**, 831).

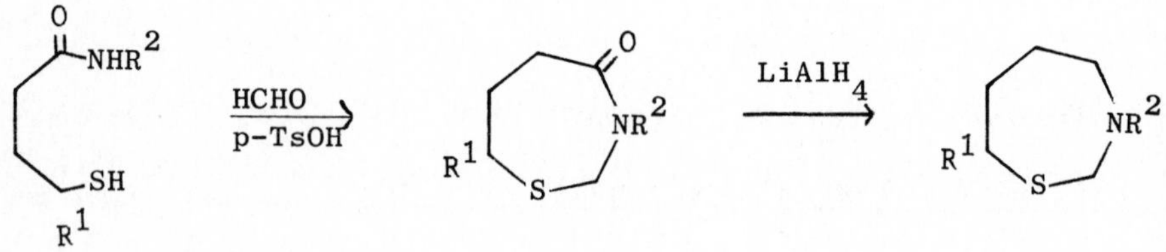

(137)

The dibenzoxanthylium salt (138) in a one pot reaction with
N-phenyl-*N*'-(4-aminobutyl)thiourea   affords   2-iminophenyl-
4,5,6,7-tetrahydro-(3*H*)-1,3-thiazepine (139).   The thiourea
is prepared from 1,4-diaminobutane and phenyl isothiocyanate
(A.R. Katritzky *et al.*, Tetrahedron, 1981, <u>37</u>, 2383).   The
^{1}H-nmr spectrum of (139) indicates the predominance of the
imino   tautomer   (D.L.   Garmaise,   G.Y.   Paris   and   G.
Efthymiadis, Canad. J. Chem., 1971, <u>49</u>, 971).

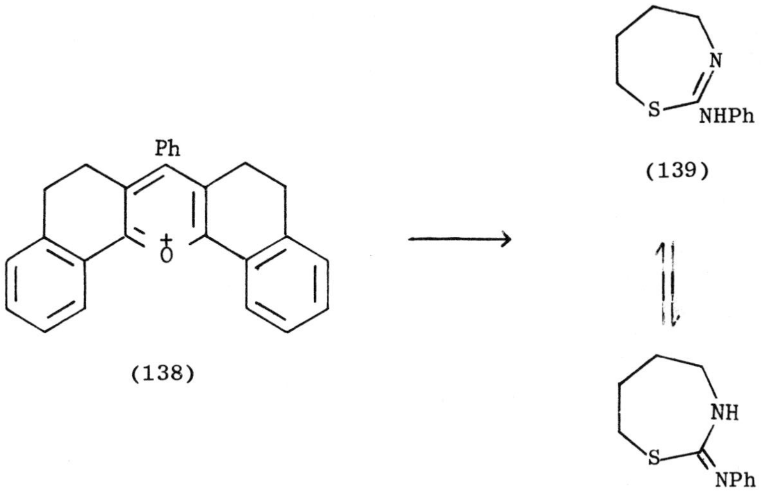

(139)

(138)

Although   other   spectroscopic   techniques   have   failed   to
establish   the   preferred   tautomer   of   2-arylamino-1,3-
thiazepines,   their ^{13}C-nmr-spectra   indicate   that   the
exocyclic double bond is favoured when R is H or Me (P.
Sohar, G. Feher and L. Toldy, Org. mag. Res., 1978, <u>11</u>, 9).

Thermolysis of the cyclic sulphilimine, S-methylaza thiaphenanthrene (140), a stable compound which may be obtained by deprotonation of the azasulphonium salt or, better, by the reaction of N-chlorosuccinimide with 2-amino-2'-methylthiobiphenyl, yields 7H-dibenzo[d,f][1,3]thiazepine (141), m.p. 168–172°C. The proposed mechanism of this reaction is a Stevens–like ring expansion of the methylide, although β-elimination is a successful competing reaction pathway in the case of the S-ethyl derivative. Acylation and oxidation of (141) leads to the sulphone (142), m.p. 223–226°C (H. Shimizu et $al.$, Chem. pharm. Bull., 1984, $\underline{32}$, 4360).

Reagents: (i) Δ, xylene; (ii) Ac$_2$O; (iii) m–CPBA

(iii) 1,4-Thiazepines

The synthesis of a stable unsaturated 1,4-thiazepine (146)
has been realised (K. Yamamoto *et al.*, Angew. Chem. intern.
Edn., 1986, <u>25</u>, 635). Stabilisation is achieved by
introducing a bulky *t*-butyl group at each position adjacent
to the sulphur heteroatom, thereby preventing its extrusion.
The synthesis is outlined below. Although a Beckmann
rearrangement of 2,6-di-*t*-butylthiopyran-4-oxime has not
been achieved, ring expansion occurs with the 2,3-dihydro
derivative and affords the lactams (143) and (141).
Oxidation of the latter to the sulphoxide and subsequent
Pummerer reaction yields the diene (145) and methylation
converts the lactam into the lactim, 2,7-di-*t*-butyl-5-
methoxy-1,4-thiazepine, which is stable up to 130°C.

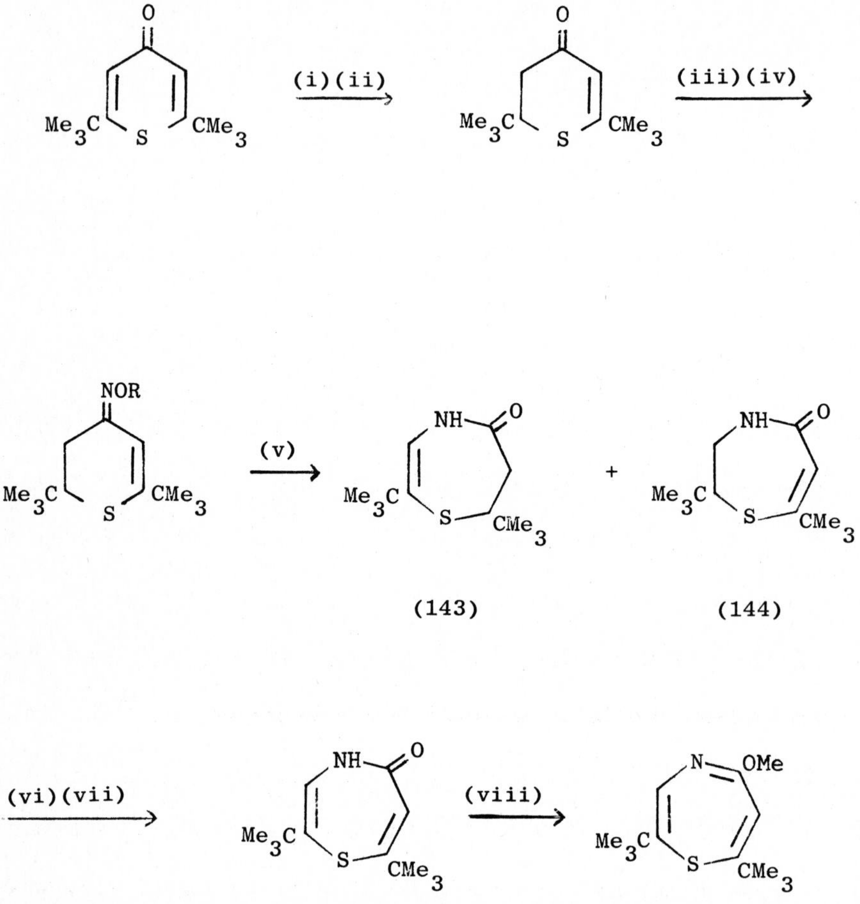

(143)   (144)

(145)   (146)

Reagents: (i) NaBH$_4$, THF; (ii) Sarett's reagent (iii) NH$_2$OH;

(iv) p-TsCl; (v) Et$_3$N, aq.dioxane; (vi) m-CPBA;

(vii) CF$_3$COOH, CH$_2$Cl$_2$; (viii) Me$_3$O$^+$ BF$_4^-$

The nmr-chemical shift data for compound (146) are given below.

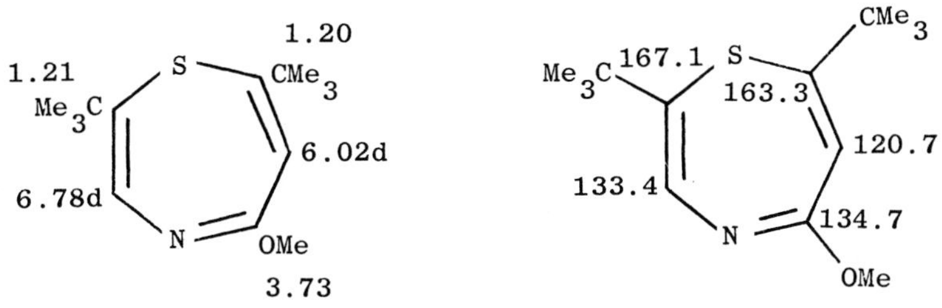

Fig. 2 ^{1}H and ^{13}C chemical shifts (δ)

There are many examples of the rearrangement of 6-substituted penicillanates and related compounds to the corresponding thiazepinone (R.J. Stoodley, Tetrahedron, 1975, 31, 2321). Thus, 6-acyl (J.C. Sheehan *et al.*, J. org. Chem., 1978, 43, 4856) and 6-acrylate esters (C.D. Foulds *et al.*, J. chem. Soc. Perkin I, 1984, 21) thermally rearrange to thiazepin-2-one, presumably as a consequence of the acidity of 6-α proton.

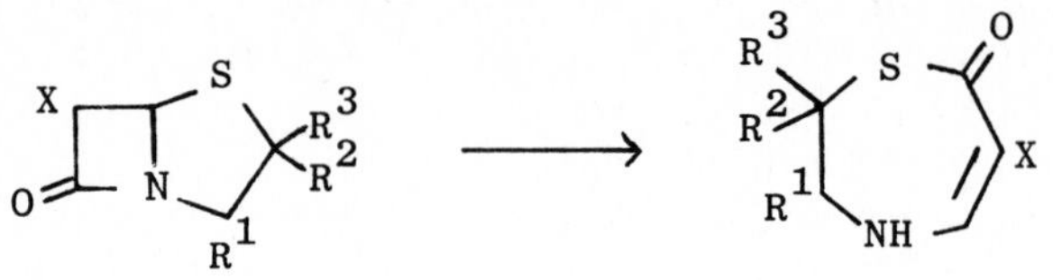

Thiazepinones also result from the thermolysis of 4-alkylsulphinylazetidinones (147), presumably *via* the 2-phenoxypenem (148) which can be isolated when the thermolysis is effected in the presence of triphenylphosphine (M.D. Cooke *et al.*, Tetrahedron Letters, 1983, 3373).

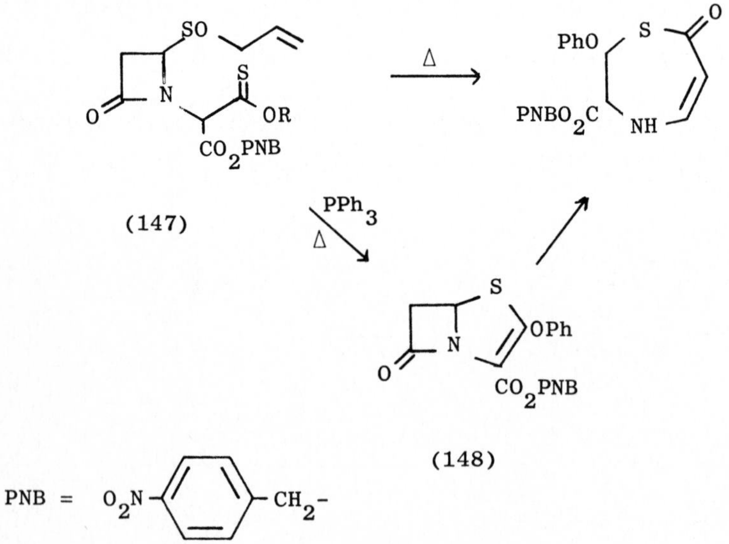

(147)

(148)

PNB = O_2N—⟨ ⟩—CH_2-

The rhodium-catalysed reaction of benzhydryl 6-diazopenicillanate (149) with alcohols (S.A. Matlin and L. Chan, Chem. Comm., 1980, 798) and the trimethyl phosphite induced desulphurisation of the thiacephem (150) (A. Henderson *et al.*, Chem. Comm., 1982, 809) are related processes.

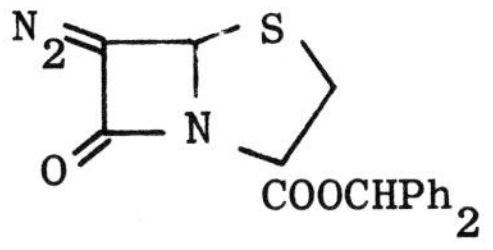

(149)

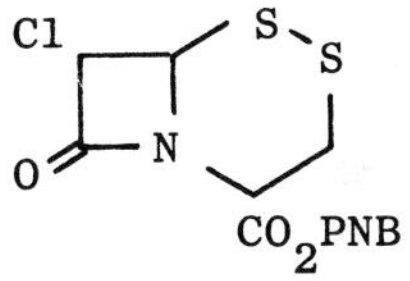

(150)

Thiopenams (152) also appear to be susceptible to this rearrangement, since attempts to generate them from the thiosilyl derivatives (151) leads to the dihydro-1,4-benzothiazepinthiones (153) (E. Schaumann, W.-R. Forster and G. Adiwidjaja, Angew. Chem. intern. Edn., 1984, 23, 439).

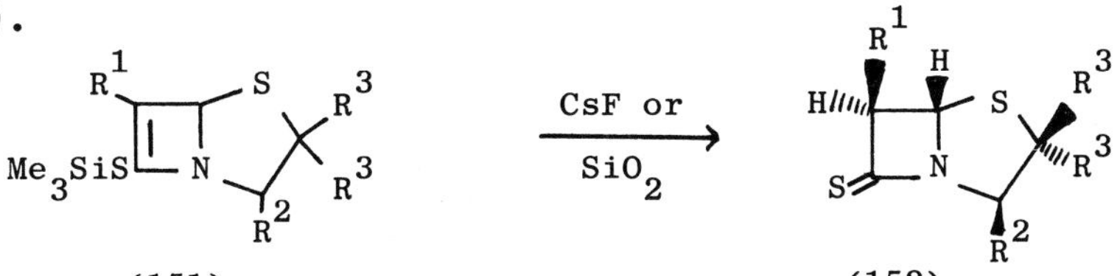

(151)

(152)

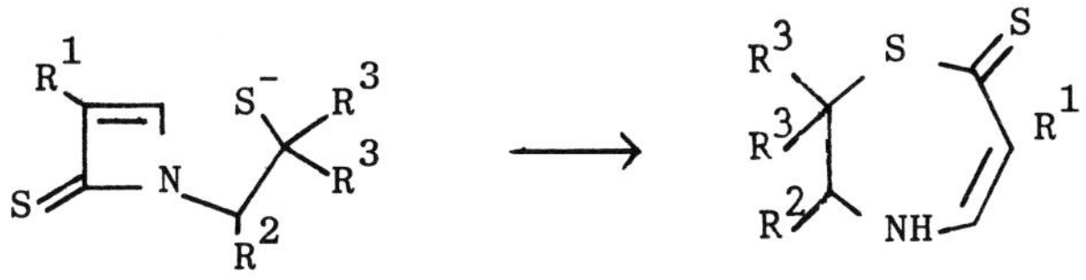

(153)

Irradiation of *N*-phthaloylmethionine (154; R=H) or its methyl ester in acetone solution unexpectedly gives the hexahydro—11b—hydroxy[1,4] thiazepino[3,4—a]isoindol—7—one (155; R=H) or the corresponding 5—carboxylate (155;R=COOMe). The ester can be separated into the *cis* and the *trans* isomer which are distinguishable as a consequence of the hydrogen bonding in the former (Y. Sato *et al.*, Chem. pharm. Bull., 1982, 30, 1263).

(154)   (155)

Reaction of *N*-benzyloxycarbonyl dehydroalanine methyl ester (156), readily obtainable from L—cystine, with D—penicillamine provides a route to 6b—carbobenzyloxyamino—3D—carboxy—2,2—dimethyl—5—oxoperhydro—1,4—thiazepine (157) which allows the incorporation of ^{14}C atoms at specific sites (S. Wolfe *et al.*, Canad. J. Chem., 1981, 59, 406). The introduction of α—aminoadipyl and glycyl—α—aminoadipyl units is possible by the reaction of the 6—amine with the N—Boc protected α—aminoacids. Oxidation of the thiazepines is best accomplished with *m*-chloroperbenzoic acid.

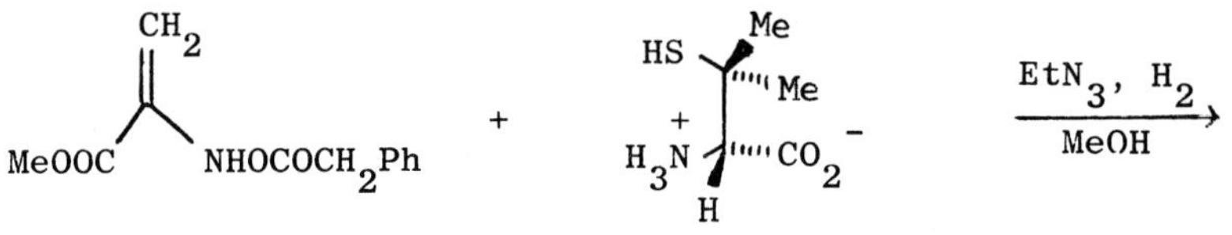

(156)

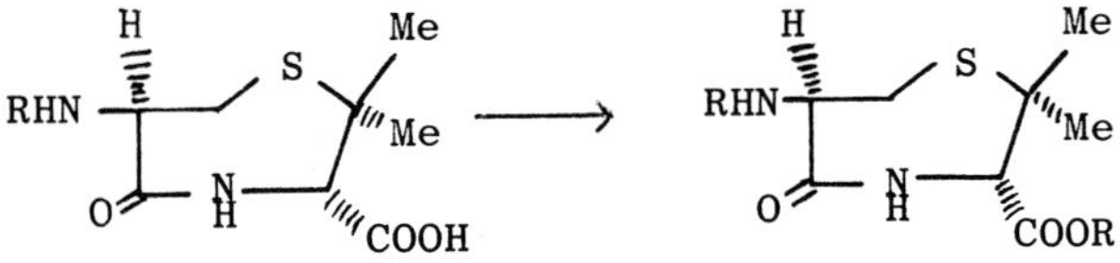

(157)

On the basis of their [1]H-nmr-spectra, the 3D,6L-thiazepines are considered to exist in a flexible boat conformation (158) whilst the 1R,3D,6L-thiazepine sulphoxides adopt the chair conformation (159), stabilised by an internal hydrogen bond between H-6 and the sulphinyl oxygen.

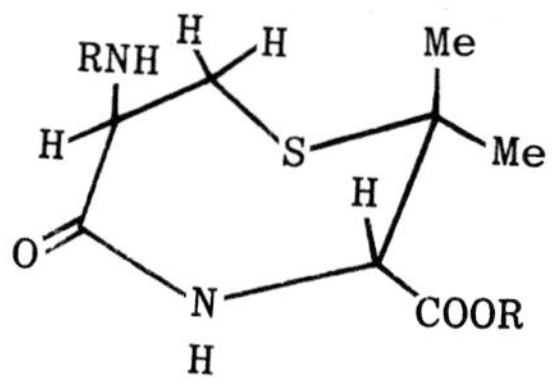

(158)

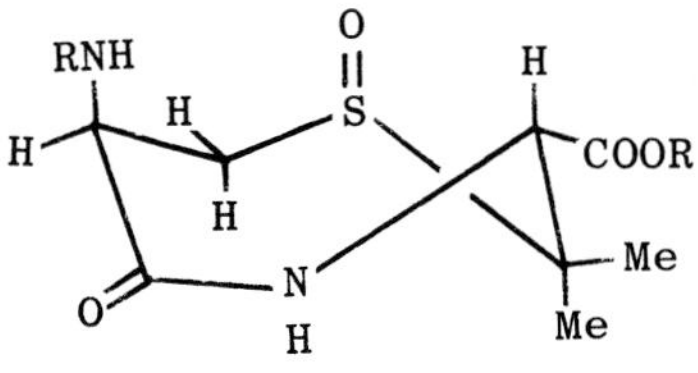

(159)

The fully saturated 1,4-thiazepine system is accessible by the reaction of 2-chloroethyl-3-chloropropyl sulphide with primary amines (M. Strzelczyk and B. Kotelko, Pol. J. Pharmacol. Pharm., 1979, __31__, 73).

Perhydro-1,4-thiazepin-5-ones react with various chlorinating agents to give the 2- or the 7-chloro derivative. The latter readily eliminate hydrogen chloride to give the 6-ene (161), but the former are only dehydrohalogenated to the 2-ene (160) in the presence of base (H. Wamholff and C.H. Theis, Ber., 1980, __113__, 995).

(160)   (161)

Fused β-lactams behave in a similar manner to the penicillanates, readily undergoing ring expansion to 1,4-benzothiazepines on treatment with base (L. Fodor *et al.*, Tetrahedron, 1984, **40**, 4089).

A ring expansion following a photochemical [2+2] cycloaddition of ethyl vinyl ether to 3-phenylbenzisothiazoles is a plausible mechanism for the observed formation of 1,4-benzothiazepines in high yield (M. Sindler-Kulyk and D.C. Neckers, Tetrahedron Letters, 1981, 529).

The cyclodehydration of the β-aminothiol derivative (162) is effected by dicyclohexylcarbodiimide, but the reaction can lead to a mixture of 1,4-benzothiazepin-5-one (163) and 4,1-benzothiazepin-5-one (164). The intervention of a Smiles rearrangement is implied. The isomers show different ^{13}C chemical shifts for the carbonyl carbon atom, ca. 195 ppm for the O=C-N moiety in the 4,1-compounds and 164 ppm for the O=C-S function in the 1,4-isomer (E.Y. Chan, R.G. Crampton and L.B. Clapp, Phosphorus Sulfur, 1979, 7, 41).

(162)   (163)   (164)

The reaction of the dinucleophilic 2-aminothiophenol with a three carbon electrophilic moiety continues to provide new examples of 1,5-benzothiazepines in a variety of oxidation states (Scheme 2).

(169)   (165)

(168)   (167)   (166)

Reagents: (i) PhC≡CCOPh, NEt$_3$, AcOH; (ii) R^{1}CH=CHCOR^2

(iii) R^{1}R^{2}C=CHCOOR^3; (iv) RCHBrCH$_2$COOH; (v) RCH(COOH)$_2$

1,3-Diphenylprop-2-yn-1-one and the thiophenol furnish the fully unsaturated 2,4-diphenyl-1,5-benzothiazepine (165), m.p. 108-110°C, although other acetylenic ketones appear to yield benzothiazole derivatives or an uncyclised product (A.S. Nakhmanovich *et al.*, Izv. Akad. Nauk. S.S.R., 1982, 1371).

Extensive studies on the use of chalcones have shown that Michael addition to the propanone is catalysed by piperidine and that subsequent cyclisation to the 2,4-diaryl-2,3-dihydro-1,5-benzothiazepine (166) is acid catalysed (A. Levai and R. Bognar, Acta chim. Acad. Sci. Hung., 1977, <u>92</u>, 415). Direct formation of the thiazepine is observed when the reaction is carried out in the presence of triethylamine (V.D. Orlov, N.N. Kolos and N.N. Ruzhitskaya, Chem. heterocyclic Compounds, 1983, 1293) or trifluoroacetic acid (A. Levai, Pharmazie, 1981, <u>36</u>, 449). A hydroxy substituent in the *ortho*-position of either phenyl ring of the chalcone also promotes the direct cyclisation (A. Levai and R. Bognar, *ibid.*, A.K. Gupta, V.K. Singh and U.C. Pant, Indian J. Chem., 1983, <u>22B</u>, 1057). The 1,1-dioxides result on oxidation of the thiazepines with peroxyacetic acid (K.P. Jadhav and D.B. Ingle, *ibid.*, p.180).

In a closely related process, cinnamic acids and acrylic acids or their esters yield 2,3-dihydro-1,5-benzothiazepin-4(5H)-ones (167) (A. Levai and H. Duddeck, Pharmazie, 1983, 38, 827; V. Balasubramaniyan, P. Balasubramaniyan and A.S. Shaikh, Tetrahedron, 1986, 42, 2731).

1-Aryl-4-arylmethylenepyrrolidine-2,3,5-triones (170) behave as α,β-unsaturated ketones in their reaction with 2-aminophenol and generate 2-aryl-2,5-dihydro-1,5-benzothiazepin-3,4-dicarboxylic N-arylimides (171) (M. Augustin and P. Jeschke, Synthesis, 1987, 937).

(170)   (171)

Derivatives of 3-bromopropanoic acid react with 2-aminothiophenol in the presence of piperidine to afford 3-(3-aminophenylthio)- propanoic acids, which cyclise in boiling xylene to give the 2,3-dihydro-1,5-benzothiazepin-4(5H)-ones (168). The use of 3-bromopropanoic acids rather than acrylic acids gives an increased yield (A. Levai and G. Puzicha, Synth. Comm., 1985, 15, 623).

Both β-propiolactone and β-butyrolactone can function as the three carbon unit and their reaction with 2-aminothiophenol provides an easy route to 2,3-dihydro-1,5-benzothiazepin-4(5H)-ones (V. Ambrogi and G. Grandolini, Synthesis, 1987, 724).

The value of allene-1,3-dicarboxylates in heterocyclic synthesis has already been mentioned (1,4-oxazepines) and a further i1llustration is provided by the reaction of dimethyl allene-1,3-dicarboxylate with 2-aminothiophenol which yields the 1,5-benzothiazepinone (172) m.p. 226–228°C (R.M. Acheson and J.D. Wallas, J. chem. Soc. Perkin I, 1982, 1905; J. Ackroyd and F. Scheinmann, J. chem. Res. (M), 1982, 1012).

(172)

1,5-Benzothiazepin-2,4-(3H,5H)-diones (169) result from the reaction of 2-aminothiophenols with either malonic acid in the presence of dicyclohexylcarbodiimide or with S,S'-diphenyl dithiomalonates alone (W. Reid and G. Sell, Ber., 1980, 113, 2314).

A more complex heterocyclic system (174) arises when β-dicarbonyl derivates of 1,2-benzothiazine 1,1-dioxide (173) react with 2-aminothiophenol (G. Steiner, Ann, 1978, 643).

(173)   (174)

The reaction of 2-aminothiophenol with alkynic esters and ketones is complex, but 1-phenylprop-2-yn-1-one reacts with an excess of the thiophenol to afford the 2,4-disubstituted 1,5-benzothiazepine (175) in good yield (G. Liso *et al.*, J. chem. Soc. Perkin I, 1983, 567).

(175)

Variations on the aminothiophenol route to 1,5-benzothiazepines include the cyclisation of 3-(2-aminophenylthio)propanonitrile which leads to 4-amino-2,3-dihydro-1,5-benzothiazepine (176), m.p. 175-177°C. Utilisation of the cyclic amidine function in this thiazepine allows the synthesis of aryl derivatives of imidazo[2,1-d][1,5]benzothiazepine (177) (J.S. Walia *et al.*, J. heterocyclic Chem., 1985, **22**, 1117).

(176)   (177)

Reagents: (i) $ArCOCH_2Br$, $NaHCO_3$, EtOH

The synthesis of benzothiazoles from 2-aminothiophenol is well documented and two quite different routes to benzothiazepines have been developed from the former compound.

The fast, base induced, ring expansion of quaternised benzothiazoles (178) provides a good route to 5-formyl-2,3,4,5-tetrahydro-1,5-benzothiazepines (H.-J. Federsel and J. Bergman, Tetrahedron Letters, 1980, 2429).

The [2+2]-cycloaddition of alkenes to 2-phenylbenzothiazole leads to 1,5-benzothiazepines in a regioselective and stereospecific manner (M. Sindler-Kulyk and D.C. Neckers, Tetrahedron Letters, 1981, 2081).

In a related synthesis, the photo-cycloaddition of ethoxyethyne to 2-phenylbenzothiazole is regiospecific affording 3-ethoxy-4-phenyl-1,5-benzothiazepine, m.p. 83–84°C. However, the reaction with ethoxypropyne gives a mixture of 3-ethoxy-2-methyl-4-phenyl-1,5-benzothiazepine, m.p. 75–77°C, together with the 2-ethoxy-3-methyl-4-phenyl- and 4-ethoxy-3-methyl-2-phenyl-, m.p. 127–128°C, isomer (M. Sindler-Kulyk and D.C. Neckers, J. org. Chem., 1982, 47, 4914). These benzothiazepines are thermally unstable, extruding sulphur and leading to the correspondingly substituted quinolines. The base peak in the mass spectrum of each of these compounds arises from loss of sulphur and an ethyl fragment.

2-Nitrothiophenol is also a useful precursor of 1,5-benzo-
thiazepines and its reaction with phenyl glycidates has been
used to prepare several major metabolites (e.g. 179) of the
anti-anginal drug Diltiazem (180) (M. Miyazaki, T. Iwakuma
and T. Tanaka, Chem. Pharm. Bull., 1978, **26**, 2889).

(179)   (180)

Ring expansion of the *trans*-1,4-benzothiazine-1-oxide (181)
occurs in boiling acetic anhydride to yield the 1,5-
benzothiazepine, although the *cis*-isomer is recovered
unchanged after similar treatment. This result is a
consequence of the stereospecific nature of the thermal 2,3-
sigmatropic rearrangement. Conversely, the ring expansion
of 1,3-benzothiazine-1-oxides to 1,4-benzothiazepines
appears to be non-stereospecific. Loss of stereospecificity
is probably a result of easy cleavage of the S-C-2 bond (H.
Shimizu *et al.*, Chem. Pharm. Bull., 1984, **32**, 2571).

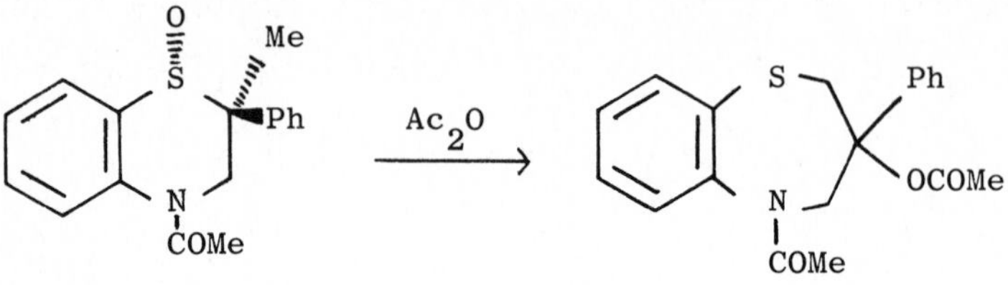

(181)

A similar ring expansion has been observed for 1,4-benzothiazin-3(4H)-ones (182) leading to the 1,5-benzothiazepin-3,4-diones (183) (Y. Maki *et al*., **Chem. Comm.**, 1983, 450).

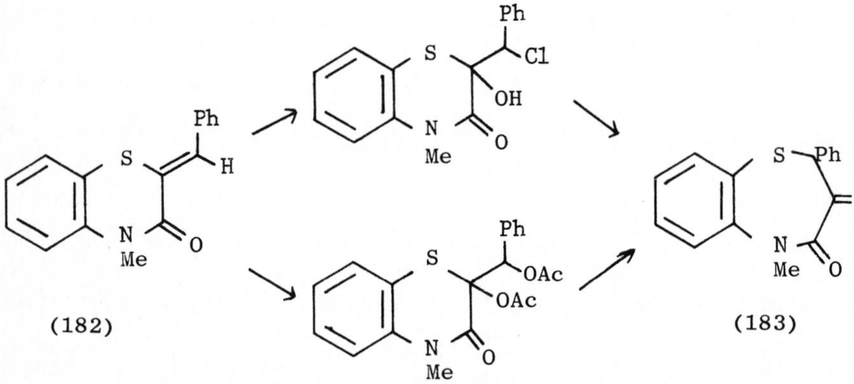

(182)   (183)

There are a number of examples of thermal ring transformation of 1,5-benzothiazepines.

Methanethiol is thermally eliminated from *trans*-2,3-dihydro-4-methylthio-2,3-diphenyl-1,5-benzothiazepine, resulting in ring contraction to a benzothiazole (G. Kaupp, Ber., 1984, 117, 1643).

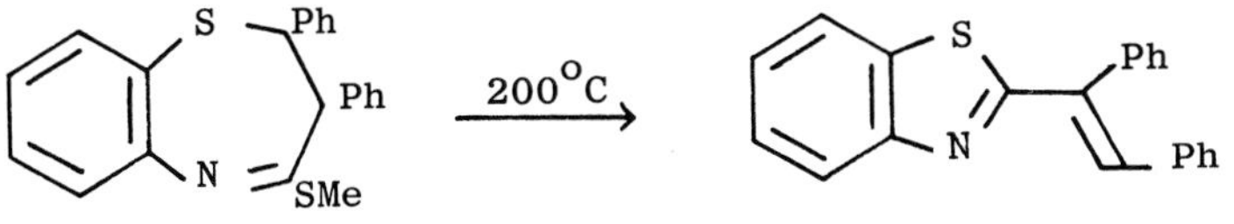

Ring contraction is also observed when 1,5-benzothiazepin-2,4(3H,5H)-dione is treated with sodium nitrite and acetic acid. Hydrolytic ring open of the 3-isonitroso compound is involved, followed by nucleophilic attack by sulphur at C-3 or C-2 (G. Kollenz and P. Seidler, Z. Naturforsch., 1984, 39B, 384).

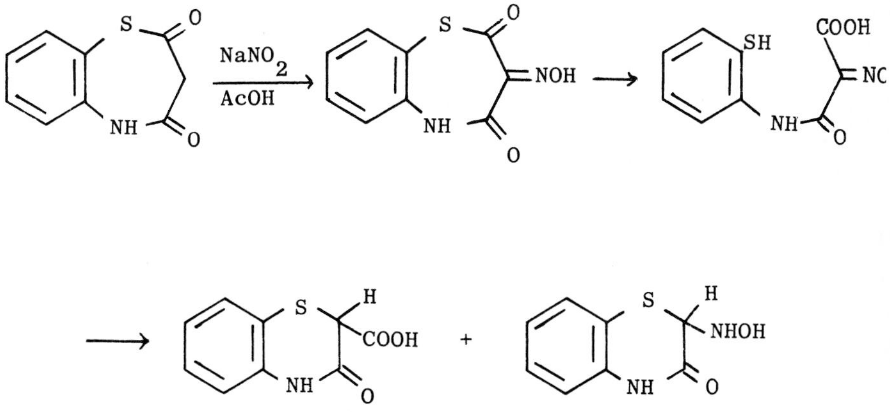

338

Ring transformations of 1,5-benzothiazepine sulphoxides proceed in a stereospecific manner. *trans*-2-Methyl-2-phenyl-5-propionyl-2,3,4,5-tetrahydro-1,5-benzothiazepine 1-oxide (184) yields the benzothiazocine (185) involving the formation of a sulphonic acid. However, the *cis*-isomer undergoes ring contraction and a small amount of the benzothiazine is formed (H. Shimizu *et al.*, Heterocycles, 1984, **22**, 1025).

(184)   (185)

The nmr-spectroscopic data for 2,3-dihydro-2,4-diphenyl-1,5-benzothiazepine are presented below (G. Toth *et al.*, Acta Chim. Hung., 1983, <u>112</u>, 167). A *para*-substituent in the 4-phenyl ring has little influence on the chemical shift of C-4, from which it is deduced that there is little conjugation between this phenyl ring and the C=N moiety.

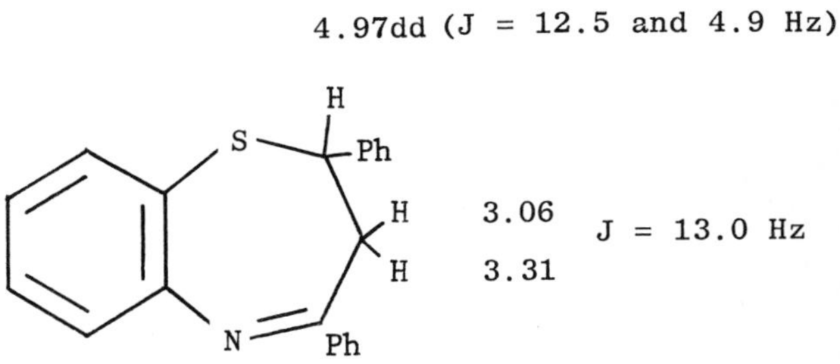

4.97dd (J = 12.5 and 4.9 Hz)

3.06
3.31
J = 13.0 Hz

(a)

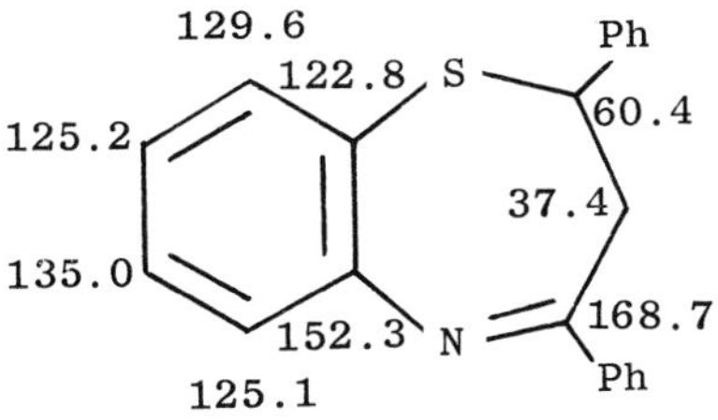

(b)

Fig. 3   (a) ^{1}H and (b) ^{13}C chemical shifts (δ) for

2,3-dihydro-2,4-diphenyl-1,5-benzothiazepine

The hetero ring is flexible and rapid inversion of conformers is observed ($\Delta G^* = 38$ kJ mol^{-1}). The conformer in which the 2-phenyl group is quasi-equatorial is preferred (P.W.W. Hunter and G.A. Webb, Tetrahedron, 1973, $\underline{29}$, 147).

In the mass spectrum of a dihydro-1,5-benzothiazepine a prominent molecular ion peak is observed, although the base peak corresponds to a 2-substituted benzothiazole ion. Fragmentation is considered to involve opening the 7-membered ring and the loss of substituents from this ring. The mass spectral fragmentation of the tetrahydro-1,5-benzothiazepines follows a similar pattern, although loss of C_6H_6SN leading to formation of a methyl allyl benzene ion is a characteristic feature of their spectra. A base peak due to the hydrobenzothiazole ion is observed. (C. Wen-Gang *et al.*, Org. mass Spec., 1980, $\underline{15}$, 643).

Dibenzo[b,f][1,4]thiazepines result from the 1,3-rearrangement of 1,2-benzisothiazole dioxides in a reaction which is akin to the carbanionic rearrangement of sulphonamides to aminosulphones. Two moles of butyl lithium are necessary, effecting *ortho*-metallation in both the fused benzene ring and the 2-phenyl ring (D. Hellwinkel, R. Lenz and F. Lämmerzahl, Tetrahedron, 1983, **39**, 2073).

The photoisomerisation of 2-aryl benzisothiazol-3(2H)-ones leads to dibenzo[b,f][1,4]thiazepin-10(11H)-ones (N. Kamigata *et al.*, Chem. Comm., 1983, 765).

N-Acylation of 10,11-dihydrodibenzo[b,f][1,4]thiazepine occurs on heating it with an acid anhydride or chloride and reaction with *O*-(mesitylsulphonyl)hydroxylamine gives the *S*-imide. 10-Acyl-10,11-dihydrodibenzo[b,f][1,4]thiazepine-5-imide-5-oxides undergo a base induced transannular migration of the acyl group giving the *S*-acylimide derivatives (P. Stoss and G. Satzinger, Ber., 1978, <u>111</u>, 1453).

7-Trifluoromethyldibenzo[b,e][1,4]thiazepin-5(11*H*)-carboxaldehyde-10-oxide undergoes a Pummerer rearrangement with trifluoroacetic anhydride, but if ethanol is present in the ether used in the work-up procedure the intermediate cation is trapped as the 5,11-bridged tetracycle (186) (H.L. Yale, J. heterocyclic Chem., 1978, <u>15</u>, 331).

(186)

(b) Thiadiazepines

From the mass spectra of a number of 3–substituted 4,5–dihydro–1,2,4–thiadiazepine–1,1–dioxides (187), it can be seen that several primary fragmentation processes are operative.   Loss of sulphur dioxide and of a hydroxyl radical are common, whilst loss of a neutral molecule CH_3NO is also prevalent.   The abundance of the molecular ion decreases as the chain length of the 3–substituent increases (U. Vettori *et al.*, Org. mass. Spec., 1984, **19**, 280.

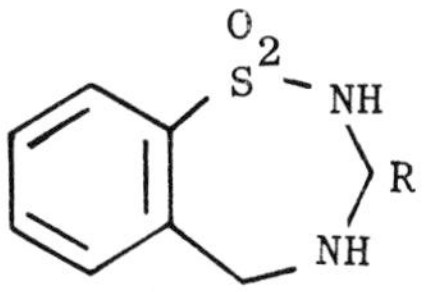

(187)

4–Phenyl–1,2,5–benzothiadiazepine–1,1–dioxide results from the reductive cyclisation of the sulphonamide derived from the reaction between 2–nitrobenzenesulphonyl chloride and ω–aminoacetophenone.   A mixture of the di– and tetra–hydrobenzothiadiazepines is produced.   Use of 4–amino–3,5–dimethylisoxazole as the 1,3–aminoketone unit leads to the 3–acetyl analogue (O. Migliari, S. Petruso and V. Sprio, J. heterocyclic Chem., 1979, **16**, 835).

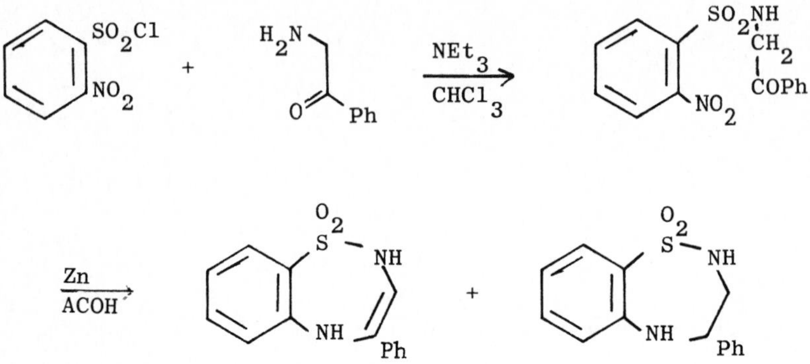

344

The ^{1}H-nmr-spectrum of the dihydro compound shows two singlets, one at δ 4.40 and the other at δ 5.90 associated with the two tautomeric species shown below.

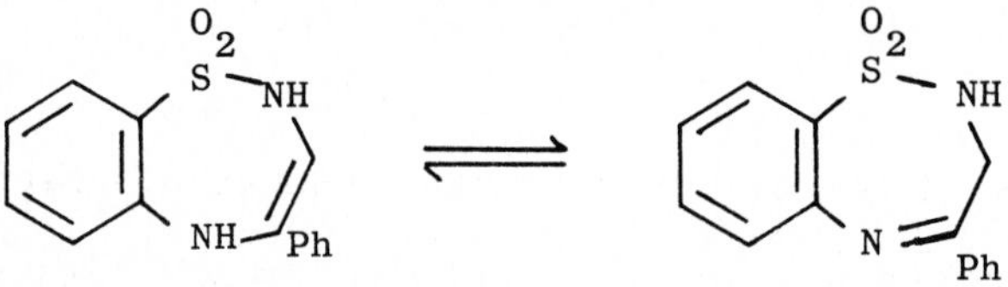

o-Phenylene diamine is a useful precursor of heterocycles containing a 1,4-diaza-arrangement and its reaction with mixed acyl sulphonyl chlorides leads to 3,3-dialkyl-1,5-dihydro-2,1,5-benzothiadiazepin-4(3H)-one 2,2-dioxides (188) (E. Bellasio, G. Pagani and E. Testa, Gazz., 1964, 94, 639).

(188)

Ethyl chlorosulphonylacetate reacts with 2-nitroanilines to give N-(2-nitrophenyl)sulphamoyl derivatives. The latter are precursors of the 1,2,5-thiadiazepine system having the nitrogen atoms and sulphur atom correctly positioned. After hydrolysis of the ester function and reduction of the nitro group, cyclisation to 2,1,5-benzothiazepine-1,1-dioxides is accomplished using dicyclohexylcarbodiimide.

Additional substitution at N-1 or N-5 may be achieved by the usual alkylation procedures or by a Mannich reaction (M. Bianchi et al., Eur. J. Med. chem. chim. Ther., 1976, 11, 101).

Reagents: (i) $ClSO_2CH_2COOEt$; (ii) OH^-; (iii) H_2, PtO_2; (iv) DCC, THF

1,4-Diaminobutane and sulphamide react in pyridine to give a high yield of the fully saturated 1,2,7-thiadiazepane-1,1-dioxide (189), m.p. 261°C (dec). In a related manner, α, α'-diamino-*o*-xylene yields 1,2,4,5-tetrahydro-3,2,4-benzothiadiazepine-3,3-dioxide (190), m.p. 271°C (V.P. Arya and S.J. Shenoy, Indian J. Chem., 1976, 14B, 766). However, the result is at variance with later work (M. Preiss, Ber., 1978, 111, 1915) which gives a m.p. 114–116°C for the compound prepared from the diaminobutane and sulphamide in diglyme. The spectroscopic properties of this product support the thiadiazepane structure.

(189)   (190)

Ring expansion resulting in the formation of the tetrahydro 1,3,4-thiadiazepine-4-carboxamide (191) occurs when 3-(2-bromopropyl)-5-trifluoromethyl-1,3,4-thiadiazol-2(3*H*)-one is treated with pyrrolidine. The product arising from simple nucleophilic displacement of the bromine accompanies the thiadiazepine (H. Kristinsson, T. Winkler and M. Mollenkopf, Helv., 1983, 66, 2714).

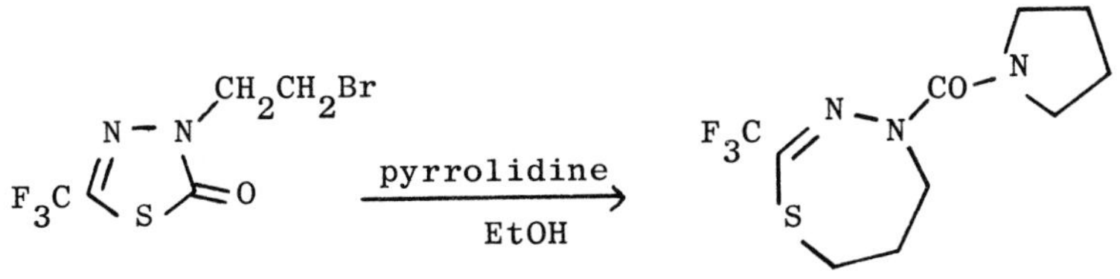

(191)

2-Phenylazo-4,6,7-triphenyl-4H-1,3,4-thiadiazepine-5-thione, m.p. 220°C (dec) is a dark purple solid which is produced when dehydrodithizone (192) reacts with diphenylcyclo-propenethione (K.T. Potts *et al.*, J. chem. Soc. Perkin I, 1981, 2692).

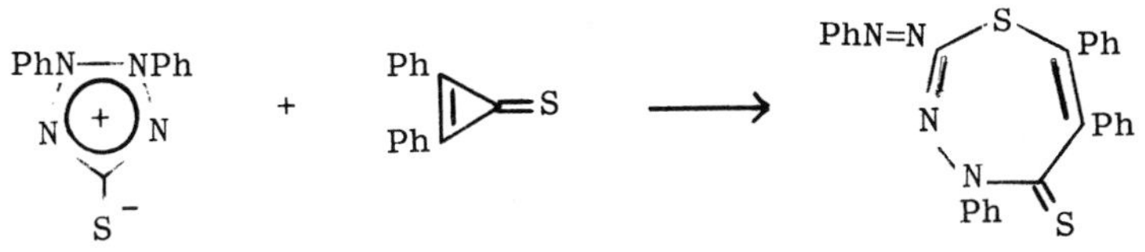

(192)

2-Chloroacetamides such as (193) react with potassium thiocyanate to yield an isothiocyanate by isomerisation of the initially formed thiocyanate. Subsequent ring closure occurs in the presence of base to yield a derivate of 1,3,5-thiadiazepin-6-one (A. Vass and G. Szaloutai, Synthesis, 1986, 817).

(193)

Reagents: (i) KSCN, acetone, 20°C; (ii) CH_3COOEt, 80°C; (iii) NaSH, H_2O, $CHCl_3$, 40°C

Thiobenzoyl isocyanate reacts with trimethylsilyl cyanide to give the 2:1 adduct, which is 2-phenyl-5-thiobenzoyl-7-trimethylsilyl-imino-4,5,6,7-tetrahydro-1,3,5-thiadiazepin-4,6-dione. It is preferential to generate the isocyanate from 2-phenylthiazoline-4,5-dione in the presence of an excess of TMSCN. Desilylation occurs on treatment with ethanolic hydrogen chloride (O. Tsuge and S. Urano, Heterocycles, 1979, 12, 1319).

Ring expansion of the ylide (194) occurs on treatment with trimethyloxonium tetrafluoroborate to yield the first example of a 1,3,5-benzothiadiazepine (195), m.p. 175.5°C, which probably arises from a Pummerer rearrangement of the intermediate aminosulphonium salt. The thiadiazepine rearranges to a benzothiazoline in the presence of acid (T.L. Gilchrist, C.W. Rees and I.W. Southon, J. chem. Res. (M), 1979, 2555).

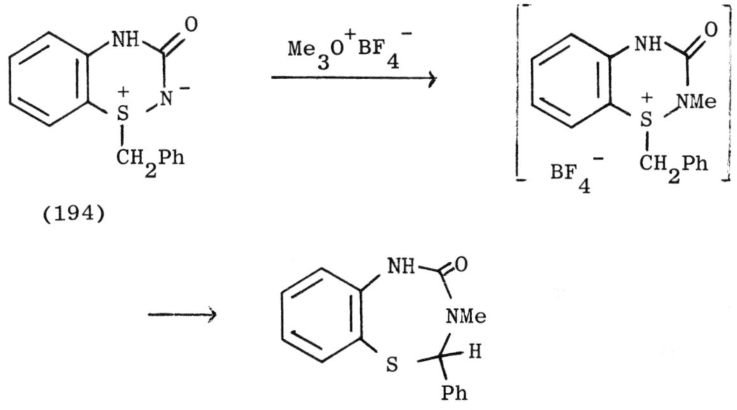

(194)

(195)

The immonium salt (196) derived from N,N-dimethylthiourea reacts with 2-aminothiophenol to give 2-dimethylamino-4-methylthio-1,3,5-benzothiadiazepine (197) admixed with the benzothiazole (198) (M. Yokoyama, K. Arai and T. Imamoto, J. chem. Soc. Perkin I, 1982, 1059).

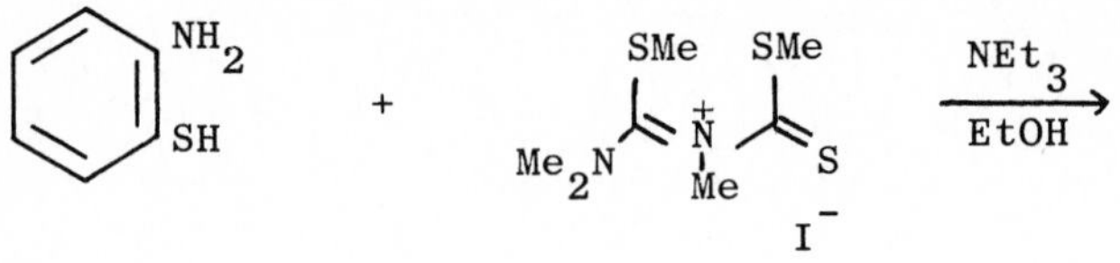

(196)

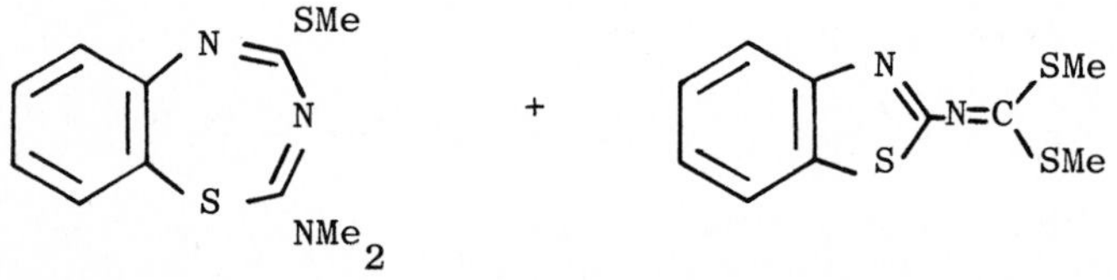

(197)   (198)

Some doubt has been cast on the 1,3,6-thiadiazepine structure (199) which has been proposed for the product resulting for the aerial oxidation of the fungicide Nabam, disodium ethylenebisdithiocarbamate (see 2nd edn., Vol. IV K, p.309). The great weight of spectral evidence, notably the non-equivalence of the two methylene groups in the [1]H-nmr-spectrum, points to the compound being 5,6-dihydro-3H-imidazo[2,1-c]-1,2,4-dithiazole-3-thione (200) (A. Senning, Sulfur Reports, 1982, 2, 33).

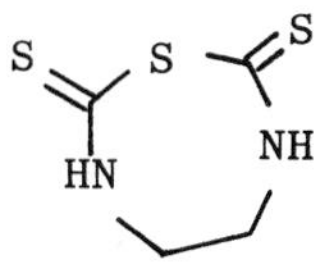

S S S
HN NH

(199)

S N N
S — S

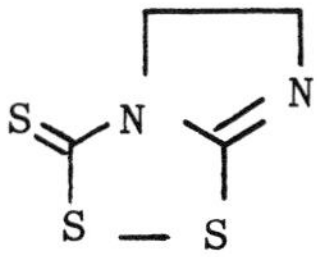

(200)

A number of 3,6-disubstituted 2,7-dihydro-1,4,5-thiadiazepine derivatives have been synthesised through the reaction of diphenacyl sulphides (201) with hydrazine hydrate (S.S. Sandhu, S.S. Tandon and H. Singh, Indian J. Chem., 1977, 15B, 664; 1980, 19B, 1023).

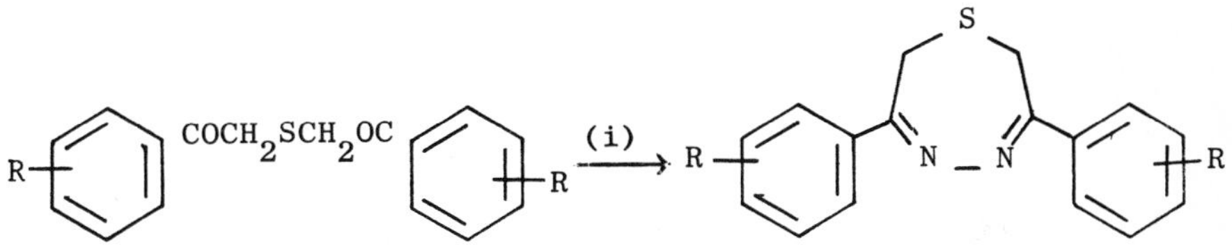

$$R \longrightarrow COCH_2SCH_2OC \quad R \xrightarrow{(i)} R \longrightarrow \overset{S}{\underset{N-N}{\bigcirc}} \longrightarrow R$$

(201)

Reagents: (i) N_2H_4, EtOH, $HSCH_2CH_2SH$

The N-heteroatoms in 1,4,5-thiadiazepines are suitably disposed to chelate to metal ions and the complexes with Cu(II), Co(II), Zn(II), Cd(II), Pd(II) and Pt(II) have been prepared and investigated (S.S. Sandhu, S.S. Tandon and H. Singh, J. inorg. nucl. Chem., 1978, 40, 1967; 1979, 41, 1239; Inorg. chim. Acta, 1979, 34, 81).

2,7-Dihydro-3,6-diphenyl-1,4,5-thiadiazepine undergoes a ring contraction to 3,6-diphenylpyridazine on treatment with bromine in acetic acid *via* a halogenation-dehalogenation sequence, involving extrusion of sulphur (J.D. Loudon and L.B. Young, J. chem. Soc., 1963, 5496). This process appears to be dependent on the reaction conditions, since in contrast bromine in methanol leads to a mixture of thiophenes and a thiadiazole. On the other hand, chlorine in methanol leads to the thiadiazole with other unidentified products (K. Kamata and O. Tsuge, Heterocycles, 1984, **22**, 1497).

Oxidation of 3,3,6,6-tetramethyl-2,3,4,5,6,7-hexahydro-1,4,5-thiadiazepine-1,1-dioxide with iodine leads to Δ^4-*trans*-3,3,6,6-tetramethyl-2,3,6,7-tetrahydro-1,4,5 thiazepine-1,1-dioxide, m.p. 138-140°C (decomp.). Irradiation at low temperature brings about isomerisation to the *cis*-isomer, m.p. 108°C (decomp.)   (H. Lind, G. Rihs and G. Rist, Tetrahedron Letters, 1980, 339).

	trans		*cis*
	δ		δ
CH$_2$	2.70 (d)	J = 13 Hz	3.46 (s)
CH$_3$	2.21 (s)	1.50 (s)	1.72 (s)

3,6-Diaryl-2,7-dihydro-1,4,5-thiadiazepines exist in enantiomeric twist-boat conformations. The barrier to ring inversion is substantially greater in the thiadiazepine and its 1,1-dioxide (Z=S or SO$_2$) than in the corresponding triazepine or diazepine   (Z=NCH$_2$Ph or CMe$_2$) (E. Cuthbertson *et al.*, Spectrochim. Acta, 1980, **36A**, 333).

An improved synthesis of 5,6-dihydrodibenzo[b,f][1,4,5]-thiadiazepines involves a Smiles rearrangement of 2-nitrophenyl 2-(α,β-diacetylhydrazino)phenyl sulphides which is brought about by potassium carbonate in dimethylformamide solution. Base hydrolysis of the acetyl compounds is facile and the resulting hydrazo compounds are oxidised by air to the dibenzothiadiazepine (C. Corral, J. Lissavetzky and G. Quintanilla, J. org. Chem., 1982, 47, 2215).

The electrochemical reduction of 2,2'-dinitrodiphenyl-sulphide in an alkaline medium leads to a mixture of products from which dibenzo[b,f][1,4,5]thiadiazepine can be isolated in *ca.* 10% yield (J. Hlavaty, J. Volke and O. Manousek, Electrochimica Acta, 1978, **23**, 589). The initial reduction product is the dihydroxylamine which undergoes a chemical oxidation to the thiadiazepine (Y. Mugnier and E. Laviron, Electrochimica Acta, 1980, **25**, 1329).

(c) *Thiatriazepines*

N-Cyanoguanidine and carbon disulphide on treatment with an alkali metal hydroxide yields the *N*-cyanoformamidino-dithiocarbimate (202). The latter reacts with sulphur in boiling methanol to give the 1,2,4,6-thiatriazepine system as the thiolate; iodomethane converts this product to the thiomethyl derivative which sublimes at 230°C. ^{1}H and ^{13}C spectral data are shown below (H. Hlawatschek and G. Gattow, Z. anorg. allg. Chem., 1983, **502**, 11).

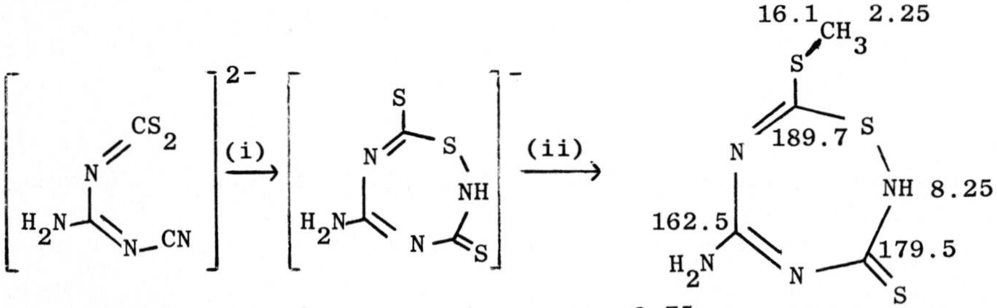

(202)

Reagents: (i) S_8; (ii) MeI, MeOH

^{1}H and ^{13}C chemical shifts (δ)

Diphenylketene-*N*-phenylimine reacts slowly with diphenylsulphur diimide to give a low yield of 5,6-benzo-2,4-diphenyl-3-diphenyl-methylenetetrahydro-1,2,4,7-thiatri-azepine (203), m.p. 160°C (N. Murai *et al.*, Chem. Letters, 1976, 1379).

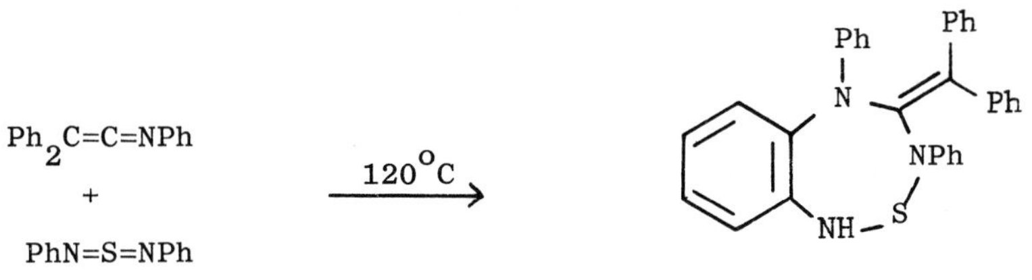

Ph$_2$C=C=NPh

+

PhN=S=NPh

$\xrightarrow{120^{\circ}C}$

(203)

Although 1,3-binucleophiles generally undergo a favoured exo-trig cyclisation with a diazabutadiene to yield a five-membered ring, the reaction of a cyclic thiourea and a 1,4-dichloro-2,3-diazabutadiene in the presence of an excess of base gives rise to an annulated 1,3,4,6-thiatriazepine (204) (S.F. Moss and D.R. Taylor, J. chem. Soc. Perkin I, 1982, 1999).

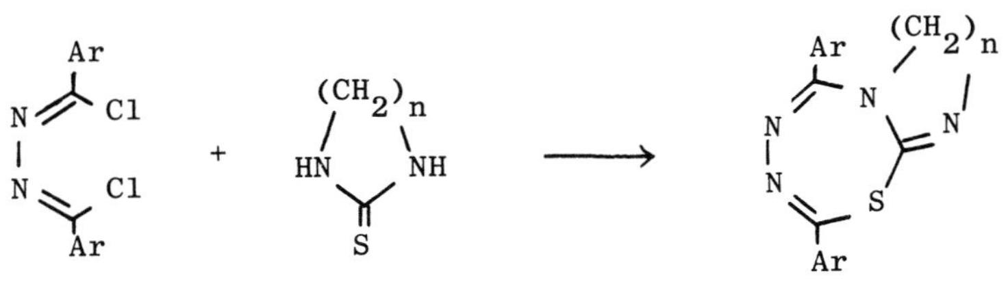

(204)

The compounds are susceptible to acid hydrolysis, which leads to a 1,3,4-thiadiazole amongst other products. Decomposition is even more facile with sodium hydroxide, affording an imidazolinyldiazabutadiene.

X-Ray crystallographic analysis indicates that the seven-membered ring is extensively puckered with one C=N bond appreciably longer than the other two (B. Beagley *et al.*, Acta Crystallog., 1981, B37, 486).

Diazotisation of *N*-(2-aminophenyl)-*N*-benzylsulphamoyl-acetamide (205) is followed by intramolecular cyclisation to the 1,2-dihydro-5H-2,1,4,5-benzothiatriazepine-2,2-dioxide derivative (206). Application of standard synthetic methods prior to diazotisation allows some variation in the nature of the substituent at the 3-position to be achieved (M. Bianchi *et al.*, Eur. J. med. Chem. Chim. Ther., 1986, 11, 101).

(205)   (206)

(d) Dithiazepines

1,2,5-Dithiazepane, b.p. 102-107°C/1.5 mm, is produced in almost 50% yield by the reaction of bis-2-chloroethyl disulphide (207) with ammonia under pressure in methanolic solution. The free base is unstable and is therefore alkylated in dimethylformamide solution (M. Strzelczyk, Pol. J. Pharmacol. Pharm., 1984, **35**, 473). The 1,2,5-dithiazepane system also results from the reaction of amines with the disulphide or by the introduction of the sulphur heteroatoms into *N*-bis-2-chloroethylamines by means of sodium sulphide in dimethylformamide.

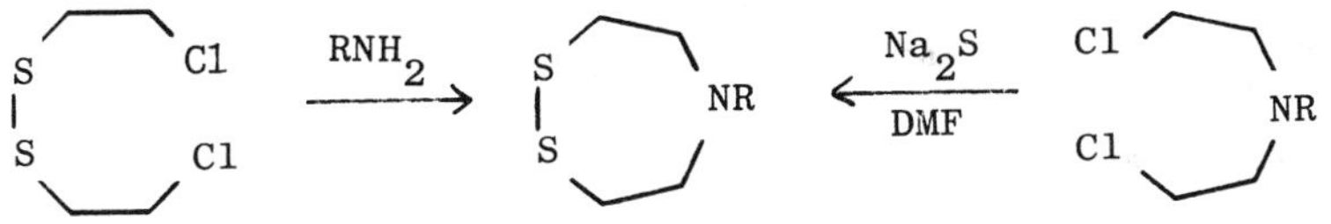

(207)

The bis(isothiouronium) salt (208) derived from the nitrogen mustard by reaction with thiourea has only been hydrolysed to the dimercaptide in a satisfactory manner in the presence of transition metal salts, of which only nickel salts promoted complete hydrolysis. Oxidation with iodine results

in cyclisation to the substituted 1,2,5-dithiazepane (209),
m.p. 129–130°C.

In the cadmium complex of (209), the dithiazepane ring is
folded such that the lone pairs of electrons on the sulphur
atoms and on the nitrogen atom are directed towards the
cadmium (G.C. Baumann *et al.*, Inorg. Chem., 1984, **23**, 3104).

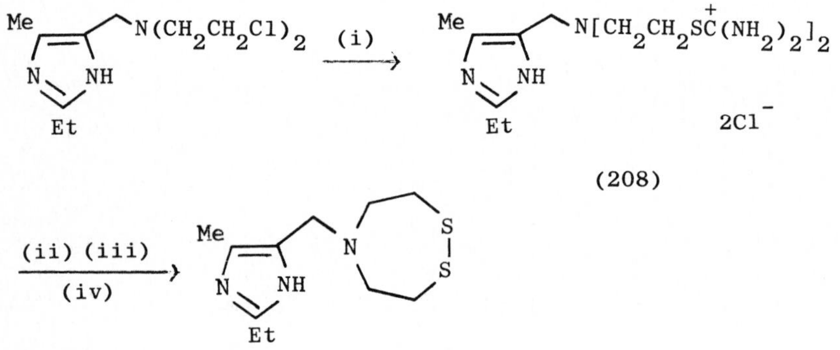

(208)

(209)

Reagents: (i) thiourea, EtOH; (ii) $NiCl_2$; (iii) H^+; (iv) I_2

3,4,6,7-Tetramethyl-2-(4-methylphenyl)-2H-1,5,2-dithiazepine,
m.p. 122–125°C, is one of the products from the reaction of
p-toluidine with the trithiepin (210) (E. Schaumann and
F.-F. Grabley, Ann., 1979, 1702).

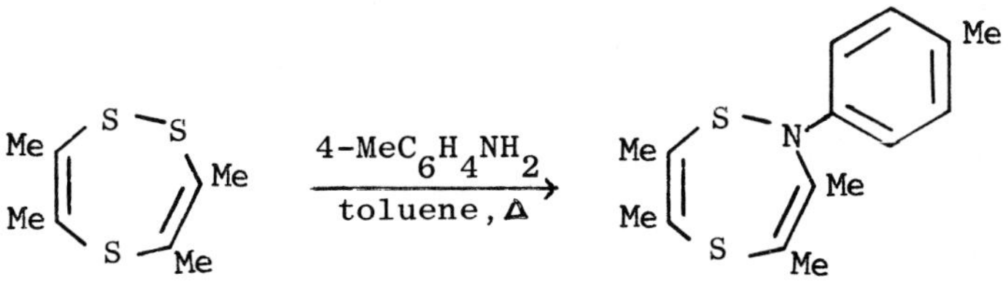

(210)

(e) *Dithiadiazepines*

2,4,1,5-Benzodithiadiazepine tetroxides (211) result from
the reaction of bischlorosulphonylmethane with *o*-
phenylenediamines (M. Vincent, G. Remond and J.-P. Volland,
J. heterocyclic Chem., 1977, **14**, 493). Dimethyl sulphate
brings about methylation of both nitrogen atoms, whilst
chlorosulphonic acid attacks at the 8-position. All four
protons in the hetero-ring are exchangeable with D_2O.

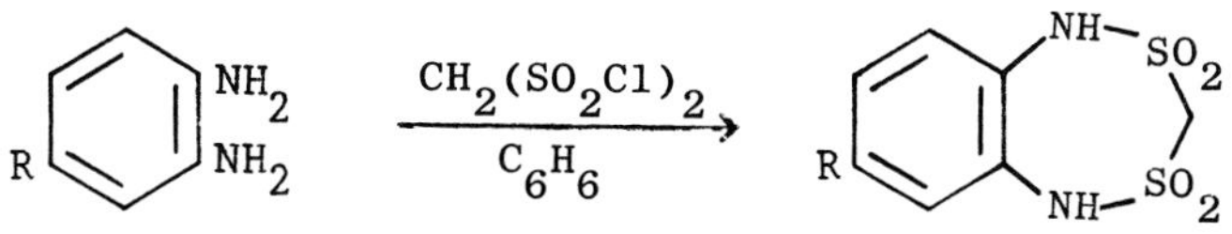

(211)

2,4,7-Trimethyl-1,3,5,6-dithiadiazepine is converted into the enamines (212) on treatment with dimethylformamide diethylacetal (W. Kantlehner, E. Haug and H. Hagen, Ann., 1982, 298).

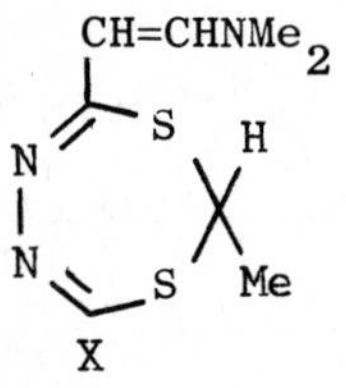

X=Me or CH=CHNMe$_2$

(212)

The disilylated urea (214) reacts with the tetrafluorosulphenyl chloride (213) to give 6,6,7,7-tetrafluoro-2,4-dimethyl-1,5,2,4-dithiadiazepin-3-one, b.p. 72°C/1 mm. (H.W. Roesky *et al.*, Ber., 1985, **118**, 2811).

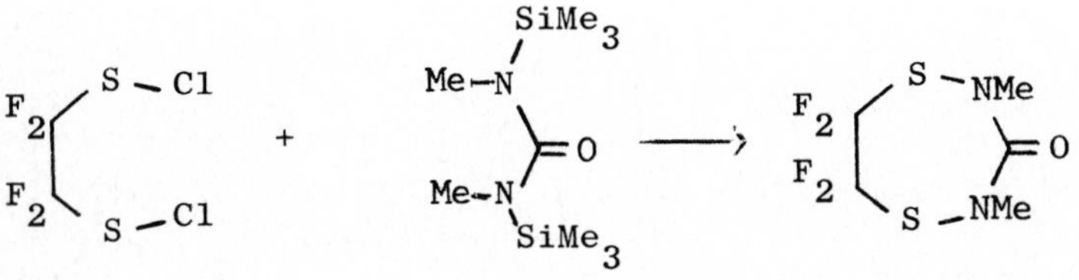

(213)   (214)

The ion derived from chloromethanesulphonamide by treatment
with an alkali metal hydroxide reacts with *o*-phenylene
diisothiocyanate to give the fused 1,6,2,4-dithiadiazepine-
3,3-dioxide (215). The structural assignment is based on
the separation of the H-7 nmr-signal from the other
aromatic-proton signals as a result of deshielding by the
thiocarbonyl function (D. Griffiths, R. Hull and T.P. Seden,
J. chem. Soc. Perkin I, 1980, 2608).

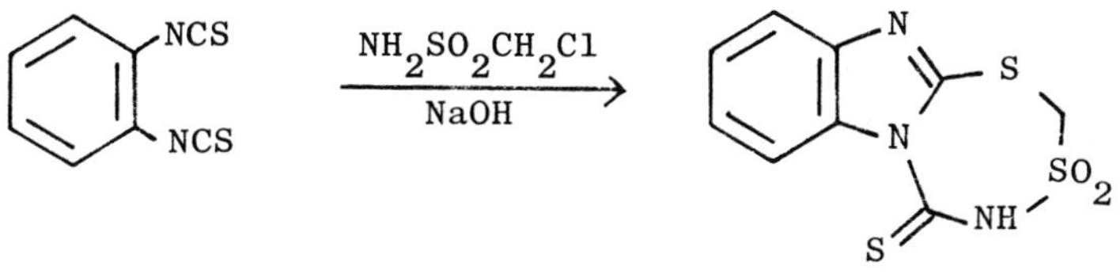

(215)

(f)   Dithiatriazepines

The bifunctional sulphur diimide (216) reacts with sulphur
dichloride to give metallic golden needles of the fused
benzodithiatriazepine (217), m.p. 254-257°C, the crystal
structure of which has been determined (P.W. Codding, H.
Koenig and R.T. Oakley, Canad. J. Chem., 1983, 61, 1562).

(216)   (217)

(g) *Trithiadiazepines*

The initial applications of tetrasulphur tetranitrogen, S_4N_4, to organic chemistry have been reviewed (J.L. Morris and C.W. Rees, Chem. Soc. Rev., 1986, 15, 1). The reagent shows promise in the preparation of heterocyclic systems which are rich in sulphur and nitrogen. For example, the reaction of S_4N_4 with dimethyl acetylenedicarboxylate in boiling toluene gives a low yield of dimethyl 1,3,5,2,4-trithiadiazepine-6,7-dicarboxylate along with other products.

The sulphurdiimide (218) reacts with the bis(sulphenyl chloride) derived from ethane-1,2-dithiol to give the dihydro trithiadiazepine (219), m.p. $30^{\circ}C$. Dehydrogenation with dichlorodicyanobenzoquinone gives 1,3,5,2,4-trithiadiazepine (220) as colourless volatile crystals. Use of the trichloride instead of the dichloride leads directly to the trithiadiazepine (J.L. Morris, C.W. Rees and D.J. Rigg, Chem. Comm., 1985, 396).

(218)   (219)   (220)

Benzene-1,2-bis(sulphenyl chloride) and bis(trimethylsilyl)-sulphurdiimide yield 1,3,5,2,4-benzotrithiadiazepine (221) as a bright yellow solid (J.L. Morris, C.W. Rees and D.J. Rigg, Chem. Comm., 1985, 396).

(221)

The trithiadiazepine (221) undergoes electrophilic aromatic substitutions at the 6-position. With an excess of reagent, a second substituent is introduced at C-7.

TABLE

1,3,5,2,4-Trithiadiazepines[*]

R^1	R^2	m.p. (^{o}C)
H	H	57
Br	H	31
Br	Br	89
NO_2	H	84
NO_2	NO_2	60
I	H	93
CN	H	71
COOMe	H	111
benzo		78

[*] Data from J.L. Morris, C.W. Rees and D.J. Rigg, Chem. Comm., 1985, 396.

Trithiadiazepyne (222) is generated when 6,7-dibromotrithia-diazepine is treated with *n*-butyl-lithium and can be trapped with furan or, better, 2,5-dimethylfuran to give the adduct (223) (J.L. Morris and C.W. Rees, Chem. Soc. Rev., 1986, 15, 1).

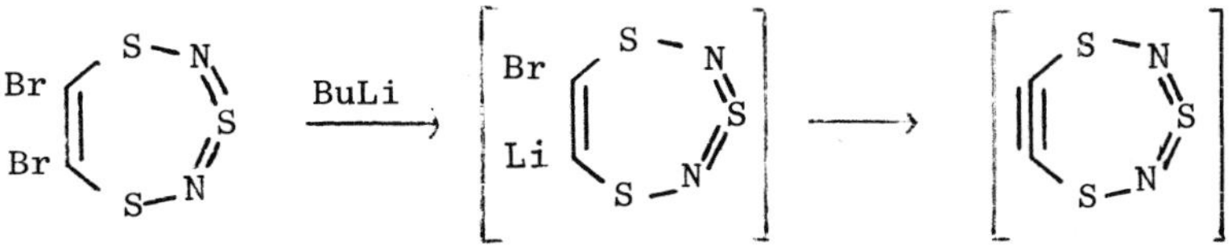

(222)

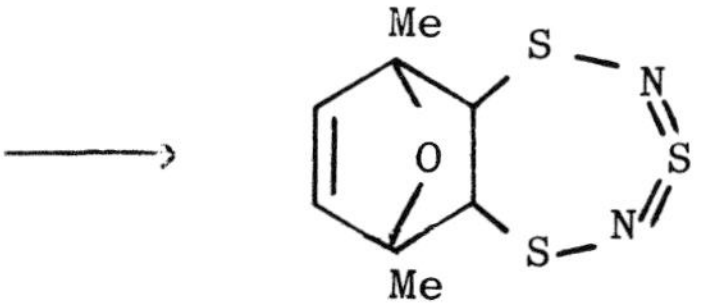

(223)

On treatment with base, trithiadiazepine yields 3,4-bis(methylthio)-1,2,5-thiadiazole, ring contraction converting two of the sulphur heteroatoms into exocyclic substituents. The 6,7-dicarboxylic ester yields the same ring system on heating with triphenylphosphine.

Ring contraction of benzotrithiadiazepine is observed on treatment with bromine, when 1,3,2-benzodithiazolium bromide is produced.

Trithiadiazepine is a planar molecule, as is the benzologue. The 10 and 14 π electrons associated with the two molecules respectively are delocalised and *ab initio* and MNDO calculations support an aromatic character (R. Jones *et al.*, Chem. Comm., 1985, 398). The stability, chemical reactivity and spectroscopic data are in accord with the theoretical predictions and the crystallographic structural assignment.

The analogous saturated seven-membered ring system results when tetrafluoro-1,2-ethanedisulphenyl dichloride reacts with the silylated sulphamide (224). The product, 6,6, 7,7-tetrafluoro-2,4-dimethyl-1,3,5,2,4-trithiadiazepane-3,3-dioxide (225, R=Me) is a colourless liquid, b.p. $139^{\circ}C/0.1$ mm, whilst the 2,4-bis(trimethylsilyl) derivative melts at $27^{\circ}C$ (H.W. Roesky *et al.*, Ber., 1985, 118, 2811). Further reaction of the trithiadiazepine (225) with the sulphenyl chloride occurs in the presence of aluminium chloride. Stepwise displacement of the trimethylsilyl groups leads to the di- and tri-cyclic systems 3,3,4,4,8,8,9,9-octafluoro-2,5,7,10,11-pentathia-1,6-diazabicyclo[4.4.1]undecan-11,11-dioxide (226) and the hexadecafluoro-2,5,7,10,12,15,17,20, 21,22-decathia-1,6,11,16-tetraazatricyclo$[4.4.1^{6,11}]$docosan-21,21,22,22-tetraoxide (227).

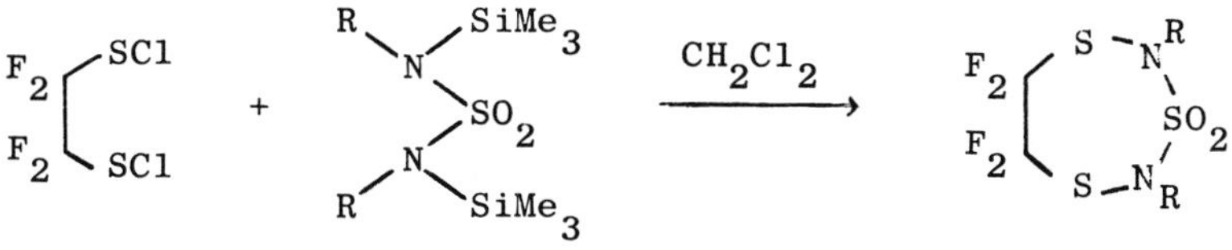

(224) (225)

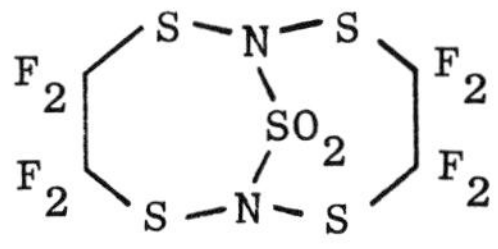

(226)

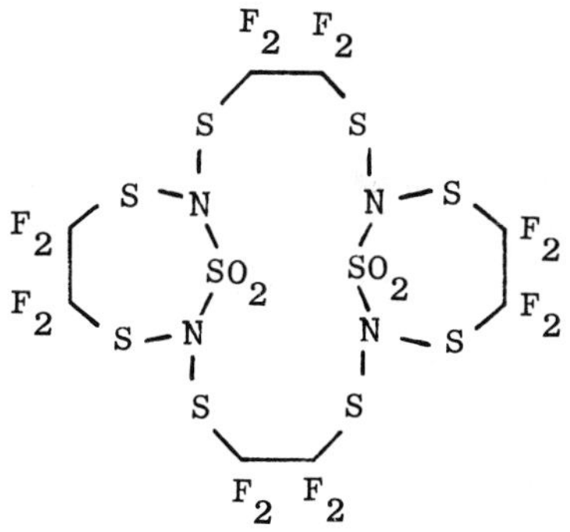

(227)

The crystal structure of the trithiadiazepane (225, R=H) tetraphenylphosphonium chloride adduct and the ^{19}F-nmr-spectrum of (225, R=SiMe$_3$) indicate a chair conformation for the heterocyclic system. At 100°C, the A,A',B,B' system collapses to a broad singlet and the energy barrier to inversion is calculated to be 67.5 kJ mol^{-1}.

(h) Trithiatriazepines

One of the products of the reaction of tetrasulphur tetranitrogen with dimethyl acetylenedicarboxylate is methyl 1,3,5,2,4,6-trithiatriazepine-7-carboxylate (228). The ester is hydrolysed by dilute hydrochloric acid and the resulting carboxylic acid is decarboxylated in boiling dioxan. X-ray crystallographic data show that the heterocycle is planar; the compound, having a 10π-system, appears to exhibit aromatic character (S.T.A.K. Daley, C.W. Rees and D.J. Williams, Chem. Comm., 1984, 55).

(228)

(i) 1,4-Selenazepines

Upon treatment with potassium acetate in acetic anhydride, benzisoselenazolin-3-one and its N-methyl derivative are converted into the 3,5-dioxo-1,4-benzoselenazepine (229). The 4-methyl derivative, (229, R=Me) m.p. 115°C, has carbonyl stretching bands at 1670 and 1625 cm^{-1}, whilst in its nmr-spectrum the 2-methylene group at δ 3.58 exhibits coupling (16 Hz) to ^{77}Se. The unsubstituted compound (229, R=H) rearranges to 3-hydroxybenzo[b]selenophene-2-carboxamide and cannot be isolated (R. Weber and M. Renson, Bull. Soc. Roy. Sci. Liege, 1979, 48, 146).

(229)

4. *Rings containing oxygen, sulphur and nitrogen*

(a) *Oxathiazepines*

Sulphene, $CH_2=SO_2$, generated from methanesulphonoyl chloride and triethylamine, adds to *N*-benzylidine-2-hydroxyaniline to give 4,5-dihydro-4-aryl-3H-1,2,5-benzoxathiazepine-2,2-dioxide (230). It is envisaged that the initial [2+2] cycloadduct rearranges with migration of the sulphone function from nitrogen to oxygen (M. Rai and B. Kaur, Chem. Comm., 1981, 971).

(230)

An intramolecular Pummerer reaction of ethyl *N*-acetyl-*S*-benzyl-*L*-cysteinylglycinate-*S*-oxide (231) or the *N*-benzoyl analogue affords the 1,3,6-oxathiazepine (232) only under very precisely controlled conditions. The same oxathiazepine results from isomerisation of the thiazolidine (233) in trifluoroacetic acid (S. Wolfe, P.M. Kazmaier and H. Auksi, Canad. J. Chem., 1979, 57, 2412).

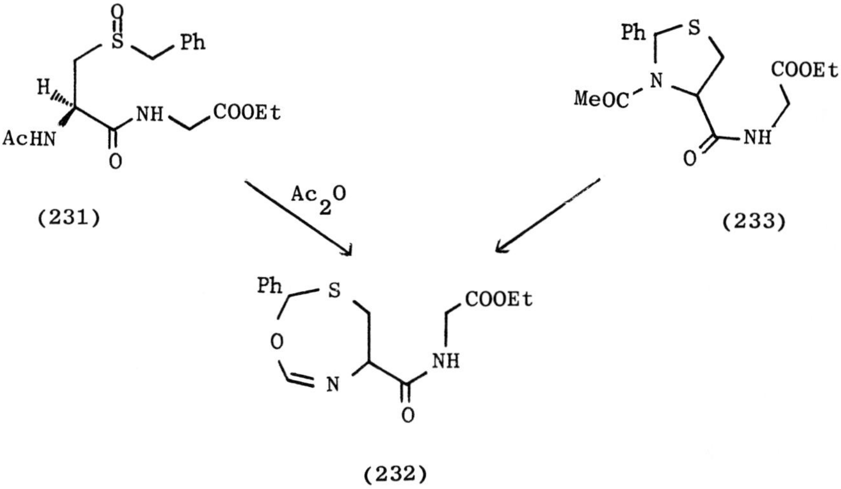

(231)

(232)

(233)

Several indolo fused 1,3,7-oxathiazepines (234) have been isolated from the Caribbean tunicate *Eudistoma olivaceum* (K.L. Rinehart *et al.*, J. Amer. Chem. Soc., 1984, 106, 1524).

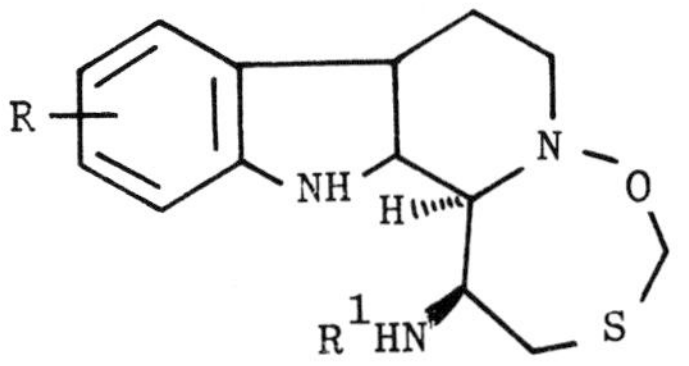

(234)

Thermolysis of 2-phenoxyethanesulphonyl azide affords 2,3,4,5-tetrahydro-1,4,5-benzoxathiazepine-4,4-dioxide (235), m.p. 153-154°C (R.A. Abramovitch *et al.*, J. org. Chem., 1984, 49, 3114).

$$PhOCH_2CH_2SO_2N_3 \xrightarrow[\text{Freon 113}]{\Delta}$$

(235)

In a similar manner, the sulphonyl nitrene generated by thermolysis of 2-phenoxybenzenesulphonyl azide (236, X=O) undergoes an intramolecular cyclisation to give 6H-dibenzo[b,f] [1,4,5]oxathiazepine-5,5-dioxide (237), m.p. 136-139°C. Neither the corresponding sulphur analogues (236, X=S, SO$_2$) nor the nitrogen analogue (236, X=NCOCH$_3$) yields a seven membered heterocycle on decomposition (R.A. Abramovitch *et al.*, J. org. Chem., 1978, **43**, 1218).

$$\xrightarrow{165^\circ C}$$

(236)   (237)

Thioacyl isocyanates usually react with bromoethanol to give N-thioacyloxazolidinones. However, the *t*-butyl derivative (238) affords 4-*t*-butyl-6,7-dihydro-1,5,3-oxathiazepin-2(2H)-one (239), m.p. 42-43°C, $\nu_{C=O}$ 1715 cm^{-1} (K. Nandi and

J. Goerdeler, Ber., 1981, 114, 1972).

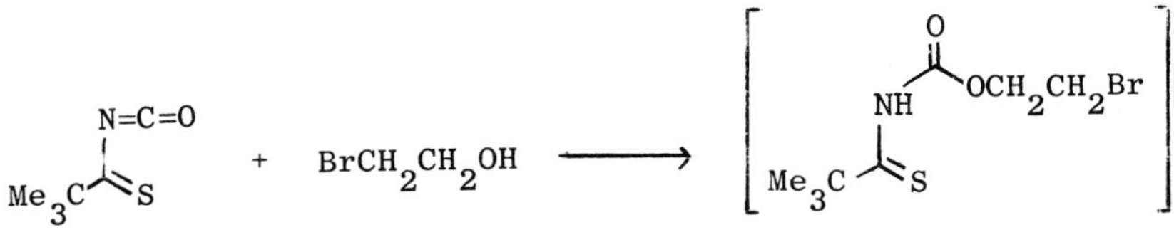

(238)

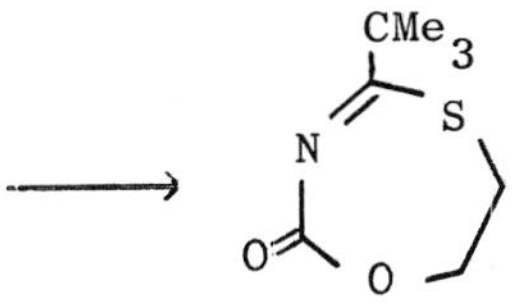

(239)

(b) *Dioxathiazepines*

Chlorosulphonyl isocyanate reacts with catechols to yield 2,2,4-trioxo-1,5,2,3-benzodioxathiazepines (240). Initial reaction occurs at 0°C and elimination of hydrogen chloride ensues on heating at 100°C (M. Hedayatullah and J.-F. Brault, Phosphorus and Sulphur, 1981, 11, 255). On boiling with formic acid, the heterocyclic ring opens to yield an *o*-hydroxyphenyl sulphamate.

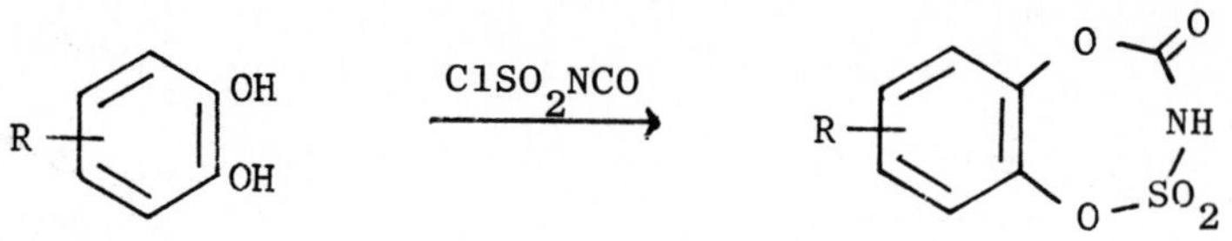

(240)

(c) *Dioxathiadiazepines*

The liquid phase reaction between perfluoro-2,5-diazahexane 2,5-dioxyl and sulphur dioxide gives an almost quantitative yield of the 1:1 cyclic adduct, 5,5,6,6-tetrafluoro-4,7-bis(trifluoromethyl)-1,3,2,4,7-dioxathiadiazepine-2,2-dioxide (241), b.p. 114°C, a seven membered ring with five heteroatoms. The compound, which is stable to hydrolysis, is reduced by triphenylphosphine to the perfluorinated six membered oxathiazine and five membered thiadiazole. Sulphur tetrafluoride and the dioxyl react in a similar manner to give the octafluoro analogue (242) (A. Arfaei and S. Smith, J. chem. Soc. Perkin I, 1984, 1791).

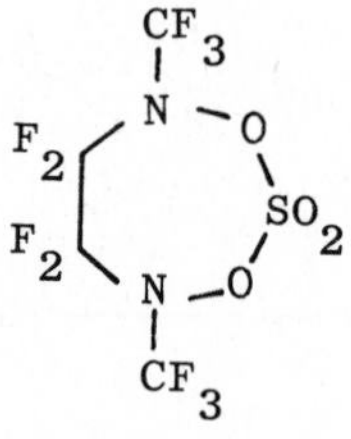

(241)

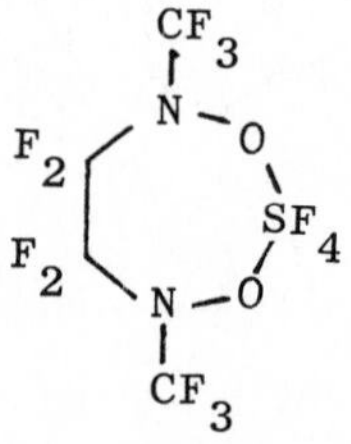

(242)

Chapter 53

COMPOUNDS WITH SEVEN-MEMBERED AND LARGER RINGS POSSESSING ONE
OR MORE UNUSUAL HETEROATOMS.

T.J.MASON

1. Magnesium and Mercury Compounds

In the crystalline state magnesacyclohexane exists as 1,7-
dimagnesacyclododecane whilst in THF solution a monomer dimer
equilibrium exists favouring the dimer. In contrast to this
the lower homologue magnesacyclopentane exists entirely in dimeric form 1,6-dimagnesacyclodecane (1) in THF solution. The
strong tendency for such species to dimerise is accounted for
in terms of the typically large C-Mg-C bond angles (128° in
crystalline 1). Such a large bond angle cannot be easily accommodated in six-membered rings and accommodation is even more
unfavourable in five membered rings. (H.C.Holtkamp, *et al.*,
J.organometallic Chem.,1982,240,1).
It has been found possible to synthesise a range of substituted phenylene mercurials by vigorously stirring sodium amalgam
with the corresponding 1,2-dibromobenzene derivatives in THF.
These are trimeric high melting solids (2) (R^1-R^4=Me, m.p.
>300°; R^1-R^4=Cl, m.p. >330°; R^1=Me,R^2-R^4=H, m.p. 323-325°;
R^1=R^4=H,R^2=R^3=Me, m.p. >320°; R^1=R^4 =H,R^2=R^3=OMe, m.p. 300°)
but the yields are very low (typically about 1-2% based on the
starting dibromo compound). Biphenylene mercurials e.g. (3)
m.p. 338° are also trimeric. (N.A.A.Al-Jabar and A.G.Massey,
J.organometallic Chem.,1984,275,9).

1   2   3

The crystal structures of the phenylene (D.S.Brown, A.G. Massey and D.A.Wickens, Inorg.Chim.Acta,1980,44,L193) and the biphenylene mercurial (K.Stender, et al., Cryst.Struct.Comm., 1981,10,613) have been reported.

2. Boron compounds

(a) Unsaturated systems containing boron: borepins

Organotin compound (1), a red brown viscous oil, has been prepared by the reaction between hex-3-en-1,5-diyne and Et₂SnBrH. The aromatic compound 1-methylborepin (2), a thermally unstable liquid, is generated when (1) is treated with MeBBr₂. Borepin (2) decomposes thermally to benzene and MeB: via 7-methyl-7-borabicyclo(4,1,0)heptadiene. (S.M.Van der Kerk, J. Boersma and G.J.M.Van der Kerk, J.organometallic Chem.,1981, 215,303, S.M.Van der Kerk,ibid.,1981,215,315). The parent borepin molecule has not yet been synthesised but a study using ab initio MO theory suggests that it would have a weakly conjugated planar structure (J.M.Schulman, R.L.Disch and M.L. Sabio, J.Amer.chem.Soc.,1982,104,3785; 1984,106, 7696). Syntheses of both the 12H-dibenzo(d,g)(1,3,2)dioxaphosphocin and -dioxaborocin ring systems have been reported each starting from bisphenol (3,R=CO₂CH₃,t-Bu). The borocin compound (4,R= t-Bu),61% m.p. 184-186° is obtained from the reaction of the bisphenol (3,R=t-Bu) and phenylboronic acid in toluene using p-toluenesulphonic acid as catalyst. The phosphocin (4,B=P,R= CO₂CH₃),73% m.p. 279-281° is obtained from (3,R=CO₂CH₃) on reaction with dichlorophenylphosphine in toluene using triethylamine as an acid scavenger (S.D.Pastor and J.D.Spivack, J. heterocyclic Chem.,1983,20,1311).

(b) Saturated systems: boropanes and related compounds.

B-Chloroborepane (5) has been used in a new strategy for the synthesis of (Z)alk-7-en-1-ols (8b) (see below). The one-pot

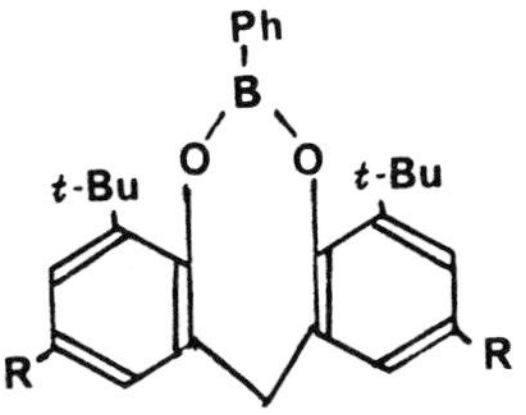

4

synthetic route involves treatment of the borepane and hex-1-yne in THF at -78° with lithium aluminium hydride followed by treatment of the vinylborane (6) with sodium methoxide and iodine in methanol at -78. The reaction proceeds *via* the 8-membered borocane species (7) and the open chain compound (8a) to yield (Z)-dodec-7-en-1-ol (8b) in 78% yield. (D.Basavaiah and H.C.Brown, J.org.Chem,1982,**47**,1792)

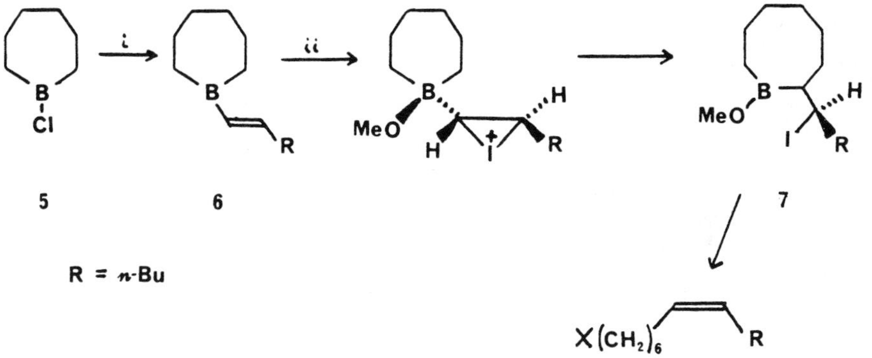

R = *n*-Bu

5   **6**   **7**   **8**

(i)   hex-1-yne / LiAlH₄ / -78°
(ii) NaOH / MeOH / I₂ / -78°

(a) X = (MeO)₂B
(b) X = HO

A new and general synthesis of boracyclanes involves the reaction of α,ω-dienes (9) with 9-borabicyclononane (9-BBN) and borane-dimethylsulphide (BMS). The reaction first produces a dumb-bell like species (10) which redistributes to 9-BBN and the product. The reaction with (9,n=2) produces borecane (11) which is inseparable from the residual 9-BBN either by recrystallisation or distillation. With higher members of the diene series polymerisation occurs on treatment of (10) with BMS. The linear polymer thus obtained can be depolymerised at 175-200° to a mixture of products including the boracyclane (12) and the 2-alkylborinane (13) (Table). The 9-BBN/BMS procedure applied to butadiene produces 1,6-diboracyclodecane (92%) b.p. 41-42°/2mm. (H.C.Brown, G.G. Pai and R.G.Naik, J.org. Chem,1984,49,1072).

Products isolated from polymer degradation

(9)(n=)	(12)%	(13)%
3	75	25
4	11	67
5	15	71
6	12	65
8	0	65
10	38	44

Tris-aminoboranes of general structure (14) are prepared from the corresponding triaza-monocyclic compounds. The symmetrical compound (14,m=n=3) iso-electronic with the guanidinium ion, is produced by the reaction of the triaza compound with trimethoxyborane and can be purified by sublimation (45-47° /0.01 mm.). Similarly prepared are (14,m=2,n=3) b.p. 80°/0.1 mm., (14,m=3,n=2) m.p. 105-108° and (14,m=n=2) m.p. 157-163°. Although (14,m=n=3) can be obtained as a monomer, due to increasing angle strain (14,m=n=2) exists only in the dimeric form (J.E.Richman, N.C.Young and L.L.Andersen, J.Amer.chem. Soc.,1980,102,5790).

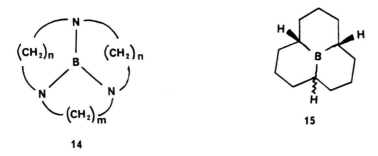

14   **15**

The original assignment for the stereochemistry of the two perhydro-9b-boraphenylenes (15,all *cis*) and (15,*cis/trans*), produced by the reaction of borane with cycododeca-1,5,9-triene (G.W.Rotermund and R.Koster, Ann.,1965,686,153) has been corrected. The isomer known as Centerbor-II has been proven to be the all-*cis* isomer and Centerbor-I is the other. (H.C.Brown, E.Negishi and W.C.Dickason, J.org.Chem,1985, 50,520).

3. *Silicon, germanium, tin and lead compounds.*

(a) *Compounds with one heteroatom*

(i) *Seven membered rings*
Treatment of 1,1-dichloro-1-germacyclopent-3-ene with hexatriene at 85° for 30h gives 1,1-dichloro-1-germacyclo-hepta-3,5-diene (1,R=Cl,76.5%), which with ethyl magnesium bromide is transformed into 1,1-diethyl-1-germacyclohepta-3,5-diene (1,R=Et) (S.P.Kolesnikov, *et al.*, Izv.Akad.Nauk.SSSR,1981, 1423). 1-Chloro-1-methyl-1-stannacycloheptane (2,R=Cl) b.p. 80-81°/0.6 mm., is prepared (65%) by the reaction of (2,R=Me) with tin(IV) chloride in nitromethane solution at -20° (E.J. Bulten and H.A.Budding, J.organometallic Chem.,1978,153,305). A series of 10,11-dihydro-5H-dibenzo(b,f)silepins (3) has been prepared by the reaction of *o,o'*-dilithiobibenzyl with chlorosilanes (Table). The addition of (3,R=H) to *N,N*-dimethyl-allylamine gives (72%) (3,R=CH$_2$CH$_2$CH$_2$NMe$_2$), b.p. 130-140°/0.1 mm., the sila analogue of the clinically effective antidepressant imipramine. The only isolated silepin product from the reaction of *o,o'*-dilithiobibenzyl with trichlorosilane however

is the spiro compound (4, 11%) m.p. 173.5-174.5° (J.Y.Corey
and R.L.Farrell, J.organometallic Chem.,1978,153,15).

1   2   3

Preparation of derivatives of
10,11-dihydro-5H-dibenzo(b,f)silepins (3)

R=	Reagents	Yield%
$(CH_2)_3Cl$	(i)$Cl_3Si(CH_2)_3Cl$,(ii)$MeMgCl$	5-14
$(CH_2)_3Br$	$Cl_2/MeSi(CH_2)_3Br$	52
H	$MeSiHCl_2$	53

Ring enlargement of silicon heterocycles can be accomplished
with compounds such as (5,n=0,1,2) which contain an exocyclic
chloromethyl substituent by reaction with aluminium trichloride or potassium fluoride. In this way (5,R=CH_2Cl,n=2) m.p.
77-79° is converted into (6,R=Cl,n=2) m.p. 77-79° on treatment
with aluminium chloride. The compound (6,R=OH,n=2),m.p.95-97°,
is obtained (47%) by the reaction of water on (6,R=Cl,n=2).
The latter with lithium aluminium hydride gives (96%) the compound (6,R=H,n=2) m.p. 72-74° (J.Y.Corey, et al., J.organomet-
allic.Chem.,1981,210,149).

4   5   6

6,7-Dihydro-5H-dibenzo(c,e)-silepine (8,R=Me,M=Si) b.p. 100-
102°/0.3 mm., yield 90%, and the stannepin (8,R=Ph,M=Sn) m.p.
189°, yield 46%, are prepared by the reaction of the bis-

Grignard reagent (7) with dichlorodimethyl and dichlorodiphenylsilane respectively. Crystal structures for some compounds
in this series have been determined (L.M.Engelhardt,et al., J.
chem.Soc.Dalton,1984,331). The spiro compounds (10,R^1=R^2=H,
30.8%) m.p. 115° and (10,R^{1}R^2=bond,10%) m.p. 66°, are prepared
by reaction of (9) with o-(ClMgCH$_2$CH$_2$)$_2$C$_6$H$_4$ (L.Birkofer and H.
Haddad, J.organometallic.Chem.,1979,164,C17).

7

8

9

10

(ii) Larger ring compounds
Cyclic hydroboration of dienes provides a simple route to bor-
ocyclic compounds which can be readily transformed into cyclic
ketones by treatment with potassium cyanide and trifluoroacetic anhydride. In this way (12,n=2) b.p. 115°/25mm., 27%, and
(12,n=3) b.p. 65°/1.5mm., 26%, are prepared from the dienes
(11,m=0) and (11,m=1) respectively. The germanium analogue of
(12,m=3) b.p. 70°/1.5mm., 23%, is similarly prepared (J.A.
Soderquist and A.Hassner, J.org.Chem.,1983,48,1801).

A range of silacrowns of general structure (13,R^1,R^2= various
alkylgroups,n=1 to 4) such as dimethylsila-17-crown-6 (13,R^1=
R^2=Me, n=3) b.p. 168-170°/0.3mm., 78%, have been prepared by
the transesterification of dialkyldiethoxysilanes R^{1}R^{2}Si(OEt)$_2$
with a polyethylene glycol HO(CH$_2$CH$_2$O)$_{n+2}$H (B.Arkles, et al.,
Organometallics,1983,2,454).

The syntheses of a number of germatranes (14,X=MeO,EtO,PrO, Me$_2$CHO,CH$_2$I,EtO$_2$CCH$_2$CH$_2$,Cl,HO,Me$_3$SiO,Me$_3$GeO,Me$_3$SnO,Ph$_3$SiO,AcO) have been reported to proceed in high yields , 72-99%, by cyclisation of N(CH$_2$CH$_2$OH)$_3$ with (MeCHO)$_3$GeX (T.K.Gar, N.Yu. Khromova, et al., Zh.obshch,Khim.,1979,49,1516). ^{1}H-nmr spectroscopy has shown that the organogermanium derivatives of (methylimino)bisacetic acid (15,R^1=R^2=Me) m.p. 215°(dec), (14, R^1=R^2=Ph) m.p. 230°(dec) and (14,R^{1}R^2=O(SiMe$_2$CH$_2$)$_2$) m.p. 239° possess 5-coordinate germanium because of transannular Ge-N interactions (S.N.Tandura, N.Yu.Khromova, et al, Zh.obshch. Khim.,1983,53,1199).

(b) Compounds with more than one heteroatom

The first oligomeric stannacycloalkane 1,1,6,6-tetraphenyl-1,6-distannacyclodecane (16) m.p. 96-98° has been prepared by the reaction of dichlorodiphenyltin with the bis-Grignard reagent of dibromobutane in THF. Single crystal X-ray diffraction shows that the ring has a boat-chair structure similar to cyclodecane. (A.G.Davies, et al., J.chem.Soc.Perkin II, 1981,369). A mixture of the macrocycles (17) and (18) containing two and four diphenylstanna units respectively has been prepared as shown in the scheme. The yields obtained and the methods used are shown in the Table (Y.Azuma and M.Newcomb, Organometallics,1984,3,9).

Preparations of tin macrocycles

n	(17)(two atoms of Sn)			(18)(four atoms of Sn)		
	Method	Yield %	m.p.	Method	Yield	m.p.
4	B	45	114-114.5	C	43	107.5-108
5	B	30	100-101	C	33	68-69
6	A	30	118-119	C	33	oil
8	A	30	92-93	A	6	70-71
10	A	26	86-87	A	7	oil
12	A	26	91.5-92	A	3	oil

Methods: A(i) Li (ii) Br$(CH_2)_n$Br; B BrMg$(CH_2)_n$MgBr; C (i) 2xBrMg$(CH_2)_n$Cl (ii) LiSnPh$_2$$(CH_2)_nPh_2$SnLi .

Dinaphthodistannocin (19,72%) m.p. 272-274° can be prepared from 1,8-dilithionaphthalene and dibromodimethylsilane. Irradiation (λ =298nm) of (19) in benzene causes extrusion of the

19   20

dimethyltin units giving perylene (20,60%)(J.Meinwald, S.Knapp
and T.Tatsuoka, *Tetrahedron Letters*.1977,2247).
Cyclic ethynyl polysilanes (21,n=1,2,3) comprise a class of
heterocycles within which the nine-membered ring compound (21,
n=3) m.p. 139° is prepared in 37% yield by the reaction of
$BrMgC\equiv C-SiMe_2SiMe_2C\equiv CMgBr$ and $Cl(SiMe_2)_3Cl$. Flash vacuum pyrolysis of this compound (650°/0.01mm.) affords (21,n=2,63%)
m.p. 55-56° and (21,n=1,3%) m.p. 64-65°. The latter has the
smallest ring known of any of the cyclic diynes. (H.Sakurai,
et al., J.Amer.chem.Soc.,1983,105,3395).
The reaction between the di-Grignard reagent from acetylene
and $ClSiMe_2(CH_2)_nSiMe_2Cl$ (n=0-6) affords mixtures of silacycloalkynes consisting mainly of the heterocycles (22,m=2,3 or
4) resulting from multiple addition. In reactions involving
the precursar with n=3 or n=4 minor traces of the 1:1 adduct
(22,m=1,n=3 or 4), detected by mass spectroscopy, occur (E.
Kloster-Jensen and G.A.Eliassen, Angew.Chem.Intern.Edn.,1985,
24,565).

21

22

Preparations of silacycloalkynes (22)

n=	Overall Yield%	m=2	m.p. m=3	m=4
4	15	86-87	-	-
3	35	162-162.5	-	-
2	25	122-123	153-154	135-137
1	15	103-104	107-108	118-119
0	5	139	146-147	-

Novel organosilicon rotane compounds (23,n=2 to 9) have been
isolated from the reaction of tetramethylenedichlorosilane
$(CH_2)_4SiCl_2$ with alkali metals in dry THF. The individual products are isolated by preparative hplc. Although formation of
5 and 6-membered rings is thermodynamically favoured, the compounds with larger rings n=7 to 12 are also formed and isolat-

ed as colourless crystalline solids. Rotanes involving the corresponding hexamethylene silicon rings are similarly prepared.(C.W.Carlson, K.J.Haller, *et al.*, J.Amer.chem.Soc.,1984,
<u>106</u>,5521).

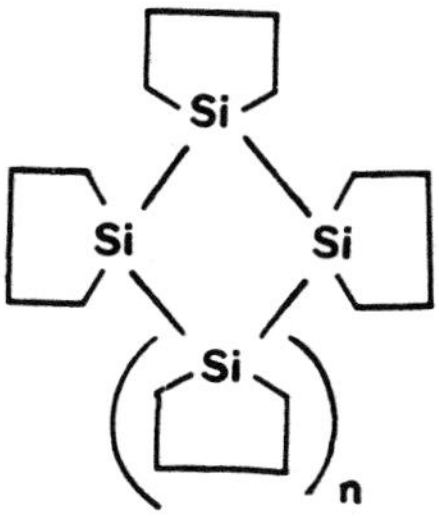

23

4. Phosphorus and arsenic compounds.

(a) Compounds containing one hetero atom

Derivatives of the novel 10,11-dihydrodibenzo(b,f)phosphepin
system (1,b to i) and (2), and of the dibenzo(b,f)phosphepin

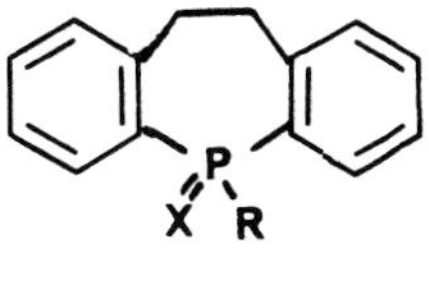

1 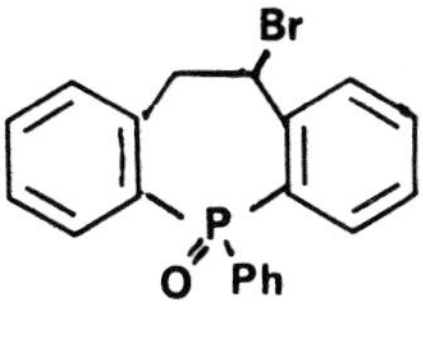   **2** 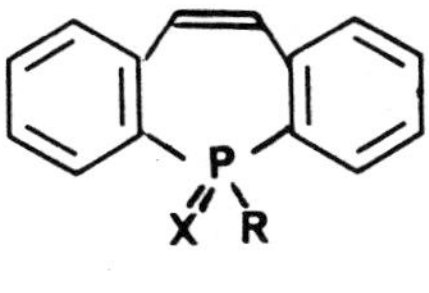   **3**

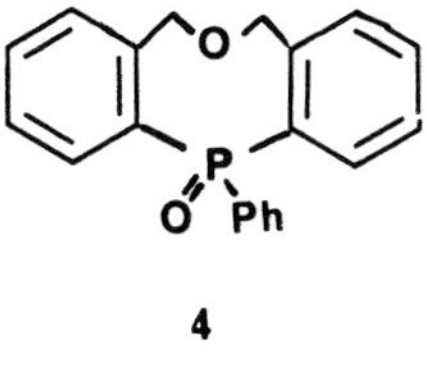

4

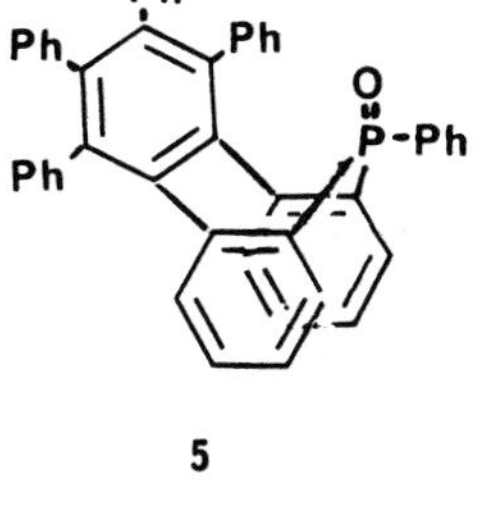

5

system (3,a to g) have been prepared from 10,11-dihydro-5-phenyl-5H-bibenzo(b,f)phosphepine-5-oxide (see Table). This key intermediate is synthesised (33%) from 2,2-dibromobibenzyl in three steps: (i) dilithiation (ii) high dilution reaction with dichlorophenylphosphine and (iii) oxidation with hydrogen peroxide. The yields of the derivatives depend critically on precise reaction conditions (Y.Segall, E.Shirin and I.Granoth, Phosphorus and Sulphur,1980,8,243). The unsaturated phosphepine (3a) has also been synthesised by heating (4) with the sodio derivative of either benzylamine or di-(trimethylsilyl)-ammonia in DMF. (K.A.Petrov, et al., Zh.obshch.Khim,1979,49, 2622). The folded "butterfly" conformation of the benzene rings in unsaturated compounds of this type has been confirmed by the X-ray structure determination of (5) (W.Winter, Ber., 1978,111,2942).

Derivatives of the dibenzo(b,f)phosphepin system

Comp.	X	R	m.p.	Comp.	X	R	m.p.
1a	O	Ph	185	2			172-3
1b	:	Ph	94	3a	O	Ph	231-2
1c	O	OH	255-60	3b	O	OH	310-14
1d	O	Cl	75-80	3c	O	Cl	
1e	O	OMe	54	3d	O	H	
1f	O	H	94	3e	O	Me	67-8
1g	O	$Me_2(CH_2)_3$	67-8	3f	:	Ph	135-6
1h	:	$Me_2(CH_2)_3$	oil	3g	S	Ph	218-9
1i	S	Ph	157-8				

The phosphinium iodide (6,R=Ph) undergoes alkaline hydrolysis to yield predominantly (1,R=Me,X=O) fifty times faster than the hydrolysis of the iodide of (6,R=Me) which mainly undergoes ring cleavage to give (7). These reactions are however much slower than the reactions of the corresponding salts with 5-membered rings (D.W.Allen, P.G.Hutley and A.C.Oades, J.chem. Soc.,Perkin I,1979,2326).

6

7

1-Phenylphosphonin oxide (9) is formed on oxidation of the bicyclic phosphorane (8,R=Ph) with hydrogen peroxide at -17° in
methanol, presumably through valence tautomerism of the first
formed phosphorane oxide. Although (9) has good stability at
low temperatures it rearranges at room temperature (half-life
= 5 minutes) to the phosphindole (10) (N.S.Rao and L.D.Quin,
J.Amer.chem.Soc.,1983,105,5960). Other approaches to the fully
unsaturated phosphonin system are based on the preparation of
the 3,8-phosphonadione system (12) formed in high yield by ozonolysis of the corresponding 3-phospholene derivative (11).
The X-ray structure analysis of (12,R^1=Me,R^2=R^3=H,) shows the
ring in a twist chair-chair form in the solid state but at
room temperature ^{13}C nmr spectroscopy indicates a rapid interconversion of conformers. (L.D.Quin, et al., J.Amer.chem.Soc.,
1982,104,1893).

8

9

10

11

12

13

Derivatives of the 3,8-phosphonadione system (12)

		Yield(%)	m.p.
R^2=R^3=H,	R^1=Me	94	130-2
	R^1=Ph	85	129-31
	R^1=OH	84	102-4
R^2=R^3=Me,	R^1=Ph	66	148-52
R^2=R^3=Br,	R^1=Me	86	100-1 (dec.)
	R^1=Ph	89	100-2 (dec.)
	R^1=OH	69	101-3 (dec.)
R^{2}R^3=Oxirane,	R^1=Ph	88	163-88(dec.)

The reaction of cyclooctatetrenyl dilithium with $RPCl_2$ produces mixtures of the *syn* and the *anti* phosphabicyclononanes (13). Examples of the compounds formed are (13,R=*i*-Pr$_2$N) b.p. 120-140°/0.01mm (13,R=Et$_2$N) b.p. 110-140°/0.01mm. The latter compound (13,R=Et$_2$N) yields the corresponding halides (13,R= Cl) and (13,R=Br) by reaction with phosphorus (5) chloride and phosphorus (3) bromide respectively. The fluoride (13,R=F) is obtained from (13,R=Cl) by treatment with potassium fluoride in 18-crown-6 ether (G.Maerkl and B.Alig, J.organometallic. Chem.,1984,273,1). *syn*-Alkyl-substituted derivatives of (13,R= *t*-Bu,cyclohexyl,menthyl) can be prepared by a similar reaction of magnesium cyclooctatetraene with $RPCl_2$ to yield (8) which undergoes thermal rearrangement in refluxing toluene. The *anti*-derivatives of (13,R=*t*-Bu,cyclohexyl) are produced in high yield by the Ni(0) catalysed rearrangement of the appropriately substituted (8) (W.J.Richter, Ber,1985,118,97).

(b) Compounds with two or more heteroatoms

The medium ring (7-11 members) diphosphonium and large-ring (16,18,20 members) tetraphosphonium halides have been prepared. Two general methods are used. Either the reaction of an α,ω-alkanebiphosphine e.g. Bz$_2$P(CH$_2$)$_n$PBz$_2$ (n=2,3,4) with an α,ω-dihaloalkane e.g. Br(CH$_2$)$_m$Br (m=2,3,4,5) to afford (14) or the linking of an α,ω-dihalobisphosphonium salt such as (15) with an α,ω-alkanebiphosphine to give (16). Dilution techniques do not affect the yields which range between 10 and 60%. The benzyl group attached to each P atom can be cleaved with lithium aluminium hydride to yield the corresponding cyclic phosphine (L.Horner, P.Walach and H.Kunz, Phosphorus and Sulphur,1978,5,171).

14   15   16

Quaternisation of methylenebis(diphenylphosphine) with 1,4-dibromobutane affords the bisphosphonium salt (17) which can be

dehydrohalogenated with Me₃P:CH₂ to give the carbodiphosphorane (18,30%) m.p. 127° dec. (H.Schmidbaur, et al., Angew.Chem.
1980, 92,557). The phosphonium salt (19,R=Me) m.p. 226° formed
(89%), by this method using α,α'dibromo-ortho-xylene undergoes elimination to the corresponding carbodiphosphorane m.p.
208° (88%) and then to the double ylid (20) m.p. 105° (53%).
The phenyl analogue (20) is similarly prepared. The latter
compound with strong base (LiBu, NaNH₂ or KH), is converted in
good yield into the alkali metal salt of the diphosphonium
triple ylide anion (21,M=Li m.p. 154°, M=Na m.p. 145°, M=K
m.p. 202°). (H.Schmidbaur, T.Costa and B.Milewski-Mahrla, Ber.
1981,114,1428). A solution of (21) in THF at -78°, when treated with MnBr₂, FeCl₂ or CoCl₂ affords air sensitive, crystalline derivatives (22) where the two triple ylide heterocyclic
ligands are folded in a boat conformation about the dipositive
metal atom (H.Schmidbaur, et al., Organometallics,1982,1,
1266).

17   18   19   20

21   22

Large ring compounds containing arsenic such as (23) b.p. 105°
/1mm. and (24) b.p. 90°/1mm. have been synthesised as shown.
The use of Me₂Si(NMe₂)₂ in the final ring closure step perm-
its the preparation of the corresponding mixed As/Si ring systems (H.T.Phung, P.B.Chi and F.Kober, Chem-Ztg.,1984,108,13).
High dilution techniques have been used to produce 11-membered
ring heterocycles such as (25) containing three heteroatoms by
the method shown in the scheme (M=H or Li, L=Cl or Br). X-Ray

(i) MeAs(NMe$_2$)$_2$
(ii) oxirane

structure determination has shown that the ring system adopts
an L-shaped arrangement with the angle between the plane of
the benzo group and the three heteroatoms being between 97°
and 114°. (E.P.Kyba, et al., J.Amer.chem.Soc.,1980,102,139).
Similar techniques have been used to prepare the symmetrical
tetrakis(tert-phosphino) macrocycle (26) two isomeric forms of
which have been isolated A(22%, m.p. 229-230°) and B(m.p. 160-
165°) either of which, on heating in refluxing xylene give a
mixture of A and B in the ratio 1.7:1 (R.E.Davis, C.W.Hudson
and E.P.Kyba, J.Amer.chem.Soc.,1978,100,3642).

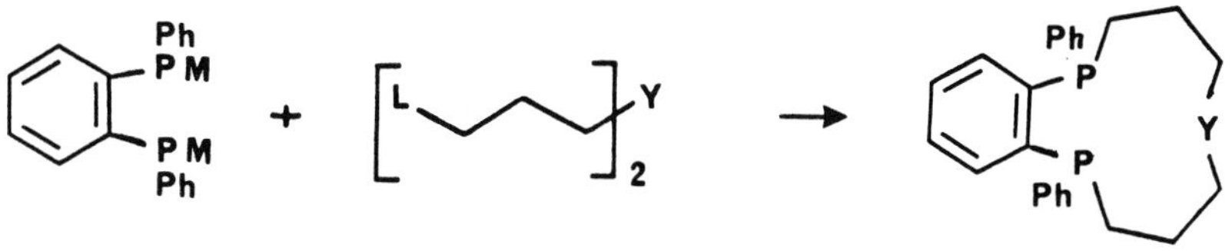

25

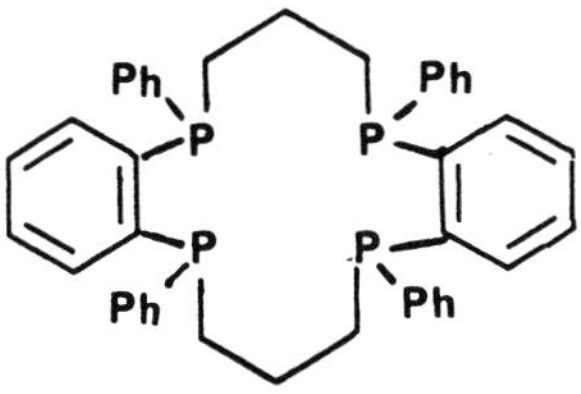

26

The preparation of compounds of structure 25

Y	Yield(%)	m.p.
PPh₃	34	158-60
S	51	112-12.5
O	32	oil
NMe	42	145-7
NPh	56	147-9
CH₂	33	120-1

Chapter 54

COMPOUNDS WITH SEVEN-MEMBERED AND LARGER RINGS CONTAINING
ONE OR MORE ATOMS OF UNUSUAL HETERO ELEMENTS TOGETHER WITH
NITROGEN OR ELEMENTS OF GROUP VI.

KAREN M. JOHNSON

1. Aluminium and boron compounds.

a) Compounds containing oxygen and/or sulphur as additional
    elements.

W.V. Dahlhoff and R. Köster (Ann., 1975, 9, 1625) have reported the preparation of several cyclic compounds
containing boron and oxygen as the additional element.
Thus, 1,3,8,10,-tetraoxa-2,9-diboracyclotetradecane
(1, n=4) and 1,3,9,11-tetraoxa-2,10-diboracyclohexadecane
(1,n=5) are prepared by heating butane-1 4-diol and pentane

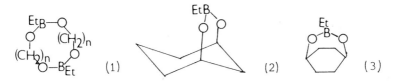

(1)   (2)   (3)

1,5-diol, respectively, with activated triethylborane at
230 - 240°C. Heating cis-1,3-bis(diethylboryloxy)cyclo-
hexane at 200°C gives a good yield of (2), while heating
cis-1,4-bis(diethylboryloxy)cyclohexane gives a low yield
of (3).6-Deoxy-1,5-O-ethylboranediyl-2,3-O-isopropylidene-
β-L-mannofuranose (4) is prepared (77%) by the reaction
between ethylboroxine and 6-deoxy-2,3-O-isopropylidene-L-
mannofuranose at 80°C. The reverse reaction occurs in
methanol at 20°C (W.V.Dahlhoff and R Koester, J. org. Chem.,
1977, 42, 3151).
   Three methods are available for the preparation of
1,6:2,3:4,5-tris-O-ethylboranediyldulcitol (5) from
dulcitol (W.V. Dahlhoff and R.Koester, J. org.Chem.,
1976, 41, 2316).

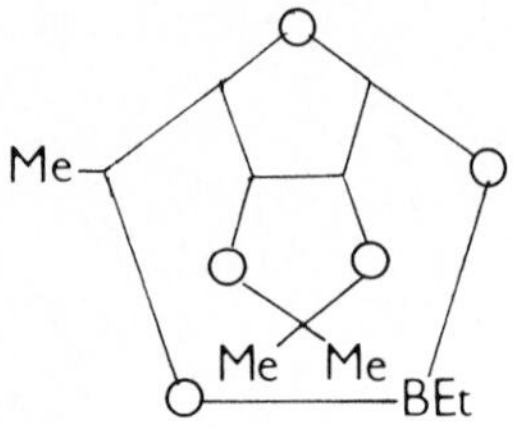

(4) 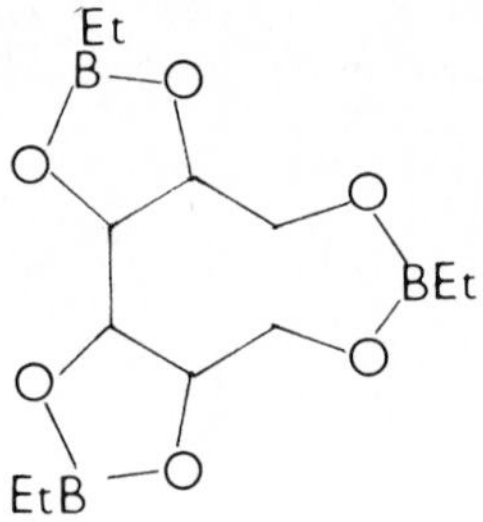   (5) 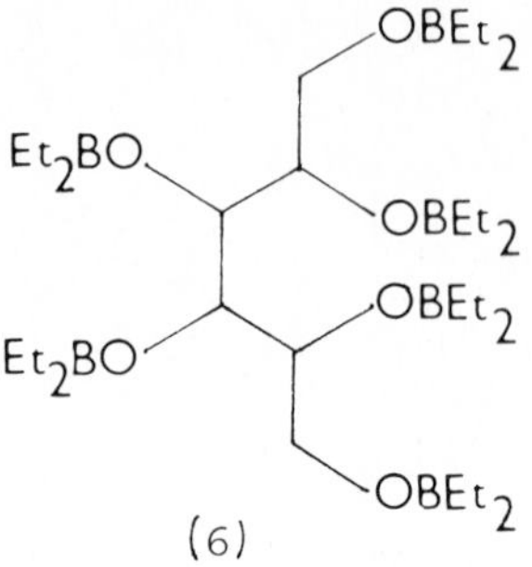   (6)

a) Addition of diethylborylpivalate ($80°C$) gives (5, 66%), this yield is considered low due to problems encountered in the separation of (5) from bis(ethylpivaloyloxy)diboroxane.
b) Addition of bis(ethylpivaloyloxy)diborane ($80°C$) gives (5, 87%).
c) Addition of triethylborane ($20°C$) produces 1,2,3,4,5,6-hexakis-0-diethylboryldulcitol (6), which when heated to 140-150°C produces 1,6-bis-0-diethylboryl-2,3:4,5-bis-0-ethylboranediyldulcitol (7).

The bicyclic compound (7) can then be converted into (5) either by heating it to 230°C or by the addition of a catalytic amount of ethyldiborane ($20°C$) giving (5) in 98% yield.

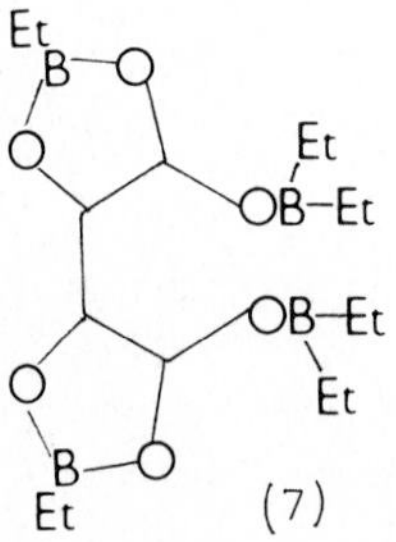

(7) 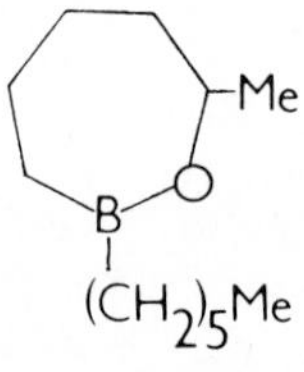   (8) 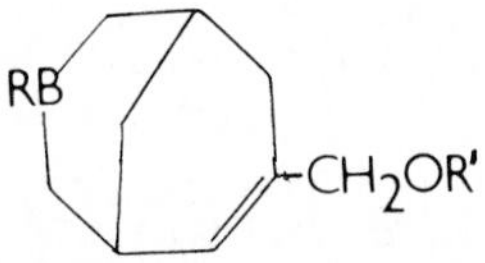   (9)

1,2-Oxeborepanes (8) are prepared by the alkaline hydrolysis and cyclisation of the corresponding e-bromoalkylboric acids and their acid halides (L.S. Vasil'ev, M.M. Vartanyan and B.M. Mikhailov, Izv. Akad. Nauk. S.S.S.R., 1976, 2308).

Thus compound (8, 73%,b.p. 66-68°C) is prepared by the hydrolysis of bromo-5-bromohexyl-n-hexylborane with 20% potassium hydroxide solution.

A number of cyclic compounds of similar structure containing boron and oxygen have also been prepared. Reduction and hydrolysis of (9, R=allyl, R'=tetrahydro-2-pyranyl) gives (10) (B.M. Mikhailov and K.L. Cherkasova, Izv. Akad. Nauk. S.S.S.R., 1979, 2729).

(11)   (10)

The cyclic compound (11) is formed when (9, R=allyl, R'=BPr$_2$) is heated to 100-130°C in vacuo. (Mikhailov and T.K. Baryshnikova, J. organometallic Chem., 1981,219, 295). It readily hydrolyses in moist air and polymerises on standing. The cyclic compounds of the general structure (12) are prepared by the reaction of carbonyl compounds

(12)   (13)

(RR'C=O, R=H,R'=Ph, 4-MeOC$_6$H$_4$, 3,4- (MeO)$_2$C$_6$H$_3$, Me$_2$CH; R=R'= Me) with 1- boraadamantane (13) at 70-80°C (Mikhailov, Baryshnikova, and A.S. Shashkov, J. organometallic Chem., 1981, 219, 301)· These compounds are colourless crystalline substances stable in air and sparingly soluble in organic solvents.

b) Compounds containing nitrogen as the additional element.

Two heterocycles containing aluminium or boron and nitrogen have been prepared. One of them is (15, Z=O,S)
(U.W. Gerwarth and K.D. Mueller, J. organometallic Chem.
1976, 110, 15; 1978, 145, 1).

(14)   (15)

2-Phenyl-1,3,2-diazaboracyclohexane, (14), reacts with phenyl-
isocyanate and phenylisothiocyanate to give (15, X=O) and
(15, X=S) respectively. The cyclohexane derivative (14)
undergoes ring expansion by insertion of two atoms of the
double bond system and a detailed description of the reaction
between the NCX (X=O,S) group and the B-N in the cyclohexane
derivative has been given.
The heterocycle (16) is prepared in 91.6% yield as shown
below, (J. Bielawski and K. Niedenzu, Synth. React. Inorg.
Met.-org. Chem., 1979, 9, 309).

(16)

c) Compounds containing nitrogen together with oxygen and/or
sulphur as additional elements.

Novel heterocyclic organoboranes may be prepared by the
reaction of an organic isocyanate with a heterocyclic
aminoborane.   (R.H. Cragg and T.J. Miller, J. organometallic
Chem., 1978, 154, C3).
Thus, 2-phenyl-3-methyloxazaborolan and toluene-p-
sulphenyl isocyanate gives (17, 92%, m.p. 80-82°C).

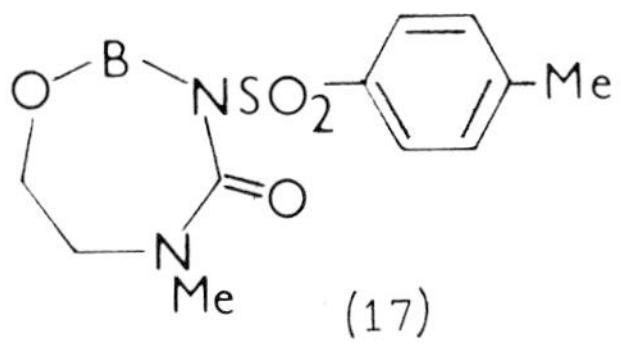

(17)

(18)

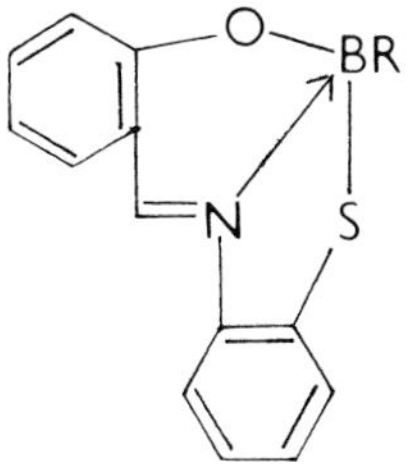

Treating $(Me_2CHO)_3B$ with o-mercaptoazomethines in 1:1 & 1:2 ratios gives the boronic ester complexes (18, R=OCHMe$_2$) and (18, R=O-SC$_6$H$_4$N=CHC$_6$H$_4$OH) respectively, both are found to be active against gram positive and gram negative bacteria (K.K. Chaturvedi, R.V. Singh and J.P. Tandon, J. prakt. Chem., 1984, 326, 817). A number of bicyclooctanes (19, X = B also H$_4$- EtC$_6$H$_4$Si , 4-MeCHC$_6$H$_4$Si, PhSi, As, Bi, P, PO) have been prepared in order to test their toxicity. (G.H. Cooper et al, Eur. J. med. Chem. - Chim. Theor., 1978, 13, 207). These compounds are prepared by cyclising triols and triamines.

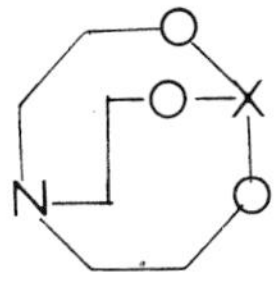

(19)

(20)

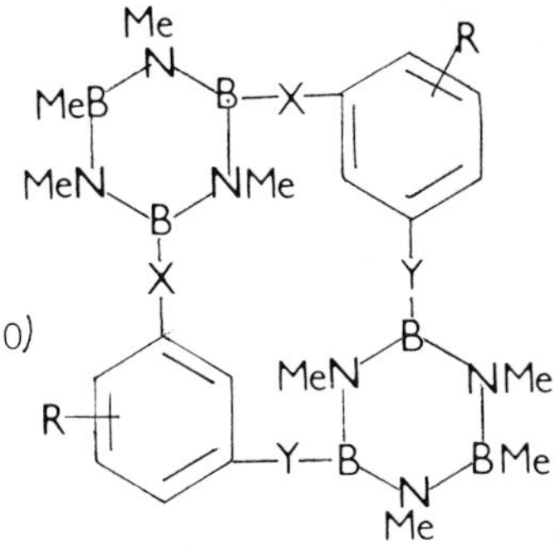

The macrocyclic oligomers (20, X=Y=O, R=H, 5-Me; X=O, Y=NH, R=H) are prepared by the interaction of a 2,4-difunctional borazine and a 1,3-dihydroxy, a 1,3-diamino or a 1-amino-3-hydroxybenzenoid compound (A. Meller, H.J.Füllgrabe and C.D. Habben, Ber., 1979, 112, 1252). They have a pseudocage structure and are less stable than the corresponding systems formed by O or N-H linked borazines.

2. Silicon, germanium, tin and lead compounds

a) Compounds containing oxygen and/or sulphur as additional elements.

Some of the numerous examples of this type of compound are listed in Table 1.

TABLE 1			
RING SYSTEM	YIELD	STARTING MATERIALS	REF.
Et_2Si, $(CH_2)_3$ $(CH_2)_3$ S		$Et_2HSi(CH_2)_3SCH_2CH:CH_2$	1
Ge O R^1 PPh R^2 R^1=Me,Ph, R^2=H; R^1=H, R^2=Me,Ph		germaphospholanes and RCHO R=Me, Ph.	2
Me Me Sn O S O		1,1-dimethyl-1-stannacyclo pentane and SO_2	3
O O O Si Me_2 O O	15%	Me_2SiCl_2 with $o-(CH_2CH_2OC_6H_4ONa-o)_2$	4
Ph Ph Ph Ph R Si O Si R R R		$R_2SiClCPh_2(CH_2)_2CPh_2SiClR_2$ (R=Ph, Me)	5

TABLE 1 contd			
(structure: eight-membered ring, Me–Si–O, Me–Si–Me; larger ring with Me, Me, Me, Me silicon-oxygen)		$(ClSiMe_2CH_2CH_2)_2O$ in dioxane/aq KOH at $-10^{\circ}C$	6
(structure: dibenzo ring with Cl–Si–Cl, Cl–Si–Cl, O)	(structure: dibenzodioxin) and $HSiCl_3$ or Si_2Cl_6 $(450-700^{\circ}C)$		7
(structure: ring with Me, Me, Me, Me, Si, O, $(CH_2)_n$, O) $n=3$ 72.5% $n=10$ 60%	siloxanes and unsaturated acids or esters. eg.$(Me_2SCH)_2O$ and $CH_2:CH\ (CH_2)_{n-2}(3,10)$ presence of H_2PtCl_6		8
(structure: XCH_2CH_2O, PhGe, Ge-Ph, OCH_2CH_2X)	$XCH_2CH_2O(Ph)\overline{GeCH_2CH_2}O$ $(X=F,\ Cl)$		9
(structure: eight-membered ring, Me Ge, S, S, Ge Me Me)	Na_2S and (structure: four-membered ring $Me\ Ge\ Me$, S)		10
(structure: ring $(CH_2)_n$, O, Me, Si–Me, O, Si, Me Me) $n=2$ $\begin{cases}-\\36.2\%\end{cases}$ $n-3$ 24.2% $n=3$ $-$	$HO(CH_2)_2OH$ and $\overline{OSiMe_2NH{-}SiMe_2OSiMe_2NHSiMe_2}$ $(ClMe_2Si)_2O$ and $HO(CH_2)_2OH$ $HO(CH_2)_3OH$ and $(Cl_2Me_2Si)_2O$ $HO(CH_2)_3OH$ and $(MeOSiMe_2)_2$		11 11 11 12

TABLE 1 contd.			
		$(Me_2CH)_2SiCl)_2O$ and glycerol	13
		$(Me_2CH)_2SiCl)_2O$ and D-erythropentosephenylosazone	13
	81.9%	hexamethylcyclotrisilthiane and $HOCH_2CHMeOH$	14
	R=H 18% R=Me 47%	$(R_2Si=CH_2)$ and $(Me_2SiO)_3$	15
		$(MeOSiMe_2)_2$ and $HO(CH_2)_2OH$	12

1. M. Draeger and H.J. Guttman, J. organometallic Chem., 1978, 212, 171.
2. C. Couret et al, Red. Trav. Chim., 1976, 95, 240.
3. E.J. Bulton and M.A. Dudding, J. organometallic Chem., 1979, 106, 339.
4. N.I. Liptuga, L.F. Irodionova and H.O. Lozinskii, Zh. obshch. Khim., 1978, 48, 1185.
5. A. Ciobanu, N. Luchian and V. Hamciuc, Rev, Roum. Chem., 1978, 23, 943.
6. J. Pola, L. Cvak and V Chvalovsky, Czech. chem. Comm. 1976, 41, 368.
7. E.A. Chernyshev et al, Zh. obshch. Khim., 1978, 48, 2798.
8. N.S. Fedotov et al, Zh. obshch. Khim., 1977, 47, 1350.
9. A. Castel et al, Compt. Rend., 1978, 287C, 205.
10. J. Barrau, G. Rima and J. Satge, Synth. React.inorg. met-org. Chem., 1984, 14, 21.
11. S.M. Saad and R. Tacke, Chem., Ztg., 1977, 101, 262.
12. E. Hengge, E. Brandstaeller and W Viegle, Monatsch., 1977, 108, 1425.
13. W.T. Markiewicz, Z. Samek and J. Smrt, Tetrahedron Letters, 1980, 21, 4523.
14. D.V. Fridland, E.P. Lebedev and V.D. Reikhsfel'd, Zh. obshch. Khim., 1976, 46, 326.
15. C.M. Golino, R.D. Bush and L.H. Sommer, J. Amer. Chem. Soc., 1975, 97, 7371.

The reaction of sila- or germaphospholanes with silicon containing carbonyl compounds produces disila- and silagermaphospholanes (C. Couret, J. Satge and J.P. Picard, J. organo- metallic Chem., 1977, 141, 35). Thus addition of dimethylsila - 2-phenyl-1-phospholane to benzoylsilane produces the diastereoisomers (1a) and (1b). Dimethylsila-2-phenyl-1-phospholane and o-methylbenzoyltrimethylsilane, and dimethyl-germa-2-phenyl-1-phospholane and benzoylsilane produce similar diastereoisomers.

(1a)   (1b)

Thermal rearrangement of $(2, R=R'=H, Me, CMe_3; R=H, R'=Me)$ occurs via one of two possible mechanisms depending on the substituents. Heating $(2, R=R'=Me, 350^{\circ}C, 16h.)$ produces

$(3, R'=Me, R'=H, M=Si)$ via the formation of a benzylic-type radical which adds to the vinyl double bond, while heating $(2, R=R'=H, 320^{\circ}C \ 72h)$ produces the six-membered analogue $(4, R=H)$. (J. Ancelle et al, Tetrahedron Letters, 1979, 3153). Both types of product are obtained $((4, R=Me)$ and $(3, R=H))$ when $(2, R=H, R'=Me)$ is heated. The relative amounts of the products (4) and (3) found is temperature dependant varying from $(4):(3)$ 92:8 at $320^{\circ}C$ to 44:56 at $380^{\circ}C$. Another method of preparing compounds of type (3) has been developed. (G. Bertran, P. Mazerolles and J. Ancelle, Tetrahedron, 1981, 37, 2459). Oxa-1-benzo-6,7-metalla-2-cycloheptanes $(3, M=Si, Ge, R=Me, R'=H)$ are formed by a radical intramolecular cycloaddition when $(2,6-$dimethyl)phenoxyvinyldimethylsilane and $(2,6-$dimethyl) phenoxyvinyldimethylgermane are heated in a sealed tube $(235-350^{\circ}C, 72h)$ giving $(3, M=Si)$ 15% and $(3, M=Ge)$ 25% respectively. Similarly $(3, M=Si, R=CMe_3, R'=Me)$ is obtained when $(2,6-$ditertiobutyl)phenoxyvinyldimethylsilane is heated at $300^{\circ}C$.

The silathiacyclooctane, (5) is prepared by the photochemical reaction of hydrogen sulphide with diallyldimethylsilane (M.G. Voronkov et al, Zh. obshch. Khim., 1984, 54, 1566). The open chain compound $H_2C:CHCH_2SiMe_2(CH_2)_3SH$ is the intermediate.

Tricyclic, compounds similar to (6), containing only O,N or

S in the central ring, with appropriate side chains often exhibit psychotropic activity. J.Y. Corey and co-workers have developed silicon analogues of such psychotropic drugs. 12,12-Dimethyl-7,12-dihydro-5H-dibenzo{c.f.} {1,5} thiasilocin (6,m.p. 123.5-125°C) is prepared from bis (O-bromomethylphenyl) dimethylsilane and sodium sulphide in ethanol. While the Grignard reagents of o-bromobenzyl-o-bromophenylsulphide and dimethyldichlorosilane give (7, b.p. 150 - 165°C 0.1 mm Hg), a light yellow oil, (B.R. Soltz and J.Y. Corey, J. organometallic Chem., 1979, $\underline{171}$,291).

$$Me \quad Me \qquad (7)$$

$$Me \quad CH_2R \qquad (8)$$

The preparation of compounds such as (7) has been developed.
Addition of potassium fluoride induces ring expansion of halomethyl derivatives such as 11-(halomethyl)-11-methyl-5, 11-dihydrodibenzo(b,e)-(1,4)oxasilepin(8, R=Cl,I,X=OCH$_2$). Treatment of (8) with potassium fluoride produces two isomers: 11-methyl-11,12-dihydro-6H-dibenzo{b,f}{1,5} oxasilocin (9,R'= F,X=OCH$_2$) and 12-methyl-11,12-dihydro-6H-dibenzo {b, f}{1,4}oxasilocin (10 R'=F,X=OCH$_2$) (Corey and V.H.T. Chang,

$$Me \qquad (9) \qquad\qquad Me \quad R' \qquad (10)$$

Organometallics, 1982, $\underline{1}$, 645). Isomers (9) and (10) are colourless oils and are separable by column chromatography but attempts to obtain a pure sample of (10) results in its decomposition. A detailed study of the influence of the fluoride source on the formation of (10,R'=F,X=O,OCH$_2$) from (8,X=O,CH$_2$,R=Cl) has been carried out (Corey et al, Organometallics, 1984, $\underline{3}$, 1051).

The cyclic acetals (12) are produced from carbonyl compounds and silicon oxygen containing heterocycles (11) with the substituents and conditions as shown in Table 2. (R.J.P. Corriu et al., J. organometallic chem., 1976, $\underline{114}$, 21);

(11)   + R^4, R^5 C=O $\longrightarrow$ (12)

TABLE 2

n	R^1	R^2	R^3	R^4	R^5	Catalyst	$T^\circ c/h$	%yield
1	CH_3	CH_3	CH_3	CCl_3	H	H_2PtCl_6	130/3	100
1	CH_3	CH_3	CH_3	CF_3	CF_3	–	exothermic	100
1	CH_3	CH_3	CH_3	CCl_3	CF_3	H_2PtCl_6	130/12	30
1	CH_3	CH_3	CH_3	C_6F_5	H	H_2PtCl_6	130/12	25
1	CH_3	CH_3	CH_3	CH_3	H	H_2PtCl_6	130/12	50
1	CH_3	CH_3	CH_3	H	H	H_2PtCl_6	80/1	70
1	α naphthyl	Ph	H	CCl_3	H	–	140–150/20	30
2	α naphthyl	Ph	H	CCl_3	H	–	140–150/20	20

Both the cyclic silicon compound (13, Z= $-(CH_2)_4-$ or $-(CH_2)_6-$) and the bicyclic compound (14, Z= $-(CH_2)_4-$ or $-(CH_2)_6-$) are formed from the corresponding diol and $MeSi(OMe)_3$, With the reactants in a 1:1 ratio the compound (13) is formed whereas with an excess of the diol (2:1) (14) is formed (R.P. Natain and A.J. Kaur, J. Indian chem. Soc., 1977, $\underline{54}$, 504).

(13)

(14)   (15)

Compounds of similar structure are prepared by the reaction
of dihydroxy aromatic compounds with various silanes (L.
Birkofer and O. Stuhl, J. organometallic Chem., 1979, 164,
Cl; Birkofer et al., J. organometallic Chem., 1980, 194,
159). The 1, 1-disubstituted-4,5-benzo-2,7,1-dioxasilacycl-
oheptanes (15) are prepared by the reaction of 1,2-di
(hydroxymethyl)benzene with a) hexamethylcyclotrisilazane to
give (15, R=Me, 36%); b) with dichlorodiphenylsilane to give
(15, R=Ph, 31%) and c) with tetramethoxysilane to give the
tetracyclic compound (16, 46%) while silocins (17) are
prepared by the reaction of 1,8-di(hydroxymethyl)naphthalene

(16)   (17)

with a) hexamethylcyclotrisilazane giving 3,3'-dimethyl-
2,4-dioxa-1,2,4,5-tetrahydronaphtho {1,8,8,a-f,g} 3H-3-silocin
(17, R=Me,25%) and b) dichlorodiphenylsilane giving 3,3'-
diphenyl--2,4--dioxa-1,2,4,5-tetrahydonaphtho {1,8,8,a-f,g}
3H-3-silocin (17, R=Rh, 44%). 3,3'-Dimethyl-2,4-dioxa-1,2,4,5-
tetrahydrodibenzo-(f,h)-3H-3-silonin (18, R=Me) and 3,3-
diphenyl-2,4-dioxa-1,2,4,5-tetrahydrodibenzo-(f,h)-3H-3-
silonin (18, R=Ph) are formed similarly from 2,2'-di(hydroxymethyl)biphenyl.

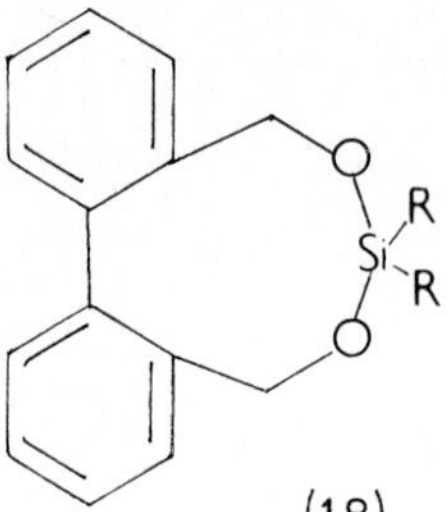

(18) 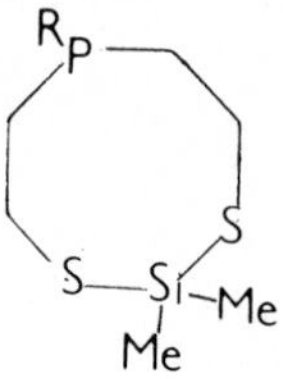   (19) 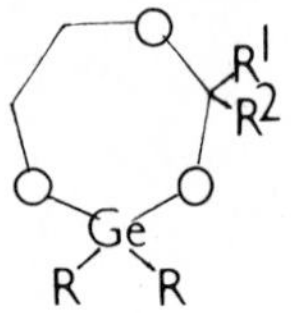   (20)

Unusual cyclic compounds, 2,2-dialkyl-1,3-dithiol-6-phospha-2-silacyclooctanes (19) are prepared by cyclising $RP(CH_2CH_2SH)_2$ (R=Me,Ph,Me$_3$C) with the silane, $(Et_2N)_2SiMe_2$. (U.Uhlig, V.P. Baryshok and M.G. Voronkov, Zh obshch. Khim., 1983, 53, 1572).

Insertion reactions of disubstituted 2-germa-1,3-dioxolanes with carbonyl compounds ($R_1^1R^2C=O_2$ R^1=H, Me, CF$_3$, R^2=Ph,Me,CCl$_3$, CF$_3$), give (20, R=Me, Et,R^1 and R^{2}as before) via a mechanism that occurs under steric control. At 160°C the substituted 1,3,5,2-trioxagermapanes (20) decomposes as shown below (G. Dousse, J. Lavayssiere, and J. Satge, Helv., 1975, 58, 2610).

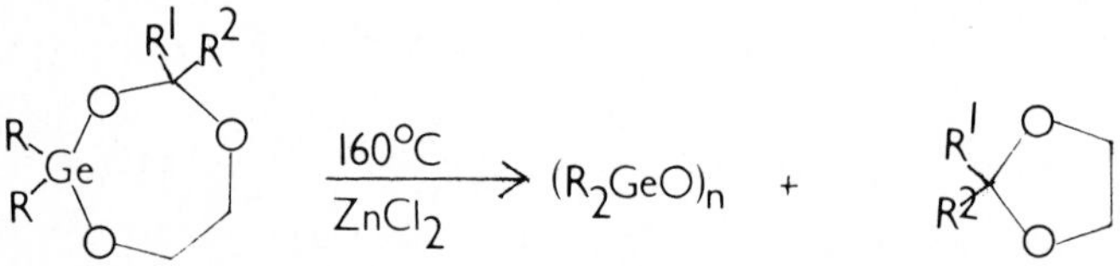

$$\xrightarrow[\text{ZnCl}_2]{160°C} (R_2GeO)_n +$$

Two methods are available for the preparation of the cyclic silicon compound 7,7-dimethyl-3-thia-7-sila-7-heptan-olide (21, b.p. 135-140°C). a) Refluxing dimethylsilyl-(allylthio)acetate in the presence of Speiers' catalyst gives (21, 64%, R=R^1=Me,n=3). b) Heating Me$_{3-m}$Cl$_m$SiCH:CH$_2$ (m=1,2) with HSCH$_2$CO$_2$SiMe$_3$ followed by treatment with trimethylchlorosilane gives (21, 57%, n=2, R=Me. R^1=Cl)and (21, 60% n=2, R=R^1=Me) (N.S. Fedotov et al., Zh. obshch. Khim., 1978, 48, 612).

(21)

(22)

(23)

Eight-membered cyclic tin (IV) compounds (22, $X,X^1=O,S$, R=Bu,Me,Ph) are prepared in 75-80% yield by cyclising di(2-thioethyl)ether ($O(CH_2CH_2SH)_2$) or di(2-hydroxyethyl) sulphide ($S(CH_2CH_2OH)_2$) with $R\bar{S}nO$ or R_2Sn $(OCHMe_2)_2$ (R. Verma, V.D. Gupter and R.C. Mehrota, Natl. Acad, Sci. Letters, (India), 1979, 2, 130).

The spiro compound (23) is prepared from tetraethoxysilane with tri(2-hydroxyethyl)amine at $170°C$, with the reactants in a 2:1 ratio (the silane in excess). (G.A. Kondrashova and L.J. Kondrashova, Chem. Abs., 1981, 97, 216453).

The silacrown ethers (24, R=R'=Me; R=Me,R'=Et,Ph,CH_2CH_2, MeO, $H_2NCH_2CH_2NH(CH_2)_3$, n=3-6) are prepared by transesterification of alkoxysilanes with polyethylene glycols. (B. Arkles et al., Organometallics, 1983, 2, 454).

$$R'RSi(OEt)_2 + HO(CH_2CH_2O)_nH \longrightarrow$$

$$+ 2EtOH$$

(24)

The silacrown ethers are generally colourless, odourless liquids of moderate vicosity.

Dibenzodisilaoxasilaheptadiene (25, $X=Me_2SiOSiMe_2$) is formed when dibenzodisilacyclohexadiene (25, $X=Me_2SiSiMe_2$) is irradiated in the presence of TCNE at room temperature (K. Sakurai, K. Sakamoto and M. Kira, Chem. Letters, 1984, 1213). When dichloromethane is used as the solvent the ratio of products ($X=Me_2SiSiMe_2$; $X=Me_2SiOSiMe_2$) is estimated to be 3:1, using n.m.r. spectroscopy and 5:2 by g.l.c. Whereas when dichloromethane /carbon

(25)   (26)   (27)

tetrachloride (2:1) is used (25 X=Me$_2$SiOSiMe$_2$, 21%) is
formed in excess.

The spiro compound 2,2,8,8-tetraphenyl-1-oxa-2,8-disila-
spiro-(7,1)-tridecanone-13 (26) together with the bicyclic
trans-7,7'-bicyclo(2,2-diphenyl-1-oxa-2-silacycloheptylidene)
(27) are obtained by photolysis of 1,1-diphenylsilacyclo-
hexan-2-one(A.G. Brook, J.B. Pierce and J.M. Duff, Canad.
J. Chem., 1975, 53, 2874). The reaction involves a siloxy-
carbene intermediate, which is then trapped by a second
acylsilane molecule. The structure of (27) has been
determined by X-ray crystallographic studies (P.T. Cheng
and S.C.Nyburg, Acta Crystallogr., 1976, B32, 930).

(28)   (29)   (30)

Cyclocondensation of MeClSiR(CH$_2$)$_4$SR'ClMe (R=R'=Me,Et;
R=Cl,R'Me) with water, ammonia, methylamine or hydrogen
sulphide gives the disilaheterocycles (28, Z=O,NH,MeN,S
respectively) and (29, Z=O,NH) (X. Chen and X. Zhou, Chem.
Abs., 1984, 101, 151920e).

The twelve-membered cyclic lead compound (30), m.p. 242-244°C is prepared (9%) by treating 2,2'-dilithiodiphenylether with dimethyllead dibromide at 0°C (D.C. Van Beelan and J. Wolters, J. organometallic Chem., 1980, 195, 185). In solution (30) exists as an equilibrium mixture of conformational isomers.

Compounds (31) and (32) are prepared as shown below

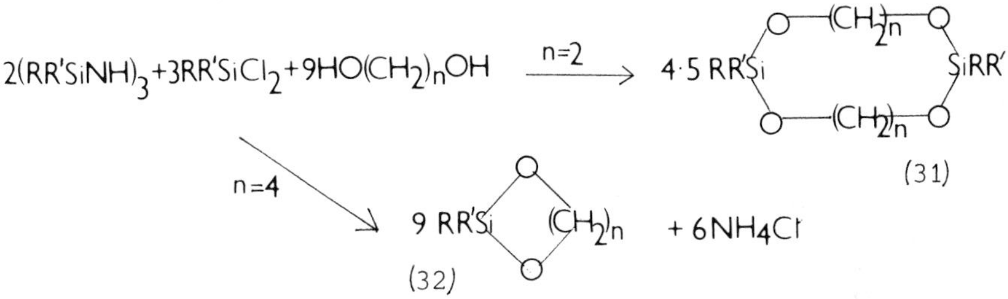

$$2(RR'SiNH)_3 + 3RR'SiCl_2 + 9HO(CH_2)_nOH \xrightarrow{n=2} 4·5 RR'Si$$

(31)

n=4

$$\searrow \quad 9 RR'Si \quad (CH_2)_n \quad + 6NH_4Cl$$

(32)

The compound (31, n=2, R=R'=CH$_3$), m.p. 57°, b.p. 194°C, is obtained in 97% yield. Compound (32, n=4.R=H,R'=CH$_3$)b.p. 125°C is obtained in 73% yield, while (32,R=R'=CH$_3$)b.p. 114°C is obtained in 96% yield. (E.P. Lebedev et al., Zh. obshch. Khim., 1975, 45, 2645).

The seven-membered heterocycle 1,1,3,3,5,5-hexamethyl-2-oxa-1,3,5-trisila-4-thiacycloheptane (33) (H.S.D. Soysa and W.P. Weber, J. organometallic Chem., 1979, 165,Cl) is

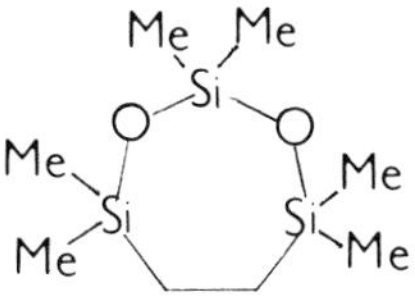

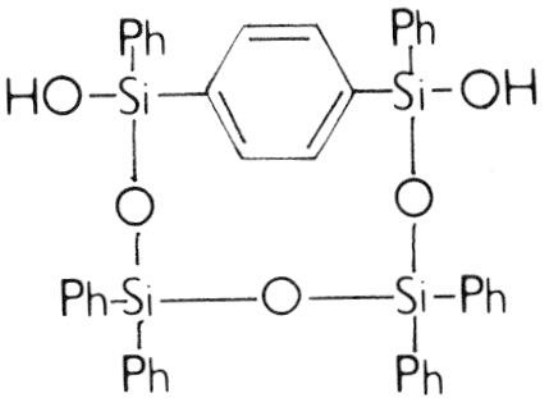

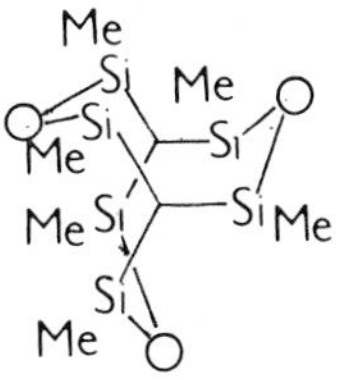

(33)   (34)   (35)

obtained by heating a 1:5 mixture of hexamethylcyclotrisilthiane and 1,1,3,3-tetramethyl-2-oxa-1,3-disilacyclopentane at 200°C for 7 hours

The silicon contained heterocycle (34), prepared by the condensation of 1,4-(Cl$_2$PhSi)$_2$C$_6$H$_4$ with HOSiPh$_2$OSiPh$_2$OH (1:1 ratio) at room temperature in the presence of pyridine, followed by hydrolysis (yield = 14.3%) (S.M. Meladze et al., Chem Abs., 1980, 92. 221358v). Finally, 2,2,4,4,6,6,8,8,9,9, 11,11-dodecamethyl-3,7,10-trioxa-2,4,6,8,9,11-hexasila {3.3.3} bicycloundecane, (35) m.p. 275-276°C is prepared in essentially quantitative yield by the thermal cyclo-condensation of tris(hydroxydimethylsilyl)methane. The structure of (35) has been confirmed by X-ray crystallo-graphic studies (C. Eaborn, P.B. Hitchcock and P.D. Lickiss, J. organometallic Chem., 1983, 252, 281).

b) Compounds with nitrogen as the additional element

Addition of potassium to tricyclic compounds such as (36, X=NMe, R=H) induces ring expansion, 1,2,-Methyl-11,12-dihydro-6H-dibenzo(b,f)(1,4)azasilepin, (37, X=NMe, R=H, R'=F) being obtained from (36, X-NMe, R'=H). (J.Y. Corey et al., organometallics, 1984, 3, 1051). Similarly the

(36)   (37)

tricyclic compound (37, R=Br, X = NEt, R' = F) is prepared from (36, R=Br, X=NEt) and potassium flouride in refluxing acetonitrile (J.Y. Corey and V.H.T. Chang, Organometallics, 1979, 174, C15).

Treating (37, X=NMe, NEt, R'=F,R=Br) with lithium alumin-ium hydride followed by CH_2=CHCH$_2$NMe$_2$/H$_2$PtCl$_6$ gives the substituted dibenzoazesilepin (37, X=NMe, NEt, R'=CH$_2$CH$_2$CH$_2$NMe$_2$, R=Br) (J.Y. Corey et al, J. organometallic Chem., 1980, 194, 15).

Compounds of similar structure to the dibenzoazasilepins (37) are prepared by the Beckmann rearrangment of suitable oximes (N.S. Prostakov et al., Khim. Geterotsikl. Soedin., 1983, 1669). Thus the oxime (38, R=Me,R'=H;R=Ph,R'=Me) gives a mixture of the benzopyridosilazepinones (39, X= C=O,

Y=NH; X=NH, Y=C=O; R=Me, R'=H; R=Ph, R'=Me). The eight-
membered cyclic compounds, the triazasilacyclootanes, (40,
R=Me)b.p. 95°C; (40,R=vinyl)b.p. 95°C; (40,R=Ph) are formed
by the reaction of the silane, $Me_3SiN(CH_2CH_2NH_2)_2$ with
appropriate $RSiCl_2Me$ (G.S.Gol'din, L.S. Baturina and T.M.
Gavrilova, Zh.obshch. Khim., 1975, 45, 2189).

The complex heterocycle (41) is prepared from $Cl_2EtSiCH_2$-
$CH_2SiEtCl_2$ and ammonia. (Q. Xie, Y. Zhou and X. Zhou, Chem.
Abs., 1983, 99, 38515d).

The heterocycles (42) containing silicon and nitrogen have
been prepared by E.A.Chernyshev and co-workers (Zh. obshch.
Khim., 1983, 53, 2725). Examples of the compounds are (42,
Z=NH,O;R=H,Me;R^1=H, alkyl , Ph, $4-MeC_6H_4$; and X = X'- X^5).

Thus cyclising 1,1-bis (dimethylamino)-1- silacyclopent -
3- ene with N- methyl diethanolamine gives (42, X=X^2,Z=O,
R=R^2=R^3=H,R^1=Me, 92%).

Finally in this section are included two tin containing
heterocycles. The first 1,7,2,-diazastannocine (43,X=
PhN-C=NPh) is prepared from the azastannacyclohexane (44)
and PhN:C:NPh and the second 1,4-azastannocine (43, X =
$MeO_2CC=CCO_2Me$) obtained from dimethyl acetylenedicarboxylate

Me—Sn—NCMe$_3$ with (CH$_2$)$_4$, Me, X

(43)

Me, Me Sn—NCMe$_3$

(44)

(CMe$_3$)$_2$Sn, N=, Ph, NMe, MeN, Sn(CMe$_3$)$_2$, Ph

(45)

and (42). (D.Haenssgen and E Odenhausen, J. organometallic Chem., 1977, 124, 143).

2,2,6,6-Tetra tert-butyl-1,2,5,6-tetrahydro-1,5-dimethyl-4,8-diphenyl-1,3,5,7,2,6-tetrazadistannocin (45) has been prepared in 49% yield by the cycloaddition of two moles of benzonitrile to one of diazadistannetidine. (D. Haenssgen and I. Pohl, Ber., 1979, 112, 2798).

c) Compounds containing nitrogen together with elements of Group VI.

The nine-membered heterocycle 1,3,5,8,2-dithiadiazagermonine (46) and the seven-membered heterocycle 1,3,6,2-thiadiaza-germepine (47) are prepared from germodiazolidine and the heterocumulene

PhN=, S, S, Ge, =NPh, MeN, NMe

(46)

Ge, S, NPh, MeN, NMe

(47)

R, R', O, Ge, O, R', N, R'

(48)

PhNCS in quantitative yields (M. Lavayssiere, G. Dousse and J. Satge, Helv., 1976, 59 1009). Germadiazolidine also reacts with other heterocumulenes, aldehydes, ketones and various other compounds to give mono and di-insertion products that are similar to compounds (46) and (47). Similarly germaoxazolidines give mono-insertion products at the germanium nitrogen bond with for example PhNCO, carbon dioxide, ketene or carbonyl compounds. The carbonyl compounds also give the di-insertion product for example the trioxazagermonine (48, R=H,R'= CCl$_3$;R=R'=CF$_3$. (G. Dousse, H. Lavayssiere and J. Satge, Helv., 1976, 59, 2961).

Germanium difluoride is used to prepare the cyclic compound 1,3,2- oxagermepine (49) as indicated in the following scheme (P.Riviere et al, J. organometallic Chem., 1978, 155, C58).

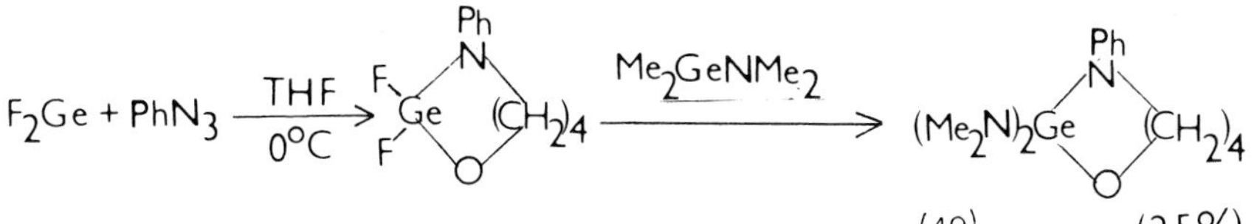

(49)   (25%)

The first of the silicon nitrogen containing heterocycles is

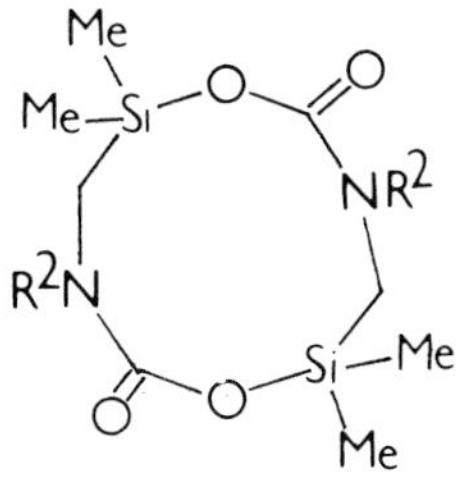

(50) 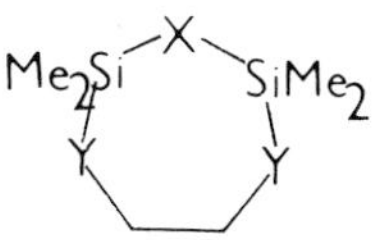   (51)

the 10-membered heterocycle (50, R^2= alkyl, H, Me_2CHCH_2Bu). It is prepared (88-98%) by cyclising Me_2 (RO) $SiCH_2NR^2(CO_2-SiMe_3)$ (V.D. Sheludyakov, et al, Zh.obshch. Khim., 1976, 46, 2712).

1-Oxa-3,6-diaza-2,7-disilacycloheptane (51, X=O, Y=NSiMe$_3$) and 1,3,5-triaza -2,4-disilacycloheptane (51,X=N, Y=NSiMe$_3$) are prepared from Li N(R)CH$_2$CH$_2$N(R)Li and X(SiMe$_2$Cl)$_2$ where X is respectively O and NH (H. Wannagat and G. Eisele, Monatsch., 1978, 109, 1059) 1,3 - Dioxa - 5, 8-diaza-2,4,9-trisilacyclononane (51), X=OSi(Me$_2$)O, Y=NSiMe$_3$) is similarly prepared from LiN(R)CH$_2$CH$_2$N(R)Li and the silicon compound-Me$_2$Si(OSiMe$_2$Cl)$_2$. The final compound of this type (51, X=N R^1, R^1=2,4,6-Me$_3$C$_6$H$_2$, Y=O) is prepared from Me$_2$SiFNC(R') SiFMe$_2$ and (NaOCH$_2$)$_2$. (J.Neeman and U. Klingebiel, J organometallic Chem., 1981, 208, 293). The unusual cyclic compound (52) has been prepared from the silylaminophosphine CH$_2$CH$_2$Si(Me)$_2$N(PMe$_2$)Si(Me)$_2$ and acetone in dichloromethane via nucleophilic attack by phosphorus and a (1,4)-silyl

(52) (53) (54)

migration from nitrogen to oxygen (D.W. Morton and R.H. Neilson, Organometallics, 1982, 1, 289). The silylamino-phosphine $CH_2CH_2Si(Me)_2N(PMe_2)Si(Me_2)$ also reacts with α,β-unsaturated carbonyl compounds and with methyl vinyl ketone giving the cyclic compound (53). Formation of the Z isomer of (53) is favoured when the methylvinylketone is slowly swept into the reaction mixture at $0^\circ C$ under nitrogen.

Condensation of trifunctional silanes $MeSiX_3$ (X=Cl, OEt, NMe_2) with N- methyldiethanolamine followed by treatment with either β-chloroethanol, β-dimethylaminoethanol or β-dimethylarsinoethanol give the cyclic compounds (54, Y= Me_2As, Me_2N or Cl). There is weak N-Si adduct bonding which is indicated by the significant shifts of the nmr-signals due to the $MeNCH_2$ and OCH_2 protons (L. Grobe and N. Voulgarakis, Z. Anorg. Allg Chem., 1984, 517, 125).

(55) (56)

The 1,2,3-thiazastannepine (55, R = p-tosyl) is formed from 1,1-dimethylstannolane and RN:S:NR. (D. Haenssgen and E. Odenhausen, J. organometallic Chem., 1977, 124, 143). 1,2,3-Thiazastannepine undergoes ring cleavage with bromine or iodine to give $Me_2Sn X (CH_2)_4X$ (X=Br or I).

The ten-membered heterocycle (56) is prepared in 74% yield by the reaction of $Sn(OBu)_2$ with $(HS(CH_2)_3)_2NMe$. It was found that there is a transannular interaction between the tin atom and the nitrogen atom (A. Tzschach, K, Jurkschat and M Scheir, Z. Anorg. Allg. Chem, 1983, 507, 196).

3.   Phosphorus, Arsenic, Antimony and Bismuth Compounds.
a) Compounds containing oxygen and/or sulphur as the
additional element.

There are numerous compounds of this type and several
examples are shown in Table 3.

TABLE 3		
RING SYSTEM	STARTING MATERIALS	REF.
	MeAs(NMe$_2$)$_2$ and MeAs(SCH$_2$CH$_2$OH)$_2$	1
	MeAs(H)CH$_2$CH$_2$OH	2
	Me$_2$C$_6$H$_2$(CH$_2$SH)$_2$ and SbCl$_3$	
	(HOCH$_2$)$_2$X, P(OPh)$_3$ and trace of sodium X=(CH$_2$)$_2$, CH:CH, 1,2-C$_6$H$_4$	4
	and NaH	5
	Ph$_2$P(O)C(N$_2$)Bz and BzPh	6
	RPOCl$_2$ and ethyleneglycol R=C$_{6-14}$ alkyl	7

TABLE 3 contd.		
	$Ph_2C{:}P(O)Ph$	8
	$2\text{-}PhC_6H_4CH_2OH$ and $PSCl_3$	9
	$2\text{-}HOC_6H_4CH_2PH$ and $PSCl_3$	9

1. F. Kober, W.J. Ruehl, Z. Anorg. Allg. Chem., 1976, 420,74.

2. P.B. Chi and F. Kober, ibid., 1983, 501, 89.

3. B.A. Arbuzov et al, Izv. Akad. Nauk. S.S.S.R., 1982, 2814.

4. A.C. Guimaraes and J.B. Robert, Tetrahedron Letters, 1976, 6, 473.

5. A.N. Hughes et al, J. Heterocyclic chem., 1976, 13, 937.

6. M. Regitz, W. Igler and G. Maas, Ber., 1978, 111, 705.

7. S.N. Aminov, A.F. Panfilova and G.A. Lankina, Chem. Abs., 1979, 91, 59212x.

8. G. Maas et al, Ber., 1982, 115, 669.

9. M.S. Bhatia and J. Lal, Indian J. Chem., 1982, 21B, 363.

(1)   (2)   (3)

Thieno (3,4-g)- 1,6,3,4,2,5-dioxadithiadiphosphocin (1, R=
Bu, cyclohexyl) is prepared in 84% yield by neutralising the
related phosphorylated sulpholene derivative (2) with
potassium hydroxide or sodium hydrogen carbonate and oxidis-
ing the resulting salt with iodine (O.N. Grishina and N.A.
Andreev, Zh. obshch. Khim., 1975, 45,2093)

The cyclisation of $(RO)_2 P(O)CH:CHOCH_2CH_2Br$ (R=Me,Et) gives
the seven-membered heterocycle (3). (V.M. Ismailov et al.,
Zh. obshch.Khim., 1984, 54, 456). while cyclising $MeAs(NMe_2)_2$
with α,ω-diols such as $HO(CH_2)nOH$ (n=5-10,12) gives the eight
- and higher membered heterocycles (4) (F.Kober and T.P.
Hoang, Chem Ztg., 1979, 103, 119).

(4)   (5)   (6)

2-tert-Butoxy-5,6-benzo-1,3,2-dioxastibepane 5) is
prepared by the transesterification of $Sb(OCMe_3)_3$ with the
corresponding glycol. A study of the nmr-spectrum of (5)
indicates that the seven-membered ring exists in a chair
like conformation (B.A. Arbuzov et al., Izv. Akad. Nauk.
S.S.S.R., 1980, 1687).

The reaction of $PhSb(OEt)_2$ with the di-carboxylic acids
$X(CO_2H)_2$ (X= CH:CH, $O-C_6H_4$) in the ratio 1:1 gives a 50%
yield of (6, Y=SbPh). (M.Weiber and I.Ketzer - Kremling, Z.
Naturforsch., 1984 39B, 754). While the analogous bismuth
heterocycles (6, Y=BiPh, X =CH₂CH₂ CH:CH) have been prepared in
15-99% yield by cyclising triphenylbismuthine with aliphatic
dicarboxylic acids $(X(CO_2H)_2$, X = CH_2CH_2, CH:CH) (A.

420

Georgiades and H.P.Latscha, ibid., 1980, 35B, 1000).

b) Compounds containing nitrogen as the additional element

Several of the compounds included in this section are based on the hexachlorocyclotriphosphazatriene ring system and the preparations of these compounds are summarized in Table 4

TABLE 4

RING SYSTEM	YIELD	STARTING MATERIALS	REF.
	28%	hexachlorocyclotri-phosphazatriene and $Me_2SiN(H)N(Me)(CH_2)_2-N(Me)N(H)$	1
	46.2%	$P_3N_3Cl_6$ and 1-amino-3-iminoisoindolenine in PhCl	2
	35.8%	$P_3N_3Cl_3$ and	3
		$P_3N_3Cl_6$ and 1,1'-dihydroxy-2,2'-binaphthyl	4

TABLE 4 contd.		
X as in ref I	$P_3N_3Cl_6$ and 2,2'-dihydroxy-1,1'-binaphthyl	4
	$P_3N_3Cl_6$ and $NH_2(CH_2)_4NH_2$	5
	methylpentachlorocyclo triphosphazene and $H_2N(CH_2)_3OH$	6

1. G.S. Goldin et al, Zh. Obshch. Khim., 1975, 45, 2566.

2. S.S. Keiznetsova and R.P. Smirnov, Chem. Abs., 1978, 88, 62840Z.

3. S.S. Keiznetsova and R.P. Smirnov, Chem. Abs., 1978, 89, 109422c.

4. K. Branat and Z. Jedlinski, J. org.Chem., 1980, 45, 1672.

5. G. Guereh et al, J. mol. struct. 1982, 96, 113.

6. P.J. Harris and K.B. Williams, Inorg. Chem., 1984, 23, 1495.

The 1,3-benzazaphosphepines (7) are prepared in 50% yields when $O-H_2NC_6H_4CH_2CH_2Ph_2$ is cyclized with carbonyl compounds $R R'CO$ ($R=Me$, $R'=Et$, Pr, $RR'=(CH_2)_5$, $R=R'=H$). (K.Issleib, H. Winklemann and H.P.Abicht, Z. Anorg. Allg. Chem., 1976, $\underline{424}$, 197).

(7) (8) (9)

The tricyclic spiro compound, spiro(cyclohexane-1,8-(1,3,5,7)-tetraza(4)phosphabicyclo(5.1.0)octane (8) is prepared in 90% yield by treating the spiro compound (9) with $Me_2CHOP(O)(NCO)_2$ (E.S. Gubnitskaya and Z.T. Semashto, Zh. Obshch. Khim., 1976, $\underline{46}$, 1183).

The cyclisation of the diketone $R'COCOR'$ ($R'=H,Me,Ph$) with the phosphadihydrazides, $RP(S)(NHNH_2)_2$ ($R=Me$, Ph,EtO, PhO) in a ratio of 1:1 gives the unusual heterocyclic compounds 1H - 1,2,4,5,3- tetrazaphosphepine sulphides (10,19-24%) (A.F. Grapov, O.B.Mikhailova and N.N.Mel'nikov, Zh.obshch. Khim., 1977, $\underline{47}$, 1704).

(10) (11) (12)

Hydrazone-azo tautomerism (10⇌11) is possible for the tetrazaphosphepine sulphides. The tetraazaphosphacyclooctene (12) is prepared in similar fashion from the reaction of a phosphadihydrazide (eg $PhOP(S)(NMeNH_2)_2$) with an unsaturated aldehyde (eg $MeCH:CHCHO$) or ketone (eg $MeCOCH:CH_2$) (M.J. Herrem, J.P. Majoral and J Navech, Phosphorus Sulphur, 1981, $\underline{11}$, 241).

(13) (14) (15)

The tricyclic compounds (13, R=Ph,Me) when added to ammonia
gives the dibenzoazaphosphepines (14, X=C=O) (Y.Segall, R.
Alkabets and I.Granoth, J.chem. Res., 1977, 310). Treatment
with lithium aluminium hydride gives the tricyclic compound
(14, X=CH$_2$)

The reaction of the phosphines $R_2PNMePR_2$ with dimethyl
acetylenedicarboxylate gives (15, R=CO$_2$Me; R^1=R^2=Me
m.p. 120-130°C; R^1=R^2=Ph , m.p. 175-178°C, R=Me, R^1=Ph m.p.
= 125-130°C). While the reaction of cyclotetraphosphazanes
with two moles of dimethyl acetylenedicarboxylate gives the
bicyclic compound 1,2,3,4,5,6,7,12-octamethyl-2,4,6,12-

(16) (17) (18)

tetraaza - 1λ^5,3,5,7,λ^5-tetraphosphabicyclo (5.4.1) dodeca -
1 (11),7,9-triene-8,9,10,11 -tetracarboxylatetetramethylester
(16) m.p. 145-150°C which contains the novel 1,2,7,-
azadiphosphepine ring (W. Zeiss and H. Henjes, Ber., 1978,
111, 1655).

The 1H-dibenzo(d,f)(1,3,2)diazaphosphepines (17, R=Me,
Et,Pr,Me$_2$CH,Ph,p-MeC$_6$H$_4$O,p-O$_2$NC$_6$H$_4$O,3,4-Me(Cl)C$_6$H$_3$O) are
prepared from equimolar quantities of the appropriate

phosphorus dichlorides and phosphorus dichloridates with
2,2'-diaminobiphenyl in refluxing benzene or toluene
(M.S.R. Niadu and C.D.Reddy, Bull. chem. Soc.Jpn. 1978, 51,
2156). The structures of these compounds have been confirmed by their ir spectra and mass spectra. The ir-spectra
indicate the existence of hydrogen bonding between the
amidic hydrogen and the phosphoryl oxygen.

A cyclo Mannich reaction between primary amines and alkyl
or aryl bishydroxymethylophosphines gives the 1,5-diaza-
2,7-diphosphacyclooctanes (18,R=Ph, R'=C_6H_5,77%, m.p. 178-
180°C; R=Et, R'=C_6H_5,43% m.p. 179-182°C). Optically active
amines give optically active cyclooctanes. (G.Maerkel, G.Y.
Jin and C.H. Schoemer, Tetrahedron letters, 1980, 21, 1409).

(19)   (20)   (21)

The synthesis of compounds containing the 2,4,3-benzodiaza-
phosphepine ring system (19) has been reported (T.L.Lemke
and D.Boring, J.heterocyclic Chem., 1980, 17, 1455). It is
prepared, containing various substituent groups, by
cyclisation of the corresponding dichlorophosphoryl carbamate
with α,α'-diamino-0-xylene. Thus (19, R=Et) m.p. 190-192°C
is prepared in 37% yield according to the reaction given
below:-

$$Cl_2P(O)N=C=O + EtOH \longrightarrow Cl_2P(O)NHC(O)OEt + o\text{-}(CH_2NH_2)\,C_6H_4$$
$$\downarrow Et_3N$$
$$(19, \ R=Et)$$

B.A. Arbuzov et al., (Izv. Akad. Nauk. S.S.S.R., 1982,440)
have prepared the bicyclic compound (20, 89%) by heating
PhP($CH_2OH)_2$ with 0-phenylenediamine in a 1:1 ratio in benzene.
If a 2:1 ratio of phosphine to amine is used (21, 97%) is
obtained.

The tricyclic compound (22) is obtained by cyclising N-phenyl-1,2-phenylenediamine with thiophosphoryl chloride (M.S. Bhatia and J. Lou, Indian J. Chem., 1982, 21B, 363).While the cyclic compounds (23) are prepared (40-63% yield) by cyclising $(CH_2)_n(CH_2NHR)_2$

(22)   (23)   (24)

(R = Me, Me_2CH, n=2-4) with $(R'_2N)_2POEt$ (R'=Me,Et)   (E.E. Nifant'ev et al, Zh obshch. Khim., 1984, 54, 1207).

The cyclic arsenic compound, tetraazadiarsinocinedione (24) is prepared (89%) by the reaction of $(Me_3SiNMe)_2CO$ with arsenic trichloride. X-ray structural analysis has established the structure and shows that the eight-membered ring has a "boat conformation". Transannular interactions are observed between the arsenic and nitrogen atoms (W.S. Sheldrick, H. Zamankhan and H.W. Roesky, Ber., 1980, 113, 3821).

c) Compounds containing nitrogen together with group VI Elements

5-(2-Benzoylhydrazino)-5,6-dihydro-2,8-diphenyl-4H-1,3,4,6,-7,5-oxa-tetraazaphosphocine (25) is prepared by treating ethyl and phenyl phosphorodichloridates with benzohydrazide and triethylamine (A.F. Grapov, O.B. Mikhailova and N.N.

(25)   (26)   (27)

Mel'nikov, Zh. obshch. Khim., 1975, 45, 2570).

The bridged diazadiphosphetidines (26, X=NMe n = 2',X=0, n=2,3) (R. Keat, and D.G. Thompson, Angew. Chem., 1977, 89, 829; R. Keat, D.S. Rycroft and D.G. Thompson, J. chem. Soc. Dalton, 1979, 1224) are prepared in 20-23% yields by heating

ClPNCMe$_2$PClNCMe$_3$ with either ethane-1,2-diol, propane-1,2-diol or the corresponding diamines. The bridged compounds have a mutual cis- orientation of the oxygen atoms. The condensation of RPOCl$_2$ (R=C$_6$-$_{14}$alkyl) with triethanolamine gives the azadioxaphosphacane (27), a biodegradable and non-toxic surfactant (S.N. Aminov, Z.F. Panfilova and G.A. Lankina, Chem. Abs., 1979, 91, 59212x).

(28)   (29)   (30)

The reactions of the phosphorus compounds ClCH$_2$CH$_2$PCl$_2$ and Cl$_2$PCH$_2$CN, with water, ethanol, PhCH$_2$NHCH$_2$CO$_2$H and methanal under acidic conditions have been studied. (L. Maier, Phosphorus Sulphur, 1981, 11, 149). It is found that ClCH$_2$CH$_2$PCl$_2$, PhCH$_2$NHCH$_2$CO$_2$H and methanal give a 90% yield of the oxazaphosphacyclooctane. (28).

S.A. Terent'eva, M.A. Pudovik and A.N. Pudovik (Izv.Akad. Nauk. S.S.S.R., 1983, 2221) have investigated the interaction of bicyclic amidophosphites with carboxylic acids. Thus cyclocondensation of O-HOC$_6$H$_4$NHCH$_2$CH$_2$(OH)R (R=H,ME) with hexaethylphosphorus triamide gives the tricyclic compund (29) which when treated with a carboxylic acid, R'CO$_2$H (R'=Me, Ph, CF$_3$, CH$_2$ = CMe$_2$) gives the eight-membered bicyclic compound (30, X = PCO$_2$R^1, R^2=H). The eight-membered heterocycle (30, R^2=Ac, X=POEt) is prepared by cyclising O-HOC$_6$H$_4$N(Ac)CH$_2$CHMe(OH) with dichloroethylphosphine in the presence of triethylamine.

A 55:45 mixture of two isomers is obtained. Heating benzo-dioxazaphosphacyclooctane (30, R^2=Ac, X=POEt) with sulphur gives (30, X=PEtS, R^2=Ac 61% yield) (S.A. Terent' eva, M.A. Pudovik and A.N. Pudovik, Zh. obshch. Khim, 1983, 53, 1420).

The adduct N,N-bis(2-(6-t-butyl-1,3,6,2-dioxazaphosphoran-2-yloxyl)ethyl-t-butylamine (31), prepared from phosphorus trichloride, hexamethyl phosphorus triamide and N-t-butyl-

2,2'-iminodiethanol, is hydrolyzed by water to give 6-t-butyl-1,3,6,2-dioxazaphosphocane-2-oxide (32, X =P(O)H) and N-t-butyl-2,2'-iminodiethanol. The heterocycle (32, X = P(O)H) reacts with o-chloranil to produce a crystalline product which in solution gives an equilibrium mixture of six-

(31)   (32)

coordinated phosphorus with a P-OH bond (33, R=OH) and the phosphate ester (34, X = P=O, R'=OH). Phenylphosphorusdichloride and N,N,N',N',-tetramethyl-P-phenylphosphorusdiamide react with N-t-butyl-2,2'-iminodiethanol to give (32, X = PPh) which adds to O-chloranil to form a tautomeric equilibrium mixture of the four-coordinate and six-coordinate phosphorus compounds (33, R = Ph) $\rightleftharpoons$ (34, X = P$^+$Ph, R'=O$^-$).

(33)   (34)

Phenylazide reacts with (32, X = PPh) to give 2-phenyl-2-Phenylimino-6-t-butyl-1,3,6,2-dioxazaphosphocane (32, X= PhN=P-P). (K.H. Osman, M.M.A Gawad and H.M.Abbasi, J.Chem. Soc. Perkin , 1984, 1189).

(35)   (36)

2,6-Dimethyl-1,3,6,2 -perhydrodioxazarsocine (36 R^1=H, R = Me, R^2= Me) is prepared from the reaction of methyldiethanolamine and $(Me_2N)_2$ AsMe. When added to o-chloranil a spiro compound 2,6 dimethyl-2,2-tetrachloropyrocatcholo-1,3,6,2-perhydro-dioxazarsocine (35, R=Me) is formed. The structure of both compound has been determined by X-ray structural analysis (P. Marone et al, Tetrahedron letters, 1976, 15, 1193). Cyclisation of $RN(CH_2CHR^1OH)_2$ with $R^2As(NMe_2)_2$ gives 2,6-dimethyl-tetrahydro-1,3,6,2-dioxazarsocine (36, $R=CH_3$, R^1=H, R^2=CH_3) and 2,4,6,8-tetramethyl-tetrahydro-1,3,6,2-dioxaz-arsocin (36, $R=R^1=R^2=CH_3$) (F. Kober and W.J. Ruhl, Z Anorg. Allg. chem., 1976, 420, 74).

d) Macrocyclic Compounds.

A number of reports have appeared that deal with the preparation and properties of compounds having a macro cyclic ether type structure. The preparations of some of these compounds are detailed below and in Table 5.

TABLE 5		
RING SYSTEM	STARTING MATERIALS	REF.
X=O, NSO$_2$, C$_6$H$_4$OH-p,S. n=0-3,5.		1

	$1,4(Br(CH_2)_3O)_2C_6H_4$, Ph_3P, $LiAlH_4$ ($PMF/CHCl_3$) and NaOH	2
	$PhAsCl_2$ with $HO(CH_2OH_2O)_2H$	3
	$PhAsCl_2$ with $HO(CH_2CH_2O)_3H$	3
	$PhAsCl(CH_2)_2AsClPh$, $HO(CH_2CH_2O)_3H$	3
	2-tert-butyl-1,3,2-dithiaphosphorinane	4
	$oBrC_6H_4OCH_2OMe$ and $PhClP(O)CH_2CH_2P(O)ClPh$	6
	$(o-BrCH_2C_6H_4)_2P(R)$ $o(R=O-C_6H_4CO_2CH_3)$ and ethylene glycol	5

TABLE 5 contd.

430

1. V I Kal'chenko et al. Zh Obshch. Khim, 1984, 54, 1754.

2. H Christol et al. Tehrahedron letters, 1979, 2591

3. V S Gamayurova and N V Shabnikova, Zh. Obshch. Khim, 1983, 53, 2796

4. J P Dutasta, J Hartin and J B Robert., J. org.Chem., 1977, 42, 1662.

5. C V Bedrin et al. Izv. Akad. Nauk. S.S.S.R., 1978, 1930.

6. L J Kaplan, G R Weisman and D J Cram, J org.Chem., 1979, 44, 2226.

The two cyclic compounds (37 and 38) have been prepared.
(V G Gruzdev, Zh. obshch. Khim., 1975, 45, 2172).

Trichlorodioxa (2,2'-thiobis(4-methyl-6-α-methylbenzylphenyl).)
phosphorane (37, R=CH(CH$_3$)C$_6$H$_5$, R^1=R^2=R^3=Cl, X=S) m.p. 195-
196°C, is prepared by the reaction of the bisphenol (2,3,5-
(HO)(R)Me C$_6$H$_2$)$_2$S with phosphorus pentachloride. It is
obtained as light green crystals, soluble in benzene and
insoluble in carbon tetrachloride. When (37, R=CH (CH$_3$)C$_6$H$_5$,
R^1=R^2=R^3=Cl, X=S)is treated with water and ethanol various
substitutions occur at the phosphorus atom. The cyclic
phosphorane (37, R=-C(CH$_3$(CH$_2$)$_4$CH$_2$, R^1=R^2=R^3=Cl, X=CH$_2$) is
prepared similarly from (2,3,5-(HO)(R)MeC$_6$H$_5$)$_2$CH$_2$, this when
treated with sulphur undergoes substitution at the phosphorus
atom. R^{2}R^3=O; R^1=Cl, R^{2}R^3=S respectively). The reaction of
the bisphenol (2,3,5-(HO)(R)Me C$_6$H$_2$)$_2$ X (X=S,CH$_2$ R=CHMe$_2$,
1-methylcyclohexyl) with phosphorus pentachloride in the
presence of ethanol gives for example, pentaoxo(ethylbis
(2,2'-thiobis(4-methyl-6-α-methylbenzylphenyl) phosphorane
(38 R=CH(CH$_3$)C$_6$H$_5$, X=S, 76% m.p. 137-138°C)

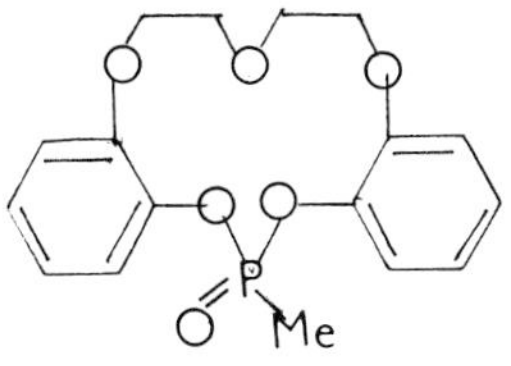

(39) 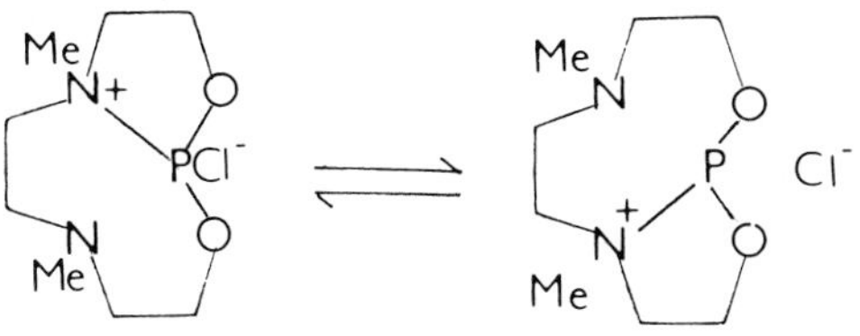 (40a) (40b)

The macrocyclic polyether (39) is prepared in a low (9%) yield by treating dichloromethylphosphine with $(O-HOC_6H_4O (CH_2)_2O)_2$ in the presence of triethylamine (K B Yatsimirskii et al, Teor. Eksp. Khim., 1976 12, 421). The polyether (39) forms complexes with lithium chloride and with magnesium chloride but not with sodium, potassium or calcium salts. The ammoniophosphonite (40) is formed by the slow disproportionation of the diazaphosphocyclopentane $(42, R-Cl(CH_2)_2O)$. The compound (40) exists as a tautomeric mixture of two

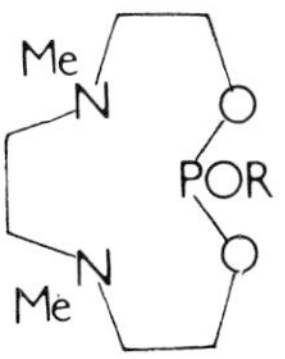

(41) 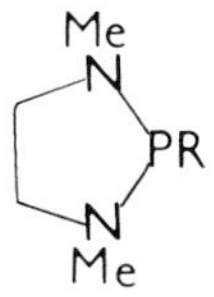 (42) 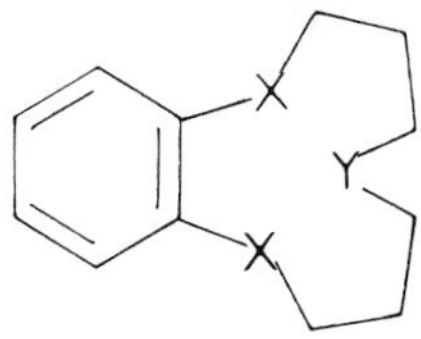 (43)

equivalent forms (40a) and (40b) while in the presence of methanol or ethanol the disproportionation of $(42, R=Cl(CH_2)_2 O)$ occurs to give the macrocycle (41, R=Me,Et, respectively) quantitatively. (H Sliwa and J Picavet, Tetrahedron Letters 1977, 18, 1583).

High dilution techniques has been employed to prepare
(43) in 32 - 56% yield where X and Y are shown in Table 6

Table 6.

$o(MX)_2-C_6H_4$ + (L⌒⌒⌒Y)$_2$ ⟶ (43).

X	M	L	Y		X	Y
PPh	Li	Br	S		PPh	S
PPh	Li	Cl	O		PPh	O
PPh	Li	Cl	NMe		PPh	NMe
PPh	Li	Cl	NPh		PPh	NPh
S	Li	Cl,Br	PPh		S	PPh

(E.P. Kybu et al, J Amer. chem. Soc., 1980, 102, 139).
X-ray analysis showed that the ring system has an L-shaped
arrangement and the phenyl substituents on the 0-phenylene
diphospha unit are cis and occupy pseudoequatorial postions
in the conformations observed in the crystal state..

An arsenic containing analogue of (43) (X=AsMe, Y=S) is
prepared from 1,2-(Li Me As)$_2$ C$_6$H$_4$ by cyclocondensation
with bis-electrophiles. The stereochemistry of these new
tertiary arsenic containing compounds follows from molybdenum carbonyl complexation experiments (E.P. Kybu and
S.S.P. Chou, chem.Comm., 1980, 449). Other examples of
this type of ring have been described (E.P. Kybu and S.S.P.
Chou, J.Amer. Chem. Soc., 1980, 102, 7012). The compounds
(43, X=AsMe, Y=S,O,NMe) are prepared from O-(LiAsMe)$_2$ C$_6$H$_4$
and $(R(CH_2)_3)_2X$ (R = Cl, X=N Me , R = mesyloxy, X=O.S) Again
thermal complexation reactions with molybdenum carbonyl
have established the sterochemistry of some of the ligands.

(44)

(45)

The macrocycle, 4,7,13,16-tetraphenyl $-1,10$-dioxa$_2$-4,7,13,16-tetraphosphacylooctadecane(44, $X^1=X^4=O$, $X^2=X^3=X^5X^6=PPh$, $Y^1=Y^2=CH_2CH_2$) is prepared by the reaction of $(ClCH_2CH_2)_2O$ and a stoichiometric mixture of 1,2-bis-(phenylphosphino)ethane and phenyllithium 'M. Ciampoline et al Chem. Comm., 1980.177) Two isomers have been isolated and each reacts with $Co(BPh_4)_2$ forming a 1:1 complex. The structures of the latter have been determined by x-ray crystallographic analysis. Cyclisation of $S(CH_2CH_2SH)_2$ with RPCl$_2$ gives the 16-membered ring, hexathiadiphospha-cyclooctadecane (44, $X^1=X^2=X^3=X^4=X^5=X^6=S$, $Y^1=Y^2=$ RP=X, X=- and R=Me, C Me$_3$) (J. Martin and J.B. Robert, Nouv. J. Chim, 1980, 4, 515). This 16-membered ring, a dimer, is in equilibrium with its 8-membered monomer, X=P(R)SCH$_2$CH$_2$SCH$_2$CH$_2$S. Oxidation with hydrogen peroxide or sulphur gives the corresponding oxo or thioxo compounds (44, X=O,S). The 16-membered ring exists in two diastereoisomeric forms which have been isolated.

The 16-membered ring compound 1,10-dipropyl-4,7,13,16-tetraphenyl-1,10-diaza-4,7,13,16-tetraphosphacylooctadecane (44, $X^1=X^4=$-N-Pr,$X^2=X^3=X^5=X^6=PPh$, $Y^1=Y^2=(CH_2)_2$) It is prepared by cyclisation (Cl CH$_2$CH$_2$)$_2$ NPr with (Ph K PCH$_2$)$_2$ in tetrahydrofuran. There are four diastereoisomers and the crystal structure of the nickel complex of one isomer has been determined (M. Ciampoline et al, Inorg. Chim. Acta., 1983, 76, 47). Five diastereoisomers exist of the cyclo-octadecane (44, $X^1=X^4=S$, $X^2=X^3=X^5=X^6=PPh$, $Y^1=Y^2=CH_2CH_2$) and are prepared from Li$_2$(PhPCH$_2$CH$_2$PPh) and 2-chloroethyl sulphide (M. Ciampoline et al Congr. Naz. Chim. Inorg., 1983, 123). These compounds form 1:1 complexes with nickel (II).

The macrocyclic compounds (45, X=S,As,Ph,$X^1=S$,$X^2=AsPhh$ n=2: Y=AsPh; X=X$^2=S$, $X^1=AsPh$, Y=AsPh, n=2; X=X$^2=S$, $X^1=S$,O Y=AsPh, n=1) and (46 X=X$^1=AsPh$,S) are prepared from organoarsenic dichlorides using cyclocondensation reactions (T. Kauffman and J Ennen, Tetrahedron Letters, 1981, 22,

(46)

(47)

5035). For example $(Cl(CH_2)_3AsPhCH_2)_2CH_2$ and disodium-
sulphide gives a 27% yield of (46, X=AsPh, X^1=S). While (Cl
$(CH_2)_3$ $AsPhCH_2)_2$ CH_2 and propanedithiol/potassium hydroxide
gives (45, X=X =S, X^2=AsPh, Y=AsPh, n=2).

The 17-membered phosphorus containing crown ethers (47,
R,X= Me,S; Ph,S; Ph,O; PhO,O) are prepared from sodium 2,2'-
(ethylene bis (oxyethyleneoxy)) bisphenolate and phosphoric and phosphonothioic dichlorides. All the products
are crystalline : the yields obtained and the melting points
observed are: R=CH$_3$, X=S, 46%, 140-141°C; R=Ph, X=S, 24%
153-154°C; R=Ph, X=O 12% 149-151°C; R=Pho, X=O 14% 90-91°C.
(T.N. Kudrya et al, Zh. Obshch. Khim, 1982, 52, 1092).

4. The Atranes

Several preparations of the atrane series of compounds
have been reported. In this section the preparation of
atranes or 1-aza-4-M-4,6,11-trioxabicyclo (3.3.3) undecanes
and related heterocycles are summarized.

(1)   (2)

The atrane, 2,9,10-trioxa-6-aza-1-silatricyclo $(4.3.3.0^{1,6})$
dodecane (1, R = CH$_2$Cl, M=Si, 60%) is prepared by heating
1-(N,N-bis(2-hydroxyethyl)amino)-3-propanol with (EtO)$_3$-
SiCH$_2$Cl in the presence of sodium ethoxide. (K. Kemme et
al, Chem.comm., 1976, 1041). An X-ray structural analysis
of (1) shows that the rigid six-membered ring exists in an
approximately semichair conformation and has shorter interaction distances than normal.

The 1-substituted metallotranes (1, R=halo, Ph, M=Ge,Sn)
and (2) are prepared by the reaction of $(R^2_3SnOCH_2CH_2)_3N$
$(R^2=alkyl)$ with halomercapto or acyloxy derivatives of
germanium and tin (K.A. Kocheshkov et al, Chem. Abs., 1978,
89, 109816j).

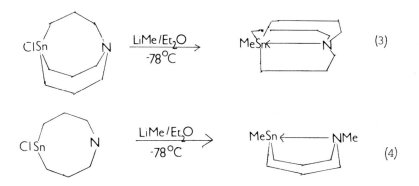

(3)

(4)

1-Aza-5-stanna-5,5-dimethyl bicyclo $(3.3.0^{1,5})$ octane (4)
and 1-aza-5-stanna-5-methyltricyclo $(3.3.3.0^{1,1,5})$ undecane
(3) are prepared in 90% yields according to the reactions
above. (K. Jurkschat and A Tzschach, J. organometallic
chem., 1984, 272, C13).

(5)

(6)

Arsatranes (5) are prepared by cyclising $HO(CH_2CH_2)_3NR$ with
$MeAs(NMe_2)_3$, when R=Me the arsatrane (5) is produced (F. Kober,
Z. Khim., 1980, 20, 49). Cyclisation of $R_2Sn(OMe)_2$ with
$MeN(CH_2CONHMe)_2$ gives 2,5,8-trimethyl-1,1-dialkyl-2-5,8-
triaza-1-stannabicyclo $(3.3.0^{1,5})$ octane-3,7-dione (6,X=Sn,
$R^1 = R^2$=Me, Et, CMe_3, Y=NMe) in 55-58% yield (A. Tzschachi
et al, Z. Anorg. Allg. chem., 1982, 488, 45). 1,1,5-Trimethyl-
2,8- dioxa-5-aza-1-germabicyclo $(3.3.0^{1,5})$ octane-3,7· dione
(6, X=Ge, Y=O) is prepared by cyclising $RR^1Ge(OR^2)_2$ (R=R'=Me,
Ph; RR^1=O$(SiMe_2CH_2)_2$, R^2 not given) with methyliminodiacetic
acid in benzene/DMF (S.N. Tandura et al, Zh. Obshch. Khim.,
1983, 53, 1199).

Chapter 55

COMPOUNDS WITH LARGER HETEROCYCLIC RINGS,
EIGHT-MEMBERED RING COMPOUNDS

J.A.H. MacBRIDE

General Considerations

(a) Synthesis

Since publication of the Second Edition of Rodd's C.C.C., Volume IVK, substantial research effort has been devoted to the synthesis of eight-membered ring heterocyclic compounds. This work has been stimulated by the isolation of a number of natural products of this type, usually with oxygen heteroatoms, and by theoretical interest in the possible aromatic character of some of these compounds or the ions derived from them.

The cyclisation of linear precursors to eight-membered and similar medium sized rings is, in general, an unfavourable process compared with formation of either smaller (e.g. five or six-membered) or larger (e.g. greater than ten-membered) rings since developing transannular interactions and loss of conformational freedom are not compensated by favourable valence angles and the proximity of reacting end-groups, as in the former case, nor by the residual flexibility and consequently more favourable entropy and absence of strain of the latter. For example the relative rates (at 50°) for formation of lactones (2) from potassium salts of w-bromoalkanoic acids (1) for 6-, 7- and 8-membered rings are 26,000, 100, and 1 respectively. Even the strained 3-membered lactone forms over 20 times faster than the 8-membered, and the 12-membered 10 times faster (G. Illuminati *et al* J. Amer. chem. Soc., 1977, 99, 2591).

Several strategies for eight-membered ring synthesis have avoided the difficulty of simple cyclisation by using ring-expansion reactions involving cleavage of bonds common to smaller fused rings, or incorporation of side-chains into

438

$$Br-(CH_2)_n-CO_2^{\ominus}.\ K^{\oplus} \longrightarrow (CH_2)_n\overset{CO}{\underset{O}{\Big<}} + KBr$$

(1)   (2)

smaller rings with initially limited conformational freedom.
On the other hand recent work has revealed reactions which
favour the closure of eight-membered (and larger) rings,
notably the formation of phenylsulphonyl lactones using
a palladium catalyst and the cyclisation of alkenyl acetals
with tin (IV) chloride (Section 2 (a)).

(b) Aromaticity

Aromatic character is well established in the 10π-electron
dianion of cyclooctatetraene and there is considerable
interest in assessing the aromaticity, if any, of corresponding π-isoelectronic heterocyclic compounds. The necessary 10π-electron system can only be realised in an eight-
membered ring if two heteroatoms (or one heteroatom and
a carbanionic centre) can each provide an electron pair
from an atomic p-orbital of appropriate geometry (as in
pyrrole, etc.) so that these orbitals must not be used for
formal π-bonds (as in pyridine, etc.) and only structures
(3) and (4) (together with those containing further heteroatoms replacing CH, and carbanions) are potentially aromatic. Of these the 1,2-isomer (3) is, in general, unlikely

(3)   (4)

to show aromatic character since essentially localised
quantum-mechanical interaction between the "non-bonding"
electron pairs (the "alpha-effect"), maximised in a planar

system, is expected to raise the energy of one of the non-bonding orbitals, perhaps to an antibonding level, and limit its participation in a delocalised π-system.

The usual experimental criteria for aromatic character are a planar ring revealed by x-ray crystallography, the effect of a diamagnetic ring current upon nmr-shifts, and the effect of π-delocalisation upon the coupling constants.

The properties of π-excessive heteroannulenes have been reviewed by A.G. Anastassiou and H.S. Kasumai (Adv. hetero-cyclic. chem., 1978, 23, 55); they have highlighted the opposing effects of ring strain with reduced flexibility versus π-electron stabilisation in a planar aromatic molecule compared with one in a buckled conformation, by writing the standard free energy difference (ΔG_o) between the two states in terms of the enthalpy differences arising from σ-electronic (ΔH_o^{σ}) and π-electronic (ΔH_o^{π}) effects, and the entropy change ($T\Delta S_o$):

$$\Delta G_o \text{ (buckled-planar)} = \Delta H_o^{\pi} + \Delta H_o^{\sigma} - T\Delta S_o$$

In eight-membered rings the ΔH_o^{σ} term, reflecting increased angle strain in the planar situation, probably outweighs the entropy effect of molecular flexibility which may be important in larger rings. The examples described in Sections 3 (b) (ii) and 4 (a) (i) show that the balance of the free energies of planar and buckled rings of type (4) is finely poised, since aromatic character is dependent not only on the heteroatoms, aromatic in the case of (4), X=Y=NH and X=NH, Y=O but not aromatic when X=Y=O or S, but on the substituent on the heterocyclic nitrogen atoms as well. Thus derivatives of (4, X=NR) with electron withdrawing R groups are non-planar and not diatropic.

1. Eight-membered ring compounds containing oxygen.

(a) With rings containing one oxygen atom.

(i) Derivatives of oxocin.

Further examples of naturally occurring reduced oxocins from the Laurencia genus of red algae include the halogenated alkyne chondrial (5) whose structure, determined by x-ray crystallography, corrects a previous bicyclic assignment (W. Fenical, K. B. Grifkins, and J. Clardy, Tetrahedron Letters, 1975, 1507) and laurallene (6).

(5)

(6)

Although the authors do not comment on the optical configuration of the allene group of (6) they describe a single compound m.p. 53-54°, $[\alpha]_D$ + 173·6°, so that unless the isolation procedure (silica chromatography of the dichloromethane extract) failed to reveal its diastereomer this must be an example of a naturally resolved allene. (A. Fukazawa, and E. Kurosawa, Tetrahedron Letters, 1979, 2797).

The predicted type of acyclic biogenetic precursor to this series, laurediol (7, and its *cis*-isomer at the alkyne conjugated double bond) has been isolated from *L. nipponica*. It is thought to cyclise to laurencia (8) and its congeners by a bromonium ion initiated process, ring size and stereochemical configuration being, presumably, under enzyme control (E. Kurosawa, and A. Fukuzawa, and T. Irie, Tetrahedron Letters, 1972, 2121 and references there cited).

Racemic laurencin (8) has been synthesised by a reaction sequence in which the furoic acid derivative (9) is reduced and ozonised to the di-aldehyde (10) which undergoes Robinson-Schoepf condensation to the amine bridged oxocin (11) leading, by an adaption of an earlier oxocin synthesis (see 2nd Edn. Vol. IVK p. 387) to the key intermediate aldehyde (12) from which the natural product (8) is elaborated (T. Masamune *et al* Tetrahedron Letters, 1977, 2507; Masamune and H. Matsue, Chem. Letters, 1975, 895).

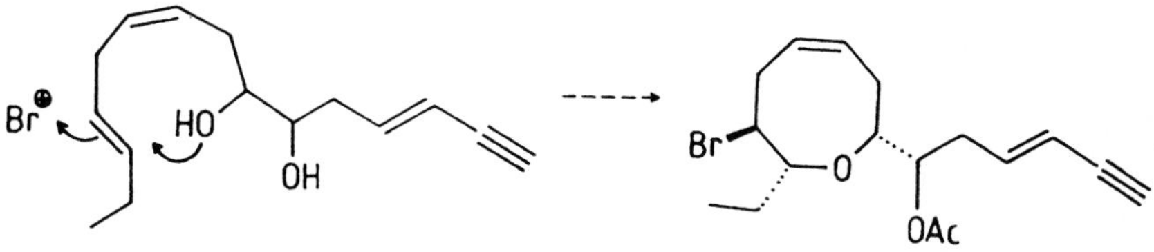
Br
HO
OH
Br
O
OAc

(7) (8)

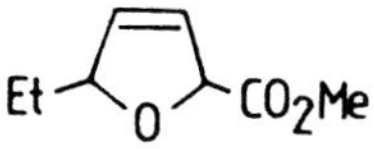
Et
O
CO2Me

(9)

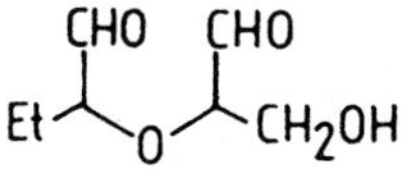
CHO
CHO
Et
O
CH2OH

(10)

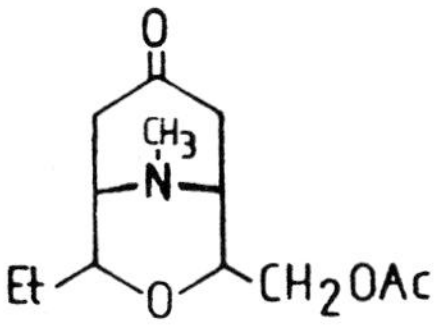
O
CH3
N
Et
O
CH2OAc

(11)

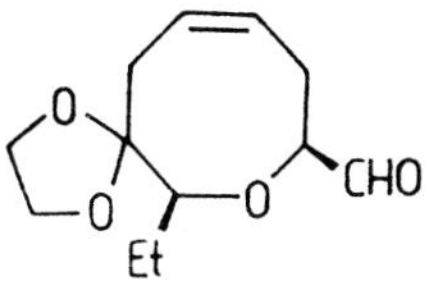
O
O
O
CHO
Et

(12)

Several research groups have sought more direct access to the substituted oxocin system of the Laurencia compounds; oxidation of 2-alkylcycloheptanones (13) with trifluoro-peracetic acid gives the lactones (14) in good yield. The

(13) (14) (15)

carbonyl group is converted into methylene by the Tebbe reagent ((C_5H_5)$_2$ Ti(μ-CH$_2$, μ-Cl)AlMe$_2$) and immediate hydro-boration-oxidation of the resulting unstable enol ethers (15) gives hydroxymethyl intermediates from which the Laurencia skeleton may be obtained. Initial incorporation of a 4,5-double bond, as in laurencin (8) etc., is achieved by spontaneous Claisen rearrangement of the vinyl ketene acetals (17) formed when diastereomeric mixtures of the selenoxides (16) are treated with base (1,8-diazabicyclo [5.4.0]undec-7-ene, DBU) in refluxing xylene. The high yields of lactones (18; R=CH$_2$OSiPh$_2$Bu^t, 84%; (CH$_2$)$_4$Me, 52%) suggest that both the *cis* and the *trans* isomer of the acetal (17) are reacting, presumably *via* transition states (17a) and (17b) (R. W. Carling and A. B. Holmes, Chem. Comm. 1986, 565 and 325).

Another approach to substituted oxocanes with a double bond between positions 4 and 5 depends on the remarkably efficient cyclisation of alkenyl acetals (18a) with tin(IV) chloride under mild conditions (See Section (1) (a)). When the double bond is not substituted (18a, R=H)4-chloro-oxocane (20) is formed in addition to the normal product (19) in 1:2 ratio respectively and a combined yield of 87%); (19, R=H) does not give (20) under the reaction conditions.

An advantage of this synthesis is that alkyl substituents can be incorporated at the positions adjacent to oxygen to give the *cis*-dialkyl configuration of the Laurencia

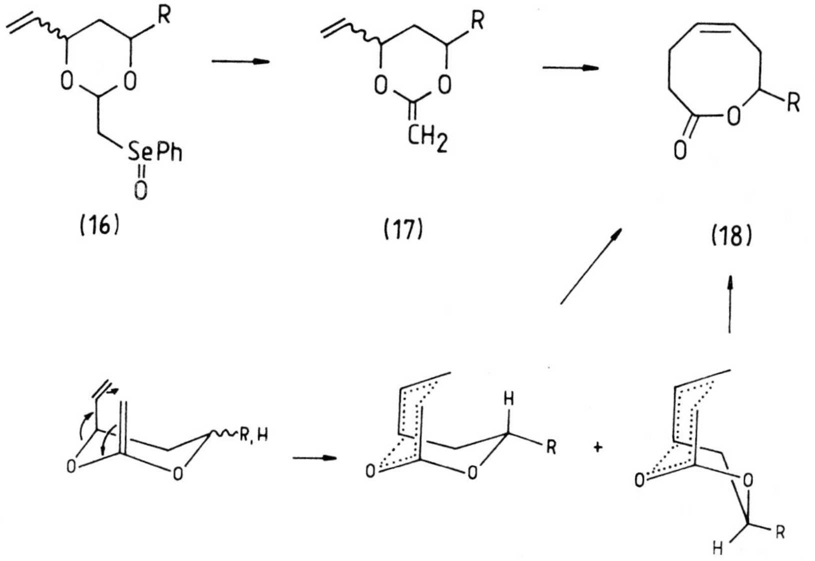

(16)   (17)   (18)

(17)   (17a)   (17b)

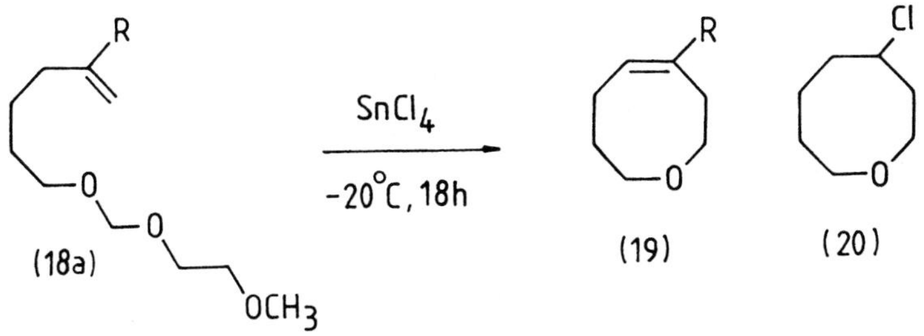

SnCl$_4$

−20°C, 18h

(18a)   (19)   (20)

(18a, R = H, SiMe$_3$, Me or ^{n}Bu.)

natural products with high stereoselectivity.

(19)   SnCl₄ ✗ -20°C, 18h   (20)

$$\text{SnCl}_4 \xrightarrow[-20°C, 18h]{\textbf{✗}}$$

(21)   (22)

Thus the dimethyl acetal (21) gives the *cis*-dimethyloxocin (22) in preponderance over the *trans*-isomer of 30 to 1. The configuration of the product has been established by de-silylation and reduction to *cis* 2,8-dimethyloxocane. A mechanism involving initial elimination of an alkoxide ion and cyclisation of the resulting oxonium species (23) by the ene reaction is put forward to explain the highly stereoselective and regioselective cyclisation to the unsubstituted carbon of the alkene, which is the Markownikoff position and electrophilic addition cannot be ruled out. (L. G. Overman *et al* J. Amer. c̈hem. Soc., 1986, 108, 3516).

A promising route to the lactone (26) depends on the seemingly surprising preference for cyclisation to an eight rather than a six-membered ring when the sulphonyl ester (24) is treated with sodium hydride in the presence of

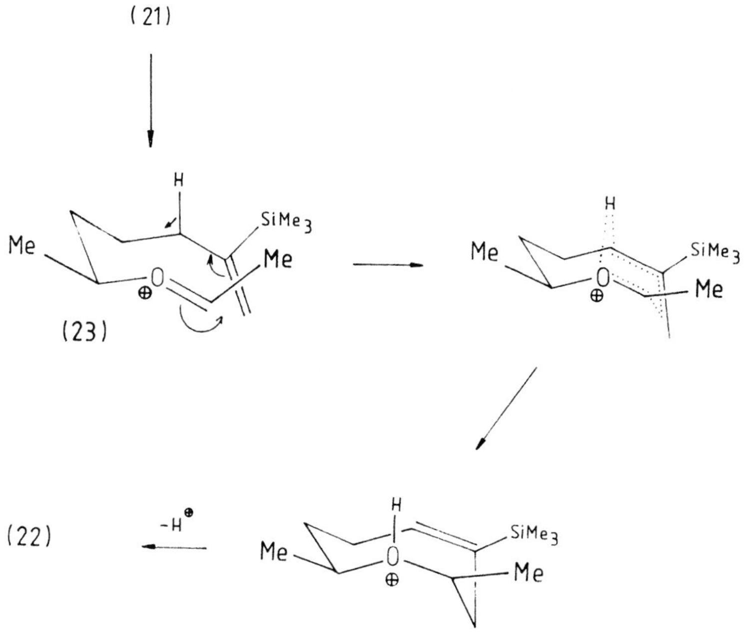

(21)

(23)

(22)   −H⊕

tetrakis(triphenylphosphine)palladium(0) and 10 mol % of 1,2-bis(diphenylphosphine)ethane (DIPHOS) (see Section (1) (a)). The reaction is thought to proceed by attack of the anionic centre, formed from the sulphonyl ester by sodium hydride, on the palladium complexed allyl cation derived by loss of acetate ion, in the intermediate (25). Similar preference for cyclisation to the less substituted end of the allyl system, independent of the size of the resulting ring, is observed with longer carbon chains   (see Chapter 56) (B. M. Trost and T. R. Verhoeven, J.   Amer. chem. Soc., 1979, 101, 1595; 1980, 102, 4743).   Since the phosphine ligands present favour the σ-bonded state of palladium in equilibrium with the π-bonded one, it may be that this reaction involves nucleophilic displacement of palladium σ-bonded preferentially at the less hindered

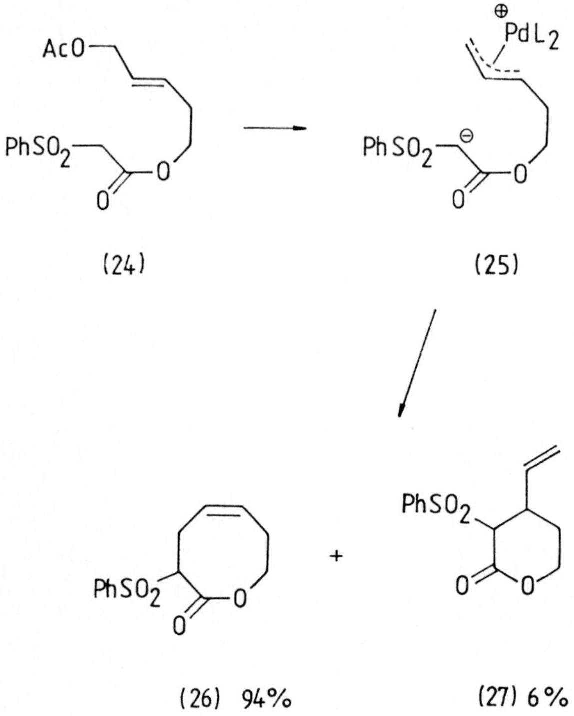

(24)   (25)

(26) 94%   (27) 6%

terminal allylic position (see R. P. Houghton, Metal Complexes in Organic Chemistry, Cambridge University Press,
1979, p. 156 *et seq.*). The phenylsulphonyl group may be
removed from the alkali labile lactone (26) using sodium
amalgam in an acetic acid buffer.

Ring expansion of the lactones (28) by two carbon atoms
derived from alkynyl-lithiums gives the oxocenones (29,
R=H, Me, or SMe) (S. L. Schreiber and S. E. Kelly, Tetrahedron Letters, 1984, 1757). The fully conjugated lactones
(31, R^1=H or Cl; R^2=Cl or CN) result from base-catalysed
ring expansion of the photo-adducts (30) of γ-pyrones with
chloroalkenes (T. Shimo, *et al.*, Chem. Lett., 1984, 1503).

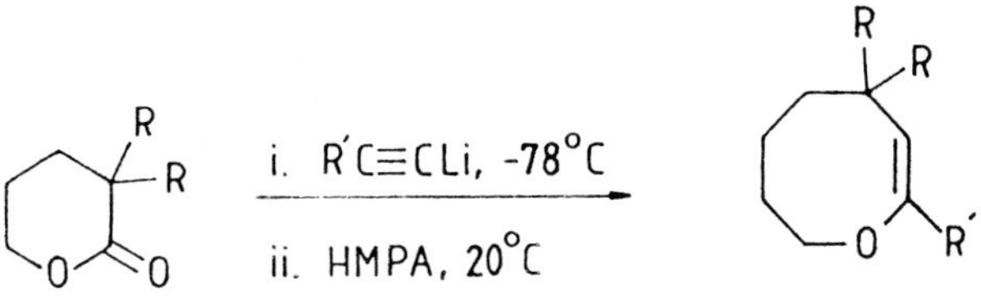

(28)　　　i. R'C≡CLi, -78°C　　　(29)

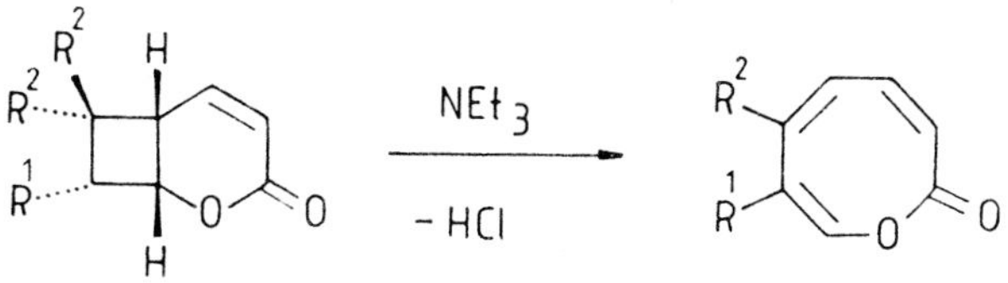

ii. HMPA, 20°C

(30)　　　NEt₃　　　(31)

− HCl

See also Section 3 (a) (ii).

(l) *Eight-membered ring compounds containing*
two oxygen atoms.

(i) *1,3-Dioxocins.*

The first unsaturated examples of this ring system lacking
benzo-fusion are obtained when the nitroso-urea derivative
(32) is treated with sodium hydrogen carbonate in methanol.
The strained *trans*-isomer　of　7-methoxy-7,8-dihydro-2H,4H-
1,3-dioxocin (33) is first produced, together with the

cyclopropyl ether (34), but it readily isomerises to the *cis*-isomer (35) in the presence of a catalytic amount of iodine.   The double bond of (33) undergoes the Diels-Alder reaction with 2,3-dimethylbutadiene at 60°C (H. Jendalla, Ber., 1982, <u>115</u>, 201).

(32)   (33)   (34)

(35)

(ii) 1,4-Dioxocins.

The γ-pyrone derivative of 1,4-dioxocin (37) (see 2nd. Ed. Vol. IVK p. 391 *et seq.*) is obtained when the natural antibiotic LL-Z 1220 (36), which contains the syn benzene dioxide group, is heated in acetic anhdyride at 100°C. The antibiotic is recovered (in racemic form) when its re-arrangement product (37) is refluxed in ethyl acetate. Both the benzene dioxide and dioxocin isomers show the same antibiotic activity against two species of fungi, suggesting equilibration under biological conditions (D. B. Borders and J. E. Lancaster, J. org. Chem., 1974, **39**, 435).

(36)

(37)

2,3-Benzo-1,4-dioxocin (41, $R^1=R^2=H$) and its methyl derivatives (41, $R^1=H$, $R^2=Me$ and $R^1=R^2=Me$) are available in 15-25% yield by flash vacuum pyrolysis (FVP) of the benzofurans (40) at 550°C, 2.10^{-5} mm Hg. The precursors (40) are obtained from the benzoxepins (38), which are formed when pyridazine *N*-oxides react with benzyne, by epoxidation of the photo-isomers (39) with .*m*.-chloroperbenzoic acid. Benzopyrans (42) and (43) are also obtained in the FVP experiment and have been shown to arise by a secondary extrusion of carbon monoxide, from the benzodioxocins (41). Like the monocyclic parent compound, the ^{1}H-nmr spectra of the benzodioxocins (41) show no evidence of aromatic delocalisation and this result (as well as the small coupling constant of only 3·4 Hz between the pro-tons of the O-CH=CH-O group in the dioxocins (41, $R^1=H$)) is attributed to the high electronegativity of oxygen inhibiting release of electrons into the π-system (T. Tsuchiya *et al* Chem. Comm., 1985, 1254). This may not be the only important factor, however, since 1,4-dithiocins are also non-aromatic (see Section 2 (b)).

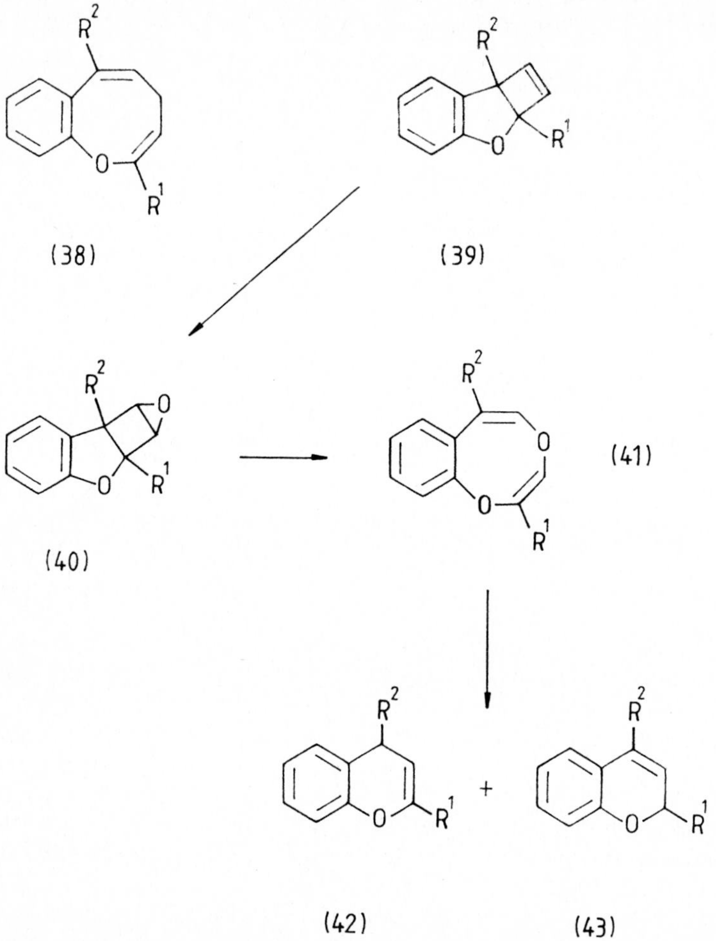

(38)

(39)

(40)

(41)

(42)   +   (43)

2. *Eight-membered ring compounds containing sulphur or selenium.*

(a) *Rings containing one sulphur atom*

(i) *Derivatives of Thiocin*

Treatment of the sulphonium salts (44), formed from 2-

vinyltetrahydrothiophen and allyl trifluoromethylsulphonate (triflate), with strong bases such as diazabicycloundecene (DBU) or lithium di-isopropylamide (LDA) leads to expansion of the thiophen ring giving the thiocin derivative (46), presumably *via* the ylid (45). Minor products result from deprotonation at position 2 of the thiophen ring. The sequence can be repeated to give 11, 14 and 17 membered ring compounds (E. Vedejs *et al.*, Tetrahedron Letters, 1978, 519; R. Schmid and H. Schmid, Helv. 1977, 60, 1361) (see also Chapter 56).

(44)   (45)   (46)

(47)   (48)   (49)   (50)

The corresponding ring enlargement which occurs when the ylid derived from the ethyl sulphonium hexafluorophosphate (47) is generated using potassium t-butoxide in THF at

452

-40°C gives the analagous methyl derivative (48:41% yield) together with two diastereomers of the *trans*-isomer (49:52%) and (50:7%). The diastereomers (49) and (50) are separable by column or preparative gas chromatography and owe their existence to the two elements of chirality inherent in the dissymetry of the *trans*-cyclooctene system and the asymmetric position of methyl substitution. Isomers (49) and (50) interconvert by a unimolecular process at temperatures over 100°C, (49) predominating at equilibrium, with activation parameters (at 110°C) ΔH 124 kJ mol^{-1} and ΔS approximately 4 kJ mol^{-1} K^{-1}. The *cis* isomer (48) is formed slowly at these temperatures by a reaction whose radical mechanism is established by its suppression in the presence of radical trapping agents. The configurations of compounds (49) and (50) were deduced from nmr measurements (A. Fava et al. J. Amer. Chem. Soc., 1978, 100, 1516; J. org. Chem., 1978, 43, 4826).

5 Steps

(51)   (52)

Inversion of the *cis*-alkene function of the unsubstituted compound (51) to the *trans* isomer (52) is achieved by a modification of the Whitham sequence in which sulphur is protected against oxidation by formation of the methyl-sulphonium salt (A. Fava et al., J. org. Chem., 1980, 45, 261).

The conformation of thiocan-5-one has been shown by variable temperature ^{13}C nmr- spectrometry to be (53) in equilibrium with its mirror image through conformational changes whose highest free energy barrier is 34·1 kJ mol^{-1}. This value, higher than that of either cyclooctanone or oxocan-5-one is attributed to increased trans-annular interaction by sulphur as well as to difference in bond lengths and

and angles (F.A.C. Anet and M. Ghiaci, J. org. Chem., 1980, 45, 1224).

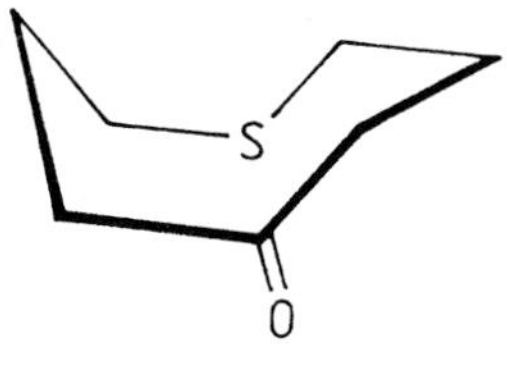

(53)

(ii) Dibenzothiocin and dibenzoselenocin derivatives

The dibenzoheterocindiones (55) where X is S, Se, O or NAr are obtained by intramolecular condensation of the keto-esters (54) and exist in distorted boat conformations with

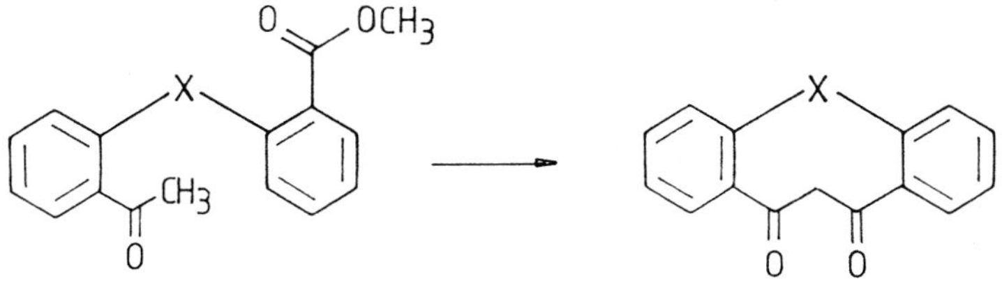

(54)   (55)

inversion barriers of 42-62 kJ mol^{-1} (D. Hellwinkel, P. Illeman, and S. Bohnet, Z. Naturforsch, 1985, 40B, 858).

454

(b) Compounds containing two sulphur atoms

(i) 1,4-Dithiocin derivatives

Attempts to prepare 1,4-dithiocin (56) by a route analogous to that used for the corresponding nitrogen and oxygen
compounds, thermolysis (temperatures above 20°C) of the
benzene bisepisulphide (57) gave only sulphur and benzene
(E. Vogel, E. Schmidbauer and H-J. Altenbach, Angew. Chem.
intern. Edn., 1974, **13**, 736).

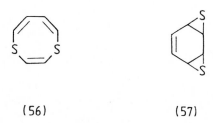

(56)   (57)

The tetrahydroderivative (59, R=H) is obtained when the
dithiol (58) is condensed with *cis*-dichloroethene   but   the
dimesyl ester (59, R=SO_2CH_3) eliminated only one sulphonate
group on treatment with a variety of bases, giving (60,
R=SO_2CH_3), and flash vacuum pyrolysis (FVP) of similar ester
and thio-ester derivatives of (59) gave only tars.   The
acetoxy and trimethylsilyloxy derivatives of the fully unsaturated system (61, R=$COCH_3$ or $Si(CH_3)_3$) are, however,
produced when the alcohol (60, R=H) is oxidised to the
corresponding ketone, converted into its unstable enolate
salt with lithium diisopropylamide (THF, -78°C) and immed-
iately quenched with acetyl (ethanoyl) or trimethylsilyl
chloride (H.J. Eggelte and F. Bickelhaupt, Angew. Chem.
intern. Edn., 1974, **13**, 345; Tetrahedron, 1977, **33**, 2151).
Comparison of the ^{1}H-nmr spectrum of 6-acetoxydithiocin
(61, R=$COCH_3$) with that of its dihydroderivative (60,
R=$COCH_3$) shows no change in shifts or coupling constants
attributable to a diamagnetic ring current. The conclusion
that 1,4-dithiocin has no aromatic character is supported
by its uv-spectrum and confirmed by observation of a far-
from-planar ring by X-ray crystallography (H.J. Eggelte,
F. Bickelhaupt, and B.O. Loopstra, Tetrahedron, 1978, **34**,

3631). This result fulfils the prediction from Huckel M.O.

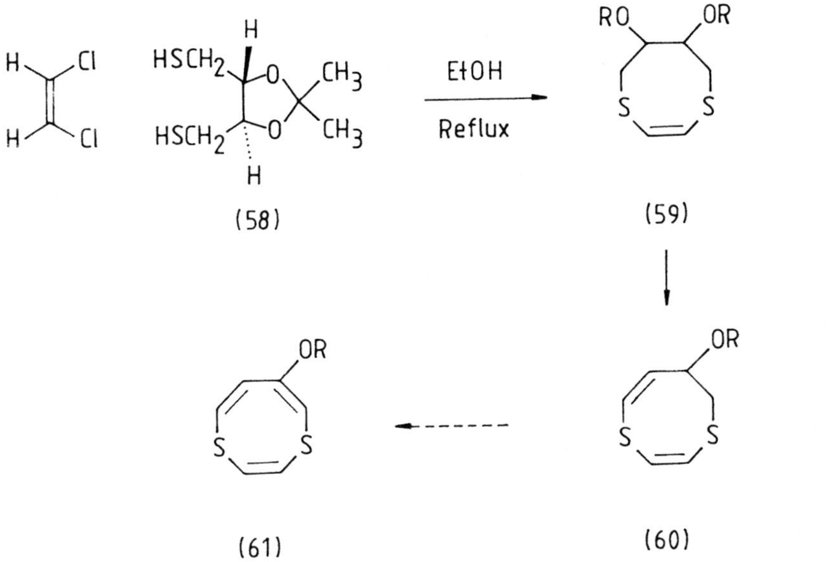

(58) (59)

(61) (60)

calculations (B.A. Hess and L.J. Schaad, J. Amer. Chem. Soc., 1973, 95, 3907) that the planar 1,4-dithiocin system would have no resonance energy to set against the strain associated with sulphur widening its valence angles, which are 118°C at S1 and 104° at S4 in the acetoxy-compound (61, R=COCH$_3$), towards the 135° of a planar regular octagon.

The thermal and photochemical interconversion of the benzo-analogue of 1,4-dithiocin (formally benzo-1,6-dithiocin) and its tricyclic isomer (see Second Edn. Vol. IVK, 396) has been further studied with the dimethyl derivatives (62) and (63).

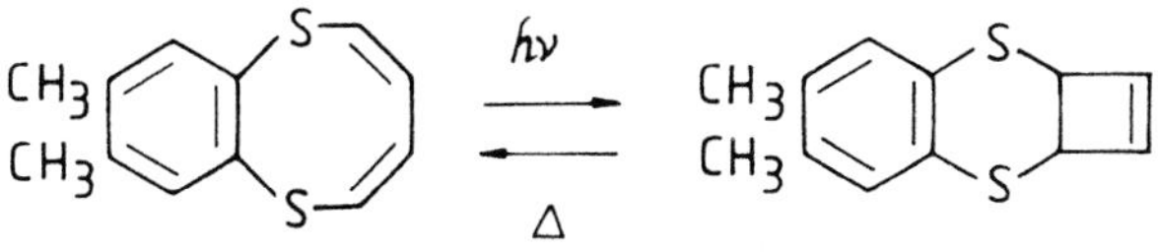

(62) (63)

3. *Eight-membered ring compounds containing nitrogen*

(a) *With rings containing one nitrogen atom*

(i) *Azocine and benzazocine derivatives*

1,2-Dihydroazocines e.g. (66, R^1=C_6H_5, R^3=CN, R^4=H) are obtained when 1,2-dihydropyridines (64) react with dimethyl acetylenedicarboxylate; an intermediate of type (65) has been detected by nmr-spectroscopy. The dihydro-azocine (66, R^1=C_6H_5, R^3=R^4=H) is resistant to *N*-bromosuccinimide, bromine, diazomethane, ammonia, and iodomethane but, like examples with other substituents, undergoes hydrogenation over palladium at the 3,4-double bond while that conjugated with the ester carbonyl group at the 5,6-position is saturated by sodium borohydride, giving the tetrahydroazocines (67) and (68) respectively. (R.M. Acheson, G. Paglietti, and P.A. Tasker, J. Chem. Soc. Perkin I, 1974, 2496). A similar reaction with 1-alkyl-1,4-dihydroquinolines (69) gives benzazocine derivatives of type (70) *via* isolable cyclobutene intermediates (D.G. Lehman, Tetrahedron Letters, 1972, 4863).

| (64) | (65) | (66) |

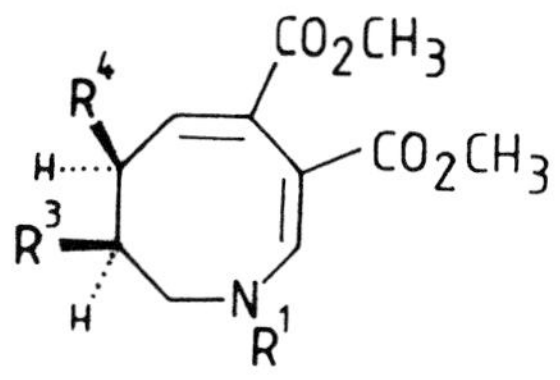

(67)

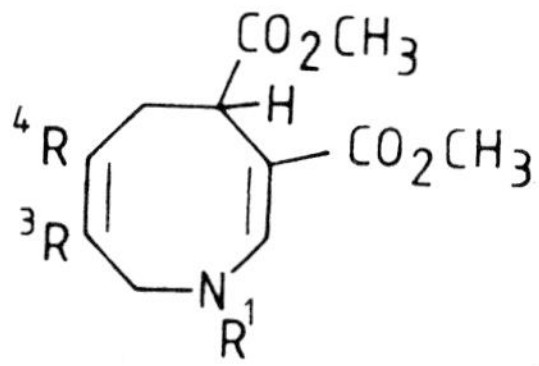

(68)

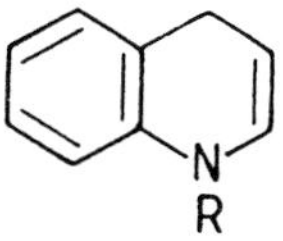

(69)

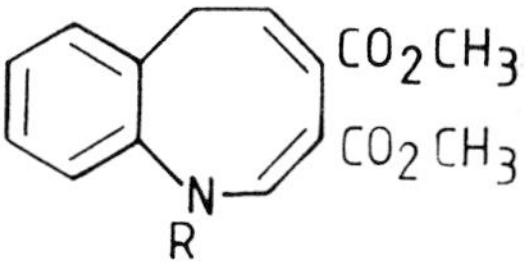

(70)

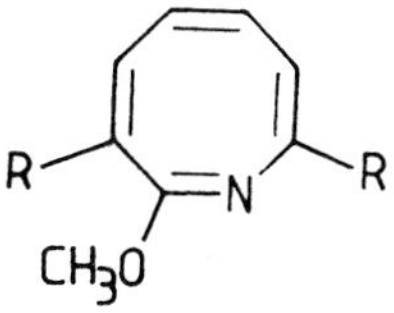

(71)

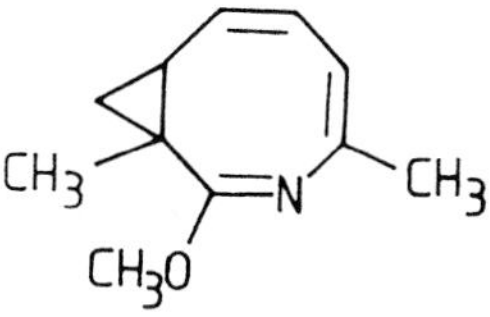

(72)

Treatment of the aromatic dianions obtained by reducing diakylmethoxyazocines e.g. (71, R=CH$_3$) with alkali metals (see Second Edn. Vol. IVK, p.401) with dichloromethane gives a mixture of isomeric cyclopropyl derivatives, principally (72). Similarly cyclopropanes (73) and (74) are formed from the corresponding isomers of benzomethoxyazocine. Reduction of these cyclopropyl derivatives with potassium in liquid ammonia gives highly coloured solutions from which

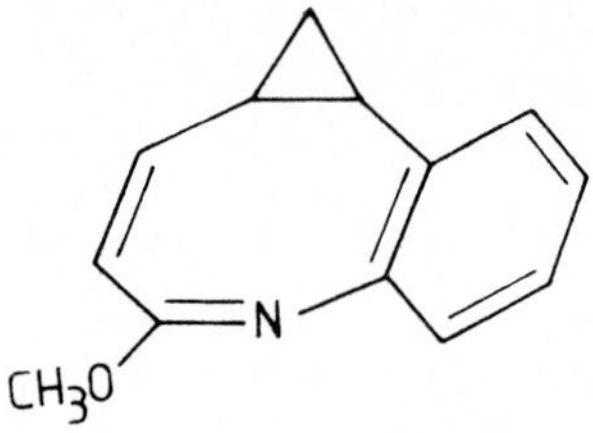

(73)

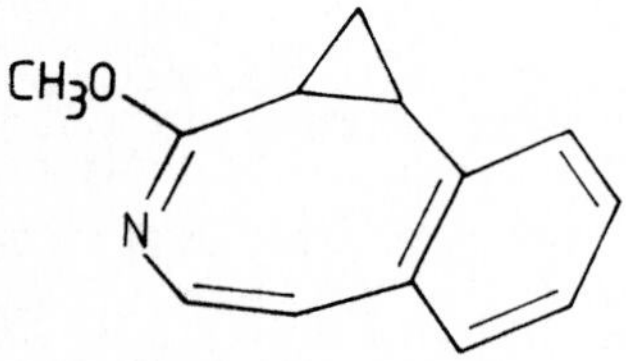

(74)

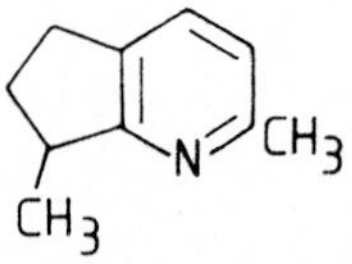

(75)

pyridines, such as (75) formed from (72), are isolated after quenching with methanol. Along with polarographic studies it is concluded that homoaromatic dianions, e.g. (76) from (72), have been formed, as with the carbocyclic analogues, but that their stabilities are not enhanced by aza-substitution (L.A. Paquette, *et al.* J. org. Chem., 1978, <u>43</u>, 4712). Thermal rearrangement of the bicyclic cyclopropane derivatives, e.g. (72), gives dihydropyridines, e.g. (77); activation parameters indicate that the mechanism is of the vinylcyclopropane rearrangement type (G.D. Ewing, S.U. Lay, and C.A. Paquette, J. Amer. chem. Soc., 1978, <u>100</u>, 2909).

Contrary to earlier interpretations the temperature dependence of the ^{1}H-nmr-spectra of some 1-alkyl-1,2,3,4-

tetrahydro-2-oxo-3-azocine carboxylic acids is due to

(76)   (77)

population changes between rapidly equilibrating rotamers
and not to dynamic nmr-effects (F.A.L. Anet, Tetrahedron
Letters, 1980, 21, 2133.

Substituted tetrahydrodibenz[c,e]azocines are available
by electrochemical oxidative cyclisation of benzylphenyl-
ethylamines with sodium perchlorate; benzylphenylacetamides
give the corresponding oxo-derivatives (M. Sainsbury and
J. Wyatt, J. chem. Soc., Perkin 1, 1976, 661; also Supplement to the Second Edn. Vol. IIIF Chapter 20 p.309 and 310).
Formation of *N*-aryl tetrahydrodibenz[b,g]azocin-4,6-diones
has been noted in Section 2 (a) (ii) of this Chapter.

*(ℓ) Eight-membered ring compounds containing two nitrogen
     atoms*

(i) 1,2-Diazocines

Further investigation of the formation of 1,2-diazocine
(79) by photolysis of the azo-compound (78) (see Second
Edn., Vol IVK, p.411) reveals that dinitrogen extrusion
products i.e. benzene (80), prismane (81) Dewar benzene
(82) and benzvalene (83) which predominate at room temperature, arise from the first singlet state of (78). Lower
temperatures and the presence of oxygen or, better, acetophenone, favour formation of the triplet state of the azo-

compound (78) leading to yields up to 67% of 1,2-diazocine (79) (N.J. Turro, *et al.*, J. Amer. chem. Soc., 1976, 98, 4320).

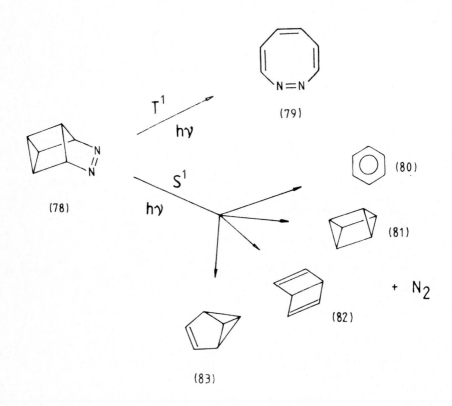

(79)

(78)

(80)

(81)

(82)

+ N₂

(83)

Thermochemical calculations show that conversion of the *cis*-hexahydrodiazocine (84, R=H) into the *trans*-isomer (85) is thermoneutral (C.G. Overberger *et al.*, Tetrahedron Letters, 1972, 4565). The four configurational isomers of the diphenyl derivative (84 and 85, R=C₆H₅) have been prepared and their photochemical interconversion studied, correcting a previous assignment (G. Vitt, E. Hadicke, and G. Quinkert, Ber., 1976, 109, 518).

(84)   (85)

(ii) 1,4-Diazocines

Derivatives of 1,4-dihydro-1,4-diazocine (88) are obtained by thermolysis in acetone of *syn*-benzene diimines (87) which are obtained in four steps from their oxygen analogues (86). The parent compound (88, R=H) is formed when the dimesyl derivative (88, R=SO$_2$CH$_3$) is treated with potassium in liquid ammonia, followed by methanol, and forms colourless air-sensitive crystals m.p. 196-198°C. This system is a potentially aromatic ten π-electron cycle (see General Considerations, Section (b)) and in the case of the unsubstituted compound and its derivatives with the electron donating substituents methyl or trimethylsilyl on nitrogen (88, R=H, CH$_3$ or Si(CH$_3$)$_3$) ^{1}H- and ^{13}C-nmr data are consistent with the presence of a diamagnetic ring current and therefore a delocalised aromatic π-electron system. In contrast derivatives with electron-withdrawing groups (88, R=SO$_2$CH$_3$ or CO$_2$CH$_3$) do not show the ring current effect and x-ray crystallography reveals non-planar geometry and pronounced alternation in bond lengths compared with the parent compound (88, R=H) which has a planar conformation (E. Vogel *et al.*, Angew. Chem. intern. Edn., 17, 1979, 962; H. Prinzbach *et al.*, *ibid*, 1979, 964).

The aldehyde derivative (89, R=CHO) of 6,11-diphenyl-dibenzo[b,f][1,4]diazocine, a chiral system because of the tub-like conformation of the heterocyclic ring, can be resolved as its oxazolidine derivative. The aldehyde (89, R=CHO) recovered after chromatography on silica gel, m.p. 184-186°C, showed [α]$_D$ + 277° (CHCl$_3$, C 1·38) and considerable thermal stability towards racemisation,

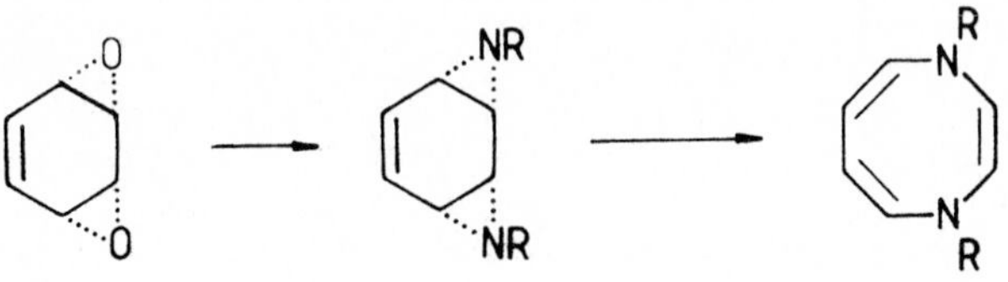

(86)   (87)   (88)

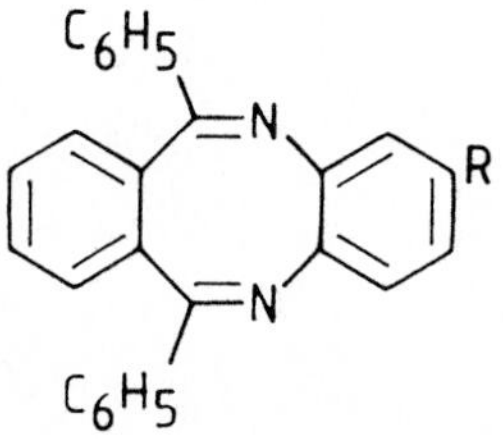

(89)

the first-order rate constant being $1 \cdot 14.10^{-5} s^{-1}$ and ΔG 37 kcal mol^{-1} at 175°C. (D. Olliero *et al.*, Chem. Comm. 1976, 276). Further studies of the synthesis and racemisation of derivatives of this compound have been described (J-M Ruxer, G. Solladie, and S. Candau, J. chem. Research (S), 1978, 408, and references there cited).

(iii) 1,5-Diazocines

A convenient two step synthesis starting from isocyanates of type (90) and aminoketones (91) gives access to unsymmetrically substituted 6,12-diaryl[b,g]-[1,5]diazocines (92), which show various types of biological activity, whereas previous routes lead only to symmetric substitution (H.K. Patney, Synthesis, 1986, 326, and references there cited).
Diazocine analogues (93) of the diazepine 'super-drugs' have been found to have antidepressant, tranquilising,

sedative, muscle relaxant, antiinflammatory and diuretic activity (Netherlands Pat., 75 07, 092; Chem. Abs., 1977, 87, 85067).

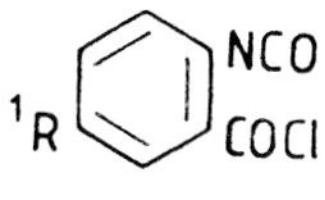

(90)

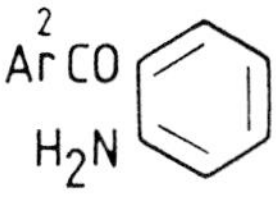

(91)

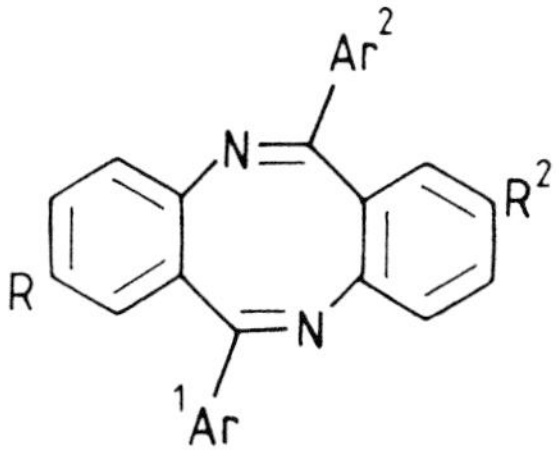

(92)

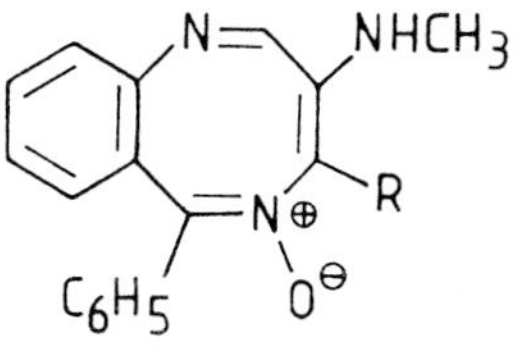

(93)

A derivative of 1,5-diazocin-2(1H) -one (98) is obtained when the condensation product (95) of the azabutadiene (94) and dichlorodiphenyl-silane is treated with an ester of acetylenedicarboxylic acid. Ring expansion to the silicon-nitrogen heterocycle (96) followed by rearrangement in which ester carbonyl replaces silicon in the ring is thought to be the reaction pathway. Ring contraction to pyridines (97, $R^1=R^2=C_6H_5$; $R^3=CH_3$ or H) accompanies de-silylation with acid, or with potassium hydroxide followed by acidification (J. Barluenga et al., Angew. Chem. intern. Edn., 1986, 25, 181).

The interesting cage diamine 1,5-diazabicyclo[3,3,3]-undecane is considered with homologous systems in the following Chapter.

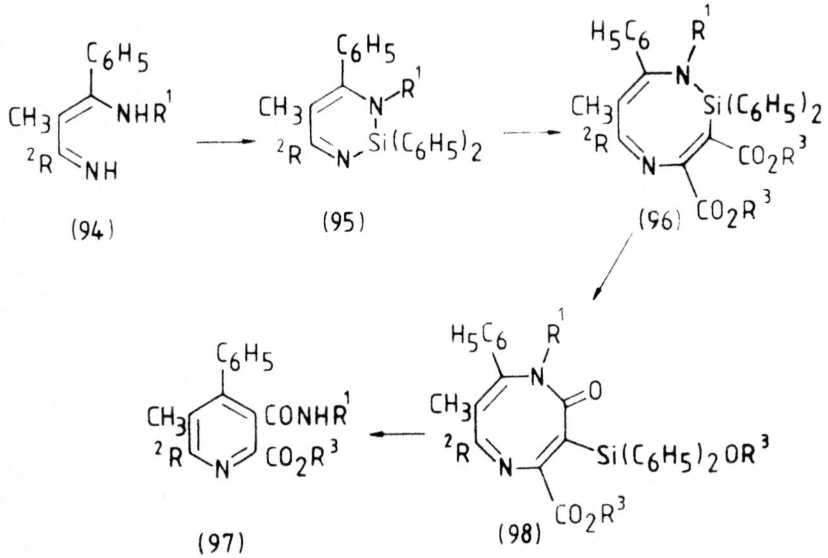

(94)　　　　　　　(95)　　　　　　　(96)

(97)　　　　　　　(98)

(iv) Tetrazocines

The products of self-condensation of the monohydrazones of the corresponding α-dicarbonyl compounds, previously reported to be 1,2,5,6-tetrazocines (99, R=CH$_3$ or C$_6$H$_5$) (see Second Edn. Vol. IVK Chapter 55 p. 424 and 425), are the mesomeric betaines (100, R=CH$_3$ or C$_6$H$_5$). Tetrazocines (99) are, however, considered to be likely intermediates in formation of the bicyclic products (100) (C.A. Ramsden, Tetrahedron, 1977, 33, 3220 and references therein). 2,6-Diethoxy-4,8-dimethyl-1,3,5,7-tetrazocine (102, R=CH$_3$), colourless crystals m.p. 26°C, is thus the first example of a tetraazacyclooctatetraene; it is obtained when the bicyclic precursor (101) is oxidised with *t*-butyl hypochlorite and potassium *t*-butoxide. ^{13}C- and ^{15}N-n.m.r. spectra show the presence of only two types of carbon and one type of nitrogen in the ring and the [1]H-nmr spectrum confirms two pairs of equivalent substituents.

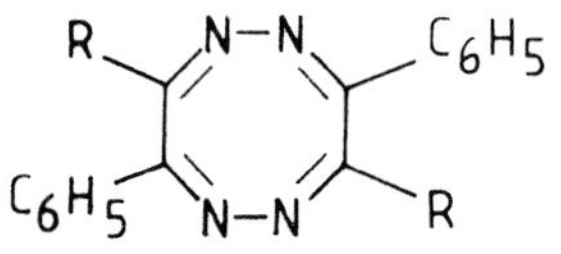

(99)

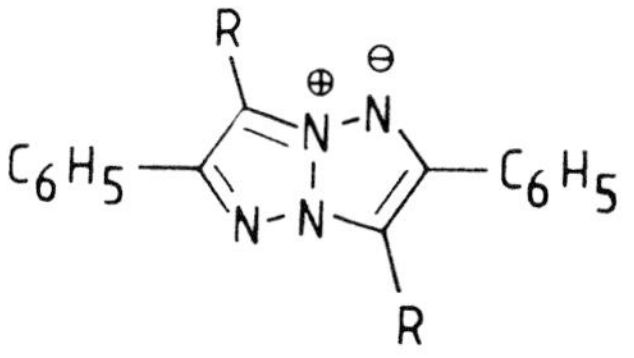

(100)

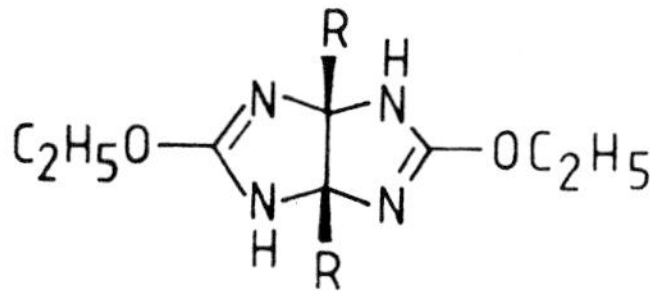

(101)

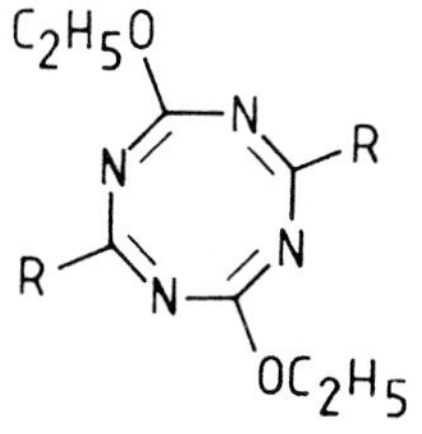

(102)

X-ray crystallography of the diphenyl analogue (102, R=C_6H_5) , similarly prepared, shows it to have a boat conformation in the crystalline state. The diethoxy-compound (102, R=CH_3) is converted into its bisdimethylamino analogue (102, N(CH_3)$_2$ replaces OC_2H_5) by heating it with dimethylamine and silica gel hydrolyses one of the ethoxy-imino groups to -CO-NH- without breaking the ring (R. Gomper and M.L. Schwartzensteiner, Angew. Chem. intern. Edn., 1983, 22, 543).

Treatment of hexamethylenetetramine (hexamine) with acetyl (ethanoyl) or trichloroacetyl chloride followed by dinitrogen tetroxide gives the tetrazocane derivatives (103, R=CH_3 or CCl_3) in a sequence similar to the formation of the explosive HMX (Y. Yoshida, G. Sen, and B.S. Thiagaragen,

J. heterocyclic Chem., 1973, 725; see also Second Edn. Vol. IVK Chapter 55 p. 425).

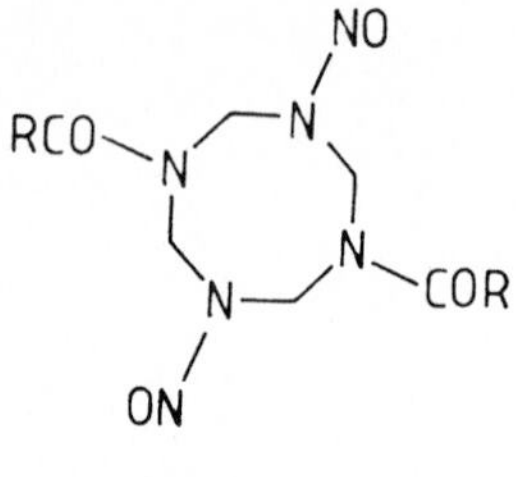

(103)

4. *Eight-membered ring compounds containing nitrogen and another heterocyclic element.*

(a) *Rings containing nitrogen and oxygen*

(i) *Oxazocines*

Thermolysis of the labile *cis*-tricyclic intermediate (104, R=SO$_2$C$_6$H$_4$CH$_3$) gives the corresponding 4H-1,4-oxazocine (105, R=SO$_2$C$_6$H$_4$CH$_3$), a route similar to that used to prepare the analogous diaza- (Section 3(b)(ii)) and dioxa-compounds (Section 1(b)(ii)). The parent compound (105, R=H), obtained (like several derivatives) from the *N*-tosyl product by

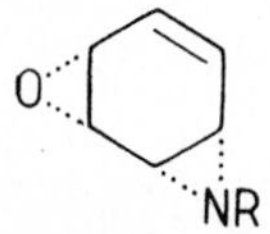

(104) 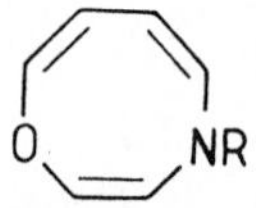   (105)

acidification (etc.) of the anion (105, negative charge replaces R with K$^+$ counter ion) formed upon treatment with potassium in liquid ammonia, crystallises from ether at -20°C in unstable yellow deliquescent platelets, m.p. 32°C. Like 1,4-dihydro-1,4-diazocine but in contrast to 1,4-diox-ocine, 4H-1,4-oxazocine (105, R=H) and its derivatives with electron donating substituents on nitrogen have ^{1}H and

[13]C-nmr-spectra indicative of a diamagnetic ring current, and therefore aromatic character, while those with electron withdrawing groups do not. The X-ray crystal structure of the trimethoxybenzyl derivative (105, $R=CH_2C_6H_2(OCH_3)_3$) confirms the planar conformation of the eight-membered ring and shows reduced bond length alternation compared with the non-planar tosyl derivative (105, $R=SO_2C_6H_4CH_3$) (H. Prinzbach, *et al* Angew Chem. intern. Edn., 1984, 23, 309).

(ii) Dioxadiazocines

(106)   (107)

1,4,6,7-Dioxadiazocine (107, $R^1=R^2=R^3=H$), m.p. 50-52°C, and its methyl derivatives, are formed in variable yields up to 37% when pyridazine di-*N*-oxides (106) are irradiated (high-pressure Hg-lamp, Pyrex filter) in dichloromethane. Tricyclic intermediates (108) are postulated. The ring protons exhibit an n.m.r. signal at 5·75δ for the dimethyl derivative (107, $R^1=R^3=CH_3$, $R^2=H$) and the absence of uv-absorption at wavelengths greater than 220nm show that the system is not aromatic (H. Arai, *et al* Chem. Comm. 1977, 133).

(108)

(b) Rings containing nitrogen and sulphur

(i) Dithiatetrazocines and related systems

Reaction between benzamidine and sulpur dichloride in
the presence of diazabicycloundecene (DBU) gives a low yield
(5-10%) of 3,7-diphenyl-1,5-ditha-2,4,6,8-tetrazocine (109,
$R=C_6H_5$) as shiny bright yellow plates m.p. 225-226°C. The
compound decomposes only slowly at 240°C forming benzonitrile and is not basic. X-ray crystallography shows that
the ring is perfectly planar with all the S-N bonds being

(109)   (110)

of equal length (1·56 A) and all the C-N bonds of equal
length (1·32 A). Taken with its characteristically struc-
tured uv-spectrum, these data suggest that the diphenyl
compound (109, $R=C_6H_5$) has a delocalised aromatic system
of, presumably, ten π-electrons, two of these being provided
by sulphur.

The unsubstituted ring system (109, R=H) could not be
obtained by the corresponding reaction of formamidine, but
the bisdimethylamino derivative (109, $R=N(CH_3)_2$) is formed
in 54% yield using *NN*-dimethylguanidine. This compound shows
quite different uv-absorption, at shorter wave-length, from
the diphenyl compound (109, $R=C_6H_5$) and x-ray crystallography reveals that the molecule (110) is not planar but
folded about the axis of the two sulphur atoms, the two
halves each forming planes, including the dimethylamino
grups, which intersect at 101°. All the S-N bonds are the
same length, as are all the C-N bonds including those to
the dimethylamino groups. The sulphur atoms are separated
by only 2·43 A, compared with 3·79 A in the diphenyl analogue (109, $R=C_6H_5$) and with 3·6 A for two van der Waals

radii of sulphur indicating a bonding interaction, so that
the structure resembles that of tetrasulphurtetranitride.
The authors suggest an unusual delocalised electronic system, represented by (110) (R.B. Woodward *et al* J. Amer.
chem. Soc. 1981, 103, 1540).

Treatment of the bis-dimethylamino compound (110) with
chlorine in acetonitrile gives the cationic S-chloro-deriv-
ative (111) with Cl_3^- as counter ion; x-ray crystallography
shows that the S-Cl···S bridge is not symmetric (R.T. Oakley
et al Chem. Comm. 1985, 655).

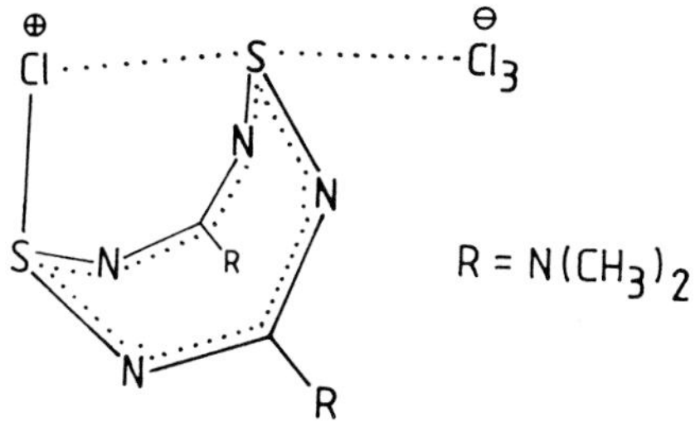

$$R = N(CH_3)_2$$

(111)

A compound having the related bridged system (112) is
formed as golden air-stable plates when tristrimethylsilyl-
benzamidine reacts with 1,3,5-trichloro-1,3,5,2,4,6-trithia-
triazine in dichloromethane at 0°C (R.T. Boere, A.W. Cordes,
and R.T. Oakley, Chem. Comm. 1985, 929).

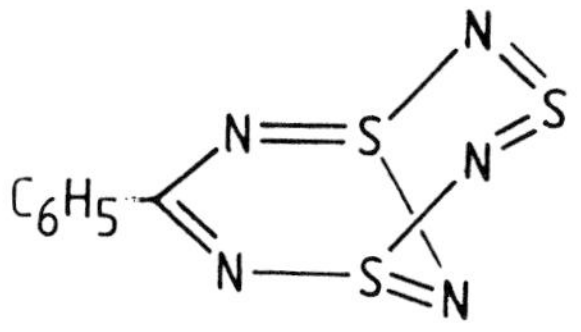

(112)

(c) Rings containing nitrogen and phosphorus

(i) Cyclophosphazenes substituted by organic groups

Detailed consideration of the cyclo-phosphazene system containing only nitrogen and phosphorus atoms in the ring is outside the scope of this work, save to note that a range of organic substituted derivatives exemplified by (113) have been prepared. Some are used as fire-retarding agents (M. Biddlestone *et al*, Phosphorus, 1973, **3**, 179).

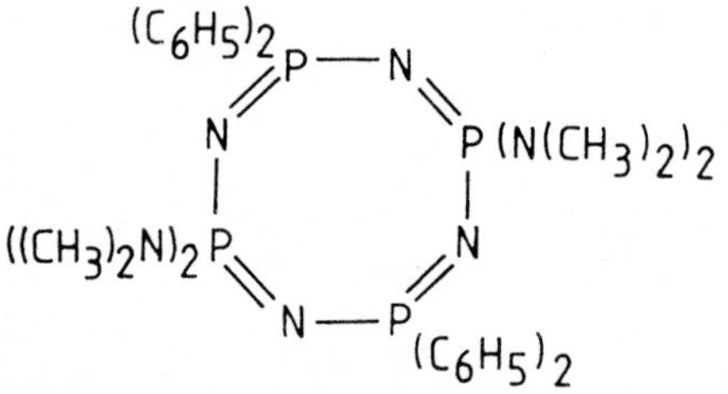

(113)

Sequential reaction of two molecules of dilithioruthenocene with octafluorocyclo-phosphazene gives the doubly bridged system (114). The structure was deduced by nmr-spect-

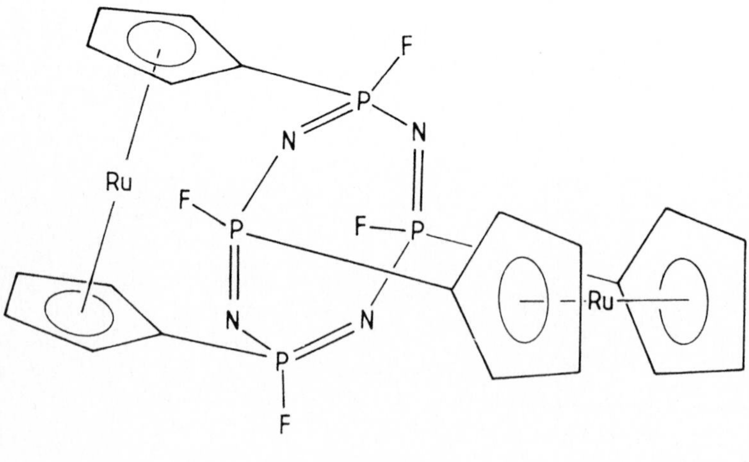

(114)

rometry and confirmed by x-ray crystallography. Only one bridging metallocene group could be affixed in an analogous experiment with ferrocene (K.D. Lavin *et al*, Chem. Comm., 1986, 117).

Cyclic metallocene derivatives such as (114) constitute a distinct class of large ring heterocyclic compound in which the cycle is completed through metal atoms π-bonded to unsaturated groups. A great variety of such compounds have been prepared in the last decade or so, but since their chemistry is more closely associated with that of organometallic compounds these are not further considered in this or the following Chapter.

(ii) Phosphatranes

Treatment of phosphatranes of type (115), in which the nitrogen atom is planar, with strongly electrophilic reagents gives derivatives of phosphorus in the hypervalent 10-P-5 state with trigonal bipyramidal geometry, such as (116) using boron trifluoride etherate. This compound dissociated upon attempted isolation but was detected in solution by changes in the nmr-spectrum upon addition of the reagent, notably development of coupling between phosphorus and the protons adjacent to nitrogen with J^3(PNCH) 3·1Hz (L.E. Carpenter II, and G.J. Verkade, J. Amer. Chem. Soc., 1985, 107, 7084 and references there cited).

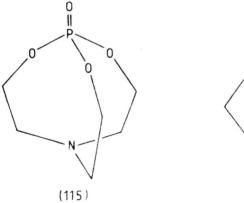

(115)

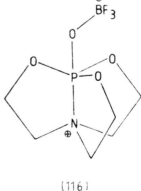

(116)

5. *Eight-membered ring compounds containing phosphorus*

(a) *Phosphocanium Salts*

The phosphocanium bromide (117) has been prepared and shown to undergo inversion at phosphorus during alkaline hydrolysis (K.L. Marsi and F.B. Burns, Phosphorus, 1974, **4**, 211).

(117)

Chapter 56

COMPOUNDS WITH LARGER HETEROCYCLIC RINGS:
COMPOUNDS WITH RINGS OF MORE THAN EIGHT ATOMS

J.A.H. MacBRIDE

Introduction

The organisation of this Chapter differs from that of Chapter 55, and to a lesser extent from that of the corresponding one in the Second Edition, where compounds are divided into sections by ring size and subdivided according to the heteroatoms present, and their numbers. The majority of recent studies of larger-ring heterocyclic systems described here are concerned with the variation of properties brought about by changing the number of atoms, or the nature of the heteroatoms, in otherwise similar rings so that the classification which has been adopted here according to structural type is more appropriate.

This is particularly true in the specialised areas of crown ethers and cryptands, cage amines, and macrolides where numerous papers disclosing major advances have appeared since publication of the Second Edition, and which now constitute the bulk of the material of this Chapter. A comprehensive review of macrocyclic complexing agents would occupy several large volumes. This topic has therefore been divided into areas of interest such as Synthesis, Ion Transport etc., and each is treated by means of a summary of the review literature with a few selected, illustrative examples.

Compounds of general interest which have not been included in the specialised sections are classified according to their heteroatoms, not ring size, in the last Section of this Chapter.

1. Crown Ethers, Cryptands, and Related Compounds.

(a) General Considerations

The range of structural types of macrocyclic complexing agent has expanded from the original crown ether (chorand) series (*e.g.* 1) to include bicyclic systems (cryptands), usually with nitrogen atoms at the teriary ring junction positions as in (2), and those with acyclic side-chains which can take part in complexing (*e.g.* 3) known as lariat ethers.

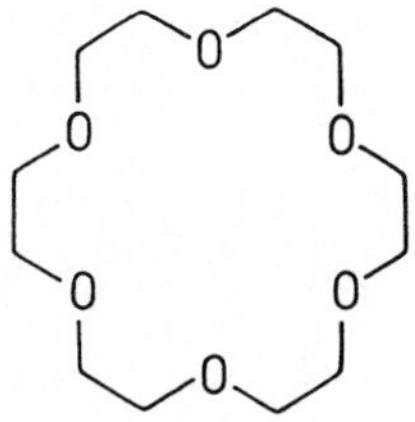

(1)

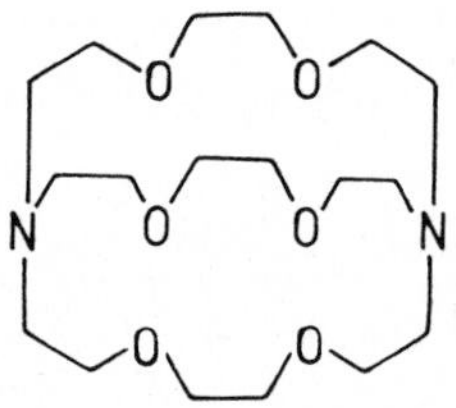

(2)

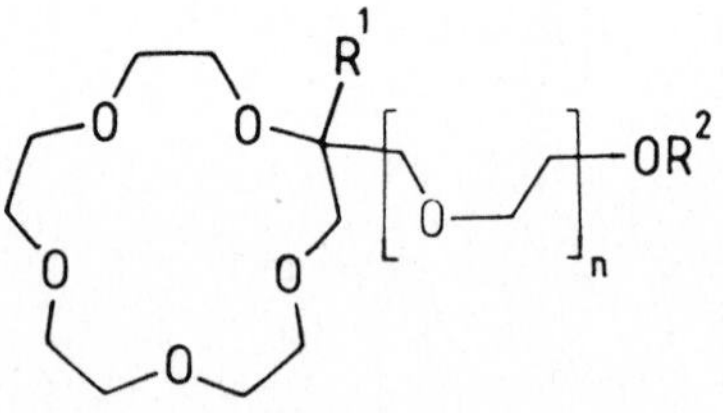

(3)

Incorporation of aromatic and heteroaromatic groups,

multiple bridging, and a variety of further groups, leads to cavitands, spherands, hemispherands and calixarenes which differ according to the size of their cavities in relation to that of the complexed species and the degree to which the size of the cavity can vary by conformational change.

The various classes of host molecule have been reviewed with particular reference to pre-organisation, the degree of conformational change which the ligand must undergo before complex formation, which has an important effect on the rates of formation and stabilities of complexes since such changes contribute an adverse entropy term to the free energy of complexation (D.J. Cram, Angew. Chem. intern. Edn., 1986, 25, 1039).

Parallel with diversification of structural type has come a rapid expansion of the variety of properties and applications of these compounds, although the molecular inclusion or 'host-guest' character of their interactions pervades the work. A 'Journal of Inclusion Phenomena' (1983; ed. J.L. Atwood and J.E.D. Davies published by Reidel, Dordrecht, Holland) has been inaugurated.

Reviews which do not specialise in particular aspects of crown ether and cryptand chemistry are:

R.M. Izatt and J.J. Christensen, (eds.) Synthetic Multi-dentate Macrocyclic Compounds, Academic Press, New York, 1978; Progress in Macrocyclic Chemistry, Wiley, New York, 1979, Vol. 1; J.F. Stoddart, Chem. Soc. Rev., 1979, 8, 85; J.M. Lehn, Acc. chem. Res., 1978, 11, 49; Pure appl. Chem., 1979, 51, 979; 1980, 52, 2303, and 2441; S. Patai, (ed.). The Chemistry of Ethers, Crown Ethers Hydroxyl Groups and their Sulphur Analogues, Part 1, Wiley, Chichester, 1980; J.S. Bradshaw and P.E. Stott, Tetrahedron, 1980, 36, 461; M. Hivaoka, Crown Compounds, Elsevier, New York, 1982; J.E.D. Davies, et al. J. incl. Phenom., 1983, 1, 3; J.F. Stoddart, Royal Society of Chemistry Ann. Rep., 1983, 80B, 353; E. Weber and F. Voegtle, (eds.), Host Guest Complex Chemistry. Macrocycles, Synthesis, Structures, Applications, Springer, Berlin, 1985, Chapter 1.

Nomenclature of macrocyclic complexing agents has been discussed particularly by W. Liebscher, et al. (Z. Chem., 1985, 25, 16, and by J.-M. Lehn (Structure and Bonding, 1973, 16, 1).

In the following pages the well established m-crown-n notation, where m is the number of atoms in the ring and n is the number of oxygen or other donor atoms, so that (1) is 18-crown-6, is used for crown ethers and their aza and thia analogues. Bicyclic systems are defined as [1,m,n]-cryptands where 1, m and n indicate the number of donor atoms in each bridge between the tertiary (usually N) centres; the donor atoms (usually 0) are almost invariably separated by two methylene groups, as noted in the following Section; thus (2) is [2,2,2]-cryptand.

(b) Synthesis

Crown ethers are commonly prepared, using the Williamson ether synthesis, by condensation of a suitable ethylene glycol oligomer, usually as its di-alkali metal alkoxide (4), with a dihalide, or ditosylate (5), also readily obtainable from a similar poly-ether diol. A remarkable

(4)   (5)   (6)

Hal, OSO_2Ar, etc.

feature of some of these reactions is that high dilution conditions are not needed to give a satisfactory yield of cyclic product rather than a linear oligomeric product. The yield of the crown compound is frequently very dependent upon the choice of alkali metal counter ion and this has been attributed to a 'template effect' in which the final cyclisation step is assisted by cyclic complexation of the cation, which needs to be of appropriate size. Thus in the analogous synthesis of the tetrathia crown (9) the yield is only 7.5% when the potassium dithiolate (7, M=K) is used but 80% with the caesium salt (7, M=Cs) (J. Buter and

Br

—S⊖M⊕

+

(7)

S—

S—

Br

(8)

S   S

S   S

(9)

R.M. Kellogg, J. org. Chem., 1981, <u>46</u>, 4481).

Crown ethers are also formed by the boron trifluoride catalysed cyclo-oligomerisation of ethylene oxide; a template effect is also evident in this reaction since the proportions of 12-crown-4, 15-crown-5 and 18-crown-6 are strongly dependent on which metal tetrafluoroborate is suspended in the dioxan medium. Thus with $LiBF_4$ the yields are 30, 70, 0% respectively, $RbBF_4$ and $CsBF_4$ give only 18-crown-6, and $Cu(BF_4)_4$ and $Zn(BF_4)_2$ form 15-crown-5 in 90% yield with 5% of each of the other oligomers (J. Dale and K. Daasvatn, Chem. Comm., 1976, 295). For discussion of the template effect see: C.J. Pedersen in Synthetic Macrocyclic Compounds, *loc. cit.*, Chapter 1; D.W. Busch, Acc. chem. Res., 1978, <u>11</u>, 392; M.D.S. Healy and A.J. Rest, Adv. inorg. Chem. Radiochem., 1978, <u>21</u>, 1; Stoddart, 1979, *loc. cit.* and references therein.

It has been pointed out that the apparent operation of a template effect can be misleading. Acid catalysed condensation of furan with acetone gives a higher yield of the macrocycle (10) in the presence of salts such as lithium lithium perchlorate (M. Chastrette and F. Chastrette, Chem. Comm., 1973, 534; with J. Sabadie, Org. Synth., 1977, <u>57</u>, 74). This result is due to increased effective acidity of the medium associated with hydration of the metal ions, rather than the template effect as initially supposed (M.D.S. Healy and A.J. Rest, Chem. Comm., 1981, 149).

(10)

In the Williamson synthesis of crown ethers the possible template effect must be weighed against the base strength of the reagent or the catalyst which is affected by the cation as well as the anion (B.R. Bowsher and A.J. Rest, J. chem. Soc. Dalton, 1981, 1157; Inorg. chem. Acta, 1981, 53, L175). The nucleophilic power of the alkoxide anion will also be affected by association with metal ions, a further factor to complicate recognition of the template effect in experiments in which the metal is varied.

Incorporation of nitrogen atoms to form aza crown or cryptand systems is generally achieved by methods similar to those used for the all-oxygen analogues (G.W. Gokel, *et al.*, Synthesis, 1982, 997). The necessary amine precursors such as (5, X=NH$_2$) are obtainable from halides such as (5, X=Cl, m=2), which is commercially available, by standard nucleophilic substitution methods with, for example, ammonia, potassium phthalimide then hydrazine, or sodium azide then lithium aluminium hydride. The additional complication of a third, possibly unwanted, alkylation on nitrogen may necessitate the use and removal of *N*-protecting groups, although the diaza crown (13) is obtained in 35% yield from the secondary amino-diol (11) and dichloride (12), compared with 66% for the all-oxygen analogue. Dilute conditions are maintained in these preparations by simultaneous dropwise addition of a 1M solution (100 cm.^{3}) to a stirred

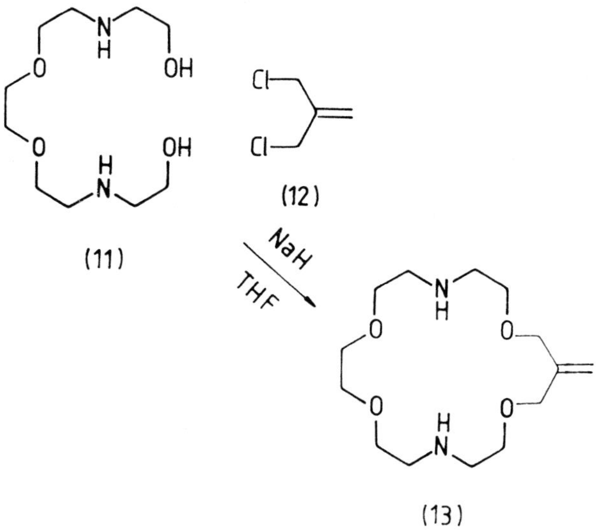

(11) (12)

NaH
THF

(13)

suspension of sodium hydride in tetrahydrofuran (300 cm.^{3}) (M. Tomoi, *et al.*, Tetrahedron Letters, 1978, 3031). In contrast, synthesis of the dibenzo-compound (16), Scheme 1, needs *N*-benzyl protecting groups but the more rigid *o*-di-substituted benzene precursor facilitates efficient ring formation (C.J. Pedersen and M.H. Bromels, USP, 3,847,949 (1974); Chem. Abs., 1975, 82, 73049).

A single step synthesis of [2,2,2]-cryptand (2) in 21% yield requires the mixed dihalide (18) and sodium carbonate in acetonitrile; the product is isolated as its sodium iodide complex and since no cryptand is obtained using potassium carbonate a template effect is probably involved (S. Kutstad and L.A. Malmsten, Tetrahedron Letters, 1980, 21, 643).

Perfluoro-derivatives of the crown ethers are formed by direct fluorination of the solid precursor mixed with sodium fluoride (to increase surface area and absorb liberated HF) at -78°C. Perfluoro 18-crown-6 is obtained in 34% yield as

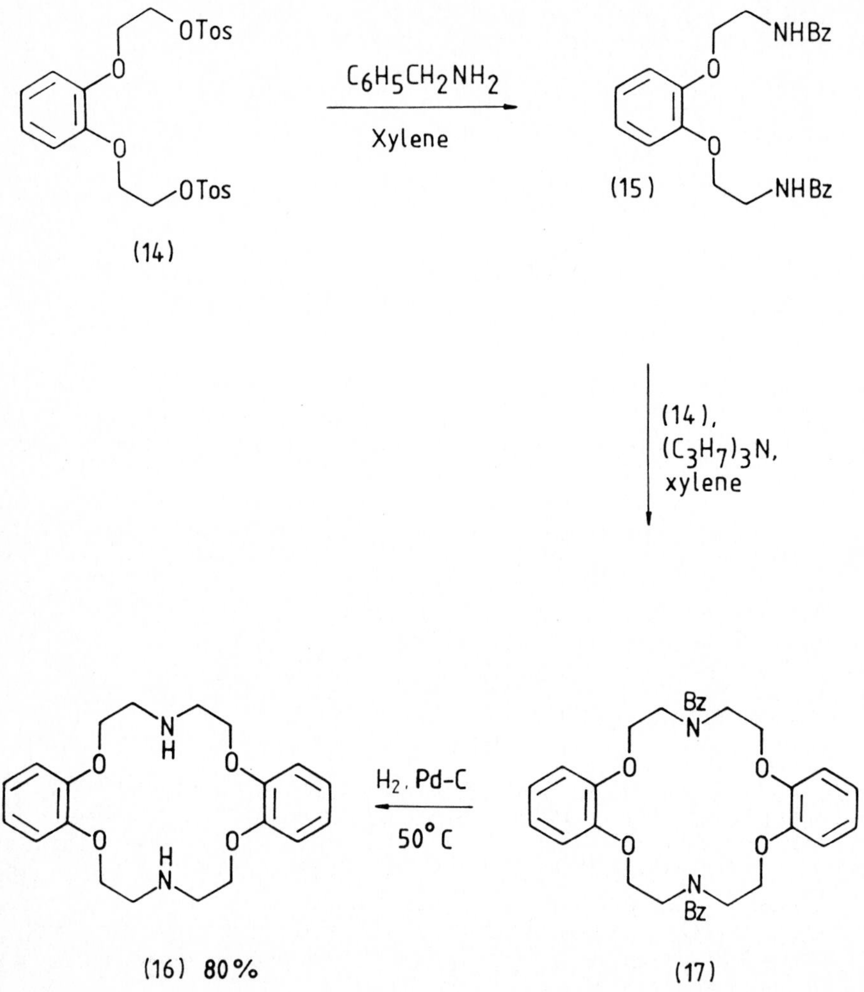

Scheme 1

colourless volatile crystals m.p. 34°C. Its X-ray crystal
structure shows that the ring is puckered so that the oxygen
atoms project towards a potential metal coordination site on
one side of the molecule. Although the oxygen atoms are

H₂N ... O ... O ... NH₂

$$\xrightarrow[\text{CH}_3\text{CN}]{\text{Na}_2\text{CO}_3} \quad (2)$$

2 Cl ... O ... O ... I

(18)

likely to have reduced donor character, the unique solvent and solubility properties of perfluoroalkyl compounds may give these materials interesting applications (W.-H. Lin, W.I. Bailey and R.J. Lagow, Chem. Comm., 1985, 1350).

For a comprehensive review of synthetic procedures up to 1980 see: G.W. Gokel and S.H. Korzeniowski, Macrocyclic Polyether Syntheses, Springer Verlag, Berlin 1982; and for synthesis of aza crowns G.W. Gokel, Synthesis, 1982, 997. Also M. Fieser, Reagents for Organic Synthesis, 10, John Wiley, New York, 1982 and other volumes in this series.

(c) Metal complexes

The outstanding features of complexation by the crown ethers, evident since the early experiments in 1967, are firstly that the alkali (and alkaline earth) metal ions are retained within the molecular cavities (few complexing agents were previously available for these metals) and secondly that the stabilities of these complexes are sharply dependent upon the match between the size of the metal ion and that of the molecular cavity.

X-ray diffraction studies show that in the alkali metal complexes of the smaller, more rigid, crown ethers such as potassium 18-crown-6, the oxygen atoms are nearly coplanar; the two trigonal faces of the oxygen array in the latter complex are separated by only 0.38Å (J.D. Dunitz and

P. Seiter, Acta Crystallog., 1974, B30, 2744). Although 30-crown-10 complexes potassium using all ten oxygen atoms in a three dimensional array the complex is far less stable than the planar ones of the smaller crowns (M.A. Bush and M.R. Truter, J. chem. Soc. Perkin II, 1972, 345). These planar alkali metal complexes contrast with the well known types of transition metal complexes where highly directional bonding to the ligands in three dimensions, with octahedral geometry, is a most prominent feature. This difference probably explains why the more rigid crown ethers are relatively ineffective complexing agents for the transition elements and it may be associated with greater ionic character of the alkali metal-oxygen bonds.

Irrespective of the overall geometry, bonding between a metal ion and adjacent donor atoms separated by two methylene groups gives a favourable five membered chelate ring, and the majority of macrocyclic ligands are built up from the synthetically convenient ethylene glycol, or diamine, oligomers.

The potentially very selective formation of highly stable complexes of macrocyclic ligands with cations of particular size arises from the interplay of the free energy of formation of metal-donor bonds against those of conformational distortion of the macrocycle and solvation of the free ion, the last term rising sharply with smaller ionic radii.

Uv photoelectron spectroscopy reveals that interaction between donor atoms of crown ethers and cryptands raises the energies of their HOMOs, which are those of the electron pairs available for complexation, making these compounds 'softer' (more easily polarised) ligands than their acyclic analogues (A.D. Baker, G.H. Armen and S. Fumaro, J. chem. Soc. Dalton, 1983, 2519). Since soft ligands tend to form more stable complexes with soft cations, and *vice versa*, this effect may be involved, for example, in the preference of hard actinide ions for the oxygen sites in water and nitrate ion rather than crown ethers (P.G. Eller and R.A. Penneman, Inorg. Chem., 1976, 15, 2439).

The formation of 'sandwich' complexes in which a metal coordinates two macrocyclic ligands whose cavities are too small to enclose the cation may also involve polarisation effects. Thus the complex of the relatively hard Mg^{++} ion with two molecules of benzo-15-crown-5 is probably stabilised by the larger number of coordination sites of the soft ligand (N.S. Poonia, *et al.*, Inorg. nucl. Chem. Letters, 1977, 13, 19).

The tendency to complex transition rather than alkali metal ions increases as the oxygen atoms of crown ethers are replaced by nitrogen or sulphur. Thus the logarithms of the stability constants (K_s) for the potassium (in methanol) and silver (in water) complexes of 18-crown-6 (19, X=0) are respectively 6.10 and 1.60 but for the diaza-compound (19,

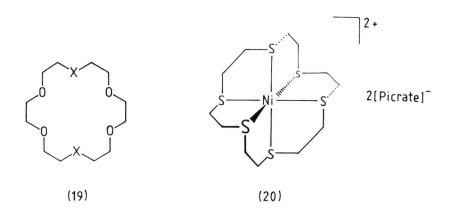

(19)   (20)

X=NH) these values are 2.04 and 7.8 and for the dithia-analogue (19, X=S) 1.15 and 4.35 (H. Frensdorff, J. Amer. chem. Soc., 1971, 93, 600).

The nickel(II) pricrate complex of hexathia-18-crown-6 (20), orange monoclinic prisms, provides an interesting example of three dimensional coordination with a monocyclic ligand. X-ray diffraction studies show that the sulphur atoms are accurately octahedrally disposed, the macrocycle being folded with achiral *meso* geometry. The ligand is a tight fit for the Ni^{++} cation, however, since the Ni-S bonds are compressed by more than 0.05Å compared with similar complexes although the ligand bond lengths are little distorted from those in the free state (in which sulphur atoms with endodentate (inward) and exodentate (outward) coordination potential are present) (E.J. Hintsa, R. Hartman and S.R. Cooper, J. Amer. chem. Soc., 1983, 105, 3738).

Since the ionic radius of Ni^{++} is between that of the

high-spin state and that of the low-spin state of cobalt(II) it was anticipated that the hexathia-18-crown-6 ligand might similarly compress Co^{++} into an octahedral complex in the smaller low spin state, an unusual situation since octahedral cobalt(II) complexes of strong field ligands (*e.g.* S) usually dissociate into low spin complexes of smaller coordination number. Expectation was confirmed, and the steric constraint of the macrocyclic ligand does indeed allow formation on octahedral low spin complex of cobalt(II) (J.R. Hartman, E.J. Hintsa and S.R. Cooper, Chem. Comm., 1984, 386).

Treatment of the acetonitrile complex of copper(I) perchlorate with hexathia-18-crown-6 gives a di-copper complex (21) in which each of the copper ions is tetrahedrally coordinated by three sulphur atoms, the fourth positions retaining acetonitrile. The tetrahedral geometry, the sulphur and nitrogen ligands, and the relative resistance to oxidation of Cu(I) in the absence of acid give this complex an interesting relationship to the active copper sites of the blue copper proteins plastocyanin and azurin (R.O. Gould, A.J. Lavery and M. Schröder, Chem. Comm., 1985, 1492). A notable structural feature of the complex (21) revealed by X-ray diffraction is the trans-

(21)   (22)

annular contact (3.64$\overset{o}{\text{A}}$) between sulphur atoms diametrically opposed in the ring which is remarkably similar to that noted in Chapter 55, Section 4(b)(i), for the unsaturated compounds numbered (109) and (110).

An unusual example of the stabilisation of particular valence states by crown ether complexation is provided by the silver complex of the surfactant long chain alkyl derivative of 18-crown-6 (22). The silver(I) complex can be reduced by sodium borohydride, photochemically, or electrochemically to the silver(0) (metallic) state but the silver does not precipitate. The redox potential of the system leads to an estimate of log K_S for the Ag(0) complex of 21, silver apparently expanding to a near-perfect fit in the cavity upon reduction (R. Humphry-Baker, *et al.*, Angew. Chem. intern. Edn., 1977, 18, 630).

The addition of one, as in the lariat ethers, or more side chains with donor atoms to these macrocycles increases their complexing ability and the cryptates are especially stable, provided the metal fits the molecular cavity. The values of log K_S for the [2,2,2]-cryptates of sodium, potassium and caesium ions are 7.2, 9.7, and 4.4 respectively showing a 300-fold preference for potassium and 4000 times the stability of its 18-crown-6 complex (M. Kivch and J.-M. Lehn, Angew. Chem. intern. Edn., 1975, 14, 555).

One area of study of such compounds has been the incorporation of groups which can change the complexing ability of the ligand in response to some external influence, a so-called 'switching' effect. For example while the azo-group has the *trans*-configuration macrocycle (23) does not complex metal ions but on uv-irradiation the *cis*-form (24) is produced which can bind metals effectively until it reverts thermally to the *trans*-form (S. Shinkai, *et al.*, J. Amer. chem. Soc., 1983, 56, 1700). Electrochemical 'switching' occurs with the nitrophenyl lariat ethers of type (25, X=N or CH) since the radical anion formed by reduction shows enhanced cation binding by factors of 15-13x10^6, the effect decreasing in the series Li > Na > K. The azo-cryptand (26), whose cavity is an almost exact fit for potassium similarly shows greater enhancement for K$^+$ than for Na$^+$ (G.W. Gokel, *et al.*, J. Amer. chem. Soc., 1985, 107, 1958).

The double armed macrocycle (27) complexes barium strongly

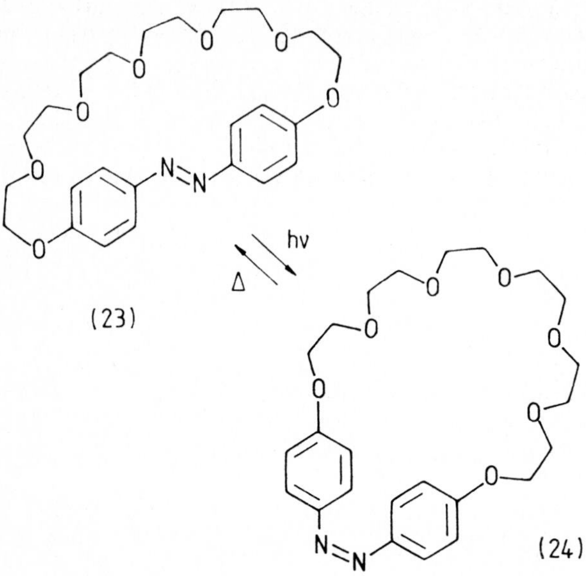

(23)

Δ ⇌ hν

(24)

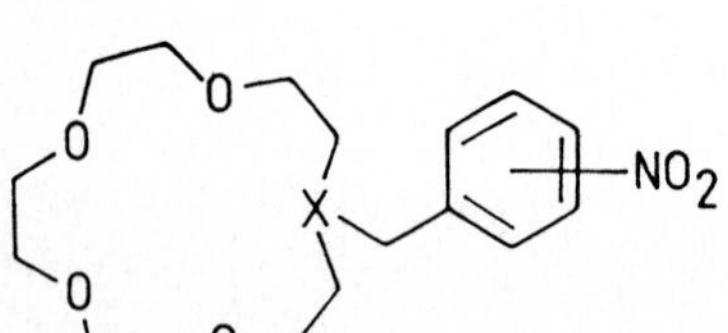

(25)

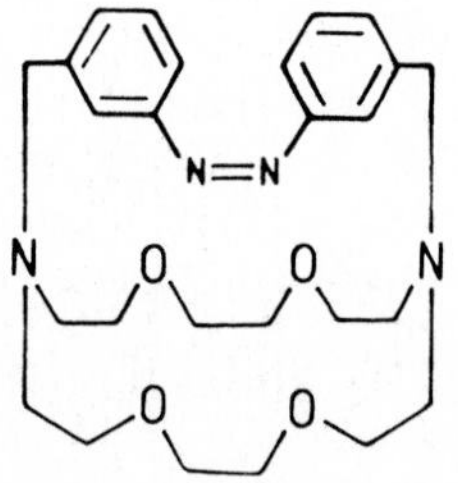

(26)

enough to dissolve barium sulphate in aqueous solution and has been considered as a de-scaling agent for oil wells in the North Sea, where barium sulphate deposits in equipment and in porous rock strata from which oil is to be displaced by pumped water can cause serious losses (F. de Jong, *et al.*, Recl.: J.R. Neth. chem. Soc., 1983, <u>102</u>, 164).

An interesting series of ligands whose donor atoms are

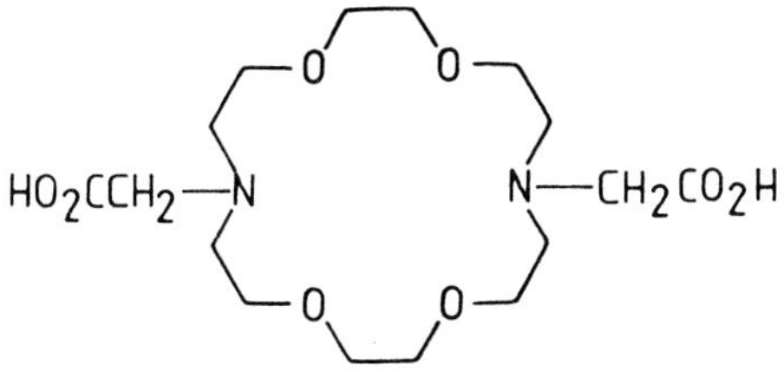

(27)

the nitrogen of pyridine have been described, including the tetrapyridine (28a) which complexes lithium selectively and changes from the red tautomer (28a) to the colourless form (28b) in the complex (S. Ogawa, R. Narushima and Y. Arai, J. Amer. chem. Soc., 1984, <u>106</u>, 5760).

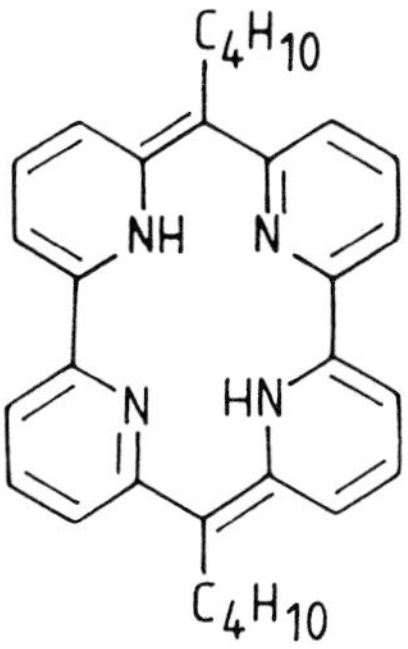

(28a) 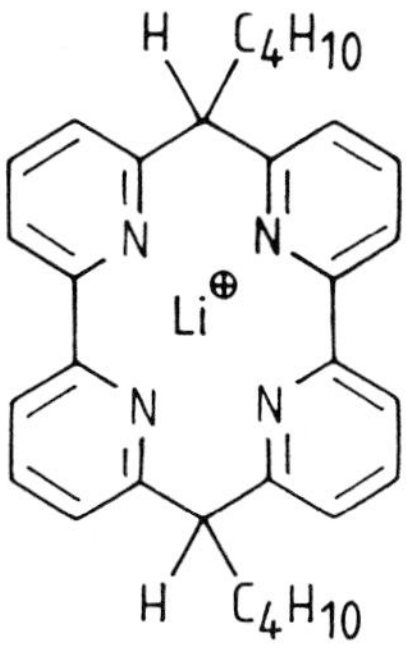   (28b)

The diamine (30a) formed as its zinc complex by treating the dichlorobipyridyl (29) with ammonium tetrachlorozincate, and somewhat ineffectively liberated from the complex by passing hydrogen chloride into its solution in sulphuric acid, shows similar tautomerism. The amine form (30a) is favoured by polar solvents such as water or acetonitrile and the imine (30b) by less polar ones like benzene and chloroform (S. Ogawa and S. Shiraishi, J. chem. Soc. Perkin I, 1980, 2527).

2,6-Cyclosexipyridine (32, R=H) m.p. 292-295°C (decomp.), whose synthesis has been discussed for fifty years, was first obtained when the tetraketone (31) was refluxed with hydroxylamine in acetic acid (G.R. Newkome and H.-W. Lee, J. Amer. chem. Soc., 1983, 105, 5956). Substituted examples of this ring system (32, R=p-C$_6$H$_4$.CH$_3$ and p-C$_6$H$_4$.C$_2$H$_5$) have been prepared from the bipyridyls (33) and (34) with ammonium acetate in refluxing acetic acid. The products precipitate as highly insoluble sodium acetate complexes, the sodium ion (ca. 15 mg from 0.51) being derived from the glassware since purified reagents were used (a property well-known for EDTA, etc.) (J.L. Toner, Tetrahedron Letters, 1983, 24, 2707).

For a comprehensive compilation of macrocyclic complexing

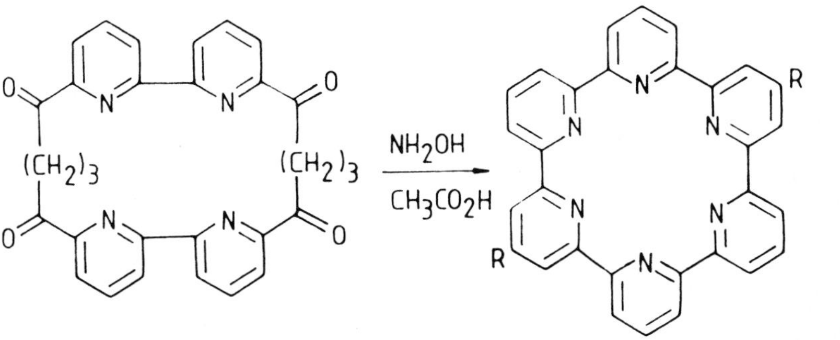

(31)

$\xrightarrow{\text{NH}_2\text{OH}}$
$\text{CH}_3\text{CO}_2\text{H}$

(32)

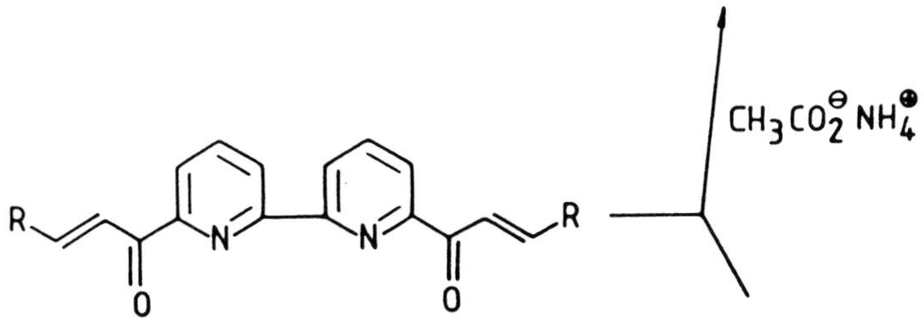

$\text{CH}_3\text{CO}_2^{\ominus}\,\text{NH}_4^{\oplus}$

(33)

$.2\text{I}^{\ominus}$

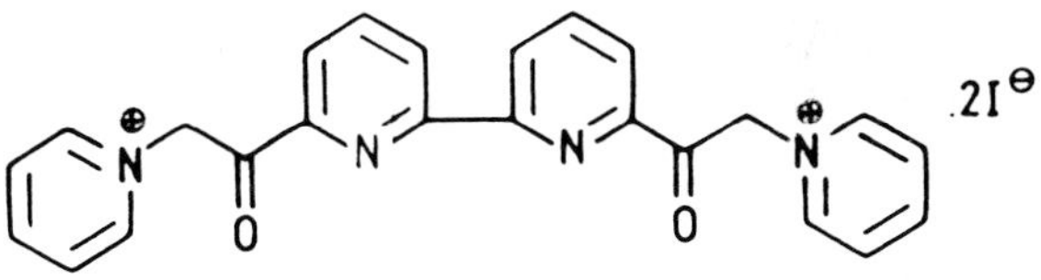

(34)

agents, the ions they bind and the associated thermodynamic parameters, which shows the enormous amount of work done in the five or six years following the first report of a crown ether, see: J.J. Christiansen, Chem. Rev., 1974, 74, 389).

Later reviews. R.M. Izatt and J.J. Christiansen, (eds.) Synthetic Multidentate Macrocyclic Compounds, Academic Press, New York, 1978; G.A. Nelson, (ed.) Coordination Chemistry of Macrocyclic Compounds, Plenum Press, New York, 1979; M. Dobler, Ionophores and Their Structures, John Wiley and Sons, New York, 1981; P. Hubberstey, Coord. Chem. Rev., 1980, 34, 27; N.S. Poonia, J. Sci. Ind. Research, 1978, 34, 202.

(d) Cation transport phenomena

The migration of metal, or substituted ammonium, ions between aqueous phases separated by lipid barriers, present for example in cell membranes, is important in biological processes and in other aspects of electrochemistry. Crown ethers, cryptands and similar compounds are particularly well suited to converting intrinsically hydrophilic cations into lipid soluble complexes which can pass such barriers, since the ions are then largely or completely surrounded by lipophilic material while the free ligands retain water solubility. These compounds have therefore been used as models for more complex biological complexing agents in studies of transport phenomena. Ion transport by the natural macrolides such as valinomycin is considered in Section 4.

Simple apparatus for comparing the transport characteristics of complexing agents consists of a U-tube containing aqueous solutions in the side-arms separated by a stirred layer of denser immiscible solvent, such as a chlorinated hydrocarbon, containing the complexing agent (in equilibrium with the aqueous solutions). The rate of transport of a chosen ion between the aqueous phases is monitored by suitable analyses.

For effective transport in this type of experiment the complexing agent must be able to enclose the ion either within the first aqueous phase or across the interface, carry it with an associated anion across the lipid region, and release it at or beyond the second interface. Since both capture and release of the ion must occur readily, complexes of intermediate stability are required and there should be no significant activation barrier to complexation. These factors tend to militate against high selectivity for a particular ion (see, for example, M. Kivch and J.-M. Lehn, Angew. Chem. intern. Edn., 1975, 14, 55; R.C. Hayward, Chem.

Soc. Rev., 1983, 12, 293).

Macrocyclic hosts bearing two (*e.g.* 35) or four (*e.g.* 36) donor arms show greater and more selective transport characteristics than simple crown or cryptand systems (H. Tsukube, *et al.*, J. Chem. Soc. Perkin I, 1986, 1033).

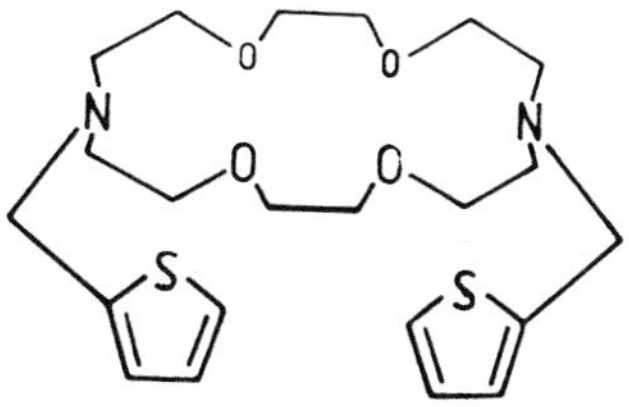

(35)

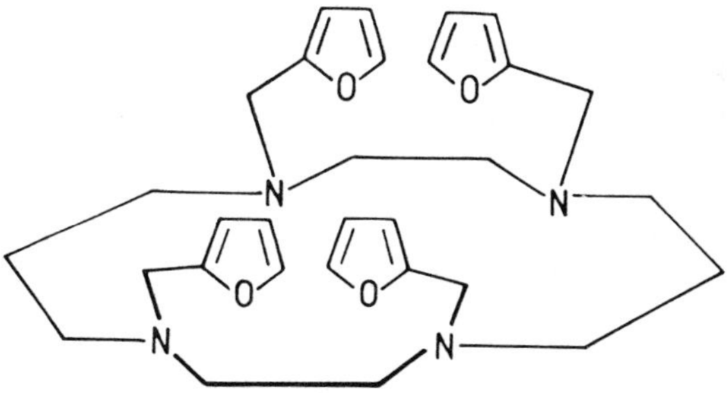

(36)

The concept of 'switchable' complexing agents, such as the photoresponsive azo macrocycles noted in the previous Section 1(c), introduces the possibility of ion transport against the concentration gradient, ion pumping, with potential applications in energy conversion, separation, *etc.*

A different mechanism believed to account for specific ion transport across biological membranes (M. Klingenberg, Nature, 1981, 290, 449) involves the formation of a hydrophilic tunnel through which the ion can pass by transfer between adjacent complexing sites. Linear peptides such as gramicidin A can form a tunnel through their helical structures and have been used in model studies of this phenomenon (B.C. Pressman, Ann. Rev. Biochem., 1976, 45,

501) but tunnels of stacked macrocycles have been suggested and approaches to the synthesis of suitable compounds (*e.g.* 37, R^1=CONHR, R^2=CO_2^-) which can form stacks by specific interactions of the substituents have been described (J.-M. Lehn, *et al.*, J. Amer. chem. Soc., 1981, 103, 701).

(37)

(e) Synthetic applications

(i) General considerations

A variety of preparative reactions are catalysed or mediated by crown ethers or cryptands; this Section highlights some of these but excludes asymmetric synthesis (Section 1(g)), although examples involving initially optically active materials are included. Synthetic applications of crown ether chemistry have been reviewed (G.W. Gokel and H.D. Durst, Synthesis, 1976, 176-182; see also M. Fieser, Reagents for Organic Synthesis, 10, John Wiley, New York, 1982 and other volumes in this series).

Most of the reactions in which crown ethers *etc.*, usually 18-crown-6 or a mixture of the *cis-* and the *trans-*isomer of dicyclohexano-18-crown-6 (38) often called dicyclohexyl-18-crown-6, have useful catalytic or similar activity involve the anion of a metal salt as the essential reagent and crown complexation of the cation may assist the reaction in one or both of two distinct ways. Firstly, the solubility of the

(38)

salt in the reaction medium may be enhanced or its transfer
across an interface with another liquid or solid phase
facilitated.   Secondly, the anion may be freed from
association with the cation, or with molecules of polar
solvent needed to dissolve the salt in the absence of crown
ether, and thus rendered more reactive ('bare' or 'naked'
anion).
When the crown ether serves as a phase transfer catalyst
it may have the advantage over quaternary ammonium cations,
which are often used to assist reaction across the interface
between aqueous and organic solvents, that a solid reagent
may be used, such as a potassium salt or KOH, avoiding
side-reactions due to water. Work-up may then be simplified
to filtration through a short column of silica to remove
metal salts and complexed crown, and evaporation of the
organic solvent.

(*ii*)   *Nucleophilic substitution and addition (see also
reactions with borohydrides, Subsection iv)*

Simple anions which show increased reactivity following
crown   complexation   of   the   cation   include   cyanide,
carboxylates and nitrite leading to nitriles, esters and
nitro   compounds   (with   nitrite   esters)   respectively   by
nucleophilic displacement of halide, tosylate and similar
leaving groups (J.W. Zubrick, B.I. Dunbar and H.D. Durst,

Tetrahedron Letters, 1975, 71).

Cyanide ion from potassium cyanide and 18-crown-6 adds to a diazonium cation to give an arene diazocyanide, Ar-N=N-CN (M.F. Ahern and G.W. Gokel, Chem. Comm., 1979, 1019). Similarly 18-crown-6 with potassium acetate assists the Gomberg-Bachmann biaryl synthesis by promoting formation of a diazo-acetate (S.H. Korzeniousky, L. Blum and G.W. Gokel, Tetrahedron Letters, 1977, 1871).

Chloroformates, RO-CO-Cl, give cyanoformates, RO-CO-CN, with potassium cyanide and 18-crown-6 in dichloromethane (M.E. Childs and W.P. Weber, J. org. Chem., 1976, 41, 3486).

The relative reactivities [n] of nucleophilic anions of potassium salts complexed by 18-crown-6 with benzyl tosylate have been compared in acetonitrile. Azide 10.0 and acetate 9.6 are the most nucleophilic but the halides are remarkably similar: fluoride 1.4, chloride and bromide 1.3, and iodide 1.0 (C.L. Liotta and E.E. Grisdale, Tetrahedron Letters, 1975, 4205).

The reactivity of hydroxide and alkoxide ions is also enhanced and methoxide will even displace chloride from un-activated aryl halides such as 1,2-dichlorobenzene (39) with crown catalysis; a benzyne intermediate (41) is precluded in this reaction since the product, 2-chloro-anisole (40) contains none of the 3-chloro-isomer (42) (D.J. Sam and H.E. Simmons, J. Amer. chem. Soc., 1972, 94, 4024; 1974, 96, 2252).

(39)   (40)

(41)   (42)

Of the halide ions, fluoride probably shows the greatest increase in both basic and nucleophilic reactivity when the gegen ion, usually potassium, is sequestered by a crown ether and solvation is reduced as in acetonitrile solution; in the absence of a catalyst the relatively expensive caesium fluoride is the most reactive alkali metal fluoride. A variety of valuable nucleophilic fluorination reactions by substitution at primary sp^3, and sp^2 carbon atoms can be achieved under mild conditions using potassium fluoride, 18-crown-6 (*etc.*), and dipolar aprotic solvents, particularly acetonitrile, but strictly anhydrous conditions are essential. Thus 2,4-dinitrochlorobenzene (43, X=Cl) gives the fluoro-analogue (43, X=F) - Sanger's Reagent - quantitatively in five hours at room temperature whereas uncatalysed fluorination with potassium fluoride, with or without solvent (*e.g.* *N*-methyl-2-pyrrolidone) requires temperatures approaching $200^{O}C$ (C.L. Liotta and H.E. Simmons, J. Amer. chem. Soc., 1974, <u>96</u>, 2250, and references therein). The highly reactive 8-fluoro-deriva-tive of tri-*O*-acetyladenosine (44, X=F) is similarly obtained from the 8-bromo-

compound (44, X=Br) with this reagent (Y. Kobayashi, *et al.*, Chem. Comm., 1976, 430).

Only the more reactive positions of the hexachloro-compound (45) are exchanged by the KF/18-crown-6/CH_3CN reagent at reflux temperature giving the tetrafluoroderivative (46, X=Cl) in high yield. Solvent-less fluorination with KF at 265°C is needed to form the hexafluoro-compound (46, X=F); the corresponding diazabiphenylenes are obtained by flash vacuum pyrolytic extrusion of nitrogen (J.A.H. MacBride, *et al.*, Chem. Comm., 1983, 425).

(45)   (46)

Benzoylsilanes (*e.g.* 47) are converted into ketones (48) by treatment with KF and 18-crown-6, followed by an alkyl halide, with loss of the silyl group. The anion (49) from nucleophilic addition of fluoride to carbonyl carbon is considered a likely intermediate (A. Degl'Innocenti, *et al.*, Chem. Comm., 1980, 1021).

$$C_6H_5COSi(CH_3)_3 \quad \xrightarrow[\text{(ii) } C_6H_5CH_2Br]{\text{(i) KF, 18-crown-6}} \quad C_6H_5COCH_2C_6H_5$$

(47)   (48)

$$\overset{O^-}{\underset{F}{\overset{|}{C_6H_5-\overset{|}{C}-Si(CH_3)_3}}}$$

(49)

Nucleophilic substitution with potassium superoxide is practicable using crown catalysis giving dialkyl peroxides from primary or secondary halides; efficient reductive cleavage of the initial product to an alcohol occurs under some conditions (J. San Filippo, C.-I. Chern and J.S. Valentine, J. org. Chem., 1975, 40, 1678; R.A. Johnson and E.G. Nidy, J. org. Chem., 1975, 40, 1680).

$$C_8H_{17}Br \xrightarrow[DMSO]{(i)} C_8H_{17}OH \ (63\%)$$
+ minor products

$$C_{18}H_{37}Br \xrightarrow[benzene]{(i)} C_{18}H_{37}\text{-}0\text{-}0\text{-}C_{18}H_{37} \ (61\%)$$
$$+ \ C_{18}H_{37}OH \ (18\%)$$

(i) KO_2, 18-crown-6

Scheme 2

(*iii*)   *Base promoted reactions*

A convenient preparation of an ethereal solution of diazomethane using hydrazine hydrate, chloroform and potassium hydroxide is catalysed by trace quantities of 18-crown-6 and obviates the need for expensive or carcinogenic nitroso-intermediates (D.T. Sepp, K.V. Scherer and W.P. Weber, Tetrahedron Letters, 1974, 2983). Other reactions involving dihalocarbenes generated by phase transfer reaction between trihalomethanes and aqueous alkali are assisted by crown ethers, which may give more selective carbene formation with a mixed halogen precursor than a quaternary ammonium catalyst. Thus chlorodibromomethane gives only adducts of bromochlorocarbene with alkali metal hydroxide and dibenzo-18-crown-6 (M. Fedorynski, Synthesis, 1977, 237).

Cyclopentadiene (50) condenses with ketones, but not aldehydes, in the presence of KOH and 18-crown-6 to give fulvenes (51) in moderate yield (H. Alper and D.E. Laycock, Synthesis, 1980, 799). Ferrocene (52) is formed when iron(II) chloride is used instead of the ketone (M. Salisova and H. Alper, Angew. Chem. intern. Edn., 1979, 18, 792).

The basic character of the fluoride ion promotes an efficient Michael addition of the nitromethane anion to a conjugated ketone in the $KF/18$-crown-$6/CH_3CN$ system (I. Belsky, Chem. Comm., 1977, 237), but results mainly in elimination of HX from secondary halides (*e.g.* bromocyclooctane gives cyclo-octene); surprisingly crown-complexed potassium iodide gives the same result. The effect of counter ions and crown complexation on the stereochemistry of β-elimination reactions has been reviewed (R.A. Bartsch, Acc. chem. Research, 1975, 8, 239).

The basicity of potassium t-butoxide is also enhanced by crown ethers; "reluctant" alkyl halides give alkenes and even 0.1 mol. % of 18-crown-6 promotes bisdehydro-halogenation of vicinal dibromides or geminal dichlorides, and syn-elimination from haloalkenes, to give alkynes (E.V. Dehmlow and M. Lissel, Synthesis, 1979, 372; Annalen, 1980, 1). The same reagent effects specific N-alkylation of pyrrole, pyrazole and similar compounds by alkyl halides (W. Guida and Dr. J. Mathre, J. org. Chem., 1980, 45, 3172) and facilitates oxidative hydrolysis of nitriles to carbo-xylic acids containing one less carbon atom (G.W. Gokel, J. org. Chem., 1980, 45, 3630).

The configurations of alkenes obtained by Wittig synthesis are strongly dependent on the solvent, tetrahydrofuran or dichloromethane, when crown-complexed potassium t-butoxide, or carbonate, is used to generate the ylid intermediate, Scheme 3 (R.M. Boden, Synthesis, 1975, 784).

Potassium carbonate with 18-crown-6 is an efficient catalyst for the Wadsworth-Emmons reaction; the cyclic enone (54), an intermediate in the synthesis of 6a-carbaprosta-

$$(C_6H_5)_3\overset{+}{P}CH_2CH_3 \quad \xrightarrow{\text{(i)}} \quad (C_6H_5)_3P{=}CHCH_3$$
$$Br^-$$

$$\downarrow C_6H_5CHO$$

$$C_6H_5CH{=}CHCH_3$$

Solvent	_cis_:_trans_	Total yield
THF	85 : 15	96%
CH$_2$Cl$_2$	22 : 78	93%

(i) KOBu^t _or_ K$_2$CO$_3$, 18-crown-6

Scheme 3

glandin I$_2$, is formed in 77% yield from the β-ketophos-phonate (53) whereas stronger bases promote elimination of the ring tetrahydropyranyloxy-group giving the trienone (55), a minor product (9%) under crown-catalysed conditions. The reaction with potassium carbonate fails in the absence of the catalyst (P.A. Aristoff, J. org. Chem., 1981, 46, 1954). The same reaction and catalyst is successful in cyclising ester intermediates in syntheses of the macrolides tylonolide and methynolide (see Section 4).

Wittig-Horner condensation of diethylmethanephosphonate (56) with carbonyl compounds is efficiently catalysed by sodium hydride and 15-crown-5; aldehydes give only the _trans_ (E) alkene (57, R' = H) (R. Baker and R.J. Sims, Synthesis, 1981, 117).

Potassium hydride with an equimolar proportion of [2.2.2.]cryptand catalyses the nearly quantitative methylation of benzoylcyclopropane (58) to the ketone (59) whereas reduction occurs in the absence of the complexing agent (H. Handel, M.-A. Pasquini and J.-L. Pierre, Bull. Soc. chim. France II, 1980, 351). Alkylation of _N_-substi-tuted trifluoroacetamides (60) with primary alkyl halides to give _NN_-disubstituted amides (61), from which secondary amines are readily obtained by hydrolysis, is efficiently promoted by KH with 18-crown-6 as catalyst (J.E. Nordlander _et al._, Tetrahedron Letters, 1978, 4987).

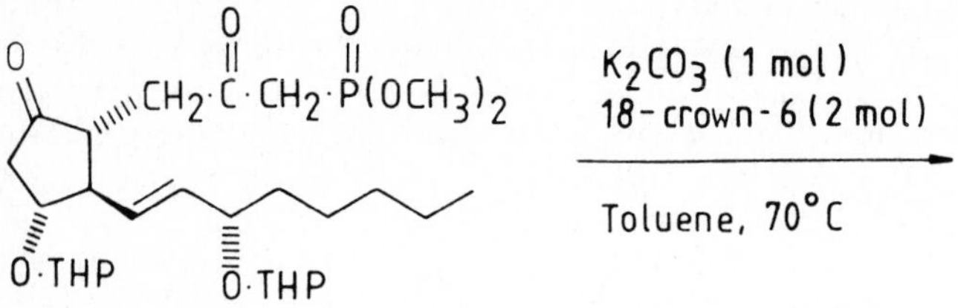

$$
K_2CO_3 \ (1 \ mol)
$$
$$
18\text{-crown-}6 \ (2 \ mol)
$$
Toluene, $70°C$

(53)

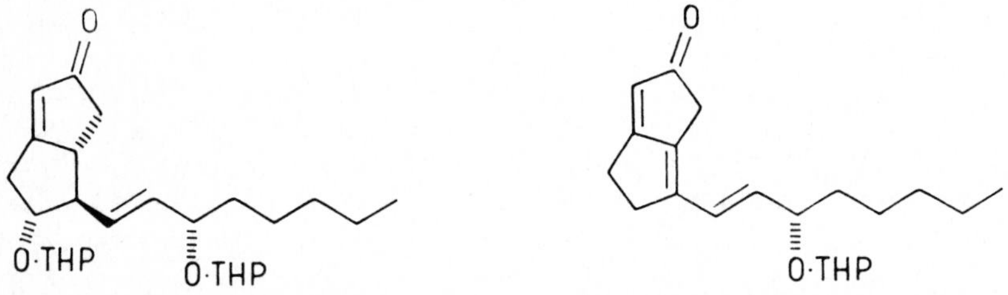

(54)

· (55)

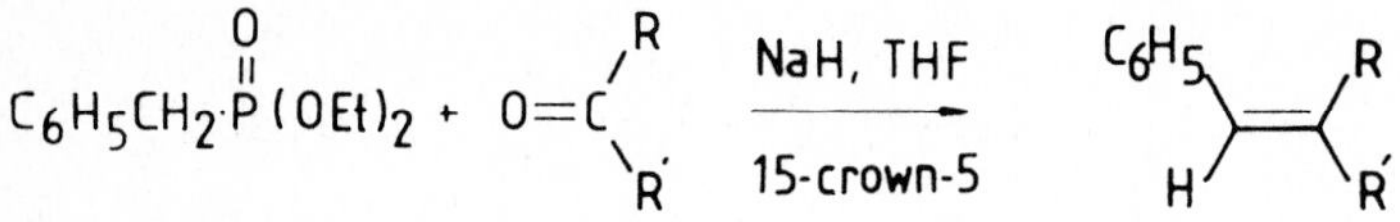

$C_6H_5CH_2\overset{O}{\overset{\|}{P}}(OEt)_2$ + $O=C\overset{R}{\underset{R'}{\big<}}$

NaH, THF

15-crown-5

$\overset{C_6H_5}{\underset{H}{\big>}}C=C\overset{R}{\underset{R'}{\big<}}$

(56)   (57)

$-CO.C_6H_5$

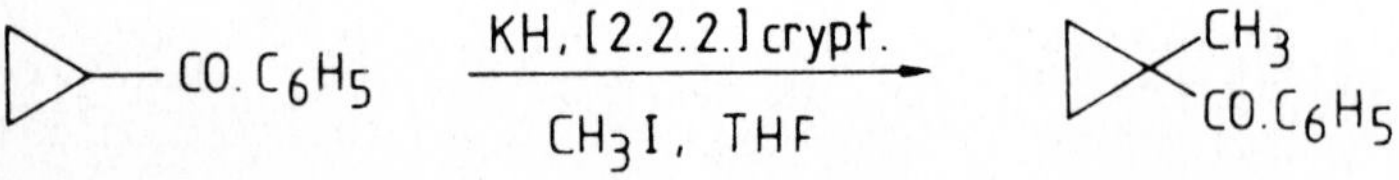

KH, [2.2.2.] crypt.

CH_3I, THF

(58)   (59)

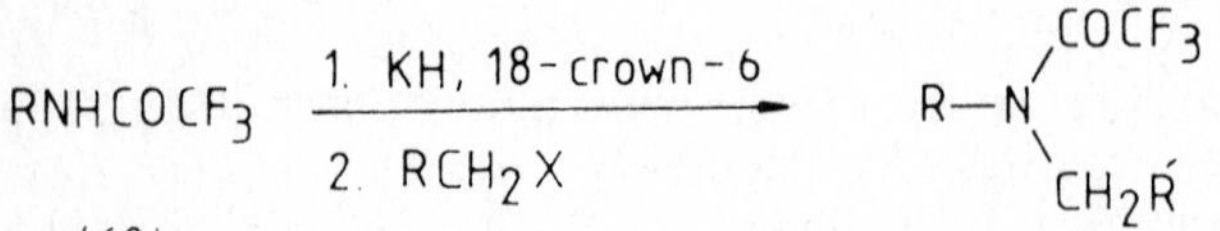

$RNHCOCF_3$

1. KH, 18-crown-6

2. RCH_2X

$R-N\overset{COCF_3}{\underset{CH_2R'}{\big<}}$ (61)

(60)

(iv) Reduction and related reactions

Reduction with alkali borohydrides in dichloromethane using 18-crown-6 greatly extends the range of application of these reagents (H.D. Durst, J.W. Zubrick and G.R. Kieczykowski, Tetrahedron Letters, 1974, 1777).

Cyclohexenone is reduced almost exclusively to 2-cyclohexenol (1,2-addition) by lithium aluminium hydride but in the presence of [2.2.1.]cryptand cyclohexanone (1,4-addition) is the major product after work-up. An increased proportion of 1,4-addition product is also formed using lithium borohydride (A. Loupy and J. Seyden-Penne, Tetrahedron Letters, 1978, 2571).

Potassium dissolved in t-butylamine to which 18-crown-6 has been added efficiently reduces thiocarbonate and particularly NN-dialkylthiocarbamate derivatives of alcohols to alkanes, giving an effective method for alcohol deoxygenation (A.G.M. Barrett, P.A. Prokopion and D.H.R. Barton, J. chem. Soc. Perkin I, 1981, 1510).

The formation of anions (equation 1) and electrides (equation 2) by the alkali metals is facilitated by stabilisation of the metal cations as crown ether complexes, displacing both equilibria towards the ionic products (J.L. Dye, Angew. Chem. intern. Edn., 1979, 20, 587).

$$2M \; \rightleftharpoons \; M^+ + M^- \qquad\qquad \text{Equation 1}$$

$$M \; \rightleftharpoons \; M^+ + e^-, \text{solv.} \qquad\qquad \text{Equation 2}$$

(v) Oxidation

Potassium permanganate dissolves (up to 0.06M concentration) in benzene in the presence of dicyclohexano-18-crown-6 and the solution ('purple benzene') is a powerful oxidant for organic compounds. Progressive dissolution of excess reagent may be effected by ball-milling to prevent the surface becoming coated by manganese dioxide; dichloromethane can also be used as solvent (D.J. Sam and H.E. Simmons, J. Amer. chem. Soc., 1972, 94, 4024).

Although the reaction temperature must not exceed $25^{\circ}C$, to avoid self oxidation of the reagent, toluene and trans-stilbene give benzoic acid quantitatively and many other functional groups can be oxidised. In contrast an equimolar amount of this reagent at room temperature aromatises γ-terpinene (62) to *p*-cymene (63) quantitatively, but conjugated cyclohexadienes do not react (A. Poulose and R. Croteau, Chem. Comm., 1979, 243).

$$CH_3 \quad \xrightarrow[\text{DC-18-crown-6}]{\text{KMnO}_4} \quad CH_3$$

(62)   (63)

Photochemical generation of singlet oxygen is commonly accomplished using aqueous solutions of rose bengal as sensitiser, but 18-crown-6 complexes of the sodium or potassium salt of this dye can be used in carbon disulphide or dichloromethane solution (R.M. Boden, Synthesis, 1975, 783).

Fluorene (64) is oxidised to fluorenone (65), and thence gives the potassium salt of biphenyl-2-carboxylic acid (66) by reaction with hydroxide ion from water liberated in the first step, when treated with potassium t-butoxide in the presence of air. 18-Crown-6 increases the yield in this process which is thought to involve formation of the superoxide anion (S.A. DiBiase and G.W. Gokel, J. org. Chem., 1978, 43, 447).

$$\xrightarrow[\substack{\text{KOBu}^t \\ \text{18-crown-6}}]{\text{O}_2} \quad \xrightarrow{} \quad CO_2^{\ominus}\ K^{\oplus}$$

(64)   (65)   (66)

(f)   Inclusion complexes of neutral molecules, anions and complex cations

The cavities of crown ethers, cryptands and related structures can enclose suitably sized molecules or ions of either charge and retain them by van der Waals interaction, hydrogen bonding with donor atoms of the host, or electrostatic attraction between ionic centres. The stability of an inclusion complex is often sensitive to relatively minor changes in molecular dimensions so that separation of similar substances is possible.

The macrocyclic ether (69) forms an inclusion complex with benzene which it can separate from *p*-xylene with a selectivity of 97%. The macrocycle (69) is obtained by condensation of two molecules each of the dibromoxylene (67) and bisphenol A (68). Benzene facilitates ring closure by

condensation with the second molecule of dibromo-xylene (67) in what appears to be a remarkable example of a template

504

effect by a neutral guest molecule (K. Saigo, *et al.*, J. Amer. chem. Soc., 1986, <u>108</u>, 996).

The macrocyclic ether (70) named cryptophane-D, one of a family of isomeric cavitands, encloses molecules such as CH_2Cl_2 whose ^{1}H-nmr-signal is shifted upfield by 4.2 ppm upon complexation. An average shift value indicating rapid inclusion-exclusion is observed at 220 K. The crystalline complex loses CH_2Cl_2 on exposure to air but its X-ray crystal structure was determined using a crystal sealed with its mother-liquor in a glass capillary (A. Collet, *et al.*, Chem. Comm., 1986, 339).

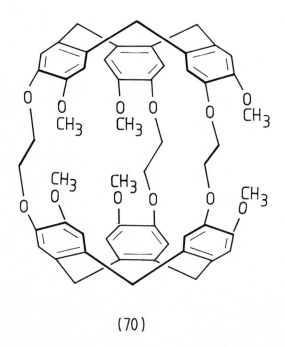

(70)

Heterocyclophanes of type (71) where X can be N, $\overset{+}{N}CH_3.BF_4^-$, or $S^+.BF_4^-$ form inclusion complexes with CH_2Cl_2 and $CHCl_3$ whose X-ray crystallographic structures reveal major crystallographic differences. The chloroform complex of (71, X = N) crystallises with monoclinic space group C2 (chiral) having cell dimensions a = 25.166Å, b = 5.668, c = 13.438, β = 111.31° and Z = 2, while that formed by methylene chloride has space group C2/c and a = 25.359Å, b = 5.485, c = 50.565, β = 96.92° with Z = 8. The difference is attributed to changes in host conformation, in particular the tilting angles of the benzene ring planes, resulting from the shapes of the included molecules (I. Tabushi, *et al.*, J. Amer. chem. Soc., 1984, 106, 2621, and references therein).

(71)

Chiral host molecules form complexes of different stability with each of the enantiomers of a chiral guest and optical resolution can be performed in this way. Chiral crown ethers may incorporate binaphthyl units (*e.g.* 72) or have asymmetric centres in the ring formed by replacing one or more ethylene glycol units by tartaric acid or similar

(polystyrene)   (72)   (73)

groups (cf. 37).   These and related macrocycles and some of their applications have been reviewed (S.T. Jolley, J.S. Bradshaw and R.M. Izatt, J. Heterocyclic Chem., 1982, 19, 3).

Resolution of amino acid ester cations (73) is achieved using resolved binaphthyl systems (72) bound via position 6 to polystyrene resin as a chromatographic stationary phase. Alternatively specific transport of one enantiomer (e.g. d-73, R = C_6H_5) through a chloroform solution of a chiral crown (e.g. RR-72) from an aqueous solution of the racemic ester into a second aqueous phase in a U-tube gives a product with optical purity > 90%.   An ingenious extension uses a W-tube containing racemic ester in the central reservoir and a chloroform solution of RR-72 in one arm and a solution of SS-(72) in the other arm; the d and the ℓ ester accumulate simultaneously in the separate upper aqueous phases (D.J. Cram, et al., J. Amer. chem. Soc., 1979, 101, 3035; 4941).

Neutral or cationic metal complexes, including ammine hydrates, can be included within macrocycles, the process being termed second sphere coordination. Thus the hydrated $Cu(NH_3)_4^{++}$ ion forms a crystalline complex with 18-crown-6 and can thus be separated from $Co(NH_3)_6^{+3}$ both salts having

PF_6^- as the counter ion (H.M. Colquhoun and J.F. Stoddart, Angew. Chem. intern. Edn., 1986, 25, 487).
Crown ethers, cryptands, and their complexes, are themselves included in the cavity of γ-cyclodextrin, a cyclic octamer of glucose (F. Vogtle and W.M. Muller, Angew. Chem. intern. Edn., 1987, 26,
Inclusion complexes, and their relationship to enzyme action, have been reviewed (D.J. Cram, Science, 1983, 219, 1177; R.C. Hayward, Chem. Soc. Rev., 1983, 12, 285).

(g) *Asymmetric synthesis and enzyme models*

The highly selective binding of crown ethers, cryptands, and related compounds to particular molecules or ions is similar to substrate binding at the active site of an enzyme and many workers have used them to investigate and imitate enzymic reactions.
The biologically important hydrolysis of adenosine triphosphate (ATP) to its diphosphate (ADP) is catalysed by (suitably protonated) hexa-aza-24-crown-8 (74, R = H) which

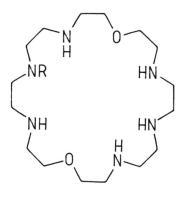

(74)

initially accepts the monophosphate group giving the phos-
phoramidate (74, R = $PO_3^=$) whence (74, R = H) is regenerated
by hydrolysis.   In the presence of Ca^{++} the intermediate
(74, R = $PO_3^=$) reacts with phosphate ion to give inorganic
pyrophosphate, which is also released during the analogous
hydrolysis of acetyl phosphate, while the rate of ATP
hydrolysis increases like that of the natural enzymic
reaction (P.G. Yohannes, M.P. Mertes and K.B. Mertes, J.
Amer. chem. Soc., 1985, 107, 8285; M.W. Hosseini and
J.-M. Lehn, Chem. Comm., 1985, 1155).

The hexa-aza macrocyle (74), and its analogue in which the
oxygen atoms   are replaced by CH_2, complex two atoms of
Cu(II) and thus serve as models for the active sites of
several biochemically significant copper proteins including
bovine erythrocyte superoxide dismutase, hemocyanine and
tyrosinase.   X-ray crystallography shows that the copper
atoms of the complex ion $[(74)\text{-}Cu_2(OH)(ClO_4)]^{2+}$ are bridged
by the OH group, and temperature dependent magnetic
susceptibility measurements show strong antiferromagnetic
coupling between them, in contrast to the very weak coupling
in the corresponding polymeric acetate bridged complex.
Comparison with natural copper proteins supports the
suggestion that the copper atoms of some of these are also
OH bridged (P.K. Coughlin and S.J. Sheppard, J. Amer. chem.
Soc., 1984, 106, 2328 and references therein).

The formation of an optically active product from a
symmetric or a racemic substrate is a common characteristic
of an enzyme reaction which various experiments have sought
to imitate.   In a relatively simple example the nickel(II)
chloride complex of the chiral tetrathia-compound (75),
derived from S,S-tartaric acid is used to catalyse the
quantitative coupling of the racemic Grignard reagent (76)
with vinyl bromide to give the chiral hydrocarbon (77) with
17% excess of the favoured (S) enantiomer.   Enantiomeric
excesses over 80% have, however, been observed when chiral
(β-aminoalkyl)phosphine ligands are used in this reaction;
a diastereomeric intermediate incorporating one enantiomer
of the Grignard reagent with a chiral complexed vinylnickel
unit has been proposed (M. Lemaire, et al., Chem. Comm.,
1984, 309; M. Kumada, et al., J. org. Chem., 1983, 48,
2195).

(75)

$$\underset{(76)}{\overset{\overset{\displaystyle MgCl}{\displaystyle |}}{C_6H_5 \cdot CH \cdot CH_3}} + BrCH=CH_2 \quad \xrightarrow{\;(75)-NiCl_2\;} \quad \underset{(77)}{\overset{\overset{\displaystyle CH=CH_2}{\displaystyle |}}{C_6H_5 \cdot CH \cdot CH_3}}$$

R.M. Kellogg and coworkers have demonstrated a more bio-chemically oriented example in which ethyl phenylglyoxylate (79) is reduced to chiral ethyl mandelate (80) by the macrocyclic derivatives of 1,4-dihydropyridine (78) in the presence of Mg^{2+}. Although this ion is loosely bound by the macrocycle it can simultaneously complex the carbonyl group of the substrate, activating it towards accepting hydride ion from the adjacent dihydropyridine group in the reacting complex (81). Enantiospecificity results from interaction of the H and R groups (CH_3, $CH_2 \cdot C_6H_5$ or C_6H_5) of the (S) amino acid residues incorporated into the macrocycle with the larger (C_6H_5) and smaller ($CO_2C_2H_5$) substituents on the carbonyl substrate. With an eighteen membered ring (78, n = 5) the enantiomeric excess of the (S) product (80) is 90%, falling to 42% with n = 12 when magnesium is less effectively held by the macrocyclic ligand (R.M. Kellogg, *et al.*, J. Amer. chem. Soc., 1985, 107, 3981).

Asymmetric syntheses and model enzyme reactions involving macrocyclic compounds have been reviewed (R.M. Kellogg, Angew. Chem. intern. Edn., 1984, 23, 782; R.C. Hayward,

510

Chem. Soc. Rev., 1983, <u>12</u>, 285; J.-M. Lehn in IUPAC
Frontiers of Chemistry, ed. K.J. Laidler, Pergamon Press,
Oxford, 1982, p. 265 *et seq.*).

(*h*) *Crown ethers of unusual geometry*

Ring closure of the ethene linked poly-ether chains (82)
at high dilution gives two products, a crystalline solid
m.p. 107-108°C identified as the cylindrical molecule (83)
and an oil having the novel Möbius strip geometry (84).

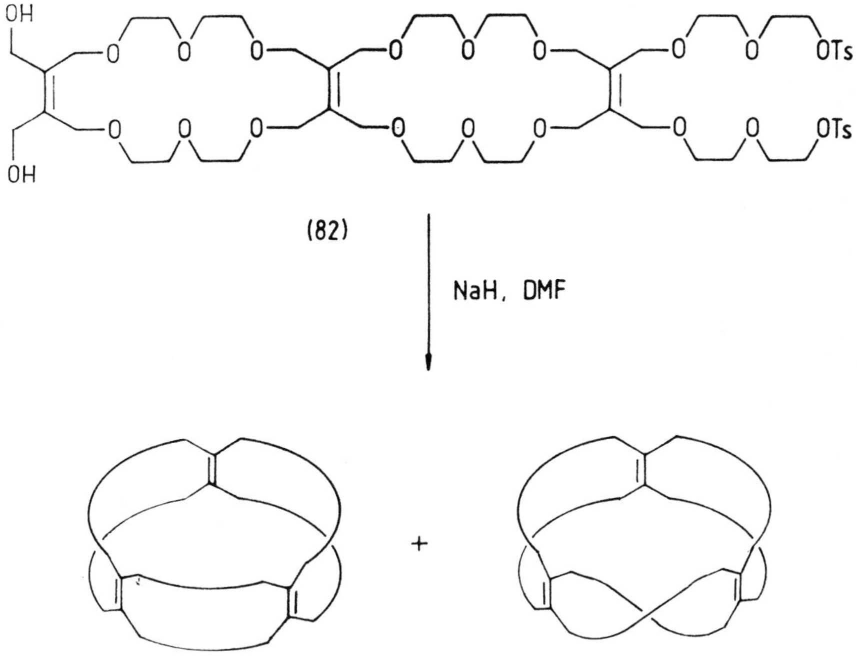

(82)

NaH, DMF

(83)   +   (84)

The two compounds are distinguished by their ^{13}C-nmr-spectra in a chiral solvent (CDCl$_3$ saturated with (+)2,2,2-tri-fluoro-9-anthrylethanol) when the achiral cylinder isomer (83) shows a single resonance of the alkene carbon while two signals are observed from the two enantiomers of the chiral Möbius isomer (D.M. Walba, R.M. Richards and R.C. Haltiwanger, J. Amer. chem. Soc., 1982, 104, 3219).

Linked macrocycles incorporating diphenyl-1,10-phenanthro-line units, Scheme 4, are formed in a template reaction when the phenol groups of the pseudo-tetrahedral copper(I) complex are condensed with di-iodopolyether. The resulting catenand complex, dark red-brown crystals m.p. 274-275°C, shows enhanced kinetic stability towards demetallation (A.-M. Albrecht-Gary et al., J. Amer. chem. Soc., 1985, 107, 3205, and references therein).

Unusual planar coordination of trigonal bridgehead, or "capping", nitrogen atoms in cryptand and related molecules is treated in Section 2.

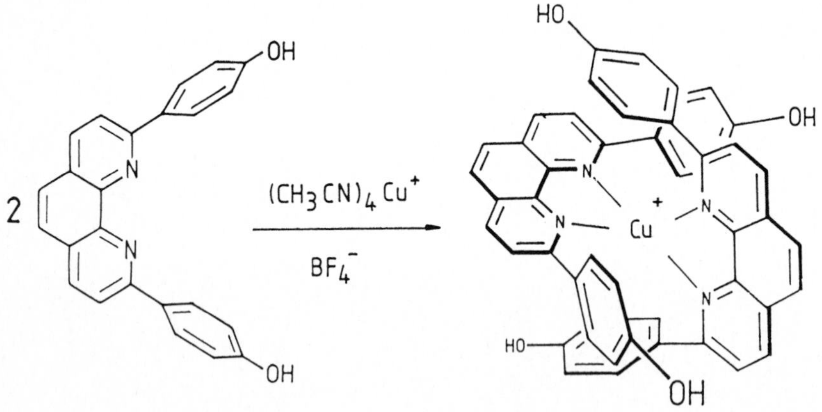

Scheme 4

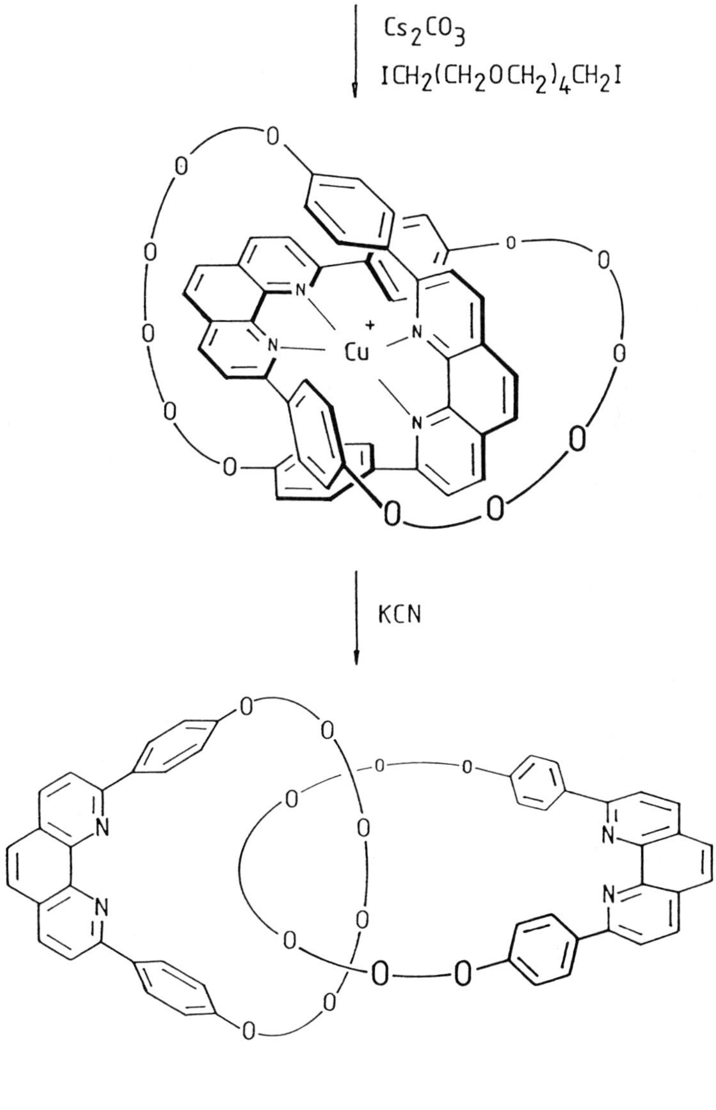

Cs_2CO_3

$ICH_2(CH_2OCH_2)_4CH_2I$

KCN

Scheme 4

2. *Protonation and stereochemistry of cage tertiary amines*

Studies of the structure and protonation of cage tertiary amines of type (85, X = CH or N) have given remarkable and

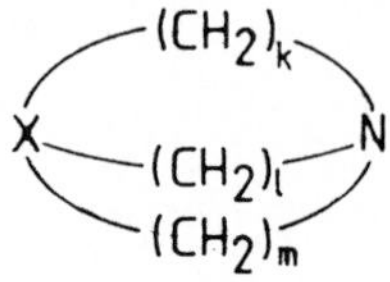

(85)

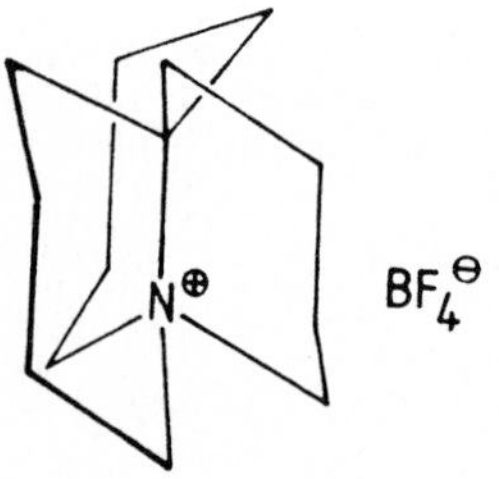

(86)

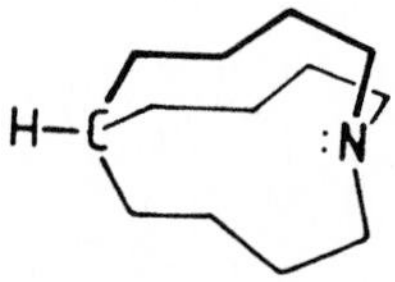

(87)

unexpected results. The isomer of 1-azabicyclo[4.4.4.]-tetra-decane (87), m.p. 168-171°C, which has its bridgehead hydrogen outside the molecular cage (the 'out-6H' isomer, but see A.H. Haines and P. Karntiang, J. Chem. Soc. Perkin 1, 1979, 2577 for discussion of the limitations of the 'in-out' notation) is formed when the quaternary salt (86) is treated with sodium in liquid ammonia. This amine (87) is soluble in organic solvents but not in water and is an exceptionally weak base, being precipitated from solution in concentrated hydrochloric acid by dilution with an equal volume of water. pK_a (of the conjugate acid) is 0.6 (48% v/v ethanol-water) and this low basicity is ascribed to steric hindrance to inside protonation coupled with an increase in strain energy of 60-85 kJ mol^{-1} associated with outward pyramidalisation of the nitrogen (R.W. Alder and R.J. Arrowsmith, J. chem. Research, (S) 1980, 163; (M) 1980, 2301-2313).

The corresponding bridgehead alkene (89), which decomposes slowly in air but has m.p. 116-122°C in a sealed tube and is prepared by treating the quinolizidinium salt (88) with

sodium and *t*-butanol in liquid ammonia, presents a remarkable contrast. The photoelectron spectrum and calculated conformation are consistent with strong interaction between the lone pair electrons of inwards pyramidal nitrogen and those of the π-bond and it is protonated on carbon, giving the quaternary propellane salt (91), even by acids as weak as ethanol! Base does not reverse this reaction, however;

Hofmann elimination occurs forming the quinolizidine (90)
(R.W. Alder, *et al.*, Chem. Comm., 1982, 940). *N*-Methyl-
tetrahydroberberine (92) undergoes similar *C*-protonation,
also by ethanol, but in this case the reaction is suppressed

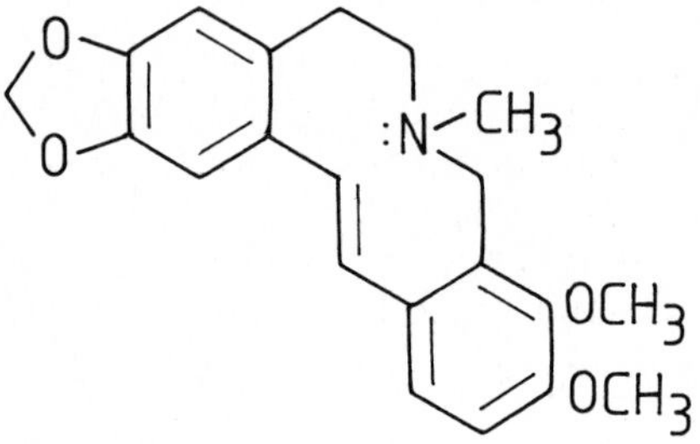

(92)

by competing *N*-protonation at higher acidity (A.J. Kirby and
C.J. Logan, J. chem. Soc. Perkin II, 1978, 642).
  In a study to find the smallest cage which can encap-
sulate a proton the reaction of 1,6-diazabicyclo[4.4.4.]-
tetradecane (93, k,1,m = 4) with acids has been

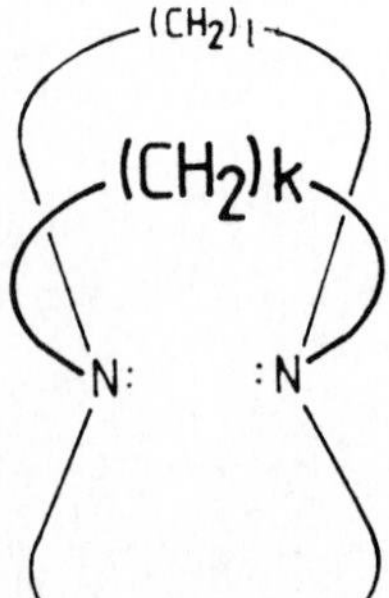

(93)

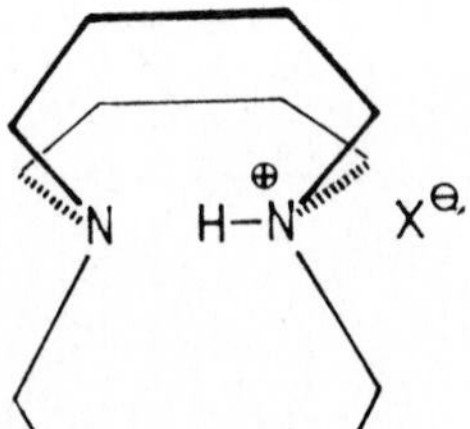

(94)

investigated. Its pK_a 1 and pK_a 2 values near 6.5 and -3.25

respectively, measured by [1]H-nmr spectroscopy in aqueous sulphuric acid, are 4 and 10 log units respectively lower than those of normal diamines with comparable N...N distances, reflecting the strain produced by outwards pyrimidalisation. When the diamine (93) is left in moderately strong acid (*e.g.* 40-70% aqueous sulphuric or trifluoromethanesulphonic acid in acetonitrile) for a week

or more the [1]H-nmr-spectrum no longer corresponds with that of outside protonated or diprotonated cation and the highly deshielded NH resonance at 17.4δ shows that the inside protonated species (94) has been formed. This ion shows extraordinary resistance to deprotonation, even by sodium amide in liquid ammonia for two hours or dry heating of the

solid hydroxide (94, X = OH) at 160°C for 30 min. A second, outside, protonation is achieved with fluoro-sulphonic acid, but only in the presence of antimony(V) fluoride (R.W. Alder, A. Casson and R.B. Sessions, J. Amer. chem. Soc., 1979, 101, 3852).

A general synthesis of bicyclic diamines of type (93) starts from bicyclic hydrazines (95). Cyclisation of the hydrazinium salts (96, Y = OH, X = Hal) is achieved with

(95) (96)

(93) (97)

40% aqueous tetrafluoroboric acid, in which medium the
easily hydrolysed hydrazinium dications (97) are stable.
Similar conditions with the addition of silver tetrafluoroborate cyclise the corresponding bromides (96, X = Br).
Reductive cleavage of the N-N bond is effected by metals,
such as iron powder, in the same dilute acid (R.W. Alder, *et
al*., J. Chem. Soc. Perkin I, 1982, 603).

The air-sensitive [4.4.4.]diamine (93; k,l,m = 4) and the
corresponding hydrazinium dication (97; k,l,m = 4) react
when mixed in equimolar proportions in acetonitrile giving
the radical cation (98) as stable dark red, almost black,
crystals; esr-spectrometry shows that the nitrogen atoms are
equivalent. The corresponding [3.3.3.] and [3.3.4.] systems
have also been obtained (R.W. Alder and R.B. Sessions, J.
Amer. chem. Soc., 1979, 101, 3651; Chem. Comm., 1977, 747).

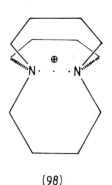

(98)

X-ray crystallographic studies of the bistrifluoromethane
sulphonate of the dication (97; k,l,m = 4) show that the
separation of the nitrogen atoms is 1.53Å for the two-
electron N-N-σ-bond, which increases in the perchlorate of
the radical cation (98) to 2.30Å corresponding with a
three-electron σ-bond between these atoms (R.W. Alder,
A.G. Orpen and J.M. White, Chem. Comm., 1985, 949).

Of twenty examples of bicyclic diamines (93) containing
bridges of two to six carbon atoms, twelve may be converted
into inside protonated cations, identified by slow exchange
with D_2O, either by direct protonation which is slow (nine
examples) and/or by redox promoted rearrangement using per-

disulphuric acid (five examples). Inside protonation is not observed with compounds of less than a total of thirteen carbon and nitrogen atoms and with this number only if a five or six-membered ring is present. 1,6-Diazabicyclo-[4.4.4.]tetradecane (93; k,l,m = 4) inside protonates by a remarkable mechanism whereby the N-proton is derived from a methylene group adjacent to nitrogen; deuteriated acid thus gives an inside N-protonated, C-deuteriated, cation (R.W. Alder, R.E. Moss and R.B. Sessions, Chem. Comm., 1983, 997).

X-ray crystallographic studies of the [4.4.4.] system (93; k,l,m = 4) show that inside protonation shortens the distance between the nitrogen atoms by 0.28Å, giving the shortest reported NHN hydrogen bond, and suggests that the proton is symmetrically placed between the nitrogen atoms (R.W. Alder, A.G. Orpen and R.B. Sessions, Chem. Comm., 1983, 999). Nmr- and ir-spectrometric studies confirm the symmetrical protonation of the symmetrically bridged compound (93; k,l,m = 4), but in examples where the bridges do not contain equal numbers of methylene groups the N-H...N system is not linear and the proton position potential has a double minimum (R.W. Alder, R.E. Moss and R.B. Sessions, Chem. Comm., 1983, 1000).

Symmetrical protonation of the [4.4.4.] diaza-compound (93; k,l,m = 4), and the mechanism of protonation of the unsaturated amine (89), bear an interesting relationship to the stable product of protonation of the hydrocarbon alkene (99) by trifluoroacetic acid whose proton- and ^{13}C-nmr-

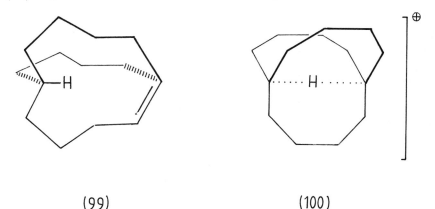

(99)   (100)

spectra reveal a static symmetrical structure (100) in which the two positively charged bridgehead carbon atoms are considered to be connected by a hydrogen atom carrying negative charge, giving a singlet resonance at -3.5δ (J.E. McMurry and C.N. Hodge, J. Amer. chem. Soc., 1984, 106, 6450).

The kinetics and thermodynamics of the proton transfer reactions of [1.1.1.]cryptand (101; n = 1) have been studied by ^{1}H-nmr spectroscopy. Using the notation i and o for in-

(101)

ward and outward pyramidal nitrogen and i^+, o^+ for the corresponding inside and outside protonated ions, five protonated species io^+, o^+o^+, ii^+, i^+o^+, and i^+i^+ have been identified while ir-spectroscopic and pH measurements show that the free ligand is ii. Outside proton transfer is predictably fast with pK_a 7.1 and $ca.$ 1 for the io^+ and o^+o^+ ions respectively, but internal protonation to ii^+ occurs only via io^+ with a half-life of about 7 minutes at room temperature and pH 1 and a high activation energy, 110 ± 8 kJ mol^{-1}, reflecting the distortion of the cage necessary for access of the proton carrier to the cavity. The second internal protonation (to i^+i^+) requires several weeks under these conditions and has the same activation energy as the first internal protonation. Deprotonation is similarly slow, the diprotonated ion i^+i^+ being stable for weeks in 5M potassium hydroxide solution at room temperature and showing similar activation energy, 105 ± 8 kJ mol^{-1}, in higher temperature experiments to give the monoprotonated ii^+ form, which could not be further deprotonated without decomposition. The ii^+ form is estimated to have pK_a not less than 17.8, thermodynamically the strongest base measured (the [4.4.4.]diaza compound (93; k,l,m = 4) is probably an even stronger base but not directly comparable owing to its complex mechanism of protonation) but with kinetically slower proton transfer, by factors of 10^8 (for i^+i^+) and 10^{10} (for ii^+), than any previously observed proton transfer (P.B. Smith, et al., J. Amer. chem. Soc., 1981, 103, 6044).

Proton transfer between carboxylic acids and the conformationally more flexible [2.1.1.]cryptand (101; n = 2) is very much faster than for the [1.1.1.] compound (101; n = 1) but still several orders of magnitude slower than that calculated for a diffusion controlled process. Substantial kinetic hydrogen isotope effects are observed and X-ray crystallographic analysis of the diperchlorate of this cryptand (101; n = 2) confirms that it has the i^+i^+ configuration; pK_a 1 for the diprotonated ion is 7.52 ± 0.05 at zero ionic strength compared to Speiss' value of 8.46 ± 0.05 at 0.05 M (B.G. Cox, N. van Truong and H. Schneider, J. chem. Soc., Perkin II, 1983, 515; B. Speiss, F. Arnaud-Neu and M.J. Schwing-Weill, Helv., 1979, 62, 1531).

X-ray crystallographic study of the (relatively sensitive) explosive hexamethylene triperoxy diamine (102) shows that the nitrogen atoms have planar coordination and that the C-N bonds are shorter than those of the carbon analogue (93; k,l,m = 4). These results are interpreted on the basis that sp^2 hybridised nitrogen is favoured over sp^3 hybridisation by reducing interaction between the electron pairs on nitrogen, inside in (93; k,l,m = 4), and those on oxygen. There is no evidence for internal protonation of the triperoxide (102) (W.P. Schaefer, J.T. Fourkas and B.G. Tiemann, J. Amer. chem. Soc., 1985, 107, 2461).

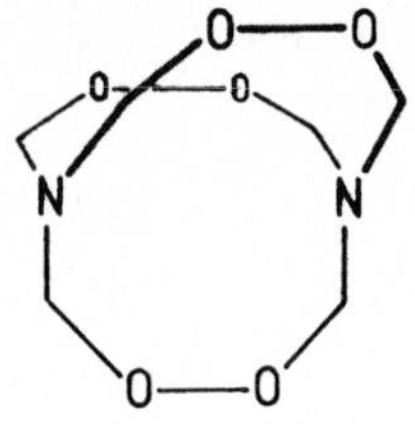

(102)

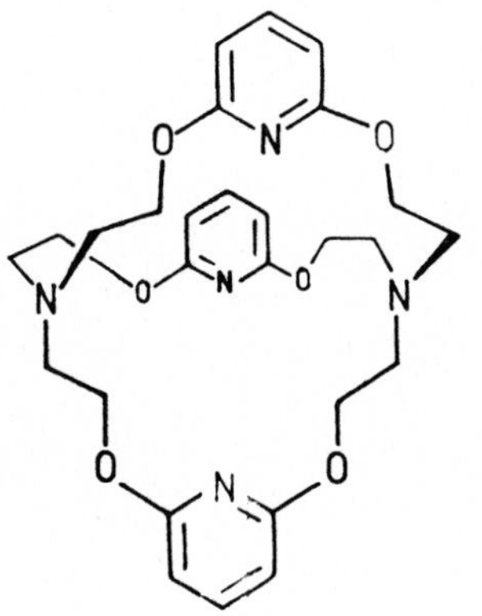

(103)

Similar electronic interaction may explain the planar coordination, and presumably sp^2 hybridisation, of the "capping" nitrogen atoms of the pyridine macrocycle (103), which is obtained by condensation of triethanolamine with 2,6-dichloropyridine, although in this case two methylene groups separate the nitrogen from the oxygen atoms (G.R. Newkome, *et al.*, J. Amer. chem. Soc., 1979, 101, 1047).

3. Conjugation and aromaticity

(a) General considerations

The experimental criterion for "aromaticity", originally described by Hückel as "enhanced electronic stability", has become almost invariably the observation of nmr-shifts indicative of a diamagnetic ring current for 4n+2 π-electron systems (diatropic compounds) while conjugated 4n π-rings show a paramagnetic effect (paratropic compounds). Atropic compounds are those giving no indication of a ring current. Usually the resonance of protons (or those of methyl groups, etc.) bonded to ring atoms, which may be inside or outside the ring, is measured. Since the magnetic field of a ring current lies in opposite directions inside and outside the ring its value at the ring atoms themselves is likely to be near to zero and their resonance, e.g. ^{13}C, is usually insensitive to these effects.

In general heterocyclic compounds can attain a particular number of π-electrons either from a completely conjugated system of double or triple bonds, i.e. annulenes in which in principle any number of =CH- groups can be replaced by =N- (or, in theory $=\overset{+}{0}-$, $=\overset{+}{S}-$, etc.), or by taking formally non-bonding electron pairs of heteroatoms into the π-system so that $-\overset{..}{X}-$ replaces -CH=CH-. The two types of heterocyclic annulenes are thus vinylogues of compounds such as pyridine and pyrrole, etc.

The constraint on macrocyclic conjugation imposed by possible preference for non-planar conformation has also been discussed in the previous Chapter. With larger rings, strain in the planar form may be reduced or eliminated by suitable combinations of cis and trans double bonds and linear alkyne, and particularly di-yne, groups can be accommodated, the latter having the synthetic convenience of ring closure by oxidative coupling of terminal alkyne functions.

Transannular bridges and fusion with partly saturated smaller rings can enhance conjugation by maintaining a planar framework, but fusion with smaller aromatic systems generally leads to reduced ring current effects when the smaller annelating system has high aromatic stabilisation energy, e.g. benzene, whereas fusion with less intrinsically stable systems such as pyrrole (at the 2,5-positions) fosters pronounced delocalisation in a macrocycle. Compounds (121, m=n=1), atropic, and (130), strongly diatropic, are among those discussed below which illustrate this trend.

Theoretical predictions for carbocyclic annulenes have suggested a limit for aromaticity at 22 or 26 π-electrons, but diatropic rings of 34 members have been described (J. Ojuna, *et al.*, Chem. Comm., 1987, 534 and references therein) while the 26 membered cycle of (130) sustains the largest ring current recorded.

Conjugated macrocyclic systems containing heteroatoms have been reviewed by A.G. Anastassiou (Acc. chem. Res., 1976, 9, 453; Pure appl. Chem., 1975, 44, 691; and, with H.S. Kasmai, Advances in Heterocyclic Chemistry, 1978, 23, 55; and by G. Schröder, Pure appl. Chem., 1975, 44, 925.

(b) *Macrocyclic vinylogues of pyridine*

Doubt, later dispelled, about the stability of monocyclic aza-annulenes made the methano-bridged examples the first synthetic targets in this class; thermal cyclisation of the isocyanate (104, R = NCO), or its acyl azide precursor (104, R = CON$_3$), gives the amide (105) whence dehydrogenation with 2,3-dichloro-5,6-dicyanobenzoquinone (DDQ), formation of the

(104)   (105)   (106)

tosyl imino-ether, and reduction with lithium aluminium hydride gives the parent compound (106) as a stable yellow liquid b.p. 63°C at 1 torr. The nmr and electronic spectra of (106) are related to those of naphthalene and high field

resonance of the methylene bridge protons, with ring protons
at low field, indicates considerable diatropic (aromatic)
character (E. Vogel, *et al.*, Angew. Chem. intern. Edn.,
1978, 17, 853).

The ring system of the isoquinoline analogue (109, R =
C_2H_5) is formed by Beckmann rearrangement of the tosyl oxime
(107) and *O*-ethylation of the resulting amide (108) gives
the imino-ether (110, R = C_2H_5); dehydrogenation with DDQ
gives the required conjugated compound (109, R = C_2H_5) as a
yellow oil which isomerises to the cycloheptatrienyl-
pyridine (111) during more prolonged dehydrogenation experi-

ments.   The ethyl ether (109, R = C_2H_5), and its methyl
analogue (109, R = CH_3) prepared independently by a similar
route, are clearly diatropic but rather less so than the
corresponding carbocyclic compound (J.M. Muchowski, *et al.*,
Angew. Chem. intern. Edn., 1978, 17, 855; G.K. Helmkamp, J.
org. Chem., 1978, 43, 3813).
    The first example of a monocyclic aza-annulene is (113)
formed by uv-irradiation at -80°C of the tetracyclic azide
(112) or of its isomer with the azidocyclopropane moiety
fused at the indicated position; a 28% yield of black-green
crystals is obtained.   The ^{1}H-nmr-spectrum of (113) shows
that a single conformer (113) with nitrogen inwards is
present and inside proton resonances at -7.0δ with outside
ones near 8.8δ and 10δ (adjacent to N) show that it has a

(112)   (113)

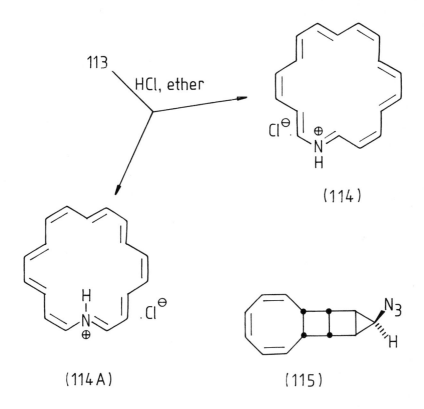

113

HCl, ether

(114)

(114 A)

(115)

substantial diamagnetic ring current which persists at relatively high temperatures. Ethereal hydrogen chloride converts (113) into its black-violet hydrochloride whose nmr-spectrum reveals that the outside protonated configuration (114) predominates over the inside protonated one (114A) by a factor of four (W. Gib and G. Schröder, Ber., 1982, 115, 240).

Similar irradiation of the azide (115) gives aza[14]-annulene (116), 50% yield, as dark violet crystals whose nmr-spectrum reveals that the solution contains the isomer (116) and conformers (117) and (118) in the ratio 83:8.5:8.5 respectively. The temperature dependent spectrum shows that all three are strongly diatropic, the more flexible conformers (117) and (118) being the more strongly so (H. Rottele and G. Schröder, Ber., 1982, 115, 248).

Ojima and coworkers have developed the synthetic route to

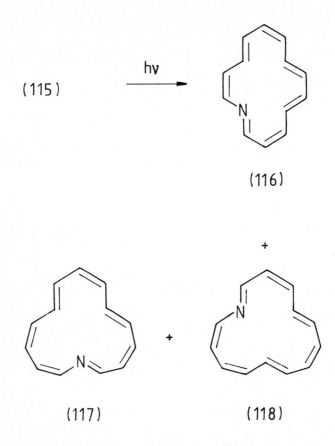

(115) → hν → (116)

+

(117) + (118)

aza-annulenes whereby nitrogen is introduced to the ring by Beckmann rearrangement of ketoximes (119) to obtain a series of examples of type (122) and their mono- and di-benzo fused derivatives, (*e.g.* 121). Tetradehydroaza-annulenes with 14, 16, 18, 20, and 22 atoms (and π-electrons) are available by this method although yields of the larger-ring members are low. The [22]annulene derivative (122, m = n = 3), the largest ring prepared in this class so far, is unstable, as is the twenty membered (4n) example (122, m = 2, n = 3).

The [1]H-nmr-spectra of these compounds (122) are tempera-

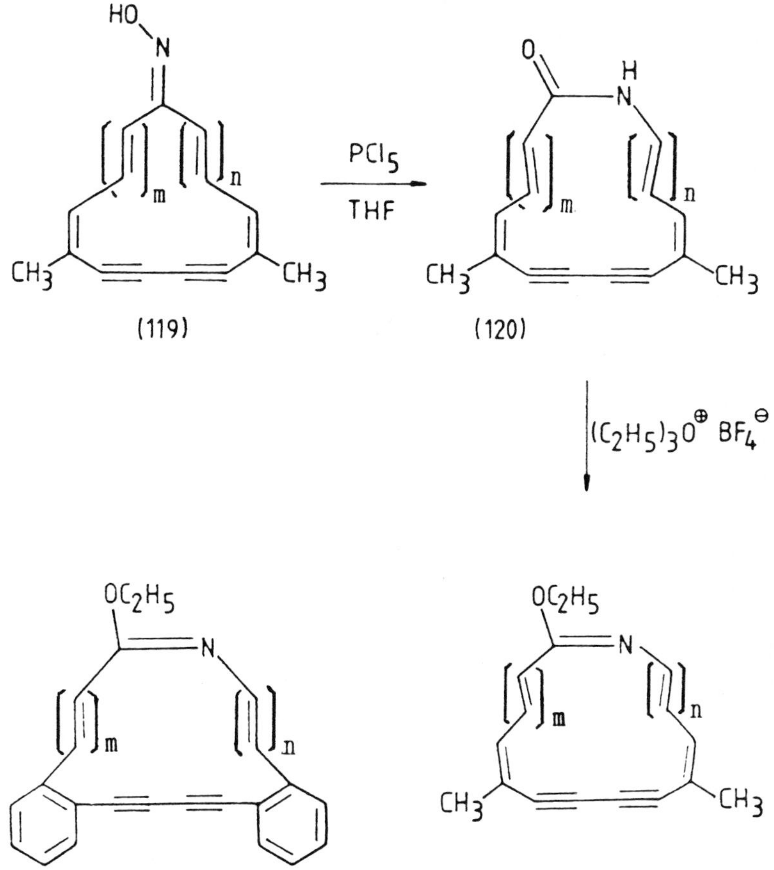

(119)   (120)

(121)   (122)

ture independent, showing conformational rigidity, and ring current effects assessed by the difference between the shifts of inner and outer protons show that the 4n+2 series including the [22]annulene (122, m = n = 3) are clearly diatropic. The [16] and [20]annulenes (4n series) are paratropic: presumably their skeletons are too rigidly planar to escape intrinsically unfavourable electronic interactions by out of plane distortion. In both the diatropic and paratropic series ring current effects fall off with increasing ring size and with benzannelation: (121, n = m = 1) is atropic and the corresponding monobenzo monomethyl compound is at most weakly diatropic.

Despite complications due to conformational differences, comparison of the nmr-spectra, clearest in the methyl resonances, of the lactams (120) and the tetradehydroazaannulenes (122) shows that the former, which are vinylogues of 2-pyridone, also show alternating diatropic and paratropic character (J. Ojima, *et al.*, J. chem. Soc. Perkin I, 1986, 933).

(123) $Pb(CH_3CO_2)_4$ (124)

The hexa-aza-annulene (124) in the 4n series is obtained by oxidation of the diamine (123); the *trans*-configuration of the azo-groups and non-planar conformation has been established by X-ray crystallography (H. Hilpert, L. Hoesch and A.S. Dreiding, Helv., 1985, 68, 325).

Bridging imino-groups maintain the planar conformation of the [14]annulene (125), scarlet prisms m.p. 158°C. The proton nmr-spectrum shows that the π-system is highly delocalised and diatropic and, with ^{13}C studies, that the unsymmetrical configurations (125A) and (125B) exchange fast at room temperature but cooling towards the slow

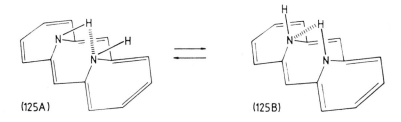

(125A)   (125B)

exchange limit at -133°C gives an unsymmetrical spectrum; activation parameters for exchange are ΔH 26.1 kJ mol^{-1}, ΔS -76.5 J mol^{-1}K^{-1} (E. Vogel, *et al.*, J. Amer. chem. Soc., 1983, <u>105</u>, 6982).

Porphycene (127), violet needles with red fluorescence in solution, is obtained in low yield from the dialdehyde (126) with low valent titanium (McMurray reaction). ^{1}H-nmr-

(128)

resonances (δ) H$_A$ 9.67, H$_B$ 9.23 and H$_C$ 9.83 indicate clear aromatic delocalisation and fast tautomerism explains the equivalence of the pyrrole rings. The tendency to form metal complexes is less than that of its porphin isomer, perhaps because of the smaller central cavity revealed by X-ray crystallography (E. Vogel, *et al.*, Angew. Chem. intern. Edn., 1986, 25, 257).

The nmr-spectra of polyalkyl sapphyrins (*e.g.* 128, R = CH$_3$), a system first observed accidentally in R.B. Woodward's laboratory, also show clear evidence of aromatic delocalisation (V.J. Bauer, *et al.*, J. Amer. chem. Soc., 1983, 105, 6429).

The high diamagnetic ring current in N,N',N'',N'''-tetra-methylporphin dication prompted the synthesis of the twenty-six membered ring vinylogue (130) by acid catalysed cyclo-tetramerisation of the hydroxy-pyrrole (129) and dehydro-genation of the initial product with bromine. The inside, H$_A$, and outside, H$_B$, proton nmr-resonances of the enlarged porphyrin (130) are at -11.64δ and +13.67δ, a difference of 25.3 ppm indicating the largest ring current observed; the N-methyl protons resonate at -9.09δ.

The dication (130) is intensely red-violet, λ_{max} 547 nm, with a molar extinction coefficient of 909,600, over twice the intensity of the corresponding (Soret) band of

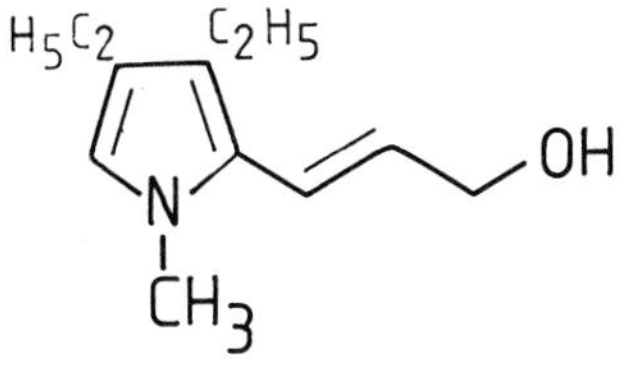

(129)

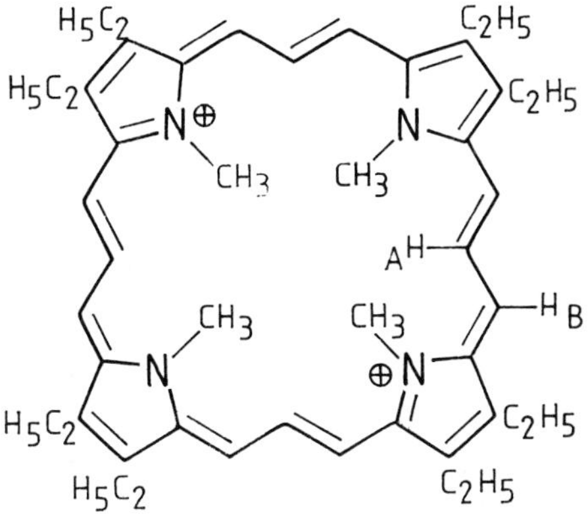

(130)

porphyrins λ_{max} 400 nm, ϵ *ca.* 400,000 (M. Gosmann and B. Franck, Angew. Chem. intern. Edn., 1986, <u>25</u>, 1100).

(c) *Vinylogues of pyrrole, furan, and thiophen*

Although azonine is clearly diatropic when the nitrogen carries hydrogen or alkyl (Second Edn. Vol. IVK p.433) the benzo-derivative (131), and even the corresponding amide anion (132) formed from it with potassium amide in liquid ammonia, are atropic localised systems. This is not due solely to the electronic effect of benzannelation, however, since on warming to $0^{o}C$ the anion (132) changes configuration to incorporate a *trans* bond (134) and the ^{1}H-nmr-resonances show that a diamagnetic ring current develops. In contrast the anion of cyclononatetraene spontaneously undergoes the reverse isomerisation so that it is probably the steric effect of the benzene *peri*-hydrogens which inhibit planar conjugation in the anion (132). The *trans*-urethane (133) remains atropic, however

(A.G. Anastassiou and E. Reichmanis, Chem. Comm., 1975, 149, and references therein).

The diphenylpyridazine fused heteronins (138, X = NCOCH$_3$, NCO$_2$C$_2$H$_5$, or O) are obtained by cycloaddition of diphenyltetrazine (135) with bicyclononanes of type (136) followed by dehydrogenation with chloranil; all are atropic (A.G. Anastassiou and E. Reichmanis, Chem. Comm., 1976, 313).

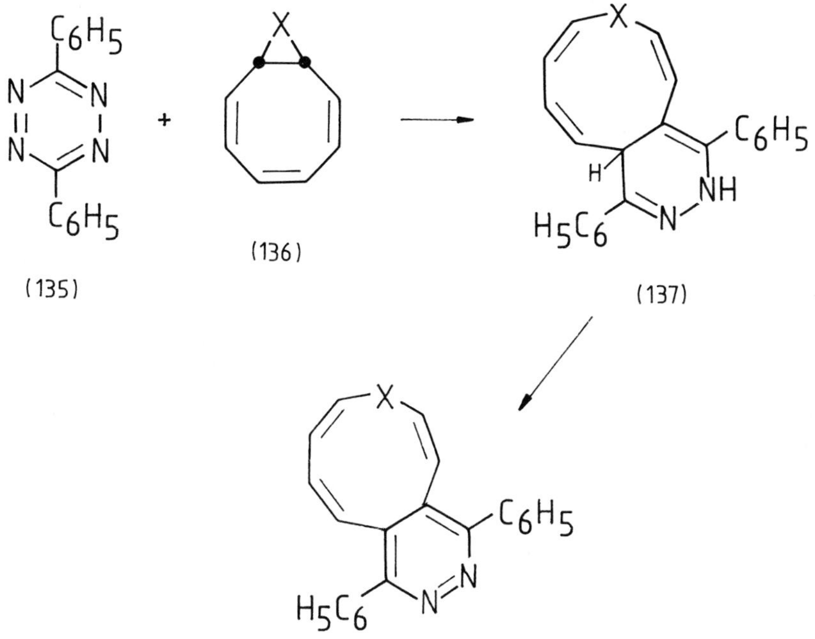

(135)   (136)   (137)

(138)

   Thirteen membered conjugated heterocycles (often called aza[13]annulene, *etc.*) are available by two photochemical routes. Generation of ethoxycarbonyl nitrene in the presence of the hydrocarbon (139) gives the adducts (140), (141), (142) and (143) with $(R = CO_2C_2H_5)$. Uv-irradiation of (142) gives the conjugated monocycle (144, $R = CO_2C_2H_5$) while (140) and (141) similarly give rings of unknown configuration; the oxaza macrocycle (145) is also formed from (140) (G. Frank and G. Schröder, Ber., 1975, 108, 3736).

   Sensitised irradiation of the azonine derivative (146) gives the monocycle (147, $R = COCH_3$) whose configuration with alternating *cis*-bonds and *trans*-bonds and three inside hydrogen atoms was deduced from its nmr-spectrum (A.G. Anastassiou and R.L. Elliott, J. Amer. chem. Soc., 1974, 96, 5257).

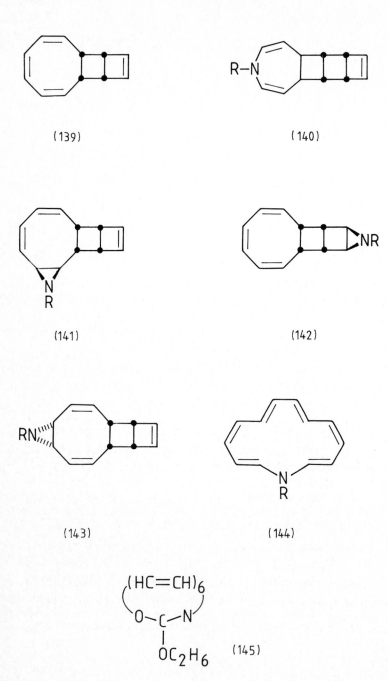

(139)

(140)

(141)

(142)

(143)

(144)

$(HC=CH)_6$

$O-C-N$

OC_2H_6

(145)

(146)   (147)

The electron withdrawing N-substituents of (144, R = $CO_2C_2H_5$) and (147, R = $COCH_3$) lead to atropic character, as in other series, but treatment of the former (or its configurational isomers) with potassium t-butoxide, and the latter with methyllithium at $-78^{\circ}C$, gives the anions (148) and (149) respectively both of which are strongly diatropic; (150) isomerises rapidly to (148) at $0^{\circ}C$. Mild protonation of Schröder's anion (146) gives the H-inside amine (149) while that of Anastassiou (150) is externally protonated to (151, R = H) but unlike the anions these amines do not interconvert at $50^{\circ}C$ indicating that the ring configuration of Schröder's system (149) is intrinsically more stable but unfavourably crowded by an additional internal proton on nitrogen. The amines (149) and (151, R = H) are clearly diatropic, as is the air-sensitive methyl derivative (151, R = CH_3) formed by quenching the anion (150) with methyl iodide. An attempt to obtain the trifluoromethyl analogue (151, R = CF_3) in the corresponding way gave the trifluoroethenyl derivative (151, R = $CF=CF_2$), probably by the usual nucleophilic attack on the iodine of CF_3I, liberating difluorocarbene and thence tetrafluoroethene from which the product may be derived. Like the acetyl (151, R = $COCH_3$) and carbethoxy (144, R = $CO_2C_2H_5$) derivatives, the trifluoroethenyl group of (151, R = $CF=CF_2$) as well as the dimethylcarbamoyl group in (151, R = $CON(CH_3)_2$) are evidently both sufficiently electron withdrawing to inhibit delocalisation of the electron pair on nitrogen since these compounds are essentially atropic, although upfield separation of the resonance of the inner protons of the latter suggests an incipient ring current (A.G. Anastassiou, R.L. Elliott and E. Reichmanis, J. Amer. chem. Soc., 1974, 96, 7825; G. Schröder, *et al.*, Angew. Chem. intern. Edn., 1974, 13, 205; Ber., 1978, 111, 84).

Two isomers, of unknown configuration, of the thirteen
membered oxygen system (153) are formed by irradiation of
the epoxides (152) and (154) but neither shows appreciable
diatropic delocalisation (W. Henne, G. Plinke and
G. Schröder, Ber., 1975, <u>108</u>, 3753).

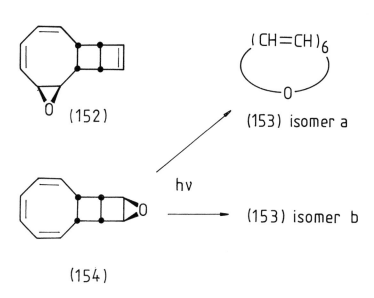

(152)

(153) isomer a

hv

(153) isomer b

(154)

The seventeen membered cycles resemble the thirteen
membered series; irradiation of the isomeric ethoxycarbonyl
nitrene adducts of *cis* cyclo-octatetraene dimer (*e.g.* 155, R
= $CO_2C_2H_5$) gives three atropic configurational isomers of
the monocycle (156, R = $CO_2C_2H_5$) (see Second Edn. Vol. IVK
p.459) each of which is converted into the H-inwards anion
(158) by potassium *t*-butoxide, whence mild protonation gives
the amine (157) without change of configuration. The
H-outwards isomer (159) is obtained by uv-irradiation of the
H-inwards isomer (157). The anion (158) shows considerable
diatropic character which is also distinct in the amines
(157) and (159) (G. Schröder, *et al.*, Angew. Chem. intern.
Edn., 1974, <u>13</u>, 205; Ber., 1978, <u>111</u>, 84).

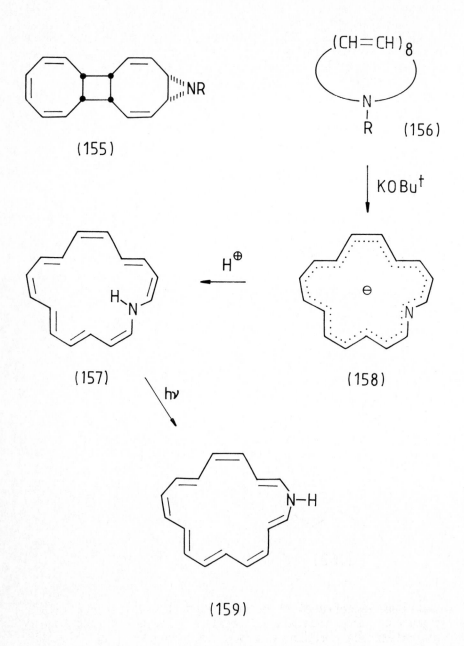

(155)

(CH=CH)$_8$

N

R   (156)

KOBu^t

(157)

H$^\oplus$

(158)

hν

(159)

Irradiation of three isomeric epoxides of *cis*-cyclo-octa-
tetraene dimer (*e.g.* 155, O replaces NR) gives the seventeen
membered oxygen ring whose probable configuration is (160)
and two inseparable isomers whose nmr-spectra are consistent
with the suggested configurations (161) and (162). The

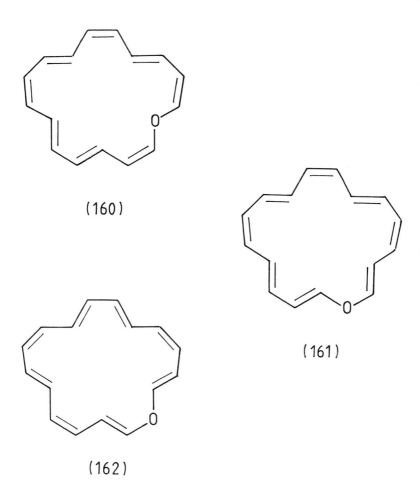

(160)

(161)

(162)

proton nmr-spectra of these compounds are weakly temperature
dependent and indicate a certain degree of diatropic
delocalisation, although no proton resonance at signifi-
cantly higher field than 4δ is present (G. Schröder,

G. Plinke and J.F.M. Oth, Ber., 1978, <u>111</u>, 99).
Delocalisation in the fifteen membered periphery of the oxygen macrocycle (165), deep red scales m.p. 255°C decomp., prepared by condensation of the dialdehyde (163) with the double Wittig reagent formed from the diphosphonium salt

(163)   (164)   (165)

(164) and lithium methoxide, is evidently enhanced by the oxygen bridges since the proton resonances show an upfield shift of 0.82-1.22 ppm compared with the model in which CH_2 replaces the macrocyclic oxygen atom, indicating a paramagnetic ring current. The sulphur analogue of (165) was not obtained in an analogous experiment, perhaps because of its potentially greater anti-aromatic character (H. Ogawa and M. Kubo, Tetrahedron, 1973, <u>29</u>, 809). The diyne systems of types (166) and (167), formed by oxidative ring closure between the alkyne groups (see also Section 3(b)) have relatively rigid frames particularly suited to delocalisation. In the nitrogen series the seventeen and twenty-one membered examples (166, X = NR', R = H, m = n = 1 or 2) show the effect of distinct diamagnetic ring currents, judged by comparison with their open chain precursors, even when the N-substituent R' is acetyl or carbethoxy, as well as hydrogen or methyl. The nineteen membered compound (166, X = $NCO_2C_2H_5$, m = 2, n = 1), deep

(166)   (167)

red crystals which decompose before melting, is clearly
paratropic (P.J. Beeby and F. Sondheimer, Angew. Chem.
intern. Edn., 1973, 12, 410 and 411; F. Sondheimer, et al.,
Tetrahedron Letters, 1974, 599 and references therein).

The oxygen and sulphur analogues in the bridged seventeen
membered series (166, X = O, R = C_2H_5, m = n = 1) and (166,
X = S, R = C_2H_5 or H, m = n = 1) are also clearly diatropic
and the proton nmr-shifts show that the size of the ring
current decreases in the sequence NCH_3 > NH > S > O >
$NCO_2C_2H_5$ > $NCOCH_3$. The corresponding sulphone (166, X =
SO_2, R = C_2H_5, m = n = 1) has remarkable paratropic
character, as equally does the syn-, but not the atropic
anti, isomer of the sulphoxide (166, X = SO, R = C_2H_5, m = n
= 1); this effect is ascribed to completion of an 18π
electron Möbius periphery using a 2p orbital of oxygen which
is suitably placed only in the syn-configuration (J.M. Brown
and F. Sondheimer, Angew. Chem. intern. Edn., 1974, 13, 337
and 339).

Of the more flexible monocycles (167) the sulphur analogue
(167, X = S) is diatropic but the oxygen one (167, X = O)
is, at most, weakly so. The sulphone (167, X = SO_2)

and the alcohol (167, X = CHOH) are predictably atropic while the ketone (167, X = CO) is paratropic. Ring current effects in the sulphur system were judged principally from the methyl proton resonance since those of the alkene groups adjacent to the heteroatom are affected by a proportion of the molecules existing in the conformation (168) in which the inside and outside protons are reversed, resulting in an upfield shift of both since the effect of the diamagnetic

(167, X = S) ⇌

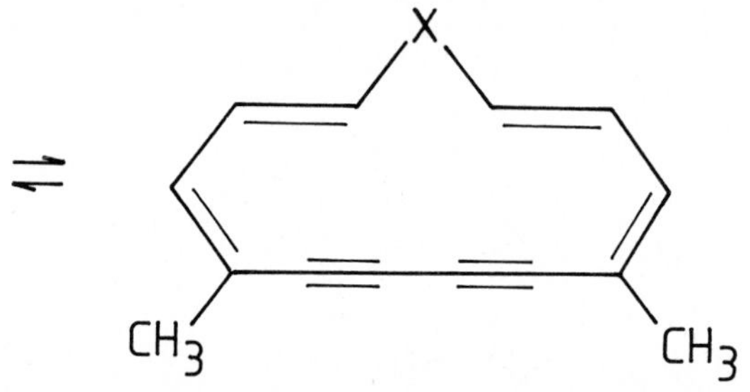

(168)

ring current is greater inside than outside the ring (R.L. Wife, P.J. Beeby and F. Sondheimer, J. Amer. chem. Soc., 1975, 97, 641).

Triazacyclononatrienes (170) are formed by heating the triaziridine (169) and have localised double bonds with a non-planar conformation. Examples with electron withdrawing

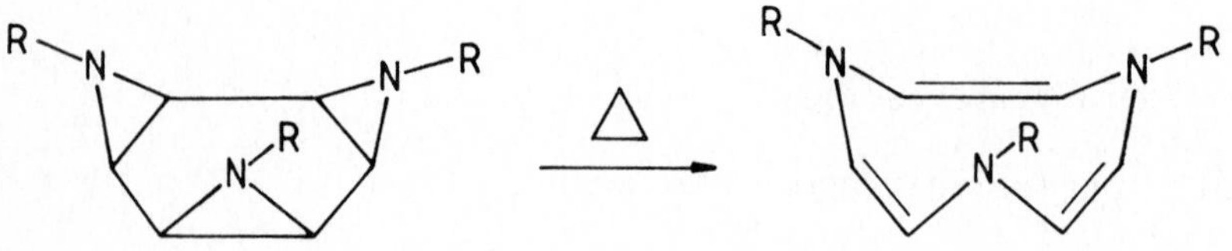

(169)   (170)

N-substituents (170, R = SO_2CH_3, SO_2CH_2-p-$C_6H_4CH_3$, or CO_2CH_3) are stable but (170, R = CH_3) is sensitive to oxygen (H. Prinzbach, *et al.*, Angew. Chem. intern. Edn., 1975, 14, 347).

The twenty-six membered ring of the tetra-imine (171) is a potential 28 π-electron system, but planarity is precluded, *inter alia*, by the σ,σ'-disubstituted biphenyl groups and its uv-spectrum shows that there is no extended conjugation (I. Agranat, Tetrahedron, 1973, 29, 1399).

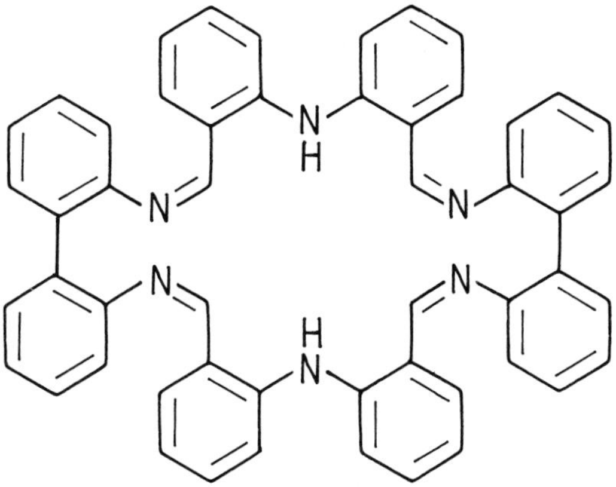

(171)

4. *Macrocyclic esters*

(a) *Synthetic macrocyclic esters*

(i) *Synthesis*

Macrocyclic lactones, rings containing a single ester linkage, have usually been studied in connection with the synthesis of natural products and this work is considered in Section 4(b), but a large number of macrocycles containing two or four ester links derived from di-carboxylic acids and di-hydroxy-compounds have been prepared. A review of synthetic methods and a comprehensive list of examples is available (J.S. Bradshaw, *et al.*, Chem. Rev., 1979, <u>79</u>, 37).

Direct synthesis of macrocyclic di- and tetra-esters (*e.g.* 174) by condensation of a dicarboxylic acid (172) and a diol (173) in the presence of an acid catalyst is sometimes effective (E.H. Hahn, H. Bohm and D. Ginsberg, Tetrahedron

Letters, 1973, 507), but the greater reactivity of acid chlorides with alcohols is more generally useful. Thus diacid chlorides of type (175) condense with ethylene glycol oligomers (176) to give macrocyclic esters (177; X, Y = CH or N), and the dicarbonyl derivative of 18-crown-6 (178; n = 3) and its thia-analogues (*e.g.* 179) are similarly obtained (R.M. Izatt, *et al.*, J. Amer. chem. Soc., 1977, <u>99</u>, 2365; J.S. Bradshaw, *et al.*, J. heterocyclic Chem., 1978, <u>15</u>, 825).

An efficient synthesis of diesters of types (186) and (187) with ring sizes from ten to twenty-one members and where X is a polymethylene bridge which may contain S, SO, or S-S, employs diacid fluorides (185) formed quantita-tively from diacids (183) by treatment with the fluoro-pyridinium salt (184) and trimethylamine, and bis-tri-phenyltin aryloxides (or alkoxides) such as (182), obtained from the diols by reaction with bis-triphenyltin oxide (181) under conditions of azeotropic dehydration with benzene. The high yields, up to 80%, of the cyclic di-esters (with minor amounts of tetra-esters and higher oligomers)

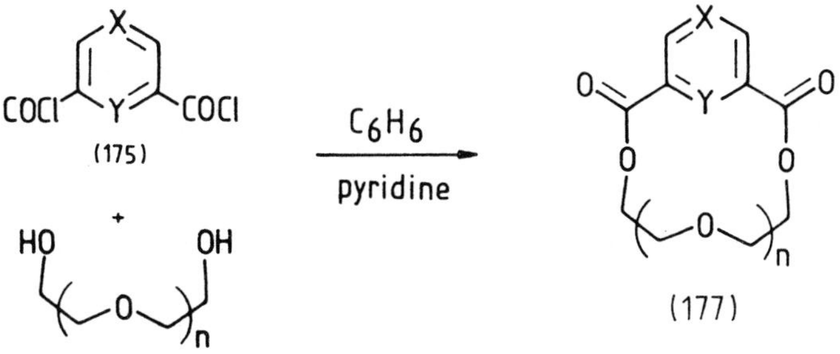

(175)

(176)

(177)

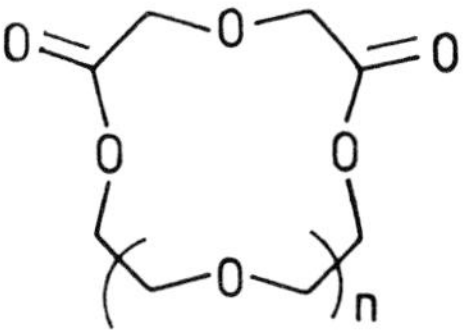

(178)

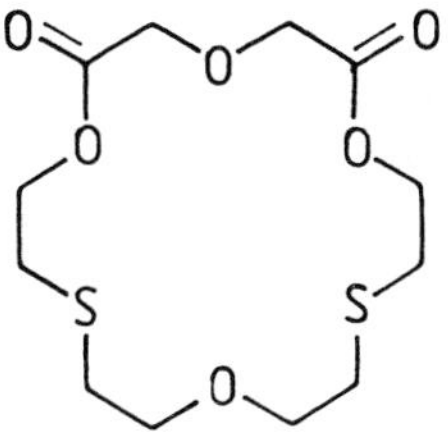

(179)

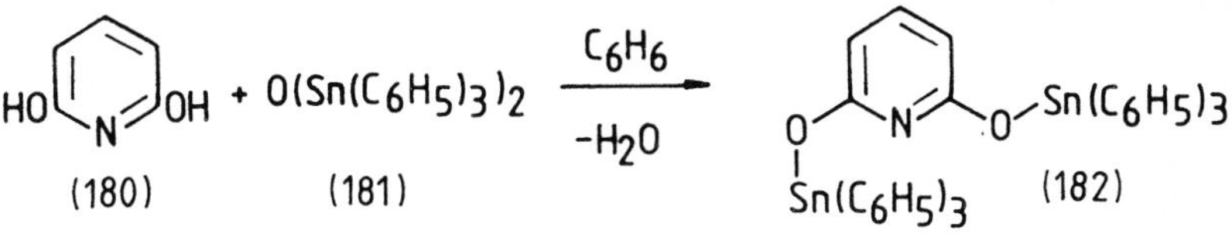

(180) (181) (182)

obtained in refluxing benzene at 10^{-2} M concentration, are attributed to template participation by tin atoms; some of the products complex calcium selectively (C. Picard, L. Cazaux and P. Tisues, Tetrahedron, 1986, 42, 3503).

Condensation of diacid salts with dihalides is also an effective method for the formation of macrocyclic esters, particularly if rigid o-disubstituted benzene units are present, as in other macrocyclisations. Condensation of dipotassium phthalate (188) with an aliphatic $\alpha\omega$-dibromide can give a macrocyclic di-ester (189), or a tetra-ester (190). Odd-membered rings are not usually produced even if n is odd (see Second Edn. Vol. IVK, p.452). Except for the eleven-membered example (190, n = 5) even-membered ring products are isolated in each experiment when n is varied from 2 to 12, in apparent agreement with earlier results and predictions. The products described are, however, those which crystallise from an extract of the total reaction

(195)

product, many in yields around 5%, so that solubility characteristics as well as cyclisation efficiency may influence the results. Similarly the condensation of *o*-bisbromomethylbenzene with dipotassium salts of dicarboxylic acids gives cyclic tetra-esters (191) in which rings of 16, 20, 28 and 32 members are apparently favoured (S.E. Drewes and P.G. Coleman, J. chem. Soc. Perkin I, 1974, 2578, previous papers in this series and references therein).

The silver salt (192) reacts with αα'dibromo-*p*-xylene (193) in acetonitrile to give the cyclic di-ester (194) m.p. 238-239°C, which forms the [2,2]paracyclophane (195) upon uv-irradiation in 1,2-dimethoxyethane in yields approaching 70% (M.L. Kaplan and E.A. Truesdale, Tetrahedron Letters,

(196)

R·N=N·R
(AZBN)

$R = -\overset{\displaystyle CH_3}{\underset{\displaystyle CH_3}{\overset{|}{\underset{|}{C}}}}- CN$

(197)

1976, 3665).

The twelve-membered ring of the bisphenolic ester (197) is relatively efficiently formed (55%) by radical addition to the diene (196); a remarkable 4% yield of the symmetric product (198) is observed, implying *either* a primary radical intermediate *or* concerted bonding to the initiating radical and between the methylene groups (H. Kammerer and J. Pachta, Makromol. Chem., 1977, 178, 1659).

Cyclisation by pyridine ring formation (Hantzch synthesis) gives the polyether pyridine di-esters (200, n = 2 or 3) and the corresponding dipyridine tetra-esters (201, n = 0-3) when the bis-acetoacetic ester (199) condenses with formaldehyde and ammonium carbonate; *N*-Methylation and reduction of the di-esters (200) to the 1,4-dihydropyridine state gives enzyme-like reducing agents, and a series of compounds capable of enantiospecific reduction of carbonyl compounds has been described in Section 1(g) (T.J. van Bergen and R.M. Kellogg, J. Amer. chem. Soc., 1977, 99, 3882).

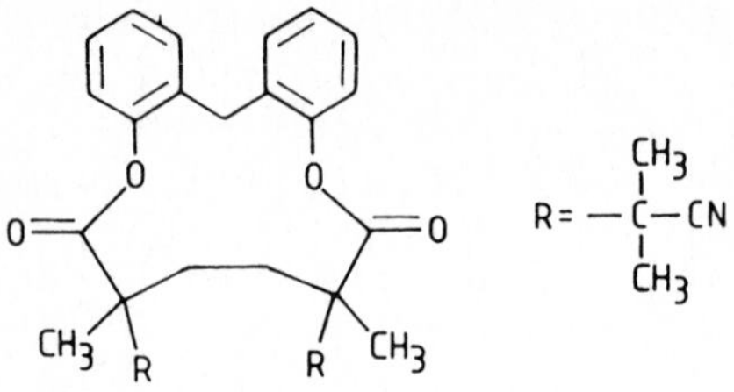

(198)

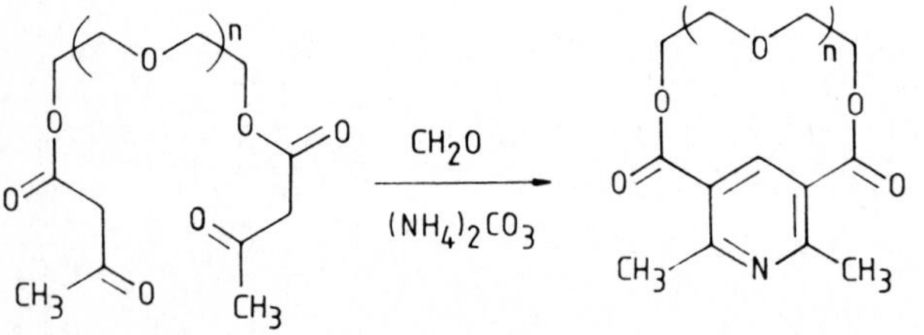

$$\xrightarrow[\text{(NH}_4)_2\text{CO}_3]{\text{CH}_2\text{O}}$$

(199)   (200)

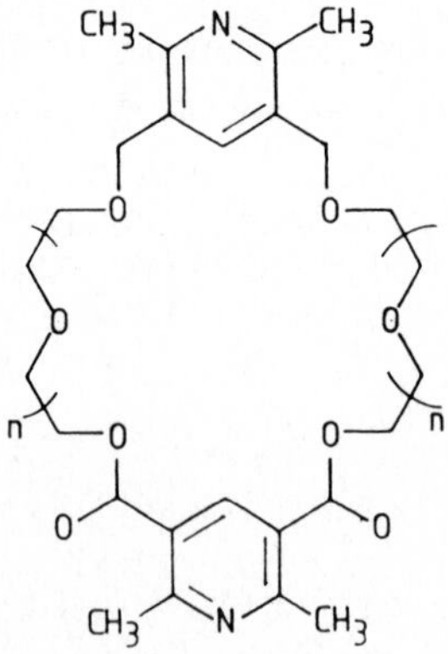

(201)

(ii) Metal complexes of synthetic macrocyclic esters

The ester analogues of the crown ethers, such as the di-ester (178) sequester alkali and alkaline earth ions in a similar way to the ethers but the complexes are less stable (R.M. Izatt, *loc. cit.*). The infrared spectrum of the crystalline di-ester (178, n = 3) shows carbonyl stretching vibrations at 1715 and 1730 cm^{-1} (which are anomalous compared with its larger and small ring analogues (178, n = 2 and 4) and thia-analogues (*e.g.* 179), all of which have $\nu_{C=0}$ 1750-1755 cm^{-1}) while that of its complex with potassium thiocyanate has one such band at 1740 cm^{-1}, interpreted on the basis that carbonyl oxygen atoms are not involved in complexation but change to a coplanar conformation (G.E. Maas, *et al.*, J. org. Chem., 1977, **42**, 3937).

The macrocyclic poly-ether ester of pyridine-2,6-dicarboxylic acid (177, X = CH, Y = N, n = 3) forms complexes with Na^{+}, K^{+}, Ba^{++} and Ag^{+} with log K 4.3-4.9, but the stability is due to the pyridine nitrogen since the benzene analogue (177, X = Y = CH, n = 3), and the macrocycle derived from pyridine-3,5-dicarboxylic acid (177, X = N, Y = CH, n = 3) where nitrogen is outside the macrocyclic ring, show no heat of reaction with any cation (R.M. Izatt, *loc. cit.*). The carbonyl stretching frequency of the pyridine-2,6-diester (177, X = CH, Y = N, n = 3), 1730 cm^{-1}, changes to a doublet at 1725 and 1715 in the potassium thiocyanate complex (J.S. Bradshaw, *et al.*, J. heterocyclic Chem., 1978, **15**, 825).

(b) Macrocyclic esters of natural origin (macrolides)

(i) Structural types

Several natural products in this class have relatively simple structures exemplified by phoracantholide I (202) excreted, along with related hydroxy and unsaturated lactones and other compounds, by the Australian beetle *Phoracantha semipunctata* for defensive and/or species recognition purposes, and brefeldin A (203) which has attracted attention because of its structural relationship to the prostaglandins (B.P. Moore and W.V. Brown, Austral., J. Chem., 1976, **29**, 1365; Y. Yamamoto, A. Hori and

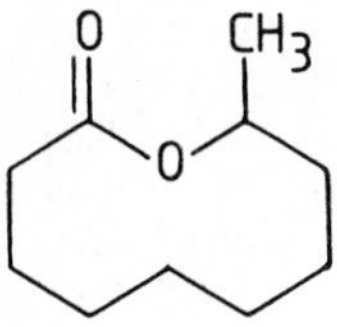

(202)

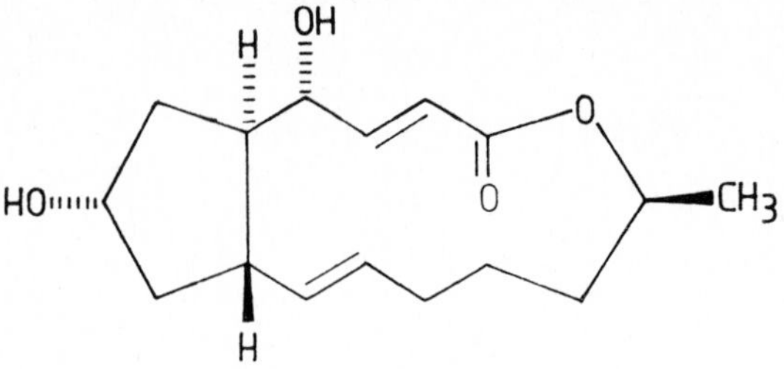

(203)

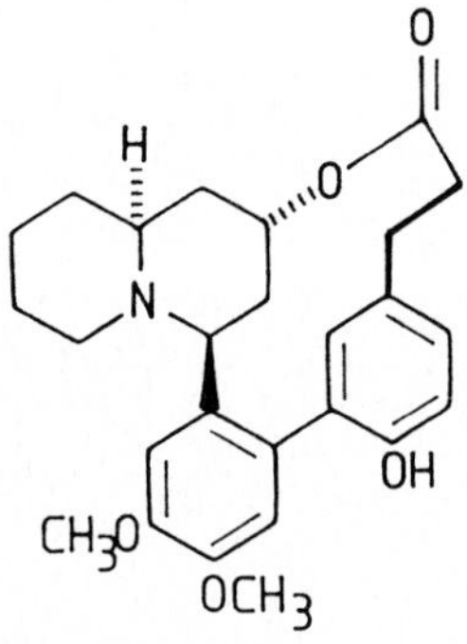

(204)

C.R. Hutchinson, J. Amer. chem. Soc., 1985, 107, 2471).

Some of the *Lythracea* alkaloids *e.g.* decinine (204) are also macrolides; other members of this series have fourteen-membered lactone rings; see Section 5 for further macrocyclic types (J.P. Ferris, *et al.*, J. Amer. chem. Soc., 1971, 93, 2963).

The majority of macrolides are, however, metabolites of micro-organisms and many have antibiotic activity; the two major structural classes are the poly-oxo macrolides usually with twelve, fourteen or sixteen-membered rings such as erythromycin (205; R^1 is the sugar desosamine and R^2 is cladinose) and mycinamycin whose aglycone is mycinolide V

(205)

(206)

(207)

(206), and the polyene type represented by amphotericin B (207) whose thirty-eight-membered ring accommodates seven conjugated *trans* double bonds (W. Mechlinski, *et al.*, Tetrahedron Letters, 1970, 3873; 3909).

The members of the ansamycin group of macrocyclic lactams, for example rifamycin B (208), are often classed with the macrolides to which they are closely related in stereochemical configuration (see Section 4(a)(iii)).

(208)

Several cyclic oligo-esters such as nonactin (209, R = CH$_3$), some of which also contain amide links as in valinomycin (210), are grouped together because of their ability to complex alkali and alkaline earth metal ions, particularly potassium.

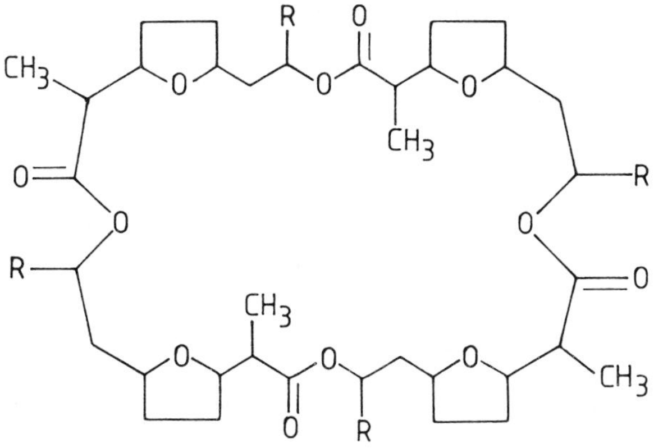

(209)

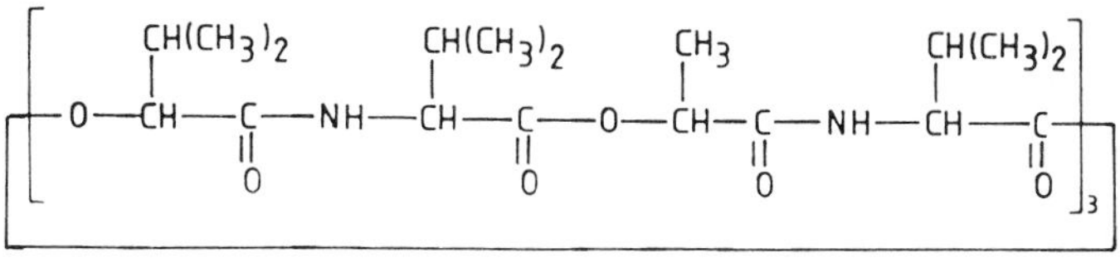

(210)

(ii) *Metal complexes*

Compounds of the above class can transport metal ions through lipophilic membranes, since the hydrocarbon groups are turned outwards in the complexes, and this aspect of their chemistry is included in a review of biological ion transport (P. Langer, Angew. Chem. intern. Edn., 1985, 24, 905). Alkali metal complexes of the ionophoric macrolides differ from those of the crown ethers and their carbonyl derivatives (to which they are frequently likened) because the metal ion is bonded to carbonyl oxygen (see Section 4(a)) and, more profoundly, because the donor atoms form a three dimensional array unlike the typically planar geometry of the smaller crown ether ligands with a similar number of donor atoms (see Section 1(e)).

X-ray crystallography shows that the potassium ion is complexed with octahedral geometry by six ester carbonyl oxygen atoms of valinomycin (210) when tri-iodide is the counter-ion; adjacent amide groups are hydrogen bonded together to form a belt round the complex cation (K. Neupert-Laves and M. Dobler, Helv., 1975, 58, 432), while a nitro-oxygen penetrates this ligand to occupy a seventh coordination position in the potassium picrate complex (J.A. Hamilton, *et al.*, J. Amer. chem. Soc., 1981, 103, 5880).

Nonactin (209, R = CH_3) complexes potassium with cubic coordination by four carbonyl and four tetrahydrofuran oxygen atoms as is shown by an X-ray crystallographic study of the thiocyanate complex (M. Dobler, J.D. Dunitz and B.T. Kilburn, Helv., 1969, 52, 2573.

(*iii*) *Stereochemistry*

There is remarkable correlation between the absolute configurations of the ten or more asymmetric centres of the various polyoxo macrolides, each of which conforms to the pattern of configurations shown for a fourteen-membered example such as erythromycin (205) by the Fischer projection (211); this sequence persists with the addition of further carbon atoms, asymmetric centres, or double bonds (W.D. Celmer, Pure appl. Chem., 1971, 28, 413). Further, the asymmetric centres of the ansamycin group of macrocyclic lactams fit Celmer's model, rifamycin B (208) differing only in configuration at C-21, for example (W. Kump and H. Bickel, Helv., 1973, 56, 2348 and references therein).

The conformations of the fourteen-membered macrolides have been studied by X-ray crystallography and by analysis of proton coupling constants and ^{13}C shifts in their nmr-spectra. Comparison of these methods, together with the small temperature and solvent dependence of the latter, show that these molecules have rather rigid conformations differing little between the crystalline and solution

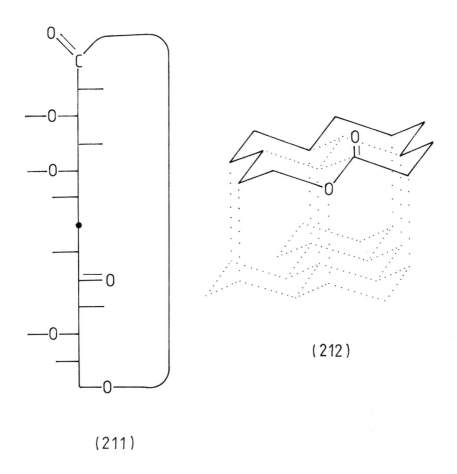

(212)

(211)

states. Conformations based on fragments of the diamond
lattice have been considered and (212) is apparently
favoured and maintained by hydrogen bonding between the
oxygen substituents (J.J. Perun, *et al.*, Tetrahedron, 1973,
<u>29</u>, 2525; J.G. Nourse and J.D. Roberts, J. Amer. chem. Soc.,
1975, <u>97</u>, 4584; Y. Terui, *et al.*, Tetrahedron Letters, 1975,
2583; S. Omura, *et al.*, *ibid.*, 1975, 2939; and references
therein). These studies combine to show that the oxygen
substituents are almost invariably located on one side of
the molecule and the alkyl groups on the other.

560

(iv) Synthesis

Synthetic approaches to macrocyclic mono-lactones and related natural products are generally divided into ring expansion methods and those depending upon cyclisation of acyclic precursors; the former are usually more suited to simpler structures.

Oxidative cleavage of bridgehead double bonds of cyclic enol ethers (see Second Edn. Vol. IVK, p.450) gives aromatic oxo-lactones of type (214) starting from benzofuran (213, n = 0, m = 4, 5, 6, or 10) or benzopyran (213, n = 1, m = 4, 5, 6, or 10) precursors (J.R. Mahajan and H.C. Aranjo, Synthesis, 1975, 54 and references therein).

(213)   (214)

Ring expansion by β-scission of alkoxy radicals is used in a synthesis of phoracantholide I (202), Scheme 5 (H. Suginome and S. Yamada, Tetrahedron Letters, 1985, 3715).

Cyclisation of linear precursors may be achieved by acid catalysed lactone formation from the hydroxy-acid, often termed the *macrolide seco-acid*, as in the synthesis of the alkaloid decinine (204) (I. Lantos and B. Loev, Tetrahedron Letters, 1975, 2011).

The rigid stereochemistry of the *cis*-enediyne hydroxy-acid (218) favours formation of the twelve-membered lactone (219) in 74% yield with *p*-toluenesulphonic acid in refluxing benzene (D. Guillerm and G. Linstrumelle, Tetrahedron Letters, 1985, 3811).

More commonly a more reactive derivative of the carboxyl group is used for lactone formation, such as a thiol ester; esters of pyridine-2-thiol are particularly effective, as in the final step of the synthesis of tetranactin (209, R =

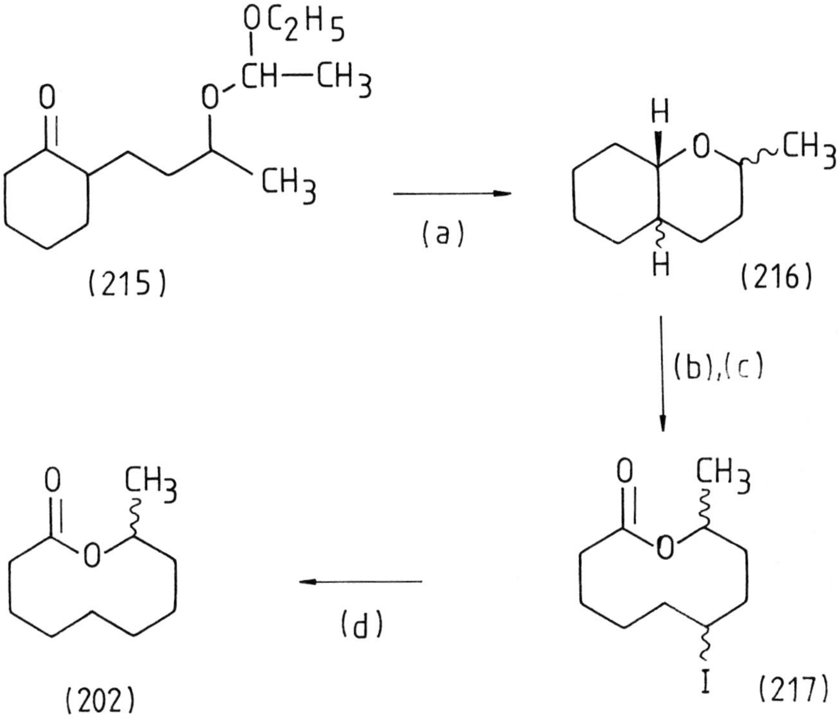

Reagents: (a) H$^+$; (b) HgO, I$_2$, pyridine; (c) uv., C$_6$H$_6$;
(d) Bu$_3$SnH.

Scheme 5

C$_2$H$_5$). This molecule, which has overall symmetry and is not optically active, is obtained in 51% yield when the carboxyl group of the linear tetrameric triester, assembled by esterification between two molecules of dimeric ester formed from pairs of enantiomeric homononactic acid units (220, 2R 3R 6S 8S and 2S 3S 6R 8R) is converted into its 3-cyano-4,6-dimethyl-2-mercaptopyridyl ester and injected into a boiling solution of p-toluene-sulphonic acid in dichloromethane at high dilution (U. Schmidt and J. Werner, Chem. Comm., 1986, 996).

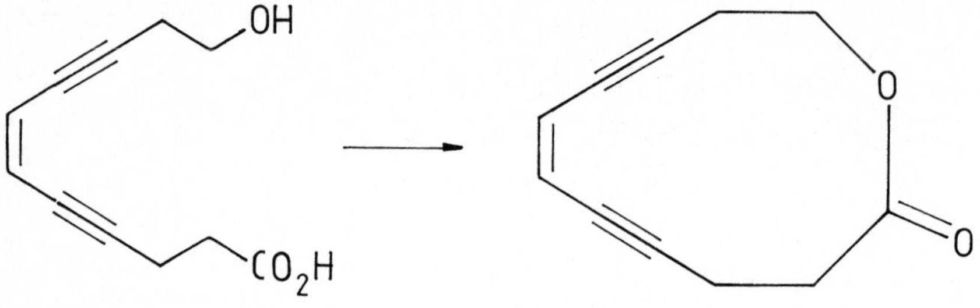

(218)

(219)

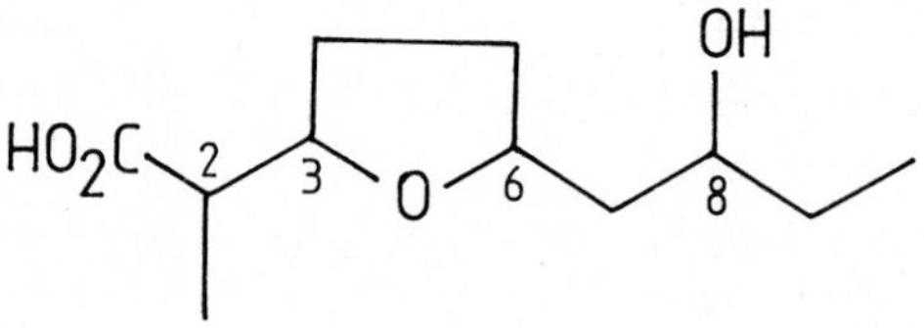

(220)

Thiopyridyl ester activated lactonisation is used to protect the *trans* double bond of the intermediate (221) from radical attack by C-10 during debromination with polymer bound tributyltin hydride in the synthesis of the highly labile thromboxane A$_2$ (222, half-life 30 sec. in aqueous medium at pH 7.4), which has not been isolated from nature but whose biochemical participation has been inferred. The lactone ring of (221) is opened by potassium

(221)

(222)

Reagents: (a) diethyl azodicarboxylate (DEAD), $(CH_3O)_3P$, CH_2Cl_2; (b) $SnH(C_4H_9)_3$, catalytic amount azo-bis-isobutyronitrile (AZBN), uv; (c) $(CH_3)_3SiOK$, THF.

trimethylsiloxide and the synthetic material gave the anticipated biochemical response in several tests (W.C. Still, *et al.*, J. Amer. chem. Soc., 1985, 107, 6372).

Lactonisation of hydroxyl groups by thiopyridyl esters is facilitated in the presence of thiophilic metal salts such as silver perchlorate, which is used, for example, in the cyclisation step in the synthesis of nonactin (209, R = CH_3) (U. Schmidt, *et al.*, Ber., 1976, 109, 2628). Tertiary butyl thiol esters react similarly in the presence of mercury(II) trifluoroacetate or, better, trifluoromethane sulphonate, which is used to close the ring of methynolide (223), the aglycone of methymycin (S. Masamune, *et al.*, J. Amer. chem. Soc., 1975, 97, 3512; 3513).

Synthesis of pyrenophorin (see Second Edn. Vol. IVK p.451) is effected by base catalysed lactonisation of the carbonylimidazole derivative of the carboxyl group (R.A. Raphael, *et al.*, J. chem. Soc., Perkin I, 1976, 1718).

Activation of the carboxyl group by treatment with tri-

(223)

chlorobenzoyl chloride followed by *NN*-dimethylaminopyridine in boiling benzene effects cyclisation of the seco-acid of mycinolide V (206), synthesised from chiral fragments. Protection of the other primary and secondary hydroxyl groups as β-trimethylsilylethoxymethyl (SEM) acetals, removed by lithium tetrafluoroborate in the last step, is used to direct lactonisation to the correct hydroxyl position (R.W. Hoffmann, *et al.*, Angew. Chem. intern. Edn., 1986, 25, 1028).

The effect of ring size on lactone formation from salts of ω-bromoalkanoic acids has been described in Chapter 55, Section (a) (G. Illuminati, *et al.*, *loc. cit.*); use of potassium carbonate and dimethylsulphoxide as solvent favours monomeric ring closure of longer chain acids. Thus slow addition of 11-bromoundecanoic acid to this reagent gives the twelve-membered lactone in 89% yield with only 9% of the twenty-four-membered cyclic dimer (C. Galli and L. Mandolini, Gazz., 1975, 105, 367).

Ring closure by carbon-carbon bond formation in linear esters has been used in several syntheses of natural macrolides. Cyclisation between anions, stabilised by phenylsulphonyl groups, and allylic acetate functions catalysed by palladium(0) complexes (see Chapter 55 Section 1(a)(i)) gives 8, 10, 12, 14, or 16 membered lactones as illustrated by synthesis of ±-recifeiolide (225, R=R'=H) from the ester

(224). The intermediate lactone (225, R = CO_2CH_3, R' = $SO_2C_6H_5$) is converted into the natural product (225, R=R'=H) by successive hydrolysis and decarboxylation with tetra-methylammonium acetate in hexamethylphosphoramide (HMPA) and reductive cleavage of the phenylsulphonyl group with sodium amalgam buffered with disodium hydrogen phosphate in THF-ethanol at $-20°C$. Phoracantholide I (202) and related compounds were similarly synthesised (B.M. Trost and T.R. Verhoeven, J. Amer. chem. Soc., 1980, 102, 4743).

Efficient regioselective coupling of dithioketals to allylstannanes in the presence of dimethyl(methylthio)-sulponium tetrafluoroborate (DMTSF) leads to a 48% yield of the fourteen-membered ketone (227) from the dithioketal (226) in 10^{-2} M solution (B.M. Trost and T. Sato, J. Amer. chem. Soc., 1985, 107, 719).

(226)   (227)

Condensation between the methylene group of β-ketophospho-
nate esters and an aldehyde or a ketone carbonyl group with
formation of a carbonyl-conjugated double bond (the
Wadsworth-Emmons reaction) catalysed by potassium carbonate
and 18-crown-6 (see Section 1(e)(iii)) is effective in the
cyclisation step in the syntheses of several natural
macrolides containing an $\alpha\beta$-unsaturated ketone group,
including that of O-mycocinosyl tylonolide (229, X = H, OH;
R = H) via the intermediate (229, X = 0, R = Si(CH$_3$)$_2$C$_4$H$_9$)
from the linear phosphonate (228). Removal of the butyl-
dimethylsilyl protecting groups with hydrogen fluoride and
pyridine, reduction of the γ-lactone and dienone carbonyl
groups with diisobutylaluminium hydride (DIBAL) and re-
oxidation of the latter selectively with DDQ gives the
natural epimer at the methyl substituted position adjacent
to ketone carbonyl (K.C. Nicolaou, S.P. Seitz and
M.R. Pavia, J. Amer. chem. Soc., 1982, 104, 2030).
The same cyclisation procedure is used in alternative
syntheses of tylonolide (229, X = H, OH; R = R' = H) and
methynolide (223) (T. Tanaka, et $al.$, Tetrahedron Letters,
1986, 3651; 3647), and of picronolide (230) where hydroxyl
protection by 4-methoxybenzyl and 3,4-dimethoxybenzyl groups
is crucial (W.C. Still and J.V. Novack, J. Amer. chem. Soc.,
1984, 106, 1148).

(228)

K₂CO₃, 18-C-6
toluene, 70°C, 10⁻³ M

(229, X = H, OH ; R = H)
with correct configuration.

(230)

A valuable review of the structural classification, stereochemical characteristics, synthesis, and biochemical features of the macrolides with a comprehensive list of examples is available (S. Masamune, G.S. Bates and J.W. Corcoran, Angew. Chem. intern. Edn., 1977, 16, 585). The macrolide antibiotics have been reviewed (Macrolide Antibiotics: Chemistry, Biology and Practice, ed. S. Omura, Academic Press, Orlando, 1984).

5. *Miscellaneous compounds of interest*

(a) *Compounds with rings containing one or more oxygen atoms*

Chromatography on silica gel of an ether extract of air-dried red alga *Laurencia obtusa* gives crystals of obtusenyne (231) whose structure and absolute configuration has been determined by X-ray diffraction (T.J. King, *et al.*, Tetrahedron Letters, 1979, 1453).

Similar extraction (with chloroform:methanol 4:1), and chromatography, of the freeze-dried Australian red marine alga *Phacelocarpus labillardieri* gives six compounds each with an acetylenic macrocycle fused to a pyrone ring. The

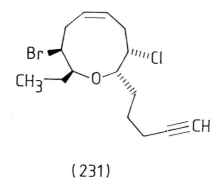

(231)

γ-pyrone (232, 12% of the extract) and the dibromo-α-pyrone (233, 0.2%) are representative examples of this new class of natural product (R. Kazlauskas, *et al.*, Austral. J. Chem., 1982, <u>35</u>, 113; J. Shin. V.J. Paul and W. Fenical, Tetrahedron Letters, 1986, 5189).

The nine-membered cyclic peroxide rhodophytin (234) is isolated from the alga *Laurencia yamada* and shows unusual stability to alkali, although in carbon tetrachloride solution the exocyclic double bond moves slowly into conjugation (W. Fenical, J. Amer. chem. Soc., 1974, <u>96</u>, 5580).

The gorgonians and soft corals contain cembranolide diterpenes including jeunicin (235) from *Eunicea mammosa* gathered near Jamaica. Related toxic compounds differing in the position of the ether bridge are found in similar species from different collection sites; they may have a protective function and be biosynthesised by a symbiotic unicellular alga which is absent, like these diterpenes, from specimens gathered in temperature waters (D.J. Faulkner, Tetrahedron, 1977, <u>33</u>, 1421 and references therein).

The most potent member of the brevetoxin group of toxins (GB toxins) from *Gymnodinium breve*, the dinoflagellate responsible for the "red tide" in the Gulf of Mexico which has caused massive fish kills and "neutoxic shellfish poisoning" in humans, is brevetoxin A, GB-1 toxin (236, R =

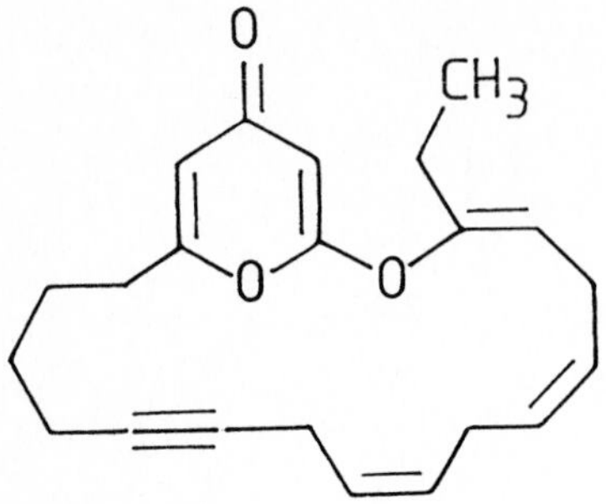

(232)

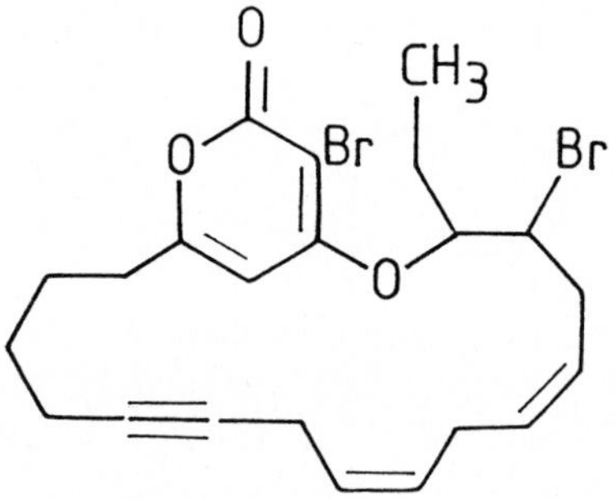

(233)

CHO), m.p. 197-199 and 218-220°C, whose structure has been elucidated by X-ray crystallographic examination of the dimethyl acetal (236, R = $CH(OCH_3)_2$) and detailed nmr spectroscopic studies. The associated toxin GB-7 is the alcohol (236, R = CH_2OH) (Y. Shimizu, *et al.*, J. Amer. chem. Soc., 1986, 108, 514).

A new synthetic approach to fused polycyclic systems of the type found in the brevetoxins (e.g. 236), and in the complex marine macrolides from the sponge *Halichondria okadai*, which have high activity against leukemia and melanoma (D. Vemura, *et al.*, J. Amer. chem. Soc., 1985, 107, 4796), depends on transannular bonding of larger macrocycles, Scheme 6. Ester coupling between hydroxyl and carboxyl protected pyrans (237) and (238), debenzylation and lactone formation *via* the pyridinethiol derivative (see

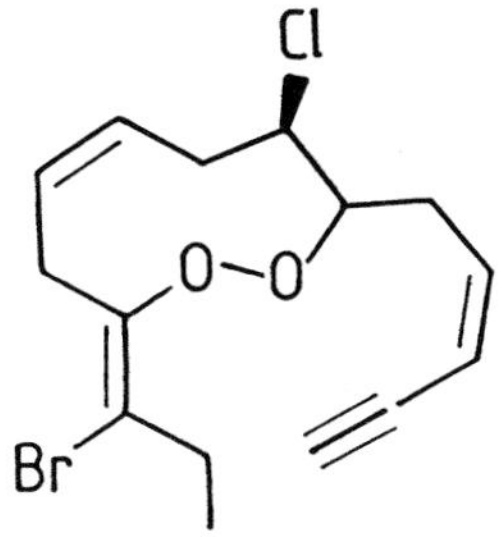

(234)

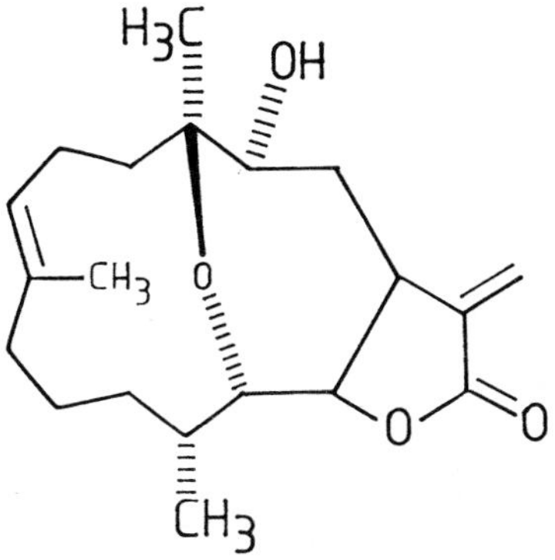

(235)

Section 4(b)(iv)) gives the di-lactone (239, X = O) which gives its dithione analogue (239, X = S) upon treatment with dimethoxyphenylthionophosphine sulphide (Laureson reagent: M. Cava and M.I. Levinson, Tetrahedron, 1985, 5087). Reductive coupling between the thione carbon atoms and quenching the resulting dithiolate anion with methyl iodide gives the bridged thioketal (240, Y = cis SCH_3) and thence

the bridgehead alkene (241). Catalytic hydrogenation or reduction with triethylsilane then gives this cis-fused system (240, Y = H), while the latter reagent in the presence of silver tetrafluoroborate gives the corresponding $trans$-fused isomer. Similar initial sequences give analogous dithioketals with six- or eight-membered central rings, $trans$ and cis fused respectively, as well as the corresponding bridgehead alkenes (K.G. Nicolaou, et $al.$, J. Amer. chem. Soc., 1986, 108, 6800, and references therein).

The N-oxide (242) of the macrocyclic diaryl ether

(236)

limacine, a dimeric benzyl isoquinoline alkaloid, is isolated from *Curarea candicans* along with its epimer (inverted at the chiral *N*-oxide centre) and an isomer bearing the *N*-oxide group on the alternative nitrogen atom. No *N*-oxides of related members of this series have been observed, suggesting that these compounds are of biosynthetic origin and not formed by N-oxidation during isolation (H. Guinaudeau, *et al.*, J. chem. Research (S), 1985, 248; (M), 1985, 2786-2793).

Cyclisation of alkenyl acetals with tin(IV) chloride, Chapter 55 Section 1(a)(i), is applicable to nine as well as to eight-membered rings containing one oxygen atom by extention of the saturated carbon chain (L.E. Oveman, *loc. cit.*).

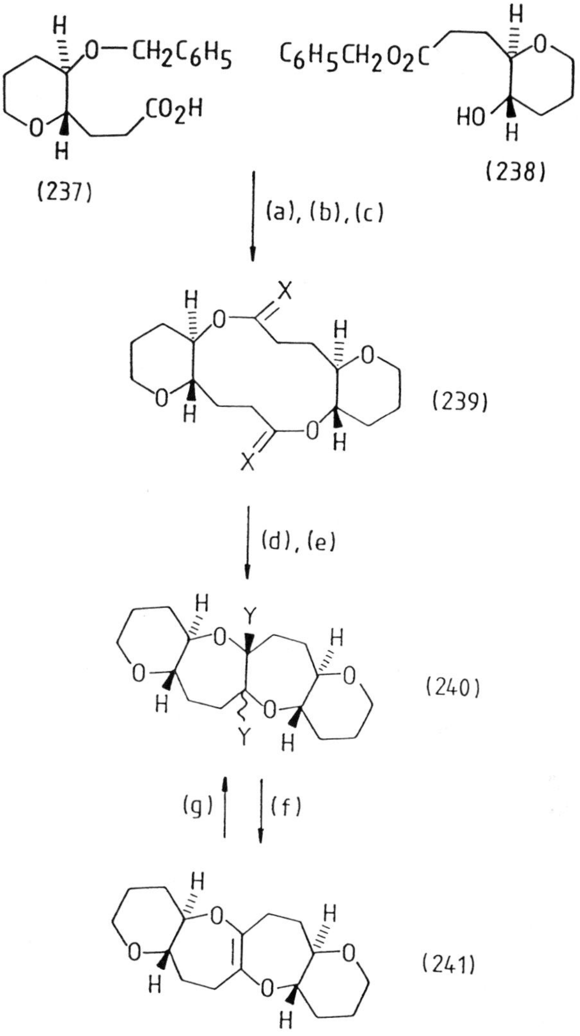

Reagents: (a) 2,2'-dipyridyl sulphide; (b) xylene, reflux; (c) dimethoxyphenylthionophosphine sulphide; (d) sodium naphthalide; (e) methyl iodide; (f) tri-butyltin hydride, AIBN or hν; (g) H$_2$-cat. or triethylsilane or triethylsilane, AgBF$_4$.

Scheme 6

(242)

(b) *Compounds with rings containing one or more sulphur atoms*

The repetitive expansion of vinylthiocine to eleven, four-teen and seventeen membered sulphur heterocycles has been noted in Chapter 55, Section 2(a)(i). An analogous sequence starting from 2-vinylthiane leads to the nine-membered ring compound (243) and thence to the twelve-membered ring analogue (244). An interesting feature of these reactions is the specific (with 243) or preponderant (with 244) formation of *trans* (E) alkene functions (E. Vedejs, *et al.*, Tetrahedron Letters, 1978, 519).

Formation of a *trans* double bond is also favoured over a *cis* double bond in the ratio 2:1 during the preparation of the thionine derivative (247) by Wittig reaction between the diphosphorane (245), derived from the corresponding bisphosphonium bromide with butyllithium, and the di-aldehyde (246); the total yield is near 2% after chromato-graphic separation. Compound (247) is an analogue of biphenylene in as much as the thionine ring is potentially aromatic, but nmr spectroscopic studies show that it is non-

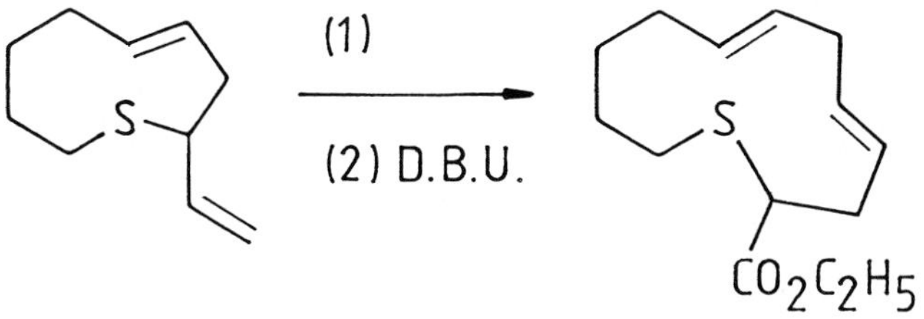

(243)   (244)

(1) = $CF_3SO_3CH_2CO_2C_2H_5$

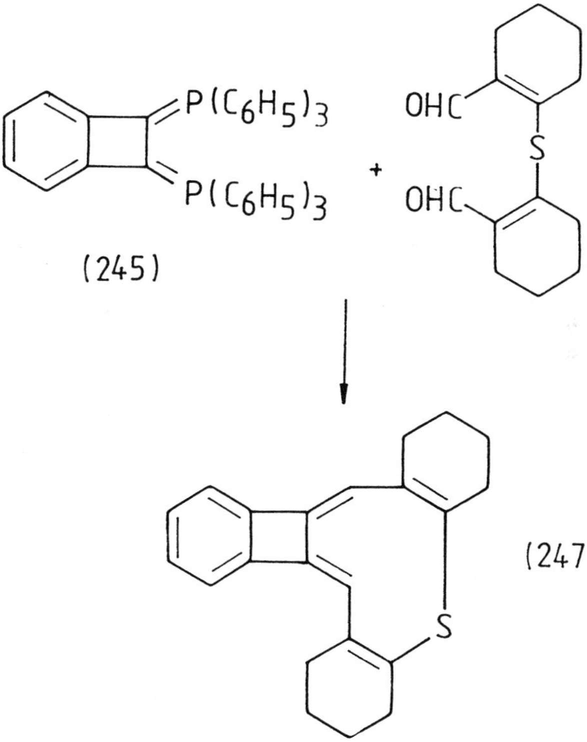

(245)   (246)

(247)

planar and atropic by comparison with the corresponding sulphoxide (247, SO replaces S) obtained from the sulphide with dilute hydrogen peroxide in acetic acid (P.J. Garratt, *et al.*, J. chem. Soc. Perkin 1, 1973, 2253).

Wittig synthesis gives relatively good yields of the cyclic sulphides (250) from dialdehyde (248) and the bis-phosphonium chlorides (249, X = O or S, Y = $(C_6H_5)_3P^+Cl^-$); the tetrafuran (251) is similarly obtained. The ring system (250, X = O or S) is also formed by Perkin condensation of

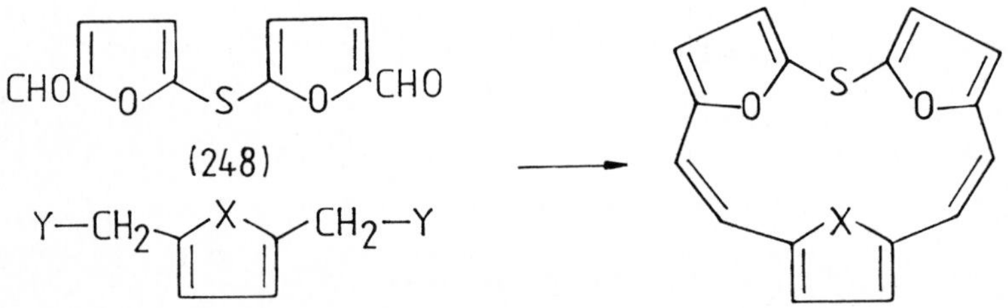

(248)

(249)

(250)

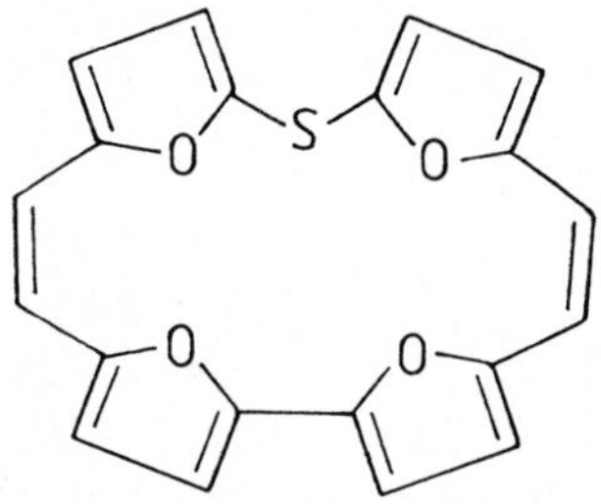

(251)

the dialdehyde (248) with the dicarboxylic acids (249, Y = CO_2H, X = O or S), but decarboxylation of the resulting di-acids is inefficient. Both systems (250) and (251) are atropic as shown by comparison of their nmr-spectra with those of the corresponding open chain sulphides (T.M. Cresp and M.V. Sargent, J. chem. Soc. Perkin 1, 1978, 1786).

Macrocyclic sulphur compounds (254) figure prominently in the synthesis of cyclophanes (255) or (256), Scheme 7, and

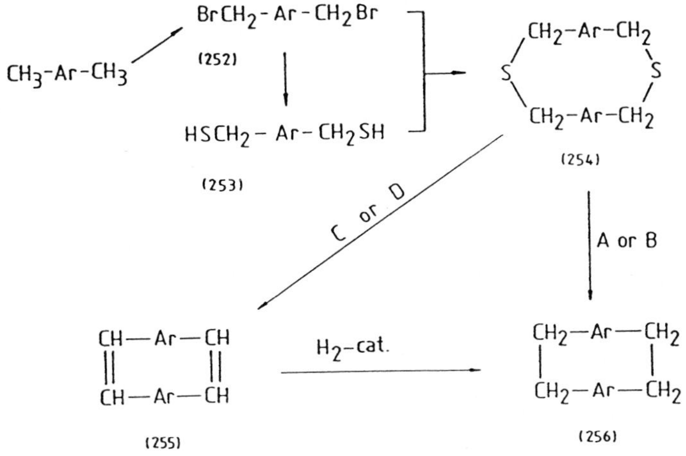

Scheme 7

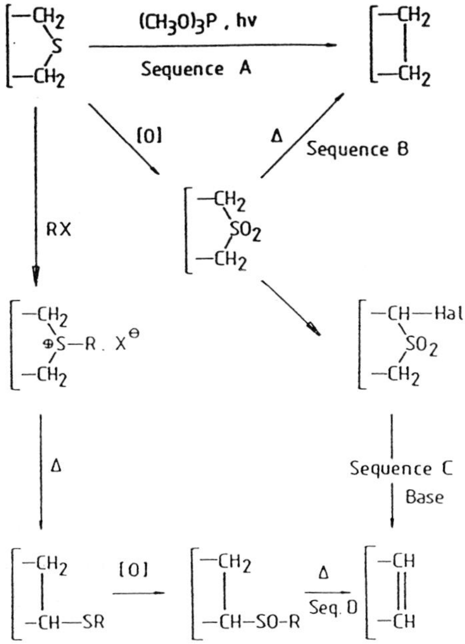

Scheme 8

these precursors are commonly formed by condensation of benzylic thiols (253, or their synthetic equivalents) with the benzylic halides (252, or tosylates, *etc.*) from which they are derived. Elimination of sulphur to complete the carbocycles, Scheme 8, is achieved by irradiation in the presence of trimethyl phosphite (sequence A), or oxidation to sulphone with thermal elimination of sulphur dioxide (sequence B), leading to carbon-carbon single bonds. Alkene links are formed by alkylation to sulphonium salts, Stephens rearrangement, and elimination of the resulting alkylthio-groups by thermolysis of their sulphoxide derivatives (sequence D), or less commonly using the Ramberg-Bäcklund reaction of α-halosulphones (sequence C).

Examples of sulphur macrocyclic intermediates in cyclophane synthesis include the *cis*-fused system (257), by sequence D (D.J.H. Funhoff and H.A. Staab, Angew. Chem. intern. Edn., 1986, 25, 742) and the pyridines (258, sequence A), (259, sequences A, B and D) and (260, sequence A) (J. Bruhin and W. Jenny, Tetrahedron Letters, 1973, 1215; V. Boekelheide, *et al.*, J. Amer. chem. Soc., 1974, 96, 1578 and 1563; J. Bruhin, W. Knewbuchler and W. Jenny, Chimia (Switz.), 1973, 27, 277).

Multiply bridged cage sulphides are prepared similarly. The ultimate example is the hexaphenylbenzene derivative (262) formed in 0.1% yield when the hexathiol (261, X = SH) condenses with the hexabromide (261, X = Br) in the presence of carbonate, at high dilution, so that all six sulphide links are formed in a single synthetic step. The use of bases having caesium as counter-ion to generate thiolate anions is uniquely effective in this and similar nucleo-philic substitutions. The cage compound (262) separates as a pale yellow precipitate, m.p. > 320°C which shows an intense molecular ion in its mass-spectrum. Chromatography on a chiral stationary phase reveals the presence of two enantiomers, expected to arise from dissymmetry associated with the tilt-angle of the *meta*-phenylene groups (F. Vogtle and W. Kissener, Angew. Chem. intern. Edn., 1985, 24, 794).

Ring strain in the eleven-membered sulphur ring of the diene (264), prepared from the dibromide (263) using sodium sulphide and caesium carbonate, leads to Cope rearrangement to the ring expanded isomer (266) from which the sulphur atom may be eliminated by irradiation in the presence of trimethyl phosphite at 60°C. The disulphide (265), in which Cope rearrangement is degenerate, is similarly obtained

(M. Eisen and F. Vogtle, Angew. Chem. intern. Edn., 1986, 25, 1026).

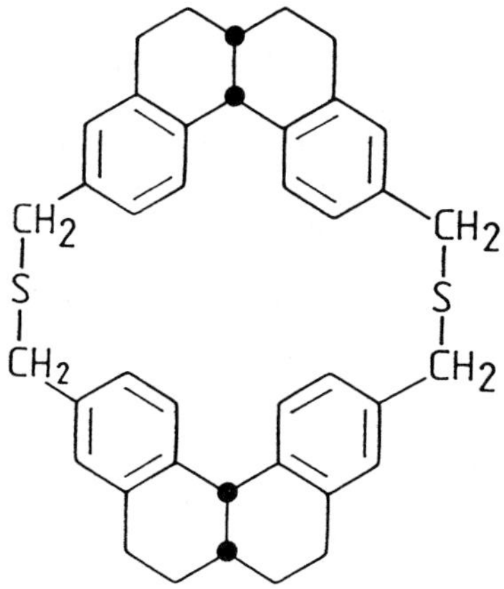

(257)

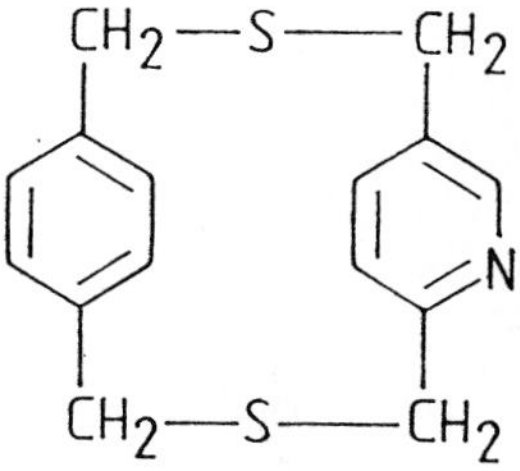

(258)

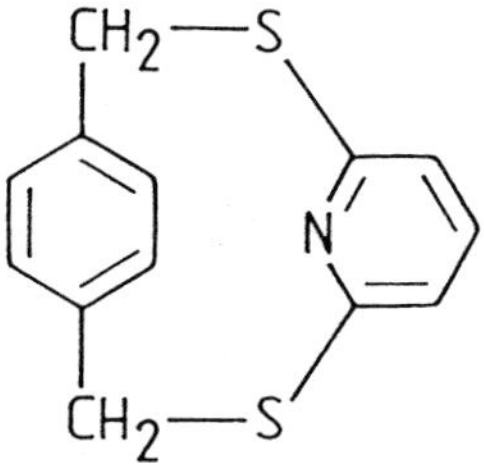

(259)

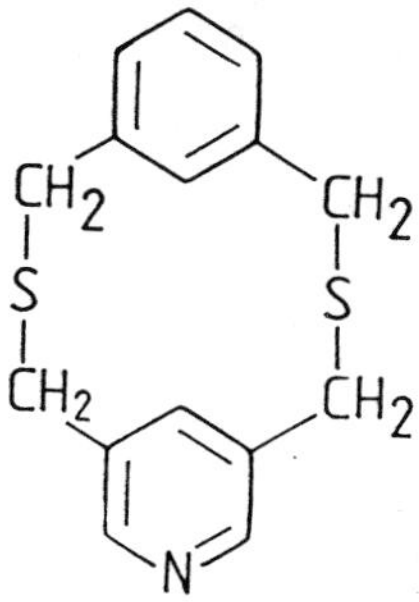

(260)

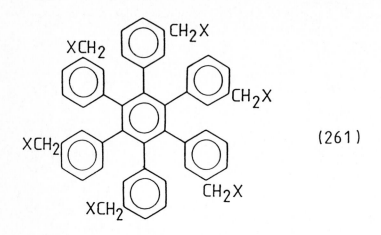

(261)

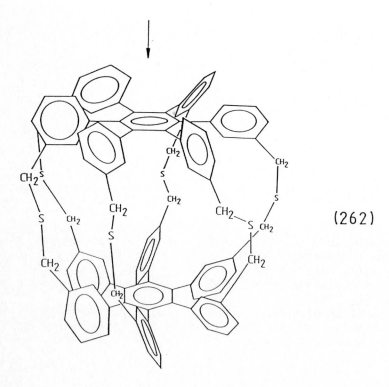

(262)

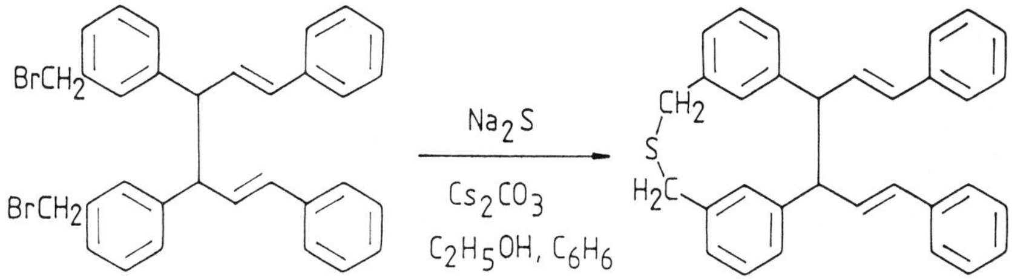

(263)

Na₂S

Cs₂CO₃

C₂H₅OH, C₆H₆

(264)

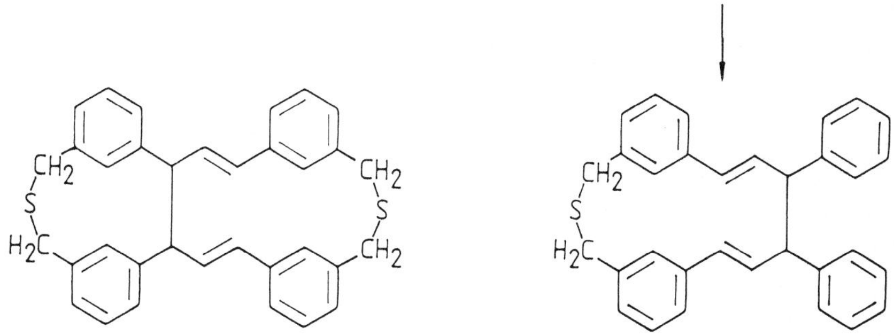

(265)

(266)

(c) *Compounds with rings containing one or more nitrogen atoms*

The conformations of the equivalent thirteen-membered rings of the diquaternary bromide, (267) determined by X-ray crystallography, are in agreement with the results of force-field calculations (B.H. Rubin, *et al*., J. Amer. chem. Soc., 1984, 106, 2088).

The nmr spectrum of the macrocyclic carbodiimide (268) has distinct signals for H_A and H_B below the coalescence

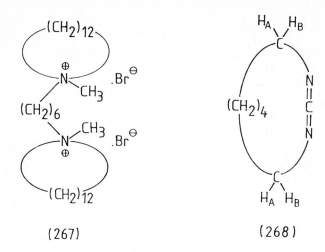

(267)   (268)

temperature of -131°C, giving an estimate of 6.7 kcal mol^{-1} for the free energy barrier to configurational inversion at nitrogen, close to that for diisopropyl carbodiimide so that the ring system has little effect (R. Damrauer, *et al.*, J. org. Chem., 1980, **45**, 1315).

Lyngbyatoxin A (269) from the alga *L.majuscula* is one of the toxins which cause dermatitis or "swimmers' itch" in Hawaiian waters. The structurally related fungal metabolite teleocidin B has similar physiological effects (J. Cardellini II, F.-J. Marner and R.E. Moore, Science, 1979, **204**, 193).

The tetrahydroazonine laufonine (270, $R^1 = R^2 = R^3 = CH_3$), with variously demethylated analogues, is extracted from the Indian plant *Cocculus Laurifolius*, attention having been drawn to the alkaloid fraction of the extract by its hypotensive activity (H. Pande and D.S. Bhakuni, J. chem. Soc. Perkin I, 1976, 2197).

Total synthesis of (±)dihydroperiphylline (271) is achieved by the elegant sequence illustrated in Scheme 9; the product is identical with that obtained by reduction of natural periphylline with sodium cyanoborohydride but natural dihydroperiphylline was not available for direct comparison. The isomeric alkaloid celacimine, in which the N-cinnamoyl group is at the alternative position indicated beside structure (271), has also been synthesised

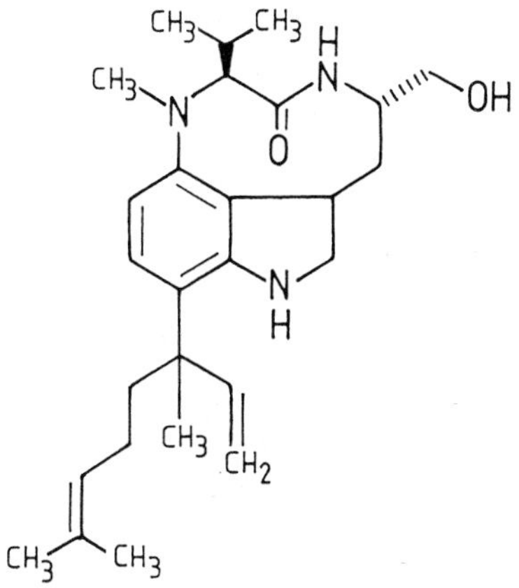

(269)

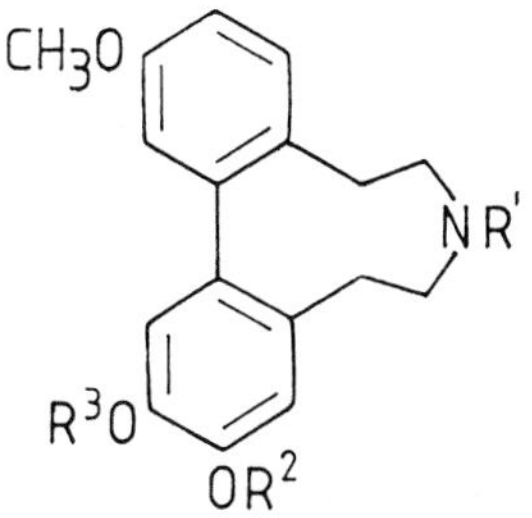

(270)

(H.H. Wasserman and H. Matsuyama, J. Amer. chem. Soc., 1981, 103, 461 and references therein).

(d) *Compounds with rings containing one or more phosphorus or arsenic atoms*

P-Phenyldibenzophosphonin (273) is obtained by the sequence shown in Scheme 10 and nmr spectroscopic studies, including the PH and HH coupling constants, show that the ring has one *trans* double-bond, is highly puckered, and atropic. The uv-spectrum of the phosphonin (273) is virtually identical to that of its oxide (272) and to that of the phosphonium methiodide (274), confirming that the electron pair on phosphorus does not contribute to the π-system of the phosphonin (273) (E.E. Middlemass and L.D. Quin, J. Amer. chem. Soc., 1980, 102, 4838).

The tetraphosphonium chlorides (277, X = H, CH_3, or Cl) are formed in near quantitative yield when the diphosphine

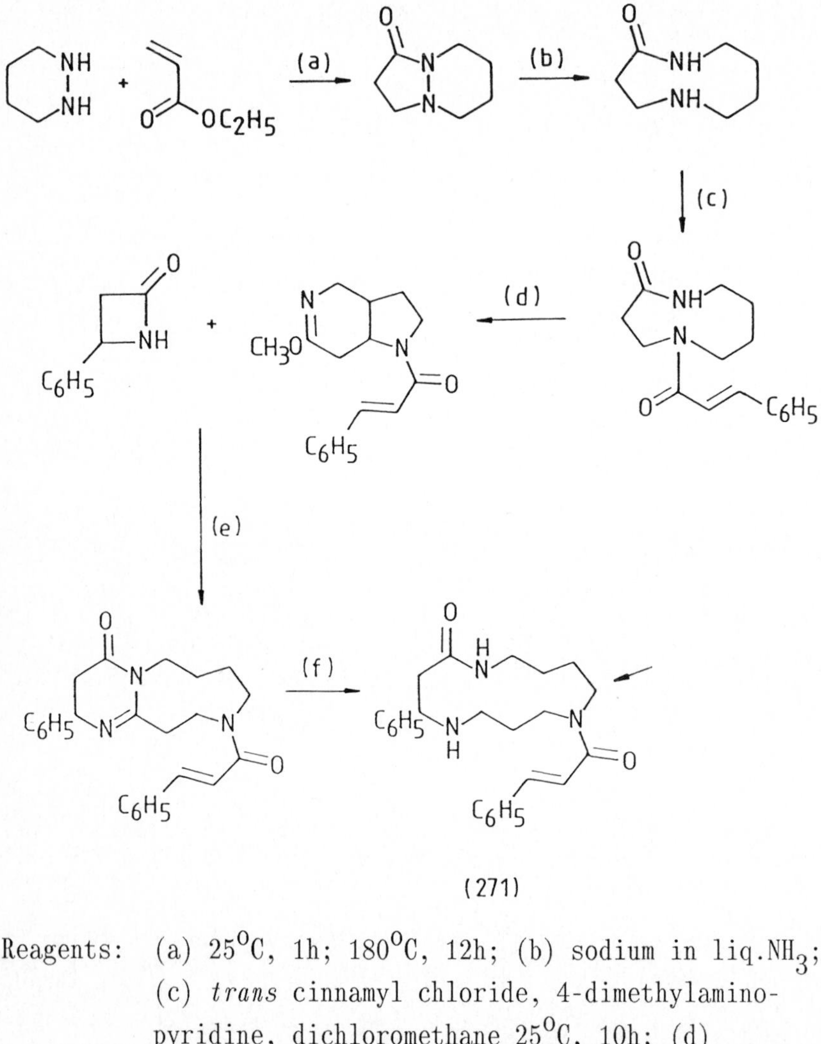

(271)

Reagents: (a) $25^{O}C$, 1h; $180^{O}C$, 12h; (b) sodium in liq.NH_3; (c) *trans* cinnamyl chloride, 4-dimethylamino-pyridine, dichloromethane $25^{O}C$, 10h; (d) trimethyloxonium tetrafluoroborate, dichloro-methane $25^{O}C$, 23h, then 50% aqueous potassium carbonate; (e) chlorobenzene, reflux 21h; (f) sodium cyanoborohydride, $25^{O}C$ 3h, 50° 2h, 25° 12h.

Scheme 9

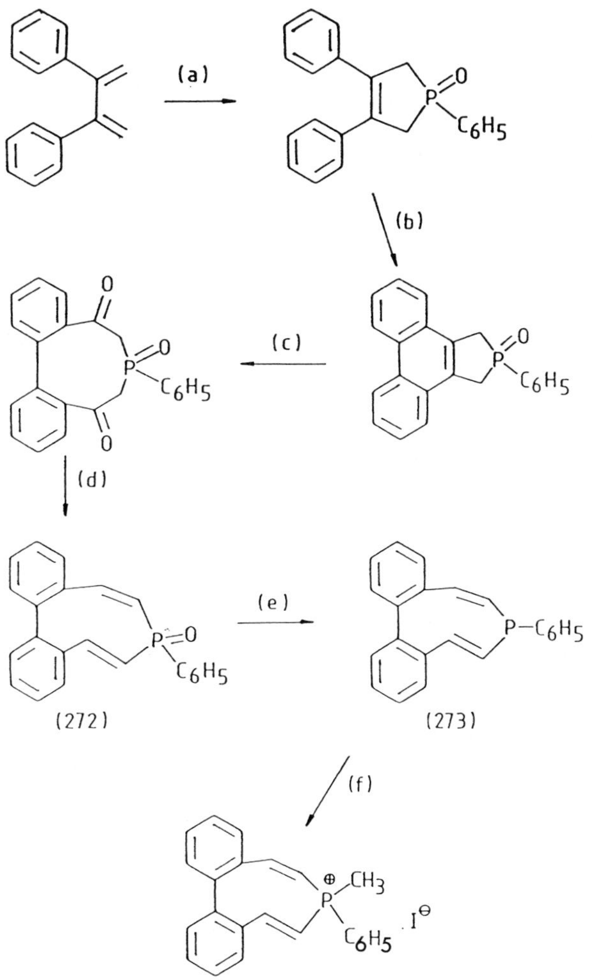

Reagents: (a) Phenylphosphine dibromide, then water; (b) iodine, oxygen, hν; (c) ozone, methanol, then potassium iodide, acetic acid; (d) sodium borohydride, then phosphoryl chloride, pyridine; (e) trichlorosilane, pyridine; (f) methyl iodide.

Scheme 10

(276) and the dichlorides (275, X = H, CH$_3$ or Cl) are heated together for four days in refluxing toluene; after three hours the intermediate formed from two molecules of the diphosphine and one of the dichloride can be isolated (S.D. Venkataramu, M. El-Deek and K.D. Berlin, Tetrahedron Letters, 1976, 3365).

The crown arsenane (280) is formed as a mixture of configurational isomers at the arsenic atoms when the lithioarsenane (278) condenses at high dilution with the dichloroarsane (279); analogous compounds with three or six arsenic atoms are formed similarly (T. Kauffmann and J. Ennen, Ber., 1985, 118, 2692; see also pp. 2703, 2714).

(e) *Compounds with rings containing more than one kind of hetero-atom*

Dichloroarsanes (e.g. 279) condense with bidentate sulphur nucleophiles, e.g. the dithiol sulphide (281) to give macrocycles (e.g. 282) with sulphur and arsenic atoms; oxygen atoms may also be incorporated (Kauffmann and Ennen, *loc. cit.*).

Tetrafluoro-1,2-ethanebis-sulphenyl chloride (283) reacts at high dilution with trimethylsilyl amines (*e.g.* 284) to give macrocyclic sulphur-nitrogen heterocyclic systems (*e.g.* 285) (H.W. Roeski, *et al.*, Ber., 1985, 118, 2811).

Sulphur and phosphorus amides give interesting exchange reactions with macrocyclic secondary tetra-amines; thus bisdimethylamino sulphide (287) reacts at reflux temperature

$$SH—S—SH \quad (281) \quad + \quad (279) \quad \longrightarrow \quad (282)$$

$$(284) \quad + \quad ClS(CF_2)_2SCl \quad (283) \quad \longrightarrow \quad (285)$$

with 1,4,7,10-tetra-azacyclododecane (286) to form sulphide bridges between diametrically opposed pairs of nitrogen atoms in the cage structure (289). X-ray diffraction studies show that the nitrogen atoms have tetrahedral positions in space, and near planar coordination as in some other cage compounds (see Section 2) (R.B. King, *et al.*, J. Amer. chem. Soc., 1986, <u>108</u>, 850).

(287) [N(CH$_3$)$_2$]$_2$S   (286)   P[N(CH$_3$)$_2$]$_3$ (288)

(289)   (290)

(291)

Hexamethylphosphorous triamide (288) reacts with the same twelve-membered cyclic tetramine (286) bridging all four nitrogens with pentavalent phosphorus to yield (290), but the analogous sixteen-membered ring compound gives the tri-valent phosphorus derivative (291) with this reagent. Symmetric and unsymmetric fourteen-membered tetramines form

mixtures of tautomeric P(III) and P(V) systems distinguishable by ^{31}P-nmr shifts, δ +104 to 116 and +53 to 61 ppm,
respectively (T.J. Atkins and J.E. Richman, Tetrahedron
Letters, 1978, 4333 and 5149).

The macrocyclic aminophosphine (292) is formed by a
template synthesis when the nickel complex of the
corresponding open chain diamine is alkylated with the
ditosylate of propan-1,3-diol, followed by removal of the
nickel using sodium cyanide in benzene. The structure and
meso configuration of the macrocycle (292) has been
established by X-ray crystallography (W.G.C. Ansell, *et al.*,
Chem. Comm., 1985, 439).

(292)

The observation that the bright red egg masses of the
nudibranch mollusc *Hexabranchus sanguineus* remain exposed on
ledges of undersea caves but have almost no predators,
prompted a chemical study. Extraction with methanol and
hplc after partitioning into carbon tetrachloride furnishes
ulapualides A and B (293) whose unusual structures, incorporating three adjacent oxazole units in the macrocycle,
were deduced by nmr-spectroscopy and mass-spectrometry on
the non-crystalline materials. Both compounds show in-
hibition of leukemia cell proliferation and growth of
Candida albicans at low concentration. A closely related
antifungal macrolide, kabiramide C, has been isolated from
the egg masses of an unidentified nudibranch (J.A. Roesener
and P.J. Scheuer, J. Amer. chem. Soc., 1986, <u>108</u>, 846;
S. Matsunaga, N. Fusetani and K. Hashimoto, *ibid.*, p.847).

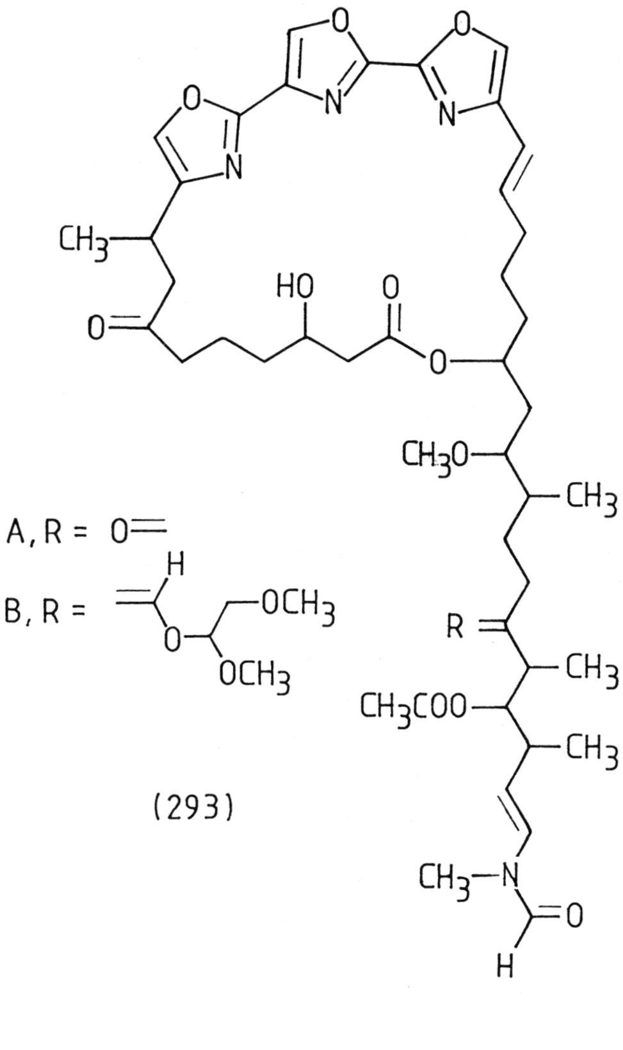

A, R = O=

B, R =
H
=CH—O—CH—OCH₃
          |
          OCH₃

(293)

Guide to the Index

This index is constructed in a similar manner to the volume indexes of the first edition of the Chemistry of Carbon Compounds. However, to make the index easier to use, more descriptive entries have been made for the commonly occurring individual, and groups of chemicals.

The indexes cover primarily the chemical compounds mentioned in the text, and also include reactions and techniques, where named, and some sources of chemical compounds such as plant and animal species, oils, etc.

Chemical compounds have been indexed alphabetically under the names used by authors, editing being restricted to ensuring uniformity of entries under the same heading. In view of the alternative nomenclature that can often be used, a limited amount of cross-referencing has been done where it is considered to be helpful, but attention is particularly drawn to Convention 2 below.

For this and the succeeding volumes, the indexing conventions listed below have been adopted.

1. *Alphabetisation*

(a) The following prefixes have not been counted for alphabetising:

n-	*o-*	*as-*	*meso-*	D	C
sec-	*m-*	*sym-*	*cis-*	DL	*O-*
tert-	*p-*	*gem-*	*trans-*	L	*N-*
	vic-				*S-*
		lin-			*Bz-*
					Py-

Some prefixes and numbering have been omitted in the index, where they do not usefully contribute to the reference.

(b) The following prefixes have been alphabetised:

Allo	Epi	Neo
Anti	Hetero	Nor
Cyclo	Homo	Pseudo
	Iso	

594

(c) A letter by letter alphabetical sequence is followed for entries, firstly for the main entry, followed by the descriptive entry. The only exception to this sequence is the placing of plural entries in front of the corresponding individual entries to prevent these being overlooked by a strict alphabetical sequence which could lead to a considerable separation of plural from individual entries. Thus "butanes" will come before *n*-butane, "butenes" before 1-butene, and 2-butene, etc.

2. Cross references

In view of the many alternative trivial and systematic names for chemical compounds, the indexes should be searched under any alternative names which may be indicated in the main body of the text. Only a limited amount of cross-referencing has been carried out, where it is considered that it would be helpful to the user.

3. Esters

In the case of lower alcohols esters are indexed only under the acid, e.g. propionic methyl ester, not methyl propionate. Ethyl is normally omitted e.g. acetic ester.

4. Derivatives

Simple derivatives are not normally indexed if they follow in the same short section of the text.

5. Collective and plural entries

In place of "— derivatives" or "— compounds" the plural entry has normally been used. Plural entries have occasionally been used where compiunds of the same name but differing numbering appear in the same section of the text.

6. Main entries

The main entry of the more common individual compounds is indicated by heavy type. Multiple entries, such as headings and sub-headings over several pages are shown by "—", e.g., 67–74, 137–139, etc.

INDEX

1*H*-2-Acetamido-1, 4-benzodiazepin-3-ones, 207

Acetophenone oxime, 69

3- Acetoxy-2-acetylbenzo[*b*]furan, 103

3-Acetoxybenzo[*b*]thiophene-1, 1-dioxide, 123

6-Acetoxydithiocine, 454

N-Acetylacetamide, 57

Acetylacetone, 64, 65

Acetylenedicarboxylic ester, 25, 53, 60, 69, 194, 223

Acetylenic macrocycles, 568

2-Acetylindene, 105

3-Acetyl-2-phenyl-3*H*-indole-3-carboxylate, 251, 252

Acridine-*N*-oxides, photorearrangement, 248

Acrolein, 42

Acrylates, 329

Acrylic acids, 329, 330

3-Acyl-3*H*-1,3-benzodiazepines, 198

3-Acyl-3*H*-2,3-benzodiazepines, 192

1-Acyl-2,3-dihydro-1,2-diazepines, 187

10-Acyl-1,11-dihydrodibenzo[*b*,*f*]-[1,4]thiazepine-5-imide-5-oxides, 342

Acylmethylenetriphenylphosphoranes, 158

1-Acyl-3-piperidones, 154

1-Acyl-3,5,7-triaryl-1,2-azepines, 180

α,*ω*-Alkane biphosphines, 390

Alkenes, reaction with silylenes, 4

Alkenyl acetals, cyclisation with tin(IV) chloride, 512

3-Alkenyl-3*H*-pyrazole-2-oxide, 183

2-Alkoxy-3*H*-azepines, 141

2-Alkoxy-3-bromo-1,4-naphthoquinones, 110

1-Alkoxycarbonyl-1*H*-azepines, 143

Alkoxy radicals, 560

Alkoxysilanes, 409

β-Alkoxyvinylphosphonic acids, 61

2-Alkyl-5-aryl-1,3,4-oxadiazepines, 277

Alkylbis(phenylaminomethyl)-phosphines, 77

2-Alkylborinanes, 380

2-Alkylcycloheptanones, 442

1-Alkyl-1,4-dihydroquinolines, 456

Alkyl halides, reaction with potassium superoxide, 496−497

Alkyl *o*-hydroxyphenyl ketones, 65

1-Alkyl-1-hydroxytetralins, 162

5-Alkylidene-2,5-dihydro-4-oxido-1,5-benzoxazepinium betaine, 265

1-Alkyl-2-oxo-1,2,3,4-tetrahydro-3-azocine, 458

α-Alkylperhydroazepines, 153

Alkylphosphonic acid esters, 61

N-Alkylphthalimides, ring expansion, 170

4-Alkylsulphinylazetidinones, thermolysis to thiazepinones, 320

Alkynes, reaction with silylenes, 4

Alkynyl-lithiums, 446

Allene-1,3-dicarboxylates, 330

Allylcyclopentadienyldimethylsilane, 11

Allylphenyl ethers, 44

Allyl trifluoromethylsulphonate, 451

Aminoacid ester cations, resolution, 506

Amino alcohols, reaction with boronic acids, 37

Aminoalkenylphosphonium salts, 76

2-Amino-3*H*-azepines, 142

2-Amino-3*H*-azepine, hydrolysis, 146

o-Aminobenzamide, 73

3-Amino-2-benzazepines, 165

2 Amino-3*H*-1,4-benzodiazepines, oxidation, 207

o-Aminobenzoic acid, reaction with arylboronic acids, 38

2-Aminobenzophenones, reactions with 2-oxazolidone, 209

2-Aminobenzoylhydrazide, 221

2-Aminiobenzoylhydrazine, adduct with dimethylacetylene dicarboxylate, 223

−, photolysis, 223

Aminoboranes, heterocyclic, 37

2-(4-Aminobutyl) cyclopentanones, 159

3-Aminocaprolactams, 153

604

610

612

614

624

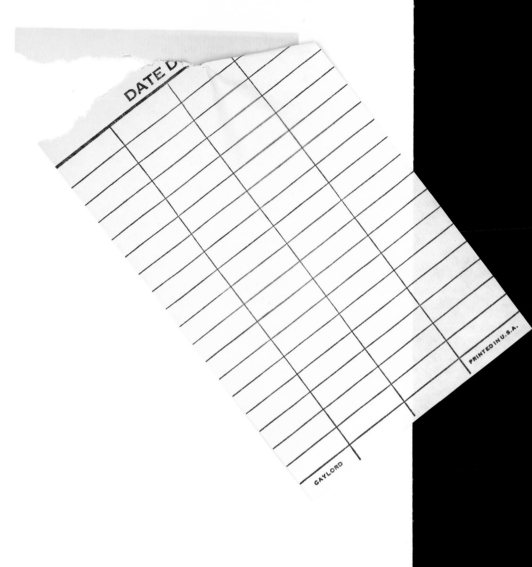